BEYOND SEPTEMBER 11TH

AN ACCOUNT OF POST-DISASTER RESEARCH

A cooperative project of the

Natural Hazards Research and Applications Information Center
University of Colorado
Boulder, Colorado
http://www.colorado.edu/hazards/

Public Entity Risk Institute
Fairfax, Virginia
http://www.riskinstitute.org

Institute for Civil Infrastructure Systems
Robert F. Wagner Graduate School of Public Service
New York University
A partnership of New York University, Cornell University, Polytechnic University, and the University of Southern California
http://www.nyu.edu/icis

Program on Environment and Behavior
Special Publication #39

Institute of Behavioral Science
Natural Hazards Research and Applications Information Center
University of Colorado
2003

This volume is the product of a cooperative project of the three entities listed on the title page. Production work was done by the Natural Hazards Research and Applications Information Center, University of Colorado–Boulder, as part of its ongoing work under National Science Foundation grant award CMS-0218780, "A Clearinghouse on Natural Hazards Research and Applications."

Financial support was provided by the Public Entity Risk Institute, a not-for-profit corporation with principal offices in Fairfax, Virginia. PERI's mission is to serve public, private, and nonprofit organizations as a dynamic resource for the practical enhancement of risk management.

Both financial and production support was provided by the Institute for Civil Infrastructure Systems (ICIS) at New York University's Robert F. Wagner Graduate School of Public Service (in partnership with Cornell University, Polytechnic University of New York, and the University of Southern California) that is funded by the National Science Foundation under Cooperative Agreement No. CMS-9728805, "The Institute for Civil Infrastructure Systems," and CMS-0203371, "Learning from Urban Disasters: National Science Foundation Response and Opportunities for Future Research Workshop," held on December 12–13, 2001, for the purpose of identifying an interdisciplinary research agenda for issues raised by the September 11th attacks. Two other grants funded by the NSF to NYU-Wagner's ICIS also provided resources for this work. These were CMS-0204660, "Urban Infrastructure Services in a Time of Crisis: Lessons from September 11th," a Small Grant for Exploratory Research (SGER), and CMS-0091482, "Bringing Information Technology to Infrastructure: A Workshop to Develop a Research Agenda," held in June 2001 to identify research areas for the use of information technology to improve infrastructure performance.

Cover images were taken from papers in this volume. The complete photograph and explanation can be found on the pages noted: back cover, p. 203; circular photo, p. 227; right side of front cover, top to bottom, p. 367, p. 365, p. 366, and p. 368.

Cite this volume as: Natural Hazards Research and Applications Information Center, Public Entity Risk Institute, and Institute for Civil Infrastructure Systems. 2003. *Beyond September 11th: An Account of Post-Disaster Research*. Special Publication No. 39. Boulder, CO: Natural Hazards Research and Applications Information Center, University of Colorado.

ACKNOWLEDGMENTS

Beginning in the early afternoon of September 11, 2001, the Natural Hazards Center, like many other organizations, focused all of its efforts on doing what it could to contribute to the response to and recovery from the terrorist attacks of that day. In our case, it was to use our expertise, under the foresighted leadership of Mary Fran Myers, Co-Director, to mobilize as many teams of scholars as possible to conduct field investigations in the immediate aftermath of the events. These investigations were essential, we believed, in order to collect data, witness what was happening, and capture the lessons that were sure to emerge.

We gratefully acknowledge all of those who responded to the call, as well as the National Science Foundation, which provided us additional funds to support the work of the scholars in the field. Further, we acknowledge the ready willingness of both the Public Entity Risk Institute and the Institute for Civil Infrastructure Systems to offer financial support to ensure that the findings of the research could be published together in this volume and widely distributed.

The production of this book was made possible with the usual invaluable input and support of every member of the Hazards Center staff and especially Mary Fran Myers, Lori Peek, and Laura Musser. The Hazards Center gratefully acknowledges the grant received from the National Science Foundation through its Research Experience for Undergraduates program, which made it possible for Musser to participate in the production of the book. Helpful substantive, editorial, and other input on various aspects of the book's development was received from Diane Adams, David Butler, Mike Campbell, Andre Filiatrault, Sarah Michaels, JoAnne Monday, and Rae Zimmerman.

Earlier drafts of this volume were improved as a result of the efforts of more than 40 anonymous reviewers, all of whom were generous with their time, critical in their thinking, and judicious with their suggestions.

The Hazards Center extends its appreciation to the authors whose work is represented here. Throughout the production process they were cooperative, cheerful, and patient.

Finally, we offer our sincere thanks to the many citizens and government officials of New York, Washington, D.C., and Pennsylvania, who allowed the authors of these papers to observe and question them during what was likely the most traumatic time of their lives.

— Jacquelyn L. Monday
Editor

PREFACE

Disasters affect our built environment as well as our social, political, economic, and cultural systems. The threats from most disasters can never be eliminated. Hazards, many of which are natural geophysical and meteorological phenomena, will always be with us. Therefore, combined with our efforts at implementing structural and non-structural mitigation measures, it is our preparedness for the known and our agility in handling the unknown that will determine the extent of damage, loss, and destruction that our systems suffer. When disaster strikes, then, it is imperative to utilize the event to inform research aimed at the mitigation of, preparedness for, response to, and recovery from hazardous events that are extreme enough to cause disaster. Our aim must be to design our physical and social structures to be "disaster resilient."

The Division of Civil and Mechanical Systems (CMS) in the Directorate for Engineering of the National Science Foundation (NSF) has a long history of responding to requests for rapid reconnaissance and research in the wake of natural and technological disasters. For over 25 years, CMS, as a key player in the National Earthquake Hazard Reduction Program, has funded both Small Grants for Exploratory Research (SGER) and long-range strategic research on the impacts of earthquakes, landslides, hurricanes, tornados, tsunamis, floods, releases of toxic materials and technological accidents, and other hazards to the built environment and those social systems governing mitigation, preparedness, emergency response, and recovery. NSF's SGER program and the University of Colorado–Boulder's Natural Hazards Research and Applications Information Center's (NHRAIC's) Quick Response (QR) program provide support for researchers to collect "perishable" data during the immediate post-impact period of disasters. The NSF-sponsored "Learning from Earthquakes" program administered through the Earthquake Engineering Research Institute has played an important role in undertaking these investigations for decades.

On September 11, 2001, each of these disaster response mechanisms was put to a new test, as were the emergency response capabilities of New York, Washington, D.C., Pennsylvania, and the rest of the nation. Key CMS Program Directors reflected on that day.

Miriam Heller, IIS Program Director—

> The traffic in Alexandria, Virginia, seemed purposeful in keeping me from the workshop I helped to support, "Mitigating the Vulnerability of Critical Infrastructures to Catastrophic Failures." The workshop represented some of NSF's targeted efforts to translate and expand research from natural disasters' effects on the built environment to critical infrastructure protection. At 8:45 a.m., the words on my radio did not clearly register in my mind, "We interrupt our regular broadcast . . . a plane appears to have crashed into the north tower of the World Trade Center." Within 30 minutes of my arrival, another workshop sponsor from the Office of Science and Technology Policy

received a phone call explaining that the Pentagon had been attacked. We were unaware that the workshop's main topic, infrastructure interdependencies and cascading failures, was playing out in New York and the Washington, D.C. metropolitan area as each of the eight critical infrastructures was disrupted. With all air travel suspended and phone lines jammed, we managed to ensure that stranded workshop participants would have lodging. We cancelled the remainder of the workshop to head to our respective zones of safety.

The Gujarat and Nisqually earthquakes would normally have presaged imminent SGER requests. Yet, getting to my family and finding my way home amidst road closures leading to the Capitol and emergency routes to the Pentagon displaced all thoughts of work. When I reached home at 3:00 p.m., I was jarred into action with the first request for additional support for QR grants from Mary Fran Myers at NHRAIC. I immediately contacted the CMS Program Officers who were in town, Joy Pauschke and Rick Fragaszy, to find how to respond. It was the end of the fiscal year and my budget was spent. Over the next two weeks, the NSF system moved with remarkable swiftness to match that of the proposing investigators, finding monies, reviewing proposals, and making awards. By September 26, pertinent information on the awardees was compiled and sent off to the Institute for Civil Infrastructure Systems in New York, whose assistance had been enlisted in the coordination of the researchers.

Joy Pauschke, NEES Program Director—

My family in Chicago contacted me via phone as soon as the first World Trade Center tower was hit. For the next hour or so, most of the NSF staff stood in front of the television to watch the tragic events of the day unfold. About mid-morning, there was collective uncertainty as to whether downtown Washington, D.C. and/or suburban Virginia had also been hit. (The Pentagon is about 8 miles from NSF). Finally, around 10:30 a.m., the Office of Personnel Management dismissed all federal employees for the day. The next day was a federal unscheduled leave day.

Making the NSF SGER rapid response awards between September 11 and 30, 2001, was the result of the great *esprit de corps* and teamwork across all of NSF. As the acting Division Director for CMS, I attended the Engineering Management Group meeting on September 12 and explained the inquiries about rapid response SGER requests that NSF had already received. Since the divisional programs at NSF were already spent out by mid-August, any remaining fiscal year 2001 NSF funding was at the Engineering Directorate or Office of Director levels. The Directorate for Engineering and the Office of the Director, through the Budget Division, quite expediently transferred \$100,000 and \$200,000, respectively, to CMS and to Engineering Education and Centers Divisions for rapid response awards. CMS Program Officers immediately processed the rapid response proposals submitted through

FastLane. Thanks to the commitment by the Division of Grants and Agreements to process all the rapid response awards, NSF was able to make these awards by September 30, 2001.

Richard Fragaszy, GGH Program Director—

I was in my office on September 11 when Joy Pauschke burst in to tell me about the first airplane crashing into the World Trade Center. My first thought was that she was pretty convincing, but it must be a joke. The next hours were spent watching the news and still not really believing what had happened. As a native of Manhattan, I was very familiar with the World Trade Center area, and had many friends and relatives in the city. Fortunately, it turned out they were all safe. The next day at work we all had the same response to the disaster—what can we do? Under Joy's leadership, resources became available and we began talking with researchers about immediate needs for data collection. We were all amazed, and proud, of the speed at which everyone at NSF worked together to fund a significant number of awards and supplements. When reading this volume, it should be kept in mind that there was no solicitation, no time to do in-depth literature reviews, no time to think and plan quietly. The work reported here was done by a dedicated group of investigators who dropped everything to do the best they could under very trying times. It is a tribute to them that so much valuable work was accomplished so quickly.

Priscilla Nelson, CMS Division Director—

I was attending U.S. Nuclear Waste Technical Review Board meetings in Las Vegas when the first plane hit. Actually, the meetings hadn't started yet, but when I came down the elevator and saw a large group of people gathered around a television, I was quickly informed as to what was happening. The initial thinking was that the first crash was an isolated occurrence—an accident. This perspective soon gave way to the horrible realization that this was not an accident, and that the world had changed. I tried to call in to NSF, but couldn't get a phone line; couldn't send an e-mail.

The meeting in Las Vegas actually started and covered most of the agenda. Presenters were in shock, and no one was focusing well on the business at hand. The rest of the day, the major topic of conversation was how to leave Las Vegas. Questions about whether "it" was over continued, and made me realize that the knowledge that something was "over" was incredibly important in determining when recovery could begin.

The next day, I was finally able to connect to NSF and with my extended family. The family were shaken but well, and the NSF program officers were superb— finding the resources for awards (the leadership at NSF really came through on that), coordinating with the Division of Grants and Agreements to get awards out quickly. The Civil and Mechanical Systems Division operated

> like a system of talented people running with smoothness, calmness, and efficiency. Joy, Rick, and Miriam did a superb job.
>
> Ultimately, I connected with a small group of Board members and we rented a car and drove back eastward, finally arriving in Washington, D.C., early Sunday morning, September 16.

Within two weeks of the attacks, NSF and NHRAIC researchers were at Ground Zero and other September 11th disaster sites. A grand total of eight NSF SGER and supplements and seventeen NHRAIC QR awards were being finalized or under discussion to support natural hazards researchers in contributing their skills and expertise toward a disaster of a *different nature.* This was a human-initiated disaster, intentionally targeting one of the densest concentrations of people and infrastructure on earth.

This disaster, by necessity, included the involvement of law enforcement and military officials, which demanded some overarching management of the research. The Institute for Civil Infrastructure Systems (ICIS) offered proximity to the site, links to local agencies, and was the natural solution. Under the direction of Dr. Rae Zimmerman, ICIS provided a point of contact for officials to verify researcher activity as well as for researchers trying to make connections with city personnel. In December 2001, ICIS also convened a workshop to bring together the researchers and government officials engaged in responding to the events to exchange experiences, findings, and data with the goal of identifying further research pursuits. After this workshop, a briefing was held for Congressional staff. The briefing included a visit to the World Trade Center site and the New York City Emergency Response Center. NSF staff, city workers, and a few NSF-supported investigators discussed their activities and answered questions posed by the Congressional staff.

This book consists of selections written by many of the researchers who received QR or SGER grants to investigate questions that arose in the September 11th disaster. Each takes a distinct view of the disaster; their topics span engineering, information technology, and behavioral sciences. We believe that the findings, lessons, and recommendations collected herein will be a worthwhile contribution to the further exploration and dissemination of information to help people, localities, and organizations make risk-informed decisions. The book should also stimulate thinking and identify areas in which additional investigation is needed to build resiliency in the face of risks from natural, technological, and malevolent causes.

Readers should note that the initial response of NSF was followed by strong interest in World Trade Center and related research from all across the

Foundation. By the end of fiscal year 2002, CMS had awarded 27 proposals for a total of $2.5 million. NSF-wide, 77 awards were made totaling approximately $24 million. More detailed information on these awards, which are listed in the Appendix of this volume, can be found at the NSF website, http://www.nsf.gov. More information about the QR awards is available on NHRAIC's website at http://www.colorado.edu/hazards.

Miriam Heller, Ph.D.
Program Director
Infrastructure and Information Systems

Joy M. Pauschke, Ph.D., P.E.
Program Director
George E. Brown, Jr. Network for Earthquake Engineering Simulation (NEES)

Richard J. Fragaszy, Ph.D.
Program Director
Geotechnical and GeoHazards Systems

Priscilla P. Nelson, Ph.D.
Directorate for Engineering
Senior Advisor (current)
Former CMS Division Director

Dennis E. Wenger, Ph.D.
Program Director
Infrastructure and Information Systems

CONTENTS

Part 3. The Engineered Environment—Buildings and Infrastructure

Part 4. Individual and Collective Behavior

Part 1

Overview

Background and Summary

Natural Hazards Center
University of Colorado

The disaster resulting from the terrorist attack on September 11, 2001, provided an unprecedented opportunity for the natural hazard research community to contribute its skills and expertise to help better understand what happened, through programs such as the University of Colorado at Boulder's Natural Hazards Research and Applications Information Center's Quick Response (QR) research program and the National Science Foundation's Small Grants for Exploratory Research (SGER). Both programs enabled several teams of scholars to enter the field quickly to collect perishable data in the days and weeks after September 11th. The information they gathered forms a unique set of data that could only have been captured in the short time frame that followed the impact of the disaster; otherwise it would have been lost. Hence, it is a valuable piece of the knowledge our society needs in order to learn how to better cope with and minimize future risks from and impacts of both natural and human-generated events.

This book consists of selections written by some of the researchers who received Quick Response grants or Small Grants for Exploratory Research to investigate questions that arose in the September 11th disaster. Each takes a distinct view of the disaster; their topics span the engineering and behavioral sciences. In their reports, the authors describe their projects and present the findings, lessons, and recommendations that grew out of this post-September 11th disaster research. This summary paper introduces the other contributions in the volume, lists the more salient findings from the research gathered here, makes overall observations about the differences and similarities between the September 11th disaster and other disasters based on these research findings, presents recommendations for improvements to practice and policy, and summarizes needs for future research.

This book should contribute to the further exploration and dissemination of information to help people, localities, and organizations make risk-informed decisions. It also should stimulate thinking to identify areas in which more investigation is needed to build resiliency in the face of risks from natural, technological, and human-induced causes.

Contents

The selections in this book are grouped loosely into six categories. Readers should note that this is a device for convenience of reading and access to the material, and is not meant to suggest that each study fits narrowly into a specific category. In fact, as is the case with much of hazards and disaster research, each of the research efforts described here tends to span several topics and disciplines, and they are all interrelated to some degree. Another important point is that this collection is in no way a comprehensive look at the September 11th disaster. For example, some aspects of the performance of some infrastructure are examined, but not all aspects of all types. Similarly, some types of human and organizational behavior have been examined in these studies, but certainly not the full range of possible behaviors in all types of organizations that participated in or were affected by the disaster. Further, the studies described here were all of the "quick response" variety, that is, they were initiated immediately after the terrorist attacks happened, sometimes without the luxury of thorough planning or time-consuming background work, and almost always conducted under less-than-optimal conditions. Most of these research projects were meant to be exploratory, pointing the way to questions for future, more intensive and extensive investigation, rather than providing definitive answers. Nevertheless, the range of topics they cover and the scientific understanding they demonstrate combine to give a solid perspective on the questions, confirmations, and puzzles the September 11th disaster has engendered.

Part one is an overview section. Besides this summary, it includes a discussion by Sarah Michaels of quick response research as a method, and a summary by Carlos Restrepo and Rae Zimmerman of a workshop on urban disasters held a few months after the September 11th event. Part two contains reports of research that focused on various issues in response to the disaster. Nancy Grant, David Hoover, Anne-Marie Scarisbrick-Hauser, and Stacy L. Muffet report on the multi-jurisdictional interactions between the federal and local government responders at the scene of the crash of United Flight 93 in Pennsylvania. David Simpson and Steven Stehr examine the process by which victims were identified and information about them managed after the collapse of the World Trade Center towers. James Kendra and Tricia Wachtendorf use the World Trade Center disaster as an exploratory case study to discuss the role and consequences of creativity in emergency response. The paper by Deborah Thomas, Susan Cutter, Michael Hodgson, Mike Gutekunst, and Steven Jones explores how geographic information systems, remote sensing, global positioning systems, and maps were used in the aftermath of the World Trade Center incident.

Part three of the volume includes four contributions dealing with affected buildings and infrastructure in New York City. William Wallace, David

Mendonça, Earl Lee, John Mitchell, and Joe Chow use a systems approach to analyze the interdependence of the critical infrastructure around the World Trade Center complex, and the impacts of their being disrupted. Gordon Warn, Jeffrey Berman, Andrew Whittaker, and Michel Bruneau report on their engineering analysis of one badly damaged building near the World Trade Center. Rae Zimmerman explains how pre-existing flexibility in design, planning, and siting of New York City infrastructures affected the provision and/or resumption of transportation, energy, environmental, and communication services after the disaster. Thomas O'Rourke, Arthur Lembo, and Linda Nozick delineate the consequences of the disaster on civil infrastructure systems in the New York City area.

Individual and collective behavior is the subject of part four. Seana Lowe and Alice Fothergill examine the influences on and motivations of volunteers in the disaster response in New York City. David Sattler reports on his investigation of the psychological distress, resiliency, and coping behaviors reported after the September 11th disaster by samples of young people elsewhere in the United States. Lori Peek describes the reactions and responses of Muslim university students in the New York City area during the aftermath of the disaster. Paul O'Brien relates his use of the traditional hazards-based model of risk communication to understand some of the behaviors of the people affected by the attack in New York City. José Holguín-Veras, Robert E. Paaswell, and Ann-Marie Yali report the findings of a research project aimed at assessing the impact the September 11th disaster had on decisions about intercity passenger travel. Jeannette Sutton applies the established typology of organizational adaptation in disasters to a selection of faith-based organizations that engaged in response activities after the September 11th disaster.

In part five are reports of two studies on the activities of the private sector in the wake of the September 11th disaster. David McEntire, Robie Robinson, and Richard Weber use the World Trade Center incident as a case study of the functions performed by businesses and the coordination that developed between the public and private sectors. Sarah Michaels' work highlights the vital role played by information technology firms to deliver previously contracted services, provide new services, and donate humanitarian aid.

The sixth and final part of the book takes a broader view of the disaster, providing some political context and a glimpse of the many public policy issues that arose. Patrick Gillham and Bob Edwards trace how the sudden transformation of the political climate after September 11th caused organizations that had scheduled social justice protests to alter their plans. Christine Rodrigue evaluates the disaster coverage in the *Los Angeles Times* for the first 12 weeks after the disaster. André LeDuc, Robert Parker, and Kathy Lynn tell how Oregon took the occurrence of the disaster as an opportunity to evaluate the effectiveness of its state agency communications for emergency response and disaster recovery. Claire Rubin and Irmak Renda-Tanali relate some of the early impacts of the disaster and its implications for federal emergency management systems, programs, and policies.

The volume ends with an appendix listing the many post-September 11th grants awarded by the National Science Foundation for topics related to homeland security.

Findings and Observations

The September 11, 2001, terrorist attack disaster has repeatedly been called "unique." Although many aspects of the disaster were unprecedented in recent U.S. experience, in most ways the aftermath of the disaster (if not the attacks themselves) was consistent with long-established knowledge and practice about appropriate response to crisis, human behavior in extreme situations, the performance of engineered systems, government disaster policy, and other issues. This conceptual framework for understanding and coping with disasters has been developed as a result of decades of past research on and experience in natural and technological disasters. The practitioners, managers, policymakers, and scientists who have been addressing natural and technological disasters over the years have expertise in responding to many types of emergencies, large-scale coordination of resources and needs, risk assessment for the built environment, designing for conditions of uncertainty, fostering community recovery, strengthening structures for multiple loadings, and simulating complex systems of infrastructure under various conditions—all of which will be needed to combat the impacts of terrorism in the future. That this disaster-based knowledge and expertise is well-suited to addressing possible future terrorism disasters is made clear by the research reported here.

That said, the September 11th disaster did have unusual characteristics that are worth noting. Some grew out of the fact that the disaster was caused by a terrorist attack; some could have developed in other types of disasters as well. Some of these factors presented "new wrinkles" in the scheme of coping with unexpected events, while others were consistent with existing attitudes, behaviors, and procedures.

- Inability to reach the disaster sites via air travel was an unprecedented problem. Organizations, agencies, and businesses could not immediately fly personnel or materials into the area. (For more on this, see the selections in this volume by Michaels; and by Sutton.)
- All three disaster sites (New York City, Washington D.C., and Shanksville, Pennsylvania) were also crime scenes. This posed additional security considerations, and the presence of security and law enforcement personnel made emergency response procedures and coordination more complicated. (For more on this, see the selection by Grant et al.)

- The scope of the disaster was unusually large by U.S. standards: millions of people were affected, multiple geographic areas were impacted, and there was a very high density of population and development in some affected areas. However, it should be noted that the loss of life was small compared to the numbers of people who die worldwide in earthquakes, typhoons, and floods every year. (See the pieces by Rodrigue; and by Rubin and Renda-Tanali.)
- The terrorist risk is unfamiliar to Americans, so neither the public nor officials knew precisely what kinds of individual protective actions could or should be taken, if any. In a flood or tornado, by contrast, most people instinctively realize they should move to high ground or seek sturdy shelter; public agencies know what kinds of warnings or advice to give. (For more on this idea, see the selection by O'Brien in this volume.)
- For decades disaster preparedness and response activities have been relatively non-politicized. But the September 11th disaster had clear political undertones and ramifications. While it is not clear yet what the relationships are among the various factors and effects, it appears that under some conditions (perhaps in the case of a disaster resulting from a deliberate act), responses may not always be as apolitical as they usually are in disasters. In those circumstances there may not be the degree of universal altruism typical of other disasters, or the same degree of respect for everyday rights (of assembly, speech, privacy). (For more on this, see the selections by Gillham and Edwards; by Rodrigue; and by Peek.)

Disaster Response

Although a number of unusual and unanticipated circumstances had to be dealt with during the course of responding to the September 11th disaster, in general the agencies, organizations, personnel, and procedures already in place proved to be workable and flexible enough to meet new and unexpected contingencies. The ability to improvise and to adapt existing knowledge, skills, and resources to an emergency/disaster situation, is one aspect of the non-linear thinking that *Disasters by Design* identifies as being essential to the effective management of future disasters (Mileti, 1999). In addition, technology for information dissemination, communications, and geospatial analysis seemed to carve out a new niche for itself in response activities, (at least in New York City). Some of the researchers' observations on these topics are listed below.

- Emergency response personnel at the Pennsylvania disaster site who had worked together beforehand had stronger and more effective interactions than those who were strangers. The emergency response was helped further by the fact that personnel had been through training and had conducted exercises of their response procedures. (See the selection by Grant et al. for more on this.)

- In New York City, emergency response personnel and organizations exhibited creativity, resiliency, and improvisation in using and providing products or services (communications, facilities, search and rescue, security, cleanup, etc.), and obtaining additional resources to carry out their activities. (See Kendra and Wachtendorf for details about this activity.)
- Creative action as exhibited by emergency response personnel and groups after the attacks yielded not only positive results but also conflicts and challenges, not unlike those documented in prior studies of the convergence phenomenon after disasters, in which volunteers, opportunists, and others converge on the scene, adding an element of uncontrollability that can complicate emergency operations, safety, and security. (See Kendra and Wachtendorf's contribution for more thinking along these lines.)
- Geographic information systems (GIS), other geo-technologies, and new mapping techniques apparently were used as decision support tools and as vehicles for risk communication in the disaster response and recovery in New York City. (See the piece by Thomas et al. for more on this.)
- The experience of some of the businesses interviewed after the disaster showed that current technology makes it unnecessary for certain types of disaster assistance to be delivered at the actual disaster site. Companies were not confined to using equipment at only one location. For example, remote sites were used to handle backup data, and some telephone services were routed through second and third locations around the country. (See the reports by Michaels; by Zimmerman; and by O'Rourke et al. for evidence of this experience.)
- Two-way pagers turned out to be one of the most reliable and secure mechanisms for contact among emergency response personnel. (See the selection by Michaels for more on this.)
- After the September 11th disaster in New York City, 33 information technology firms formed a team to use their network monitoring techniques to track signals from wireless devices. This enabled the monitoring of cellular or pager communications coming from the World Trade Center debris, helped dispel rumors about wireless communications, helped guide rescue personnel, and confirmed the safety of some people. (Consult the piece by Michaels for more on this.)

The Engineered Environment—Structures and Infrastructure

The three studies reported in this volume that examined the engineered environment suggest that many existing practices for and knowledge about building in damage resistance and coping with unexpected interruptions in normal operations are applicable for protecting against terrorism-caused damage as well, and have benefitted from prior research and experience in natural disasters. Highlights of the findings are listed below.

- In the immediate aftermath of the September 11 attacks, both telecommunication equipment and the emergency procedures enacted by the telecommunications companies performed well. Redundant, dispersed facilities and the ability to quickly bypass damaged nodes were important factors in maintaining systems and restoring lost ones. The best practices and designs have benefitted from lessons learned during previous natural disasters, including the Loma Prieta and Northridge earthquakes. (See the research by O'Rourke et al. for more details.)
- Virtually all damage to underground water, gas, and steam pipelines and telecommunications conduits in New York City on September 11th resulted from direct impact from falling debris or penetration of the ground by projectiles. Thus underground infrastructure likely will be relatively undamaged in future disasters, except in the immediate vicinity of collapsed buildings or failed structures. (See O'Rourke et al.)
- Disruptions in the water distribution system that affected firefighting were overcome in part by using fireboats that pumped water out of the Hudson River. This technique is also part of the earthquake emergency plans of San Francisco, Berkeley, and Vancouver. (See the piece by O'Rourke et al. for more on this technique.)
- The September 11th disaster in New York City confirmed that the degree of flexibility incorporated into the design of infrastructure systems influences the extent to which they are able to perform extraordinary functions in a crisis. Many distribution networks proved extremely flexible in providing users with emergency power, water, communication, and transportation. Both public and private utilities were able to quickly identify, acquire, and manage resources to provide short-term emergency service in New York City. (See the contribution by Zimmerman for a full description of this idea.)
- Despite the loss of a perimeter column over a 17-story height, a 39-story building just south of the World Trade Center tower 2 that was seriously damaged by the tower's collapse did not collapse itself, because its lateral- and gravity-load-resisting systems were highly redundant. (See the selection by Warn et al. for additional thoughts on this topic.)

- Standard tools for the two- and three-dimensional analysis of buildings subject to earthquake shaking can be used to make preliminary judgements about the damage-tolerance of buildings. (See Warn et al.)

Individual and Collective Behavior

Several aspects of individual and group behavior were investigated in the studies reported in this volume. In some instances the findings are in accord with prior knowledge about disasters while in others differences were found. For example, the fact that the September 11th disaster was human-induced and deliberate appears to have resulted in some behaviors not necessarily typical of those that develop in natural disasters.

- Symptoms of psychological distress in response to the September 11th disasters that were reported by university students surveyed at four locations around the United States included having concern about their own and family members' safety, making changes in daily routines and activities as a result of the attacks, feeling angry or depressed, and having trouble sleeping. The level of this distress was relatively low overall (in the single digits and teens), but somewhat higher in New York and South Carolina than in Colorado and Washington. (See Sattler for more on this.)
- Nearly three-quarters of the university students surveyed at each of four locations in the United States reported instances of posttraumatic growth, cognitive adaptation, resource gain, and resiliency after the attacks, such as reassessing their priorities, relationships, and philosophy of life; discovering they were stronger than they thought; and taking positive action by showing concern for strangers, displaying the American flag, or donating money to charity. These reactions fit in with prior theories of responses to psychological distress during and after disasters. (For a full description of these findings, see the piece by Sattler.)
- After the airplanes hit the World Trade Center towers, people in New York City were inundated with official and unofficial advice from multiple sources and through multiple channels to leave the area, take protective action, watch for falling debris, and other information. Official warning messages and media coverage became intermingled. (For more about warnings, see the selection by O'Brien.)
- Some of the variables traditionally thought to influence people's response to warnings were operative in the September 11th disaster (environmental and social cues) and some were not (pre-event influences such as prior disaster experience and warnings). These findings confirm recent thinking by disaster researchers that the popular model for risk communication needs updating. (For more on warnings, see the paper by O'Brien.)

- Some organizations appear to be able to adapt in more complex ways in a disaster situation than has been accounted for in the typology of organizational structure and change most widely used in prior disaster research. For example, faith-based organizations in the vicinity of the New York City disaster used a variety of adaptive strategies to meet the emerging needs of their congregations and the community. (For more on this, see the paper by Sutton.)
- The familiar pattern of convergence behavior (that is, people want to help out, so they converge on the disaster site) was repeated in the September 11th disaster. Volunteers in New York City reported that their motivations for helping were both internal and altruistic, and they also experienced feelings of healing and empowerment as a result of their volunteer work, along with a sense of solidarity with the community. (For more on these observations, see the selections by Lowe and Fothergill; by Peek; and by Grant et al.)
- People not otherwise directly affected by the attacks appeared nevertheless to share a sense of having been victimized (presumably due to the perception that the attacks had been directed at all Americans). Heightened feelings of victimization appeared to play a strong role in the people's decisions to converge on the disaster scenes to render assistance. Another sort of victimization was experienced by Muslim students at universities in the New York City area who were faced with anti-Muslim backlash in public opinion. (See the selections by Lowe and Fothergill; and by Peek.)
- There were indications that the "therapeutic community" that has been repeatedly shown to form in the wake of natural and technological disasters as people pull together to cope with the event may not be as strong or as all-encompassing in human-induced disasters. For example, Muslim students at universities in the New York City area felt themselves to be excluded from the sense of community that other Americans shared. On the other hand, these same Muslim students did report feeling increased solidarity with other Muslims after the disaster. (See the selections by Lowe and Fothergill; and by Peek.)
- A study conducted six months after the disaster found that September 11th had a noticeable impact on the process people used to decide whether to travel by airplane, train, or automobile. Two factors were significant in that choice process: the effects of the September 11th disaster on the individual person, and the person's overall level of stress. In general, the higher the overall level of stress, the less likely the person was to choose air travel. (See Holguín-Veras et al. for details on the project, model, and variables).

Private Sector Roles and Activities

Particularly in the New York City area, the September 11th disaster illustrated several interesting findings about businesses and the relationships between the public and private sectors.

- On its own, the private sector played varied roles in the response to the September 11th disaster in New York City, including building restoration, sanitation services, business relocation and resumption, communications, insurance coverage, transportation, donation management, and equipment repair. (For more, see the selections by McEntire et al.; and by Michaels).
- Businesses interacted with the public sector to fulfill such disaster functions as warning and evacuation, relocation and management of the emergency operations center, perimeter control and law enforcement, security and medical staffing, logistical support of search and rescue teams, information dissemination, and the repair of communications and other infrastructure (McEntire et al. describe these activities.)
- Both the close collaboration among public and private sector entities and the effectiveness of some companies in continuing to serve their clients and also contributing to the rescue and disaster response were found to be a result of such factors as the participants' having had previous disaster experience, having had pre-existing disaster plans or held mock exercises, the presence of communications equipment, willingness to cooperate, and post-event ingenuity. (For more on this, see the selections by Grant et al.; by Kendra and Wachtendorf; by McEntire et al.; and by Michaels).

Public Policy and Political Context

- Since 1965, the federal government has followed a pattern after key disasters of generating after-action studies; enacting new laws; modifying disaster response plans; and making organizational changes. The aftermath of the September 11th disaster fit this basic pattern of action for attempts to improve the federal approach to disaster management. However, there were some unusual aspects in the federal reaction to the September 11th disaster. The legislative process was unusually bipartisan. Both the legislative and executive branches moved quickly to pass new laws and and to re-organize. As a result, dramatic changes were made in the federal government's emergency management and counter-terrorism approaches and attitudes. (See the selection by Rubin and Renda-Tanali.)
- The September 11th disaster transformed the political climate in the United States, at least temporarily, so that dissent with government positions became unpopular. Most of the organizations that had arranged

late-September protests against economic globalization and trade liberalization in Washington, D.C., withdrew their support from the events, rather than risk losing legitimacy in the eyes of the media and the public. (See the research report by Gillham and Edwards.)

- In response to the September 11 attacks, state and local governments across the United States activated their emergency response plans and communication processes, in some cases revealing deficiencies in those preparations. Oregon, for example, used the opportunity provided by the increased awareness of risk to conduct an evaluation of its state-level mechanism for communication and coordination in disasters. (See the selection by LeDuc et al. for more on this procedure.)
- For the first 12 weeks after the September 11th disaster, the front page coverage of one metropolitan daily newspaper leaned toward the sensational aspects of the disaster, and presented a fairly narrow political point of view. For 9 of those weeks, 100% of the newspaper's front-page photographs related to the disaster. For the first three weeks, the front page coverage centered on the "disaster" aspects of the event, but in the fourth week it switched to a focus on the related war story. (See the study reported by Rodrigue for details on this evaluation of media coverage.)

Recommendations for Future Research

The research recommendations listed here are some of the more salient of the many questions for further investigation discussed by the contributors to this volume. (Note that a separate set of research recommendations is presented in the selection by Restrepo and Zimmerman, which catalogs the thinking of participants in a workshop held a few months after the disaster.) The research needs summarized below fall into three broad categories. One category calls for further investigation of the applicability of hazards-based models, theories, and tools to the occurrence of terrorism disasters. The second group includes calls for the expansion of hazards/disasters research based on hazards concepts and theories that were borne out by the September 11th disaster, just as they likely would have been in any U.S. disaster. The third category focuses on needs related to quick response research.

Research for Coping with Terrorism Disasters

- Future research should examine the long-term psychological response to terrorist attacks and threats. (See Sattler for a more complete explanation.)
- Further exploration is needed of the ways in which post-disaster therapeutic communities develop. Under what circumstances are people included or excluded? Does it make a difference if the disaster is

human-induced and/or deliberate? (See Peek for related important questions.)

- Research will be needed to determine how to overcome the uncertainty of predicting future attacks, particularly for the benefit of the commercial insurance industry. (Rubin and Renda-Tanali point out this need.)
- The risk communication model used in formulating and disseminating warnings for other types of disasters needs to be refined for a better fit with the risk of terrorism. (See the report by O'Brien on this issue.)
- There is a need to assess the full range of successes and shortcomings associated with mapping technologies, based on the experience in New York City. How did people actually use the maps? How were they incorporated into decision making? (See Thomas et al. for more on this.)

Research Needed for Understanding all Types of Disasters

- Future disaster research should explore creativity in a more systematic way than it has to date: the structure of the creative process, types of creativity, and possible discord. (See Kendra and Wachtendorf.)
- Research should explore whether the same organizational factors that impede or facilitate creativity in business settings can have an impact on the disaster response environment and emergency response organizations. (Kendra and Wachtendorf explore this idea in their paper.)
- The scale, scope, and location of the September 11 disaster in New York City demonstrated the need to better understand the interdependencies among critical infrastructure systems and to model them under different conditions. This is a step toward integrating such models with computer-based decision support systems (as called for in *Disasters by Design* (Mileti, 1999)) that can help with the design of such systems and their management during future disruptions. (See the observations made by Wallace et al.; and by Zimmerman.)
- Research should examine the interaction between the private and public sectors in disaster management. What factors hinder and foster coordination? (McEntire et al. point out questions related to this issue.)
- How applicable to the private sector are the models of disaster management, assistance, and cooperation that have been based on governmental and non-profit operations? (See the report by Michaels.)
- Further research is needed on the way communications technology (such as the wide use of the internet) is changing the formulation, dissemination of, and response to warnings. (See O'Brien.)

- Research should explore the ways in which existing typologies of organizational structure and adaptation need to be modified to take into account the more complex organizational changes that were observed in the September 11 disaster. Do certain types of organizational structures, or adaptations have more success in disaster response? (See Sutton.)
- More research is needed on the factors that facilitate volunteering during disasters, including the motivations of volunteers and their socio-demographic characteristics. (See Lowe and Fothergill.)
- Additional research is needed into how to manage large numbers of victims, from patient information to the process of DNA collection and matching. (See Simpson and Stehr.)
- Thorough exploration is needed of how geo-technologies, particularly mapping, can be integrated effectively into the emergency management cycle for all hazards. (See Thomas et al.)

Research on Quick Response

- Researchers should continue to make progress on developing a systematic approach to conducting quick response research, archiving its information and data, making findings accessible, and refining techniques. (See Michaels.)

Recommendations for Practice and Policy

Based on the findings and observations made during the conduct of the research reported here, the authors have offered numerous ideas for how public policies can be improved to better cope with future disasters, and how the practice of disaster response likewise can be enhanced. Some highlights are summarized below.

Terrorism Disasters

- Law enforcement and investigative personnel need to be integrated into disaster planning, training, and exercises, because they will have a central role in terrorist disasters. (See Grant et al. for more on this.)
- More media attention to the broader political, social, religious, and other aspects of the September 11th and similar disasters could help Americans better understand the terrorism risk and the consequences of mitigative actions the country might take. (See the paper by Rodrigue.)
- Researchers and practitioners need to determine the best protective actions that people can take in response to terrorism events, so that proper warnings and instructions can be formulated. (See O'Brien.)

- A consistent policy is needed that balances the need to know vs. the need to keep information and databases about critical infrastructure systems secure. (See O'Rourke et al.; and Zimmerman.)

All Disasters

- Disaster response plans and exercises should try to enhance the creativity and improvisational skills of people and organizations. (See Kendra and Wachtendorf.)
- Public officials and agencies should involve businesses, corporations, and faith-based organizations in all types of planning for disaster prevention, response, and recovery. (See McEntire et al.; and Sutton.)
- Practitioners should continue to emphasize networking and partnering, in order to facilitate needed cooperation between the public and private sectors in disaster management. (See McEntire et al.)
- Community service may be an effective recovery strategy for those who feel victimized by the disaster, because it enables them to transform helplessness into feelings of efficacy. (See Lowe and Fothergill.)
- Public policy should address the potential for religious and ethnic discrimination and other forms of social exclusion after human-induced disasters. (See Peek.)
- In their efforts to build and improve levels of preparedness, states and communities across the nation should consider the State of Oregon's approach, which is to help communities do long-term planning, help build local capacity, mitigate all risks, and set up partnerships among government and the private sector. (See LeDuc et al.)
- Emergency responders may need to give more attention in their plans, training, and exercises to handling massive numbers of missing people, casualties, fatalities, and family notifications. (See Simpson and Stehr.)

Quick Response

- There is a need for broader, deeper, and more institutionalized funding for research that can be carried out in the immediate aftermath of disasters. (See Michaels.)

References

Mileti, Dennis S. 1999. *Disasters by Design*. Washington, D.C.: Joseph Henry Press.

Perishable Information, Enduring Insights? Understanding Quick Response Research

Sarah Michaels
School of Planning
University of Waterloo

Introduction

The hallmark of quick response research is collecting information that would be lost if it were not gathered in the immediate aftermath of a damaging event. The intent of quick response research is to understand circumstances that exist only fleetingly and/or to document evidence created as a result of a damaging event that will not survive clean-up operations. Disaster-induced damage, ephemeral situations, and people's attitudes and actions toward them quickly become overlaid with the next layer of the drive towards a more stable, sustainable situation. Consequently, investigators in a wide range of disciplines, as diverse as structural engineering and sociology, seek to investigate the temporary world of disaster incidents and what happens in their immediate aftermath. In the social and policy sciences, researchers are likely to describe their investigations during or immediately after a damaging event as "quick response" research, while those in science, engineering, and architecture may use other terms, such as "field reconnaissance" (Frost and Deaton, 2000), or simply "reconnaissance" (Singh, 1997; Rojahn, 1997; Arnold, 1997). Regardless of the terms used to describe their efforts, investigators undertake their research to advance both our ability to prepare, respond, and mitigate future damaging events and to further the creation of new knowledge within their disciplines and beyond them.

Quick response research has become widely thought of as a research method. Yet it is not a set of procedures or techniques for undertaking systematic inquiry. Erroneously considering quick response research as a research method leads to the misconception that it is somehow a subset of social science research, particularly qualitative field work. As a result, it becomes more difficult to see that quick response research is practiced with great benefit across disciplines. Another danger of conceptualizing quick response research as a distinct methodology is that it becomes conceivable to justify lowering standards of rigor in executing all phases of an investigation. The inappropriate rationale would be that blanket exceptionalism is warranted by the phenomenon under investigation.

Consequently, rather than thinking about quick response research as research method, it is more constructive to understand quick response research as research distinguished by the time period during which the information collection phase is undertaken—during or immediately after a damaging event. The circumstances of data collection often require compromising or suspending established protocols for methods of data collection that are not inherently flexible. Methods of data collection that are flexible and do yield information in states of flux, such as interviews and observation, become the instruments of choice across disciplines in the disaster and post-disaster environment. Compromises in the non-data collecting phases of inquiry are justified only to the extent dictated by the compromises required in data collection.

The next two sections consider some of the motivations behind undertaking quick response research, including its applied utility, contribution to theory, and research objectives for undertaking it. This is followed by a discussion of selected research methods and data collection strategies and techniques useful to quick response researchers. Eleven trends shaping quick response research are then outlined, followed by a selective discussion of program support for quick response research. In conclusion, recommendations are made with the intent of stimulating discussion about developing a more systematic consideration of quick response research.

Applied Utility and Contribution to Theory

Quick response research can be important practically, often fulfilling researchers' desire that their investigations directly or indirectly bring about an improvement in a situation (Palys, 1997). Results may suggest how to minimize loss or enhance recovery (Cisin and Clark, 1962). Examples of the former are post-earthquake reconnaissance leading to improvements in the practice of structural engineering (Rojahn, 1997) or identifying the contribution architects can make to reducing structure vulnerability (Arnold, 1997). The work of Wallace et al. (2003) exemplify the latter. They focus their research on how to assist emergency managers in responding to service disruptions arising from events that affect infrastructure interdependencies.

When it comes to understanding policy research conducted in the quick response mode, it is helpful to distinguish between analysis *for* policy and analysis *of* policy (Carley, 1980). Analysis *for* policy is applied policy research intended to be used by decision makers to improve policies or programs. It is requested by and provided for decision makers (Haas and Springer, 1998). The intent is not to make a general contribution to disaster-related knowledge; rather it is intended to enhance the capability of organizations who must plan for and/or cope with disasters. As such, it focuses on specific, practical concerns (Tierney, 2002). It is exemplified by the rapid assessments that the National Weather Service (NWS) has undertaken for the last 50 years. Teams are dispatched immediately after an extreme weather event to evaluate the performance of the NWS as it relates to the particular event being investigated. This includes assessing the utility of the weather forecasts to the media and individuals (Lerner, 2002). Likewise, the Federal Emergency Management Agency carries out its own research. Routinely it administers surveys on customer satisfaction focusing on questions that the agency wants answered. The City of Los Angeles exemplifies a jurisdiction that conducts its own post-disaster reconnaissance to learn first-hand from how other communities handle disaster (Tierney, 2002).

As analysis *for* policy expands, the concern is that organizations may decline involvement in analysis *of* policy. Assisting academic researchers may come to be seen as nothing more than a nuisance if organizations are able to support investigations that focus on their specific needs (Tierney, 2002).

Investigations into policy research conducted through the Natural Hazards Research and Applications Information Center's Quick Response Research Program, funded through the National Science Foundation, typically exemplify analysis *of* policy. They are usually academic investigations of policy problems concerned primarily with illuminating causes and effects of policies and programs. The Natural Hazards Center's program is not institutionalized in such a way that the research it funds has a direct conduit for affecting policy decisions.

While quick response research has practical importance, it also can contribute to purely theoretical work. Stallings (2002a) points out that an established way of better understanding how personality and social structures are created and recreated comes from understanding exceptions to rules, such as disruptions and threats of disruptions to routines. Aguirre et al. (1998) were able to extend emergent norm theory by demonstrating the ongoing importance of lasting social relationships in shaping collective behavior. This theory presumes that nontraditional collective behavior arises out of a normative crisis. The February 26, 1993, terrorist attack on the World Trade Center was the precipitating event that created the normative crisis that Aguirre et al. (1998) investigated using data collected from a survey of 415 people who worked at the World Trade Center in the first week of May 1993.

Research Objectives

Quick response research may be conducted for exploratory purposes, descriptive purposes, or explanatory purposes. A research project may have elements of more than one of these research objectives. The distinctions between purposes are not clear cut, although research can be categorized based on its primary emphasis (Selltiz, 1976).

Exploratory

Research is done to gain an understanding or to gain new insights into a particular phenomenon. It can be used to derive precise research questions or to develop hypotheses. Exploratory research requires a flexible and wide-sweeping strategy, open-ended techniques, and the use of atypical samples. It is undertaken for heuristic purposes, to get closer to an explanation (Palys, 1997; Selltiz et al., 1976).

Kendra and Wachtendorf (2003) describe their quick response research as an exploratory case study into the role of creativity in emergency response. By becoming familiar with the phenomenon being investigated, through exploratory research, the researcher can begin to identify important variables and questions of interest. During exploratory research, the researcher must avoid foreclosing what may prove to be worthwhile avenues of investigation and remain open to various perspectives. Exploratory research is a good source of ideas and helps ensure that when more systematic research questions and designs are formulated they will be meaningful. It is important to acknowledge that the results of exploratory research may be little more than anecdotal (Palys, 1997). They may also be wrong. At a minimum, users of information gathered through exploratory research must seek out subsequent work to find out whether preliminary results have been overturned or not.

“Preliminary” is an adjective frequently used to describe the results of quick response research (Bolin, 1990, p. vii). O’Brien (2003) emphasizes that his quick response research is the first phase of a larger project. In phase two, he will use quantitative methods to help achieve his long-term objective of understanding risk communication model dynamics in the context of terrorist attacks. O’Rourke et al. (2003) describe the lessons learned from their quick response research into critical utility systems after the September 11th terrorist attacks as preliminary. Work is proceeding on collecting additional information that will lead to refining the database and clarifying issues. As a follow up to his quick response research that generated an initial assessment, Sattler (2003) is conducting a longitudinal study to examine delayed mental health problems and growth and resiliency.

Quick response researchers are encouraged to use the initial findings they generate to lay the groundwork for larger and longer-term research projects.

Twice the Natural Hazards Center has surveyed researchers it funded to undertake quick response research to find out if they had applied for funding to pursue work begun as quick response research. The first survey, completed in 1993, was of 67 researchers who performed studies between 1986 and 1992. Of those, 53 responded for a response rate of 79%. In 1996, 33 researchers who completed quick response research between 1992 and 1995 were surveyed, and 22 responded. The first survey revealed that quick response researchers submitted 31 proposals for followup research, of which 29 were funded. At the time of the survey, three additional proposals were pending and four were in the process of development. The 1996 survey revealed that 13 longer-term proposals were submitted by respondents. (Myers, 2002; Natural Hazards Research and Applications Information Center, 1996).

Descriptive

Descriptive research attempts to accurately portray the characteristics of whatever entity is being studied, be it an individual or a population (Palys, 1997; Selltiz et al., 1976). Descriptive research is particularly important in disaster and immediate post-disaster settings because it is a critical means of preserving and interpreting perishable data. It may not be possible to revisit or replicate the situations that quick response researchers describe. Grant et al. (2003) provides a thorough account of intergovernmental cooperation in a multi-jurisdictional response to the United Flight 93 air crash in a small community, Shanksville, Pennsylvania. They discuss such specifics as mass feeding and technical communication. Simpson and Stehr (2003) explain the process of identifying individuals missing after the collapse of the World Trade Center. Kendra and Wachtendorf (2003) describe instances of creativity in the recovery from the September 11th terrorist attacks on the World Trade Center. Warn et al. (2003) depict the structural and non-structural damage to 130 Liberty Street, located immediately south of World Trade Center tower 2, which collapsed in the September 11th disaster.

Explanatory

Explanatory research attempts to study causal relationships (Palys, 1997). The purpose of such research is to test a hypothesis about a causal relationship between variables (Selltiz et al., 1976). For example, Holguín-Veras et al. (2003) set out to discover if there were linkages in behavioral changes in intercity travel and the impact of the terrorist attacks of September 11th. They did discover statistically significant linkages. However, they employed an unorthodox experimental design and data collection process because they were concerned that delaying any further the collection of stated preference data would result in further dissipating the transient behavioral effects associated with the September 11th disaster. Consequently, they made pragmatic decisions about the scenarios to include in the experiment and about the people from whom they would collect stated preference data.

Research Methods

Quick response research employs approaches, methods, and data collection techniques familiar to social science researchers (Jacobsen, 1956; Cisin and Clark, 1962; Mileti, 1987; Stallings, 2002b) and by no means alien to investigators in policy, science, and engineering. They make possible analyses that have "grounded relevance," essential for analysis to make its way into the decision-making process (Haas and Springer, 1998).

Collecting high quality, primary data during and immediately after a damaging event is a challenging enterprise (Stallings, 2002b). It is essential to be successful if the value of the information obtained in the immediate post-disaster phase is to be maximized (Frost and Deaton, 2000). The actuality of quick response research highlights the gap between the ideal and the practice of conducting research in the immediate aftermath of a disaster. It may be one reason why, as Stallings (2002b) notes, disaster research methods per se have not been discussed much.

This section begins with a brief discussion of the case study approach, a standby of quick response research. Next, field work, the most commonly used method of data collection in quick response research (Tierney, 2002), is discussed in connection with allied techniques such as observation, elite interviews, and document review. Then, sampling is discussed because it highlights the often opportunistic nature of quick response research.

Case Study Approach

When the primary task is to understand how action is taken to manage specific situations, a qualitative case study approach is helpful. A case study approach provides for a complete understanding of the complexity of a situation by examining the phenomenon within the context that it occurs (Yin, 1984). Kendra and Wachtendorf (2003) describe their quick response research as an exploratory case study into the role of creativity in emergency response. Holguín-Veras et al. (2003) treat the terrorist attacks on the World Trade Center as a case study of how extreme events affect intercity travel behavior.

Field Research

Field work remains the most commonly used method of collecting data during quick response research (Tierney, 2002). Field research is conducted in science, engineering, and the social sciences. For example, in science and engineering, expert field reconnaissance immediately after earthquakes is valuable for determining the mechanisms responsible for poor performance of structures during an earthquake (Frost and Deaton, 2000) and to improve understanding of ground motion characteristics and the damage potential to different structural systems (Singh, 1997).

One example of social science field work undertaken immediately after the September 11th disaster terrorist attacks is Kendra and Wachtendorf's (2003) exploratory case study into the role of creativity in emergency response. They undertook 750 hours of systematic field observation over two months beginning on September 13, 2001. They observed key planning meetings at highly secured facilities, such as New York City's Emergency Operations Center, incident command posts, and the federal Disaster Field Office. They spent time observing operations at volunteer, supply and food staging areas, "Ground Zero," assistance centers for victims' families, and rescue workers' respite centers. They observed activities at major security checkpoints in lower Manhattan and at locations central to emergency response. They generated extensive notes, took over 500 photographs and sketches, and collected floor plans of various facilities. The primary interest of those gathering the data were the activities of formal and informal organizations and the emergence of multi-organizational networks. Systematic field observation facilitated informal interviews.

Field experience enables researchers to get a sense of initial reactions and attitudes, observe events as they unfold, access local news coverage, visit various response facilities, walk the streets, generate contacts for future research, demonstrate commitment to key informants, and suggest additional research (Simpson and Stehr, 2002). If in the social sciences "field research is the systematic study of ordinary events and activities in the settings in which they occur" (Bailey, 1996, p. 1), then what distinguishes disaster research is the circumstances in which it takes place (Jacobsen, 1956; Stallings, 2002b). If field research is typically about understanding "the meaning of daily events from the perspectives of those being studied" (Bailey, 1996, p. 3), quick response research is about understanding the meaning of exceptional events or daily events in exceptional circumstances from the perspectives of those being studied. Whereas Bailey (1996) classifies field research as a longitudinal design, because data collection usually takes months or years, quick response research employing field study is characterized by a shorter time in the field.

Field research is distinguished by the collection of data in the setting of the phenomenon of interest. In general, in field research researchers do not control events and interactions. Data collection is undertaken primarily through interactions and observations. Other techniques, such as semi-structured and structured interviews, may be used to supplement observation. In addition, the contents of documents may be analyzed or surveys distributed selectively. While field research is primarily classified as qualitative research, it may contain quantitative analysis as well. Whereas normally the interactions and observations of the field researcher are systematic (Bailey, 1996), in quick response research settings, interactions in the field may be both systematic and serendipitous (Fitzpatrick and Mileti, 1990). O'Brien (2003) describes administering interviews with City of New York workers still on emergency alert after the terrorist attacks on an "as-available basis."

Field investigators may be challenged in gaining access to key informants (Bailey, 1996). This is true for quick response researchers who are interested in minimizing the loss of details respondents will be able to provide. The sooner a researcher can speak with informants, the less opportunity informants will have had to develop a canned response (Killian, 1956). It is not always possible for researchers, however, to gain access to key informants while the latter are still fully engaged in response and initial recovery activities (Killian, 1956; Grant et al., 2003; Simpson and Stehr, 2002). Arriving in Shanksville, Pennsylvania, three days after the crash of United Flight 93, Grant et al. (2003) recognized that it would not be possible to conduct interviews with key personnel engaged in emergency response. Two weeks after the World Trade Center collapsed, Simpson and Stehr (2002) were unable to access key informants in the New York City Office of Emergency Services, the Office of the Medical Examiner, the New York State Office of Emergency Services, and the Federal Emergency Management Agency, who were engaged in victim identification.

Ultimately, it is the field researcher's understanding and interpretation that determines what is learned from field research. Field research is highly flexible. It can be undertaken by an individual or by a team. It is well suited, but not limited to, exploratory and descriptive research. It is often used for generating theory and hypotheses. It rarely involves hypothesis testing, standardized questions, or manipulation and control of variables. The field work experience is rarely predictable prior to undertaking it (Bailey, 1996).

Quasi-experimental Field Studies

Guetzkow (1962) and Drabek (2002) both advocate quasi-experimental field studies. A quasi-experimental approach adapts the logic underlying traditional laboratory experimentation to a field setting. Since, unlike in the laboratory, a field researcher does not completely control all aspects of the situation, analytic rather than manipulative control is used to control rival, plausible explanations. This involves the researcher's taking an existing situation and using her or his analytical powers to make sense of the observed causal patterns (Palys, 1997). For Guetzkow (1962), a quasi-experimental approach would expand the traditional single-case post-disaster study. While there are exceptions, e.g., Morgan's (1990) investigation of the psychological impact of the Loma Prieta earthquake on law enforcement officers and health care providers, this approach has not been widely used in quick response research.

Observation

Since researchers generally do not control events and interactions in field research, observation becomes a critical means of collecting data (Bailey, 1996). Observation is important as the basis for careful analysis of disaster conditions (French, 1990). It can be used (1) when documentation is unavailable, (2) when key informants are unavailable, (3) when estimates and statistical data are spotty, (4) to verify information obtained through other means, (5) when it is the most appropriate means of obtaining the data, and (6) to generate ideas about future research.

Documentation is Unavailable

To collect information about the structural and nonstructural damage to a 39-story office building in lower Manhattan that was damaged in the collapse of the World Trade Center, one member of the Multidisciplinary Center for Earthquake Engineering (MCEER) research team accompanied a structural engineer from LZA/Thornton Tomasetti on a detailed inspection of the interior of the building on September 23, 2001 (Warn et al., 2003). The building is located immediately south of tower 2 of the World Trade Center. The reconnaissance of the interior followed an exterior inspection of the building from ground level two days previously. The interior inspection involved climbing from the entry level to the roof via a stairwell located in the building's core and then conducting a floor-by-floor inspection, paying particular attention on each floor to a particular northerly portion. The MCEER research team was motivated to analyze the building to understand why the building did not collapse, despite the observed loss of key structural elements and severe damage. Since the research team did not have access to detailed information on the structural framing system of the study building, they made do with the approximate sizes noted during the interior building reconnaissance (Warn et al., 2003).

Key Informants are not Available for Interviews

Upon arriving at the crash site of United Flight 93 in Shanksville, Pennsylvania, three days after the flight had been downed, Grant et al. (2003) recognized that it would not be possible to conduct interviews with key personnel engaged in emergency response as they had originally intended. Consequently, they amended their method to undertake, in the first instance, observation, supplemented with informal conversations.

Estimates and Statistical Data are Spotty

French (1990) and a team of four graduate students made a preliminary assessment of damage to urban infrastructure after the Loma Prieta earthquake, relying primarily on first-hand observation. To supplement their observations they used available estimates of physical and financial damage provided by state or local officials.

To Verify Information

In her examination of the Muslim student experience after September 11th, Peek (2003) undertook participant observation to verify the experiences that students had discussed in interviews with her. For example, while walking through subway stations and sitting on trains with the young Muslim women she had interviewed, Peek (2003) confirmed that these young women received looks from passersby that they had reported in the interviews.

Most Appropriate Means of Obtaining the Data

Gillham and Edwards (2003) observed strategy meetings and protest events as one means to collect data for investigating changes in the organization field and changes in events by those who had been planning an "episode of contention" against the World Bank and the International Monetary Fund September 23–30, 2001. They observed three different decision-making meetings; an American Federation of Labor and Congress of Industrial Organizations (AFL–CIO) "disaster relief blitz;" two activist gatherings at the Anti-Capitalist Convergence Center; a number of protest events, such as a teach-in, a people's summit, and an interfaith prayer gathering; two legally sanctioned sets of rallies and marches; and an illegal march.

To Suggest Future Research

Using informal observation, Wallace et al. (2003) found that visualization tools, such as geographic information systems, were widely employed during the response to the attacks on the World Trade Center. Consequently, they suggest conducting work that will contribute to the capabilities for visualizing assumptions and implications of models of infrastructure interdependence.

Elite Interviews

Elite or specialized interviewing involves collecting non-standardized information from selected individuals having specialized knowledge about an event or process (Patton and Sawicki, 1993). After the terrorist attacks of September 11th, a number of quick response researchers used variations of specialized interviewing, both in person and over the phone (for example, McEntire et al., 2003; Sutton, 2003, Gillham and Edwards, 2003; Grant et al., 2003; Thomas et al., 2003; O'Rourke et al., 2003; Wallace et al., 2003). Three examples illustrate the types of persons investigators approached to conduct specialized interviews. First, Thomas et al. (2003) spoke with providers and users involved in the geotechnological response efforts in New York City after the September 11th disaster. Second, O'Rourke et al. (2003) interviewed deputy fire chiefs in command of the eastern and western sides of the World Trade Center complex to assess the amount of water drawn from the water distribution system. Third, Wallace et al. (2003) interviewed two public service providers to assist in choosing and developing cases featuring interorganizational aspects of disruptions in critical infrastructure systems.

Thomas et al. (2003) found that, in addition to face-to-face interviews, telephone interviews were also successful for gathering information to understand the broad use of geo-technologies. Michaels (2003) used telephone interviews to speak with representatives of information technology firms that had provided assistance in the immediate aftermath of the September 11th disaster. Using the telephone made it possible to conduct interactive interviews with individuals located throughout the United States and overseas.

Document Review

After the terrorist attacks of September 11th, quick response researchers used document analysis when access to key informants was not possible (Rubin and Renda-Tanali, 2003), as an investigative focus (Rodrigue, 2003), to identify specific events (Wallace et al., 2003; Zimmerman, 2003), and to complement interactive methods, such as interviews (O'Rourke, 2003; Gillham and Edwards, 2003; Thomas et al., 2003; Simpson and Stehr, 2003; Grant et al., 2003).

When access to key actors for personal interviews was not possible, Rubin and Renda-Tanali (2003) relied primarily on secondary sources in their investigation of the effects of the September 11th disaster on federal emergency management in the United States.

Rodrigue (2003) undertook a content analysis of *Los Angeles Times* front screen stories on the September 11th disaster. She began by recording the headline, lead sentences, date, key theme, and up to three modifiers of that key theme. The system she created yielded 17 key themes, 10 of which included more than five stories each.

Wallace et al. (2003) had independent coders identify and classify instances in the borough of Manhattan in which critical infrastructure was disrupted based on reports of such incidents in the *New York Times* Metro edition between September 12, 2001, and December 12, 2001. To identify events that relied heavily on infrastructure and to uncover how infrastructure was used in the immediate aftermath of the attacks on the World Trade Center, Zimmerman (2003) reviewed documents, reports, and media coverage supplemented by presentations from public and private owners, operators, and regulators of infrastructure in the New York City region.

O'Rourke et al. (2003) reviewed pumping records and examined high-resolution aerial photography in addition to interviewing Marine Division personnel from the New York City Fire Department. They did so to estimate how much water was supplied directly from the Hudson River to the World Trade Center complex and surrounding buildings for fire fighting after the September 11th attacks.

Gillham and Edwards (2003) gathered documents and examined the websites and e-mail list-servs of specific advocacy organizations that, immediately before the September 11th disaster, had been organizing a demonstration of support for global justice. To supplement the elite interviews they conducted, Thomas et al. (2003) tracked the use of maps in the *New York Times* since communicating to the public contributes to supporting the emergency management cycle. In their investigation of victim management and identification after the World Trade Center collapse, Simpson and Stehr (2003) collected additional data from reports in the *New York Times* and other publications.

Grant et al. (2003) supplemented observation and focused interviews with content analysis of the newspaper coverage of the response and recovery operation. While they utilized national news services, they focused on local coverage because it contained more details. Local coverage reflected that members of the local media were familiar with the local people engaged in the response effort. Grant et al. (2003) also utilized six different formal presentations made by responders about their experiences.

Kendra and Wachtendorf (2003) collected a wide range of documents produced by local, state, and federal agencies and individuals and organizations less formally tied to response efforts than government entities. Examples of documents they collected included internal and public reports, information or resource requests, information handouts, internal memos and directives, schedules, meeting minutes and agendas, and maps. They also took advantage of the extensive electronic database of articles and web-based information compiled by the Disaster Research Center. For six months after the attack, newspaper articles from major New York City papers were collected. The database also included articles from major periodicals, selected international newspaper articles, and information from government, charity, community-based, private sector, and individual internet sites that were created after and about the terrorist attacks of September 11th.

As illustrated in a number of examples above (Rodrigue, 2003; Wallace, 2003; Zimmerman, 2003; Thomas et al., 2003; Simpson and Stehr, 2003; Grant et al., 2003; and Kendra and Wachtendorf, 2003), newspapers were an important source of information. Quick response researchers used them to track specific incidents and to understand the broader context of events.

Sampling

In Bolin's (1990, p. vii) edited collection of quick response research into the 1989 Loma Prieta Earthquake, he comments how "research methods become opportunistic and flexible" in the emergency phase of the post-disaster environment where so much is in a state of flux. While this is evident in the choice of quick response researchers to use convenience, purposeful, and snowball sampling, Bourque et al. (2002) make the case that it is worth reconsidering the use of standardized, population-based surveys.

Convenience Sampling

While convenience sampling is normally discouraged, its value is recognized when it is the only feasible way to obtain information (Mason, 1996). In their investigation of what organizations and individuals did in response to aftershock and secondary hazard warnings after the Loma Prieta earthquake, Fitzpatrick and Mileti (1990, p. 76) describe how they employed a "convenience sampling technique" to gather data. Organizational respondents were selected based on their knowledge about warnings, while citizens were chosen based on "availability and convenience" (Fitzpatrick and Mileti, 1990, p. 77). Whenever possible, the researchers conducted group interviews in "affected areas where people were known to be gathering" to maximize information collected (Fitzpatrick and Mileti, 1990, p. 77).

O'Brien (2003) employed a comparable approach in New York City after the September 11th disaster. He interviewed individuals in the direct chain of responsibility for emergency response. Also, he interviewed people anywhere they congregated to discuss events; this included such venues as the street, restaurants, and subway stations.

The work of Holguín-Veras et al. (2003) provides an example of using convenience sampling in a survey conducted after September 11th. To gain insights into the effects that extreme events have upon intercity travel behavior, Holguín-Veras et al. (2003) used random utility models to provide a methodological framework for assessing behavioral changes. To generate the data to use in the model estimation process, they collected stated preference data about transportation options from volunteer research participants six months after September 11th. Between March 14, 2002, and April 4, 2002, 192 volunteers completed self-administered questionnaires. Initially the questionnaire was administered to a set of undergraduate and graduate students at the City College of New York. Undergraduate students were asked to respond to the questionnaire. To maximize the variability in the socioeconomic characteristics of the sample, the graduate students were each asked to administer the questionnaire to three other individuals. Holguín-Veras et al.'s (2003) findings are not generalizable to the population of

New York City's five boroughs because the researchers employed convenience sampling rather than representative sampling.

Purposeful Sampling

Purposeful sampling involves deliberately choosing particular people, events, or situations to obtain information that could not be obtained otherwise (Maxwell, 1996). In their qualitative investigation of volunteer behavior, Lowe and Fothergill (2003) employed a purposive sampling technique as they sought out people of diverse backgrounds. They succeeded in gaining a fairly diverse sample of 23 individuals to interview in-depth about their helping behaviors within 96 hours of the 2001 terrorist attacks on the World Trade Center.

Sattler (2003) employed purposeful sampling in conducting a study to examine if psychological distress, posttraumatic growth, resiliency, and coping varied as a function of spatial distance from the terrorist attacks of September 11th. Three weeks after the disaster, Sattler (2003) had 1283 students in four states complete a confidential and anonymous self-administered questionnaire. The response rate was 99%. He designed his sample of college students from the four regions of the country such that they had similar demographic characteristics. The rationale was that this would enable comparison across the regions.

LeDuc and Parker (2003) report on a survey designed to evaluate interagency communication and coordination in the post-disaster environment. The survey was distributed in October 2001 to the directors of the 22 member agencies of the Oregon Emergency Response System. All units in the population were targeted for study. An 87% response rate was achieved. In its capacity as state response coordinator, the Department of Oregon Emergency Management asked the Oregon Natural Hazards Workgroup, University of Oregon, to review the nature of state agency communication in light of the need to activate emergency response plans after the September 11th attacks.

Focus groups normally bring together a purposive sample of informants to discuss a phenomenon selected by the researcher (Palys, 1997). Participants value the interaction they have with their peers during focus group interviews (Michaels, 1993). Peek (2003) conducted 19 focus groups with self-selected participants as the primary means of collecting data during the initial phases of her study on Muslim student experiences after the terrorist attacks of September 11th. Such an approach enabled her to utilize effectively limited time and resources to gather a range of perspectives. She supplemented her focus group work by using one-on-one semi-structured interviews and participant observation.

Snowball Sampling

Snowball sampling starts with one or two people and then uses their connections and the connections of their connections to generate a sample (Palys, 1997). Its purpose is to identify cases of interest by using the expertise of people who know others who know which cases are rich in information (Miles and Huberman, 1994). Snowball sampling is a useful means of generating elite interviews where the actors know each other. McEntire et al. (2003), Thomas et al. (2003), Sutton (2003), and Lowe and Fothergill (2003) all describe using snowball sampling in their quick response research after the September 11th disaster.

Representative Sampling

Bourque et al. (2002) argue that in light of new, technologically sophisticated methods for conducting surveys, it is worth reconsidering the use of well-designed, standardized population-based surveys for certain quick response research topics. They use the term "survey research" to refer to a sample of respondents that represents an underlying population. They suggest that the representativeness of any random digit dialing sample after a disaster in the United States, particularly in an urban area, will be as good as if not better than any other method of collecting data, provided that the researcher is able to meet five conditions that they explain.

Selected Trends Shaping Quick Response Research

Quick response research is not a static enterprise in a constant world. Eleven trends demonstrate this.

(1) Beginning with a deductive approach to research, quick response research has evolved to include both deductive and inductive approaches.

(2) From its roots primarily in single disciplines, quick response research now is undertaken by researchers in a plethora of disciplines and has become multi-disciplinary and interdisciplinary.

(3) After a significant disaster, there is a growing probability of researcher convergence at the impact site.

(4) Those who conduct quick response research are an increasingly diverse group of individuals.

(5) The oversight of institutional review boards has become pervasive when it comes to human subjects.

(6) There is growing concern over litigation.

(7) Increasing concern over security may limit access to decision makers and documentation.

(8) The specter of biological and chemical attacks on their own or in conjunction with another disaster heightens potential health risks to quick response researchers.

(9) Information technology is making possible new means of collecting perishable data.

(10) The potential for conducting quick response research away from the immediate vicinity of a disaster site is expanding.

(11) Quick response research now functions in a context where more people know more quickly about what has happened than in times past.

From a Deductive Approach to Deductive and Inductive Approaches

The implicit approach of the early social science quick response researchers was deductive. Deductive research is theory-guided. Research is focused by what a theory suggests should be looked at and where relevant information may be gathered (Palys, 1997). Killian (1956, p. 14) talks about the need "to insure that the subjects selected are adequate for testing the hypotheses of the study." The early work emphasizes the mismatch between the rigors dictated by the hypothetico-deductive approach and the reality of disaster field work. The strategy employed to address this discord was to introduce as much discipline into data collection as is practicable.

The deductive tradition in quick response research continues. O'Brien's (2003) work exemplifies theory-guided, deductive, exploratory quick response research. He undertook to test and extend a model of risk communication to understand the warnings disseminated immediately after the September 11th terrorist attacks. Sattler (2003) also employed a deductive approach to his qualitative research. He developed a questionnaire to test his speculation (based on research, theory, anecdotal evidence, and the nature of the threat) that citizens were experiencing fear and distress and were reassessing their own lives and showing resiliency. Those researchers who operate in a deductive mode use exploratory research as pilot studies (Palys, 1997).

Other contemporary quick response researchers operate in an inductive research mode, emphasizing the generation of theory that emerges from research. For example, Rodrigue (2003) conducted what she described as an inductive and iterative approach to classifying articles discussing the terrorist attacks of September 11th in the *Los Angeles Times* in the first 12 weeks after the attacks. Sutton (2003) described how she examined the descriptive material she collected to identify themes she wished to consider in light of the Disaster Research Center's model of organizational adaptation. Exploratory research is an integral and focal part of research in theory-generating, inductive research. Initial exploration is essential as a source from which theory might emerge. The exploratory phase of research allows inductivists to gain new insights. Particularly important are the perceptions of those who inhabit the research site (Palys, 1997).

From Single Disciplines to All Comers

Speaking at a June 1952 conference on earthquake and blast effects on structures in Los Angeles, Engle posed the question, "Why is it that the practicing engineers on this [the West] coast find it impossible to collaborate on a clear, factual, complete and concise report following a major earthquake?" (Engle, 1952, p. 185). Less than a month later, earthquakes in Kern County became the foci of the first detailed analyses of the structural behavior of earthquake-resistant construction. While the emphasis on structural behavior in these initial investigations put structural aspects of building earthquake engineering ahead of other aspects, the Earthquake Engineering Research Institute (EERI) (1977) soon came to recognize that all engineering, scientific, and socio-economic effects needed to be considered in its field investigations of damaging earthquakes. In developing its multi-disciplinary approach to post-earthquake reconnaissance, EERI identified building engineering, lifeline earthquake engineering, geoscience, and social science as the concerned professions.

Likewise, while the field study of social relations after a disaster has become diversified, initially, quick response research into the immediate human response to a damaging event was almost exclusively the domain of sociologists (Quarantelli, 1994). This began with the pioneering field team that initiated disaster studies in the social sciences. The first was the National Opinion Research Center (NORC) that operated at the University of Chicago between 1950 and 1954 that studied natural and industrial disasters where circumstance might approximate a disaster in war. The Chemical Corps Medical Laboratories of the Army Chemical Center commissioned and supported its research (Quarantelli, 1987). The tradition of team field work by sociologists was continued at the Disaster Research Center, first at Ohio State University and then at the University of Delaware. Graduate sociology students received training in conducting field work during the crisis phase of disasters (Quarantelli, 1997).

The initial insistence on viewing disaster phenomena within disciplinary frameworks within the social sciences cast a long shadow throughout the late 20th century. Early efforts to launch multi-disciplinary or interdisciplinary social science investigations into disasters failed. Recruitment of representatives from other disciplines, such as anthropology and psychology, into sociology-dominated teams proved disruptive of team work (Quarantelli, 1994).

In an effort to broaden participation in post-disaster research, the Quick Response Research Program at the Natural Hazards Center was initiated in 1982. About half a dozen efforts were funded the first year. The program was inspired by the Disaster Research Center's investigations into the immediate aftermath of disasters (Tubbesing, 2002).

The multi-disciplinary National Science Foundation/Natural Hazards Center program to support quick response research was structured to avoid the more typical and lengthy process of applying for research grants, which impeded capturing perishable information. The National Science Foundation explicitly funded the quick response research program portion of the Natural Hazards Center budget. The very limited funding that is available through this program is strictly for reimbursement of the most basic out-of-pocket expenses, such as travel and lodging. Consequently, scholars may not be able to take students with them, and professors cannot afford to buy out class time for data collection and analysis or travel to destinations where travel costs are expensive, such as remote or distant locations. Limited funding means that quick response research is undertaken more on an opportunistic than a systematic basis. Still, the Natural Hazards Center program has enabled researchers throughout the United States, regardless of discipline, who might not have done so otherwise, to undertake hazards research (Tubbesing, 2002).

Trickle to Flood of Quick Response Researchers

When Killian (1956) warned researchers to be mindful of coordinating their investigations, he was concerned that two teams working in the same locale would arouse resentment among their research subjects by overworking them. Now, dozens of researchers from around the world come to study the post-event impacts of high-profile disasters. Because communities and organizations affected by a disaster can easily become overtaxed responding to multiple research efforts, a coordinated strategy for managing the needs of researchers is beneficial to both those affected by a disaster and the investigators. For example, to manage researcher convergence after the 1994 Northridge, California, earthquake, a clearinghouse was established by EERI and the California Governor's Office of Emergency Services to enable researchers to get information, make contacts, and to be briefed daily (Tierney, 2002).

Increasing Diversity in Who Is Conducting Quick Response Research

For its first two decades, quick response research was dominated by researchers who looked a lot like those they were investigating—white males. The field is becoming more diverse. While there were a few women engaged as researchers in the first 20 years, their numbers began to increase noticeably in the mid 1970s. With notable exceptions, people of color have not been attracted to the field. It is now recognized that men and women experience disaster differently, that minority populations experience disasters differently from the majority, and that these differences are a legitimate subject of investigation (Tierney, 2002).

Emergence of Human Subject Issues and Institutional Review Boards

Quick response researchers have always been subject to the same ethical considerations as their contemporaries working in other fields. Human subject issues and institutional review boards were not a factor in the early days of field research in disaster settings (Tierney, 2002). Now, compliance with Institutional Review Board process common to most academic and other research settings needs to be addressed before a disaster (Dodds and Nuehring, 1996). For example, the Natural Hazards Center's Quick Response Research Program requires pre-approved human subjects protocol from a researcher's home institution. This was not a requirement when the program began in the early 1980s. Where the opportunity for quick response research has not been anticipated in advance, Institutional Review Boards or Human Subjects Committees may be able to act quickly, as exemplified by Simpson and Stehr's (2002) experience.

The appropriateness and necessity of the standard approach to consent, geared to experimental research and studies on at-risk populations, is questionable for many forms of quick response research. Highly formalized approaches to gaining consent are a mismatch with the fluid, informal data-collecting strategies and techniques often required in quick response research (Tierney, 2002).

Growing Fear of Litigation

The fear of litigation and appearing less than competent is increasingly acting as a deterrent to officials for participating in quick response research. The emergence of quick response research was predicated on the assumption that those caught up in a disaster would be willing to be candid in expressing their thoughts and describing their actions. No longer is it possible to provide blanket assurances of privacy and anonymity. Assurances must come with the qualification that confidentiality can be provided "unless pursuant to a court order" (Tierney, 2002, p. 355).

Increasing Concern over Security

Considerably heightened concern over security in the United States and other countries is a legacy of the September 11th disaster. O'Rourke et al. (2003) describe the understandable cautiousness of infrastructure managers to provide information about critical facilities after September 11th. They describe how they collected detailed information that it is not appropriate for them to disseminate until review and clearance for it can be obtained. Taking data off-line, as occurred after the attacks of September 11th, has broad implications for the geographic information system and emergency management communities (Thomas et al., 2003), both for practice and research. Quick response researchers may face more limited access to key informants and information than they did before September 11th. Having pre-established relationships with organizations to be investigated in the aftermath of a specific event may prove decisive in enabling a researcher to access key decision makers immediately after a disaster.

Increase in Potential Health Risks to Quick Response Researchers

Since September 11th, there is the increasing specter of biological and chemical terrorist attacks either as independent occurrences or combined with other forms of disaster. In the past, quick response researchers were usually unconcerned about the immediate or long-term consequences to their health of investigating a disaster in the United States that had been triggered by a natural phenomenon, such as a tornado or flood.

Increasing Potential of Information Technology

O'Rourke et al. (2003) describe unparalleled opportunities to make use of advanced geographical information systems, remote sensing, condition monitoring, model-based simulation, and the capability to produce digital base maps that integrate the spatial characteristics of infrastructure. These tools make it possible to advance the state of the art and practice for characterizing and modeling complex infrastructure systems. A reliable system for rapidly recording querying and visualizing qualitative and quantitative information is being developed through integrating global positioning systems, geographic information systems, digital photography, and hand-held computers. Recent advances enhance the utility of data recorded during post-earthquake (Frost and Deaton, 2000) and other field reconnaissance.

Information technology makes it easier to manipulate electronic documents than paper copy. Rodrigue (2003) chose to analyze the on-line edition of the *Los Angeles Times* rather than the paper version for representation of the September 11th disaster because it made collecting data easier. She was able to copy and paste headlines and lead sentences into a spreadsheet, eliminating the need to manually retype what was being coded.

Information technology is making it possible to conduct quick response research away from the disaster impact site. First, it can expand the frame of impact and response to disaster. For example, the current state of information technology enables firms to deliver technical, communications-related disaster assistance without necessarily being in the vicinity of the physically affected area. Consequently, to understand how selected information technology firms contributed to recovery immediately after the attacks of September 11th did not require the investigator to be in the environs of the physically impacted site (Michaels, 2003). Second, information technology can provide a researcher with the same access to on-line information as someone directly affected by the disaster. For example, it would have been possible for a researcher to investigate the use of the World Wide Web in locating missing individuals after the September 11th disaster. A researcher located outside of the New York City area could use the Greater New York Hospital Association web page, "Patient Locator Service," put on-line 72 hours after the collapse of the World Trade Center, and find out—by entering a missing person's name—whether that person was being treated in an area hospital, and if so, which hospital (Simpson and Stehr, 2003). They could also explore the several websites that hosted "virtual" posters of missing people where those concerned about a particular individual could upload a scanned photograph and basic information about the missing individual (Simpson and Stehr, 2003).

On-scene to On-scene and Off-site

Where to conduct quick response research is a function of what phenomenon is being investigated. It is understandable that most quick response research has been conducted at or in the immediate vicinity of the disaster impact. Examples include Grant et al. (2003) investigating multijurisdictional response to the crash of United Flight 93 at the crash site in Shanksville, Pennsylvania, or Warn et al. (2003) investigating structural and nonstructural damage of a high-rise office building immediately south of the collapsed tower 2 at the World Trade Center.

Quick response research can also be conducted away from the site of impact. For example, three weeks after the terrorist attacks, Sattler (2003) conducted a study to examine psychological distress, posttraumatic growth, resiliency, and coping among college students in New York, South Carolina,

Colorado, and Washington State. The study was designed to examine if and how responses varied as a function of distance from the September 11th terrorist attacks. Rodrigue's (2003) examination of the representation of the September 11th attacks in the online edition of the *Los Angeles Times* provides another example of off-site quick response research. She chose the *Los Angeles Times* specifically because it was removed from where the terrorists struck and because its coverage could then be expected to focus on stories that had national appeal. Michaels (2003) investigated how information technology firms were able to deliver technical, communications-related disaster assistance without being at the sites impacted by the September 11th disaster.

More People Know More Quickly

One way in which the context for quick response research has become more complex is that there are more means than ever for the speedy dissemination of news as a disaster unfolds. One particularly striking facet of September 11th was that news of what was going on was reported instantaneously to friends or family members by people inside the World Trade Center and onboard the planes that were hijacked. The evolution of disaster-related memory is being shaped by many messages conveyed through disparate television and radio channels and through phones and the internet. Rodrigue (2003) describes how e-mail, list-servs, and news groups effectively reach audiences of a size and geographical scope previously the exclusive domain of national media conglomerates.

Program Support for Quick Response Research

The Quick Response Research Program at the Natural Hazards Center funded by the National Science Foundation is one of a number of programs that supports field investigations immediately after a disaster. Briefly noted here are examples of federal initiatives past, present, and future, and an initiative undertaken by a professional association. While the National Research Council has supported quick response research in the past, the National Science Foundation and the National Institute of Mental Health have ongoing programs. Federal legislation passed in 2002 authorizes the Director of the National Institute for Standards and Technology to establish and deploy a team within 48 hours of a major building failure. EERI has organized post-event investigations for over half a century (EERI, 1977).

A formal program existed at the National Research Council of the National Academy of Science to dispatch teams after disasters. Dozens of teams were dispatched, many of them interdisciplinary, before the formal program was disbanded (Gaus, 2002a).

In the 1960s, the National Science Foundation (NSF) made provisions to collect perishable information because its staff recognized that natural hazard events triggered full-scale tests of constructed and social systems that could not be replicated in the laboratory. The Small Grants for Exploratory Research (SGER) program of the NSF is probably the most significant source of quick response research support in the United States today (Wenger, 2002). The SGER program is well suited to quick response research because of the brevity of the proposal required (five pages) and because proposals can be approved quickly, within a day or so, at the discretion of the NSF Program Director. SGER grants can be for amounts up to $100,000 per year.

The National Institute of Mental Health (NIMH) has issued program announcement PAR-02-133 in 2002 for Rapid Assessment Post-Impact of Disaster (RAPID). It replaces an earlier announcement issued in 1990 and revised in 1995. RAPID grants provide limited funding to enable investigators, who intend to follow up with a full research application, to undertake an early assessment to generate preliminary data that can be used as a basis for their longer, full application. The intent is to enable access to a disaster site immediately after an event has occurred by providing a rapid funding mechanism for research. NIMH's Traumatic Stress Research Program recognizes that for many types of mental health studies, such as those that consider service seeking, evaluate outreach or prevention initiatives, and identify high-risk individuals based on early response, it may be critical to undertake rapid assessment (NIMH, 2002).

The National Construction Safety Team Act, which became Public Law 107-231 on October 1, 2002, is an example of legislation designed to facilitate analysis for policy, applied policy research intended to be used by decision makers to make programmatic and policy improvements. The full title of the Act provides a succinct summary of it: An Act to Provide for the Establishment of Investigative Teams to Assess Building Performance and Emergency Response and Evacuation Procedures in the Wake of Any Building Failure that has Resulted in Substantial Loss of Life or that Posed Significant Potential of Substantial Loss of Life. The law has the potential to improve the quality of failure investigations, to establish clearly who is responsible for dispatching investigative teams, and to increase the number of failure investigations (Gaus, 2002b). As of spring 2003, the National Institute of Standards and Technology, which is authorized by the Act to establish the investigative teams, had not received an increase in its funding to carry out the Act's objectives (National Institute of Standards and Technology, 2003).

EERI's program to support post-earthquake investigations is noteworthy because its evolution reflects and advances the increasing sophistication and inclusivity of quick response research. Since its inception in 1949, EERI has undertaken post-earthquake investigations. The intent of these investigations remains "to observe and document the effects of earthquakes on the built environment and the resulting social, economic, and policy impacts" (Earthquake Engineering Research Institute, 2003). The founders of EERI saw earthquakes as field laboratories—opportunities to investigate in ways that would not be feasible at test sites or true laboratories. Earthquakes allowed them to test theories and view performance under real conditions. The post-earthquake investigations have stimulated further research, changed practice in earthquake design and emergency management, and influenced building codes and standards (Tubbesing, 2002).

At the outset, EERI-affiliated engineers and earth scientists paid their own way to undertake post-event investigations. Then in the early 1970s, the National Science Foundation began funding the ongoing Learning from Earthquakes (LFE) program. Funding allowed the program to become more systematic and inclusive of a range of disciplinary expertise. It also enabled publication of results that could then be disseminated widely. The first published reconnaissance report was in 1972 on the Managua earthquake. The dissemination of observations is one of three main activities of the program. The other two are conducting investigations and developing guidelines for field data collection (Tubbesing, 2002).

The EERI program continues to evolve (Tubbesing, 2002). For example, in 1998 EERI initiated its Lessons Learned Over Time series. This series recognizes the need for a mechanism to capture lessons from earthquakes that may not become apparent until years after an event and to re-evaluate initial post-event findings in light of new understanding and knowledge (EERI, 2002). In terms of EERI's immediate post-event investigations, there is an increasing emphasis on coordinating with other teams and individuals that converge on earthquake sites. New communication and data collection technologies are being employed for more systematic data collection. Training is being provided for field investigators and team leaders. Greater use is being made of the internet to disseminate information quickly from the field (Tubbesing, 2002).

Programs to enable quick response research are not confined to the United States (Koshida, 2002a). Emergency Management Australia has piloted a program adapted from the United States' Natural Hazards Center model. The Institute for Catastrophic Loss Reduction is in the planning stages of a Canadian version of the Natural Hazards Center's Quick Response Research Program (Falkiner, 2003). Environment Canada is exploring how to develop a protocol that would enable researchers within that government organization to investigate the immediate aftermath of extreme weather events (Koshida, 2002b).

Programs that enable quick response research play a significant role in shaping how quick response research is undertaken. Program managers are in a strong position to shepherd the development of quick response research.

They can encourage the development and deployment of multi-disciplinary teams. They can also promote bringing together researchers based in the vicinity of the disaster site and those from away. They can foster teams composed of researchers at different career stages.

Recommendations

The following seven recommendations stem from the consideration of quick response research provided in this paper. The primary purpose of this incomplete listing is to stimulate discussion about developing a more systematic consideration of quick response research.

(1) Assess quick response research undertaken since World War II as a basis for setting the agenda for future quick response research.

A large body of quick response research has been generated for over half a century. What types of quick response research have been influential in advancing disaster preparedness, response, and mitigation? What has advanced disciplinary and interdisciplinary theory building? It would be valuable to look at the legacy of specific programs, such as the National Science Foundation-sponsored, Natural Hazards Center-administered quick response program, as well as gaining an overview of all programs and quick response research supported fully or partially by U.S. government funding.

What is true about maximizing post-earthquake learning—that we need to be specific about "what we do not know" (EERI, 1977, p. 15)—is true generally for maximizing learning from quick response research. A research agenda for quick response research needs to be built on identifying the gaps in the existing quick response literature. Particular attention needs to be paid to that aspect of quick response research that distinguishes it from other research, namely the collection of perishable information generated during or immediately after a disaster.

(2) Undertake periodic disciplinary and interdisciplinary assessments of quick response research.

A key goal for undertaking quick response research is to advance knowledge within a discrete discipline or interdisciplinary domain that provides the conceptual framework for specific research projects. Therefore, it would be valuable for specialists in particular disciplines or in interdisciplinary studies to periodically assess the contribution of quick response research to enhance theory in that domain. This should be the basis for proposing a research agenda for quick response research that would advance conceptual thinking. At the same time, acknowledging that quick response research employs methods utilized by scientists and social scientists outside the sphere of disasters should lead to engaging in methodological debate to advance the conduct of empirical research.

(3) Develop mechanisms that facilitate quick response research.

Given the unanticipated nature of many disasters, attention needs to be paid to how to facilitate timely and appropriate collection of perishable information. It would be worth exploring the potential of adapting the EERI model for post-earthquake reconnaissance to non-earthquake events.

(4) Designate and support one center to collect and catalog selected media coverage of major events.

Newspaper documentation, whether hard copy or on-line, has proven valuable to quick response researchers as a source of both background information and material for analysis. Having one center take the lead role in being the archival home for an event would facilitate quick response research and longitudinal research by capturing and processing popular sources of information.

(5) Establish a consortium of programs that support quick response research.

As quick response research programs proliferate, it becomes more important to establish a mechanism whereby these programs can exchange information, and as appropriate, pool resources.

(6) Provide fellowships for students to complete theses and dissertations that evolve out of quick response research.

The experience of participating in quick response research as a graduate student can stimulate a career-long research interest in disasters. The National Science Foundation/Center Quick Response Research Program encourages graduate students to apply for grants. A logical next step would be to create a program that provides financial support for graduate students to build on the quick response research they conduct.

(7) Develop quick response research capability internationally.

The impacts and responses to many disasters are not confined to one jurisdiction and there is potential to apply what is learned from one event to others. Consequently it would be valuable to think about how to support quick response researchers from different countries working together.

Conclusion

After half a century, quick response research continues to be a vibrant dimension of disaster inquiry. What is striking is how many of the issues that concerned the founders of the field continue to challenge present-day researchers. Calls for a systematic approach to field research after a disaster are not new (Killian, 1956 in general; EERI, 1977 specifically for earthquakes). Added to the historical concerns are ones that were not anticipated by the field's pioneers, such as increasing reticence of agencies to supply information. While the early "strike team" approach proved highly effective, there is growing interest in integrating quick response research into place-based research and developing the capacity of local researchers.

The term "quick response research" has been used in the past as if there were a clear conceptualization of what it is. We may have exhausted what can be accomplished, understood, and accepted from a field that has been understood intuitively rather than being defined transparently. One way forward is to explore the ramifications of appreciating quick response research as research defined by the information collection phase occurring during or immediately after a damaging event.

By recognizing that the common denominator in all quick response research is the collection of perishable data during or immediately after a disaster, it is possible to consider setting a research agenda for quick response research. This requires thinking about some fundamental questions. What are pressing quick response research needs? How should such needs be determined, prioritized, and then addressed? How can advances be made in the research methods used by quick response researchers? How can the use and functionality of multi-disciplinary reconnaissance teams be advanced in investigating a wide range of disasters? How should quick response research be evaluated? In earthquake engineering, post-earthquake reconnaissance studies collectively have had more impact on developing Applied Technology Council projects than any single investigation into specific earthquakes (Rojahn, 1997). We may find that developing a framework to understand quick response research will enable us to see that the sum of quick response research experience is greater than the parts.

References

Aguirre, Benigno E., Dennis Wenger, and Gabriela Vigo. 1998. "A Test of the Emergent Norm Theory of Collective Behavior. *Sociological Forum* 13(2): 301–320.

Arnold, Christopher. 1997. "Contributions of Earthquake Reconnaissance to Improved Understanding of Architectural Issues." Pp. 1551–1558 in *Building to Last: Proceedings of Structures Congress XV, Portland, Oregon, April 13–16, 1997, Vol. 2*. New York, American Society of Civil Engineers.

Bailey, Carol A. 1996. *A Guide to Field Research*. Thousand Oaks, CA: Pine Forge Press.

Bolin, Robert, ed. 1990. *The Loma Prieta Earthquake: Studies of Short-Term Impacts*. Environment and Behavior Monograph #50. Boulder, CO: Institute of Behavioral Science, University of Colorado.

Bourque, Linda B., Kimberley I. Shoaf, and Loc H. Nguyen. 2002. "Survey Research." Pp. 157–193 in Robert A. Stallings, ed., *Methods of Disaster Research*. Philadelphia, PA: Xlbris.

Carley, Michael. 1980. *Rational Techniques in Policy Analysis*. London: Heinemann Educational Books.

Cisin, Ira H. and Walter B. Clark. 1962. "The Methodological Challenge of Disaster Research." Pp. 23–49 in George W. Baker and Dwight W. Chapman, eds., *Man and Society in Disaster*. New York: Basic Books.

Dodds, Sally and Elane Nuehring. 1996. "A Primer for Social Work Research on Disaster." *Journal of Social Service Research* 22(1/2): 27–56.

Drabek, Thomas E. 2002. "Following some Dreams: Recognizing Opportunities, Posing Interesting Questions, and Implementing Alternative Methods." Pp. 127–153 in Robert A. Stallings, ed., *Methods of Disaster Research*. Philadelphia, PA: Xlbris.

Earthquake Engineering Research Institute. 1977. *Learning from Earthquakes*. Oakland, CA: EERI.

Earthquake Engineering Research Institute. 2002. *Request for Proposals: a New Round of Lessons Learned Grants*. http://www.eeri.org/earthquakes/LFE/rfp.html.

Earthquake Engineering Research Institute. 2003. *Learning from Earthquakes*. http://www.eeri.org/earthquakes/LFE/LFE1.html.

Engle, Harold M. 1952. "Lessons from the San Francisco Earthquake of April 18, 1906." Pp. 181–185 in Martin C. Duke and Morris Feigen, eds., *Proceedings on Earthquake and Blast Effects on Structures Los Angeles, California June 26–28, 1952*. San Francisco, CA: Earthquake Engineering Research Institute.

Falkiner, Leanna. 2003. Co–ordinator, Institute for Catastrophic Loss Reduction. London, Ontario, Canada. Personal communication with author on February 26.

Fitzpatrick, Colleen and Dennis S. Mileti. 1990. "Perceptions and Response to Aftershock Warnings During the Emergency Period." Pp. 75–83 in R. Bolin, ed., *The Loma Prieta Earthquake: Studies of Short-Term Impacts*. Environment and Behavior Monograph #50. Boulder, CO: Institute of Behavioral Science, University of Colorado.

French, Steven P. 1990. "A Preliminary Assessment of Damage to Urban Infrastructure." Pp. 17–24 in R. Bolin, ed., *The Loma Prieta Earthquake: Studies of Short-Term Impacts*. Environment and Behavior Monograph #50. Boulder, CO: Institute of Behavioral Science, University of Colorado.

Frost, J. David and Deaton, Scott L. 2000. "An Integrated System for Earthquake Damage Reconnaissance." *Sixth International Conference on Seismic Zonation: Managing Earthquake Risk in the 21st Century*. Oakland, CA: Earthquake Engineering Research Institute.

Gaus, Michael P. 2002a. President, American Association for Wind Engineering. Williamsburg, Virginia. Personal communication with author on October 25.

Gaus, Michael P. 2002b. "National Construction Safety Team Act Passed." *American Association for Wind Engineers Newsletter* (September): 9.

Gillham, Patrick and Bob Edwards. 2003. "Global Justice Protesters Respond to the September 11th Terrorist Attacks: The Impact of an Intentional Disaster on Demonstrations in Washington, D.C." Pp. 483–520 in *Beyond September 11th: An Account of Post-Disaster Research*. Special Publication #39. Boulder, CO: Natural Hazards Research and Applications Information Center, University of Colorado.

Grant, Nancy K., David H. Hoover, Anne-Marie Scarisbrick-Hauser, Mary E. Myers, and Stacy L. Muffet. 2003. "The Crash of United Flight 93 in Shanksville, Pennsylvania." Pp. 83–108 in *Beyond September 11th: An Account of Post-Disaster Research*. Special Publication #39. Boulder, CO: Natural Hazards Research and Applications Information Center, University of Colorado.

Guetzkow, Harold. 1962. "Joining Field and Laboratory Work in Disaster Research." Pp. 337–355 in George W. Baker and Dwight W. Chapman. eds., *Man and Society in Disaster*. New York: Basic Books.

Haas, Peter J. and Fred J. Springer. 1998. *Applied Policy Research*. New York: Garland Publishing.

Holguín-Veras, José, Robert E. Paaswell, and Ann-Marie Yali. 2003. "Impacts of Extreme Events on Intercity Passenger Travel Behavior: The September 11th Experience." Pp. 373–404 in *Beyond September 11th: An Account of Post-Disaster Research.* Special Publication #39. Boulder, CO: Natural Hazards Research and Applications Information Center, University of Colorado.

Jacobsen, Carlyle F. 1956. "A Special Report Prepared for The Committee on Disaster Studies. Preface." P. iii in Killian, Lewis M., *An Introduction to Methodological Problems of Field Studies in Disasters*. Disaster Study Number 8 of Committee on Disaster Studies, Division of Anthropology and Psychology. Washington, D.C.: National Research Council, National Academy of Sciences.

Kendra, James and Tricia Wachtendorf. 2003. "Creativity in Emergency Response to the World Trade Center Attack." Pp. 121–146 in *Beyond September 11th: An Account of Post-Disaster Research.* Special Publication #39. Boulder, CO: Natural Hazards Research and Applications Information Center, University of Colorado.

Killian, Lewis M. 1956. *An Introduction to Methodological Problems of Field Studies in Disasters*. A Special Report prepared for the Committee on Disaster Studies. Disaster Study Number 8 of Committee on Disaster Studies, Division of Anthropology and Psychology. Washington, D.C.: National Research Council, National Academy of Sciences.

Koshida, Grace. 2002a. *Literature Review and Evaluation of Existing Quick Response Models*. Ottawa, Ontario, Canada: Office of Critical Infrastructure Protection and Emergency Preparedness.

Koshida, Grace. 2002b. Researcher, Environment Canada. Toronto, Ontario, Canada. Personal communication with author on October 23.

LeDuc, Andre and Robert Parker. 2003. "Natural Hazard Mitigation in Oregon: A Case Study." Pp. 559–583 in *Beyond September 11th: An Account of Post-Disaster Research.* Special Publication #39. Boulder, CO: Natural Hazards Research and Applications Information Center. University of Colorado.

Lerner, William. 2002. U.S. National Weather Service, Silver Spring, Maryland. Personal communication with author on October 28.

Lowe, Seanna and Alice Fothergill. 2003. "A Need to Help: Emergent Volunteer Behavior After September 11th." Pp. 293–314 in *Beyond September 11th: An Account of Post-Disaster Research.* Special Publication #39. Boulder, CO: Natural Hazards Research and Applications Information Center, University of Colorado.

Mason, Jennifer. 1996. *Qualitative Researching*. London: Sage.

Maxwell, Joseph A. 1996. *Qualitative Research Design*. Thousand Oaks, CA: Sage.

McEntire, David A., Robie J. Robinson, and Richard T. Weber. 2003. "Business Responses to the World Trade Center Disaster: A Study of Corporate Roles, Functions and Interactions With the Public Sector." Pp. 431–457 in *Beyond September 11th: An Account of Post-Disaster Research*. Special Publication #39. Boulder, CO: Natural Hazards Research and Applications Information Center, University of Colorado.

Michaels, Sarah. 1993. "Weighing Science and Politics in Local Decision Making About Hazards." *Knowledge and Policy* 6(2): 3–22.

Michaels, Sarah. 2003. "Information Technology Firms Respond to the Immediate Aftermath of the September 11, 2001 Terrorist Attacks." Pp. 459–480 in *Beyond September 11th: An Account of Post-Disaster Research*. Special Publication #39. Boulder, CO: Natural Hazards Research and Applications Information Center, University of Colorado.

Miles, Matthew B. and A. Michael Huberman. 1994. *Qualitative Data Analysis*. 2nd edition. Thousand Oaks, CA: Sage.

Mileti, Dennis S. 1987. "Sociological Methods and Disaster Research." Pp. 57–70 in Russell R. Dynes, B. de Marchi and C. Pelanda, eds., *Sociology of Disasters: Contributions of Sociology to Disaster Research*. Milan, Italy: Franco Angeli Libri.

Morgan, Jerri Laube. 1990. "Psychological Impacts on Law Enforcement Officers and Health Care Providers." Pp. 65–74 in R. Bolin, ed., *The Loma Prieta Earthquake: Studies of Short-Term Impacts*. Environment and Behavior Monograph #50. Boulder, CO: Institute of Behavioral Science, University of Colorado.

Myers, Mary Fran. 2002. Co-Director, Natural Hazards Research and Applications Information Center, University of Colorado–Boulder. Personal communication with author on October 23.

National Institute of Mental Health. 2002. *Rapid Assessment Post-Impact of Disasters*. http://grants.nih.gov/grants/guide/pa-files/PAR-02-133.html.

National Institute of Standards and Technology. 2003. *NIST's Responsibilities under the National Construction Safety Team Act*. http://www.nist.gov/public_affairs/factsheet/constructionact.htm.

Natural Hazards Research and Applications Information Center. 1996. "1995–1996 Annual Report." Boulder, CO: NHRAIC.

O'Brien, Paul. 2003. "Risk Communication and Public Warning Response to the September 11th World Trade Center Attack." Pp. 355–372 in *Beyond September 11th: An Account of Post-Disaster Research*. Special Publication #39. Boulder, CO: Natural Hazards Research and Applications Information Center, University of Colorado.

O'Rourke, Thomas D., Arthur J. Lembo, and Linda K. Nozick. 2003. "Lessons Learned from the World Trade Center Disaster About Critical Utility Systems." Pp. 269–290 in *Beyond September 11th: An Account of Post-Disaster Research.* Special Publication #39. Boulder, CO: Natural Hazards Research and Applications Information Center, University of Colorado.

Palys, Ted. 1997. *Research Decisions*, 2nd edition. Toronto: Harcourt.

Patton, Carl V. and David S. Sawicki. 1993. *Basic Methods of Policy Analysis and Planning*. Englewood Cliffs, NJ: Prentice Hall.

Peek, Lori A. 2003. "Community Isolation and Group Solidarity: Examining the Muslim Student Experience After September 11th, 2001." Pp. 333–354 in *Beyond September 11th: An Account of Post-Disaster Research.* Special Publication #39. Boulder, CO: Natural Hazards Research and Applications Information Center, University of Colorado.

Quarantelli, Enrico L. 1987. "Disaster Studies: An Analysis of the Social Historical Factors Affecting the Development of Research in the Area." *International Journal of Mass Emergencies and Disasters* 5(3): 285–310.

Quarantelli, Enrico L. 1994. "Disaster Studies: The Consequences of the Historical Use of a Sociological Approach in the Development of Research." *International Journal of Mass Emergencies and Disasters* 12(1): 25–49.

Quarantelli, Enrico L. 1997. "Disaster Studies: The Disaster Research Center Field Studies of Organized Behavior in the Crisis Time Period of Disasters." *International Journal of Mass Emergencies and Disasters* 15(1): 47–69.

Rodrigue, Christine M. 2003. "Representation of the Terrorist Attacks of September, 2001 in the Online Edition of the *Los Angeles Times.*" Pp. 521–558 in *Beyond September 11th: An Account of Post-Disaster Research.* Special Publication #39. Boulder, CO: Natural Hazards Research and Applications Information Center, University of Colorado.

Rojahn, Christopher. 1997. "Contributions of Earthquake Reconnaissance in Developing Improved Recommendations and Guidelines for Seismic Design." Pp. 1605–1615 in *Building to Last: Proceedings of Structures Congress XV, Portland, Oregon.* New York: American Society of Civil Engineers.

Rubin, Claire B. and Irmak Renda-Tanali. 2003. "The Implications for Federal Emergency Management of the September 11, 2001 Terrorist Attacks." Pp. 585–604 in *Beyond September 11th: An Account of Post-Disaster Research.* Special Publication #39. Boulder, CO: Natural Hazards Research and Applications Information Center, University of Colorado.

Sattler, David N. 2003. "Resiliency, Posttraumatic Growth, and Psychological Distress After the Attacks on America." Pp. 315–332 in *Beyond September 11th: An Account of Post-Disaster Research.* Special Publication #39. Boulder, CO: Natural Hazards Research and Applications Information Center, University of Colorado.

Selltiz, Claire, Lawrence S. Wrightsman, and Stuart W. Cook. 1976. *Research Methods in Social Relations*, 3rd edition. New York: Holt, Rinehart and Winston. Simpson, David M. and Steven D. Stehr. 2002. "Investigating the Disaster in New York City: Conducting Field Research following the Collapse of the World Trade Center." *Natural Hazards Review* 3(2): 34–35.

Simpson, David M. and Steven D. Stehr. 2003. "Victim Management and Identification After the World Trade Center Collapse." Pp. 109–120 in *Beyond September 11th: An Account of Post-Disaster Research.* Special Publication #39. Boulder, CO: Natural Hazards Research and Applications Information Center, University of Colorado.

Singh, Jogeshwar P. 1997. "Contributions of Earthquake Reconnaissance to Improved Understanding of Ground Motions." Pp. 1600–1604 in *Building to last: Proceedings of Structures Congress XV, Portland, Oregon.* New York: American Society of Civil Engineers.

Stallings, Robert A. 2002a. "Weberian Political Sociology and Sociological Disaster Studies. *Sociological Forum* 17(2): 281–305.

Stallings, Robert A. 2002b. "Methods of Disaster Research: Unique or Not?" Pp. 21–44 in Robert A. Stallings, ed., *Methods of Disaster Research.* Philadelphia, PA: Xlbris.

Sutton, Jeannette. 2003. "A Complex Organizational Adaptation to the World Trade Center Disaster: An Analysis of Faith-based Organizations." Pp. 405–428 in *Beyond September 11th: An Account of Post-Disaster Research.* Special Publication #39. Boulder, CO: Natural Hazards Research and Applications Information Center, University of Colorado.

Thomas, Deborah K., Susan L. Cutter, Michael Hodgson, Mike Gutekunst, and Steven Jones. 2003. "Use of Spatial Data and Geographic Technologies in Response to the September 11th Terrorist Attack on the World Trade Center." Pp. 147–162 in *Beyond September 11th: An Account of Post-Disaster Research.* Special Publication #39. Boulder, CO: Natural Hazards Research and Applications Information Center, University of Colorado.

Tierney, Kathleen J. 2002. "The Field Turns Fifty: Social Change and the Practice of Disaster Fieldwork." Pp. 349–374 in Robert A. Stallings, ed., *Methods of Disaster Research.* Philadelphia, PA: Xlbris.

Tubbesing, Susan K. 2002. Executive Director, Earthquake Engineering Research Institute, Oakland, CA. Personal communication with author, November 6.

Wallace, William A., David Mendonca, Earl Lee, John E. Mitchell, and Joe E. Chow. 2003. "Managing Disruptions to Critical Interdependent Infrastructures in the Context of the 2001World Trade Center Attack." Pp. 165–198 in *Beyond September 11th: An Account of Post-Disaster Research.* Special Publication #39. Boulder, CO: Natural Hazards Research and Applications Information Center, University of Colorado.

Warn, Gordon, Jeffrey Berman, Andrew Whittaker, and Michel Bruneau. 2003. "Investigation of a Damaged High-rise Building Near Ground Zero." Pp. 199–240 in *Beyond September 11th: An Account of Post-Disaster Research.*Special Publication #39. Boulder, CO: Natural Hazards Research and Applications Information Center, University of Colorado.

Wenger, Dennis. 2002. Program Director, National Science Foundation. Personal communication with Jacquelyn Monday, Natural Hazards Research and Applications Information Center, December 31.

Yin, Robert K. 1984. *Case Study Research.* Thousand Oaks: Sage.

Zimmerman, Rae. 2003. "Public Infrastructure Service Flexibility for Response and Recovery in the September 11th Attack." Pp. 241–268 in *Beyond September 11th: An Account of Post-Disaster Research.* Special Publication #39. Boulder, CO: Natural Hazards Research and Applications Information Center, University of Colorado.

Acknowledgments

I would like to thank Dennis S. Mileti, Mary Fran Myers, and Jacquelyn Monday, Natural Hazards Center, for their endorsement of my embryonic idea that led to this paper. I would also like to thank Robert A. Stallings, University of Southern California; Enrico L. Quarantelli and Kathleen J. Tierney, University of Delaware; Michael P. Gaus, American Association of Wind Engineering; William Lerner, National Weather Service; Susan K. Tubbesing, Earthquake Engineering Research Institute; Grace Koshida, Environment Canada; Mary Fran Myers, Wanda Headley, and Lori Peek, Natural Hazards Center, for providing information and sources. Dennis Wenger, National Science Foundation, and two anonymous reviewers contributed valuable comments on an earlier draft of this paper.

Correspondence should be directed to Sarah Michaels, School of Planning, University of Waterloo, Waterloo, Ontario Canada N2L 3G1; (519) 888-4567 ext. 6863; fax: (519) 725-2827; e-mail: michaels@fes.uwaterloo.ca.

Learning from Urban Disasters: Summary of a Workshop

Carlos Restrepo
Institute for Civil Infrastructure Systems
Wagner Graduate School of Public Service
New York University

Rae Zimmerman
Institute for Civil Infrastructure Systems
Wagner Graduate School of Public Service
New York University

The collapse of the World Trade Center in New York City on September 11, 2001, was one of the worst urban disasters in the history of the United States. Almost 3,000 people perished as a result of the disaster. The economy was dealt a severe blow, the consequences of which are still felt today. When the World Trade Center was first built, its approximately 1.25 million square meters of office space accommodated about 40–50,000 people (Extreme Events Mitigation Task Force, 2002, p. 52). The number of telephone lines installed in the towers was similar to that found in cities such as Cincinnati or Copenhagen[1]. The collapse of the World Trade Center raised a large number of research questions related to understanding what happened on that day, why the buildings collapsed, how agencies and individuals responded to the event, how civil infrastructure systems were affected, and how the lessons learned can be used to prevent similar disasters from happening.

To address the many issues raised by the September 11th disaster, the National Science Foundation (NSF) awarded Small Grants for Exploratory Research and, through the Natural Hazards Research and Applications Information Center at the University of Colorado, "Quick Response" grants, to obtain information that might not be available later. NSF provided funding to the Institute for Civil Infrastructure Systems (ICIS) to convene researchers and others directly involved in response activities at a workshop to exchange experiences and knowledge. The workshop's objective was to promote connectivity among members of the research community for future collaboration and between researchers and emergency managers to advance research on catastrophic events of the kind that occurred on September 11th.

On December 12 and 13, 2001, ICIS hosted these researchers, NSF, and officials and other managers involved in the aftermath of the September 11th disaster to identify some of the key problems and issues associated with the attack and potential research themes to address these issues. Research participants came from many disciplines, and included engineers, urban planners, geographers, sociologists, political scientists, and other academicians. They came from about two dozen universities in over a dozen states in all regions of the country. Managers who were involved in response to the September 11th disaster participated as speakers at the workshop. They included professionals from federal, state, and local government; officials in emergency management and key service sectors such as environmental protection and transportation; the utilities; health institutions; the National Institute of Standards and Technology (NIST) and the American Society of Civil Engineers investigative team.

This report describes the major themes and lessons discussed at the workshop. The research areas included are structural engineering, building performance, and fire safety; civil infrastructure systems; data and information gathering; agency coordination during response; and behavioral responses. After the presentations made at the workshop, a number of participants discussed future research issues and cross-cutting themes. The results of these discussions are presented in the last section of this paper.

Readers should note that the summaries given here reflect presentations made at the time of the workshop, and that the researchers' observations and conclusions may have changed since the workshop was held. Many of the presentations at the workshop were developed into papers for this volume, and more detailed descriptions of the projects and their updated conclusions can be found by referring to other selections in this book.

Structural Engineering, Building Performance, and Fire Safety

Several structural engineering research issues are associated with the collapse of the World Trade Center. These include understanding what happened, why the structures collapsed, and how the buildings performed during extreme conditions. A number of structural engineers and related specialists began to work immediately at the site of the disaster to address these issues. Although the World Trade Center collapsed, a number of neighboring buildings that sustained serious damage did not, including the Verizon building (which was damaged by debris). Engineers were interested in finding out why those structures did not collapse. Moreover, the World Trade Center towers remained standing for a sufficiently long period of time to allow around 20,000 people to evacuate the towers. This suggests that future designs could provide for even greater egress during such conditions, allowing more people to survive a similar catastrophe.

Preliminary Impressions of the World Trade Center Disaster, Post-disaster Reconnaissance, and Perishable Structural Engineering Data Collection[2]

Researchers Abolhassan Astaneh-Asl, W. Gene Corley, and Michael Bruneau, who were working in this area, were faced with a serious challenge. The cleanup efforts and response at the site began to proceed at a rapid pace. A large amount of steel from the collapse was taken to New Jersey where some of it was being recycled. By the end of 2001 over 150,000 tons of steel were processed and recycled. Hence, the clean-up efforts did not allow some researchers to collect as much data as they would have liked. Structural engineers tried to save samples of the steel to attempt to understand the collapse. The column trees and core columns were coded and had stamps or engravings that identified their location within the structure. In addition, photographs and videos were collected at the site to aid in the analysis. Such data was necessary to find out what components of the structure were significant from a structural point of view.

One of the main research efforts was to establish a sequence of events for the collapse. The two towers were very similar but not identical. The typical floor plan included roughly 1-meter spacings. The core was made up of steel columns and was there for the purposes of providing elevators, stairs, and other services. The exterior of the buildings provided all lateral load resistance. Was the collapse of the towers initiated by the collapse of the floors, caused by heat, or due to buckling? What happened on different floors? It appears that joints were the first to go where fire was intense and then the outside columns buckled. It also appeared that the inside columns did not buckle but instead collapsed after the outside columns collapsed.

Early reconnaissance work at the site was also carried out with the goal of allowing researchers to better understand how techniques from earthquake design and engineering could be used to prevent this kind of urban disaster. Two kinds of debris were analyzed for this purpose: impact debris collected after the collapse, and falling debris resulting from the collapse of the towers or collateral damage. In order to gain insight in this area, it is necessary to understand how the load is distributed when certain components of a structure are removed. For instance, the façade of 130 Liberty Plaza was pierced over many floors yet the structure remained standing. What happens when you remove a column from a building? Why did this structure survive?

Buildings that use floor diagram systems can arrest projectiles because every beam is continually connected to columns. Such systems have a desirable feature that really helps, and it is that loads can find alternate load paths. There is enough redundancy in those buildings so that it would probably take the loss of three columns before they would collapse. If elastic capacity is exceeded it is still possible to have plastic capacity. One lesson learned from the Northridge earthquake is that better beam connections are needed. These lessons are also applicable to the prevention of future urban disasters.

Fire Safety Aspects of the World Trade Center Disaster[3]

A critical part of understanding the World Trade Center collapse is the effect of fire, as explored by Frederick W. Mowrer. A number of significant research questions about this aspect of the disaster have emerged, including the effect of jet impact on fire safety systems; the role of jet fuel on the ensuing fire; the effect of fire on the structural integrity of long span steel joists; human behavior throughout the event; and the question of why the World Trade Center collapsed but the First Interstate Bank Building in Los Angeles and the One Meridian Plaza Building in Philadelphia did not, despite being subjected to severe multi-floor fires. In the case of the World Trade Center collapse, both planes that struck the towers went in at an angle and as a result the fire started in several floors simultaneously.

Conditions in the World Trade Center before September 11th were important in trying to understand what happened. The condition of fireproofing applied to steel structural members needs to be considered. The use of long-span steel bar joists to support the floors as well as the columns requires evaluation, particularly because these floor joists were part of the primary structural frame of the buildings.

Another question for fire safety research is the effect of the jet impacts on fireproofing, structural stability, exit stairs and elevators, ventilation systems, and automatic sprinklers. The World Trade Center towers were retrofitted with sprinklers but were not originally designed with them. Similarly, the role of the jet fuel is of interest. Much of the jet fuel burned outside the buildings and the remainder served as a very large accelerant. Most of the jet fuel must have burned in about 10 minutes or shortly thereafter. The results of such research will allow future designers to improve building performance in order to make this kind of outcome less likely.

Research Needs in Structural Engineering and Fire Safety[4]

More research is necessary in several areas, according to W. Gene Corley. Research will be carried out in the short and medium term in areas such as structural fire effects and protection; building egress; and search and rescue. The main research needs for structures are robustness, redistribution of loads, and connections. There is simply not enough knowledge in these areas. Connections are always considered a problem for structural engineers. More research is needed about fireproofing, how fire spreads, temperature rise, combustibles, and connections. According to some experts, this is where the largest part of the research funds should go. Fireproofing allows for enough time to put the fire out or to let it burn itself out. While what we have in buildings may burn like wood it's no longer wood and that's what we need information about. Most of the available research on the potential effect of combustibles was done on wood materials. There is little or no information in the literature about fireproofing connections. In terms of search and rescue, more research is needed in detection and locators, access, and rescuer safety.

Some preliminary research cost estimates are: $12 million for structural engineering; $28 million for fire effects and protection, where much modeling is needed; $8 million in the area of egress, where many interviews should be carried out to ask people how they came out of the towers; and $12 million for search and rescue.

Impact on Civil Infrastructure Systems

The collapse of the World Trade Center had a profound impact on civil infrastructure systems in the area. Despite the losses, most systems proved to be resilient and service was restored within short periods of time. Transportation sustained the greatest damage. The PATH (Port Authority–Trans-Hudson) train and the 1/9 and N/R subway lines sustained severe damage and service interruptions. The Holland Tunnel was also disrupted. A key subject for future investigation in the area of infrastructure systems is the role played by the various agencies responsible for these services in responding to the disaster.

Improved Security and Management of Underground Infrastructure Systems[5]

A broad area of research on infrastructure systems investigated by Thomas O'Rourke relates to improved security and management of underground systems such as electric power, water distribution, telecoms, transportation, natural gas, steam, and wastewater conveyance. Specific research goals include documentation of September 11th infrastructure systems performance; comparative analysis with infrastructure system performance after earthquakes; and developing a methodology for assessing interdependencies. Lessons drawn from infrastructure system performance during events such as September 11th involve the water distribution and electric power systems. New York City has 6,123 miles of pipeline. This is comparable to the system in Los Angeles, but is packed in 1/5 of the area. Heavy damage to underground water pipelines resulted in serious loss of pressure from hydrants surrounding the World Trade Center site. Substantial water was provided by fireboats pumping from the Hudson River. The way in which electric power is configured into local distribution systems provided a very resilient overall network that allowed for the preservation and rapid restoration of power in the World Trade Center area. New York City has the densest electrical load in the world, and it is a highly compartmentalized system.

The dense nature of infrastructure systems in New York City also suggests that disruptions can cause major damage through a cascading effect. Such incidents are comparable to what happens during extreme events such as earthquakes. Several research projects funded by the National Science Foundation related to interdependencies in infrastructure systems. The interdependencies and their performance during extreme events can be evaluated using information technology. The tragedy of September 11th resulted in a willingness to share information that should benefit research in this area. Unfortunately, such information is also viewed as having a detrimental value, as people believe it can be used for terrorist activities. If such information is not shared, researchers will not be able to fully understand what happens during an urban disaster such as the collapse of the World Trade Center.

Developing and Planning Infrastructure Systems in the Aftermath of Extreme Urban Events[6]

Today, planners and developers are trying to promote even more density. For many reasons it is not clear how to approach the issue of having large concentrations of workers along with a transportation network under one building. Discussions about shorter buildings and less density have not taken place. Implicit to this discourse, according to Cruz Russell, is the understanding that downtown needs a renewed commitment to developing the World Trade Center site, creating the critical mass of class A office space required to sustain a substantial workforce and attract future business development.

Agencies such as the Port Authority of New York & New Jersey found it difficult to plan ahead and find the time to think about what to do next in the immediate aftermath of the September 11th disaster. Urgent activities such as debris removal required a tremendous amount of work and interdisciplinary attention. Engineers estimated that it would take 24 months to restore PATH service with some kind of temporary service, assuming a very rapid schedule of work.

Such daunting tasks made it difficult to think about how to replace and restore downtown's office infrastructure in a timely fashion. There are very few places that can be developed in Manhattan that consist of 6x6 contiguous blocks situated over transportation networks, especially places that have the economic impetus that the World Trade Center had. This site represents important development challenges and opportunities.

The Port Authority of New York & New Jersey and agencies like it will be looking at fortifying resources such as tunnels and bridges, ferries and rail crossings, and attempting to create both the physical and human infrastructure systems that will withstand the reality that society now will have to accept, after September 11th.

Transportation Systems

The transportation sector suffered severe disruption after the collapse of the World Trade Center. Research efforts in this area center on understanding the disruptions to travelers caused by damage to transportation infrastructure, the coordination of response efforts in the sector, and the behavioral response of transportation users.

Impact of the World Trade Center Collapse on PATH Tunnels and Subway Lines[7]

Damage to public transportation systems in the vicinity of the World Trade Center was extensive, as noted by Joe Englot and Joseph N. Siano. The PATH station beneath the complex was compromised by the collapse of the World Trade Center. There was flooding in the PATH tunnels and two of them had to be plugged with concrete to avoid flooding the rest of the system. Users responded by going to Christopher Street via the 33rd Street line and changing to the Metropolitan Transportation Authority's subway service. However, these stations were not designed to be used by such high numbers of people. The 1/9 subway train also sustained severe damage, as it was essentially destroyed by the collapse of the buildings. Approximately 1,200 linear feet of tunnel were damaged or filled with rubble.

Some areas of the PATH station were intact after the towers' collapse, but damage was expected to be total because of the removal operations above and the heavy machinery being moved around above the PATH stations. The presence of many utilities in the area complicated restoration efforts. The steps needed to restore the PATH included recovery of human remains, debris removal, stabilization of the bathtub wall, removal of PATH tunnel plugs at Exchange Place, rehabilitation of PATH tunnels under the Hudson River, construction of a temporary PATH station at the World Trade Center site, planning and coordination for a new permanent PATH station, and actual construction of a permanent PATH station.

Disruptions in transportation posed serious problems for people who needed to commute to work from areas such as Jersey City and Brooklyn. As a result, restoration of service to the N/R subway line, and planning for reconstruction of the 1/9 line were high transportation priorities.

Coordination Efforts in the Transportation Sector[8]

In considering the transportation coordination efforts after the World Trade Center collapse, as David Woloch did, it is important to remember that responsibilities were diffused among many different players. Even within the City of New York Mayoralty, there were many different agencies with transportation functions, including the Department of Transportation, the Department of City Planning, the Department of Design and Construction, and the Police Department. Other entities included the State Department of Transportation, the Port Authority of New York & New Jersey, and the Metropolitan Transportation Authority. Coordinating these agencies is a very difficult task under the best of circumstances—but in the wake of September 11 this coordination was crucial.

The City Department of Transportation is primarily responsible for both maintaining and regulating the uses on the City's streets and sidewalks. With the primary transportation responsibilities after September 11th, the Department of Transportation needed to immediately develop and implement new solutions to allow for mobility to and within lower Manhattan—and in doing so needed to work effectively with its many partners.

On September 25, as the lower Manhattan business district tried to get back on its feet, motorists began to return to the roads. But while traffic actually appeared to be moving in Manhattan there was absolute gridlock in Brooklyn, Queens, and New Jersey. In addition to the reduced roadway capacity in the World Trade Center area and the closure of crucial corridors and crossings (including the West Side Highway, the Holland Tunnel, and the Brooklyn Battery Tunnel), security checks at the remaining crossings were vigilant and created long vehicular queues outside of Manhattan. As a result, that morning motorists faced a commute that took hours. It quickly became clear to the Department of Transportation Commissioner Iris Weinshall and her staff that all bets were off and unusual measures were called for.

The most important step recommended by the Commissioner—and accepted by Mayor Giuliani—was to implement a single-occupancy-vehicle restriction during the morning rush hours into midtown and lower Manhattan. (Although illegal under normal circumstances, such a ban was possible through the post-September 11th Mayoral Emergency Order). The ban reduced vehicular volume during the morning peak period as many solo drivers appeared to car pool or take transit. There was also a spike in volume before the peak (5 a.m. to 6 a.m.) and after the peak, helping to spread the traffic volume throughout the day. The ban's success would not have been possible without extraordinary interagency cooperation. It required the participation not only of the Department of Transportation and the Police Department but also the Metropolitan Transportation Authority (responsible for the Midtown Tunnel) and the Port Authority (responsible for the Lincoln Tunnel).

A year and a half later the ban continues (with shorter hours and only into lower Manhattan). Other measures taken by the City Department of Transportation, including establishing a Brooklyn Ferry route and providing bus priority space in lower Manhattan, also continue. The aftermath of September 11th brought out creativity and coordination among agencies in a manner that fostered long-term transportation and transit improvements.

Impacts of Extreme Events on Passenger Travel Behavior[9]

The events of September 11th affected not only the transportation system but also its users. One research area is the assessment of the behavioral changes that took place as a response to the disaster. A research project conducted by José Holguín-Veras included analysis of data, assembling and selecting a panel, definition of surveys, and estimation of behavioral models. The behavioral models estimated and quantified the role of September 11th on the mode choice process for air travel. This research should contribute to a better understanding of the changes in travelers' behavior after extreme events, and how transportation agencies should deal with them.

Energy and Electric Power

Damage to Energy Infrastructure during the World Trade Center Collapse[10]

The provision of electric power to approximately 13,000 customers in the World Trade Center area was affected by the September 11th disaster, according to Elie Chebli. A 13-kilovolt (kV) substation was lost when the first tower was hit. The summer demand of these customers was approximately 450 megawatts (MW). This load area is the equivalent demand of the city of Hartford, Connecticut. The share of this total used by the World Trade Center buildings was 90 MW, which is equivalent to the amount of power used by the city of Albany, New York. Despite the magnitude of the disaster, Con Edison managed to stay within its design standard because there is a lot of redundancy built into its system. The five 138-kV transmission lines feeding the substation in lower Manhattan were de-energized as a result of the collapse of World Trade Center buildings 1, 2, and 7. Measures were taken to re-energize those feeders. They were first isolated from the destroyed buildings 1, 2, and 7 so that they could be re-energized from another substation in the lower Manhattan area. In addition to providing electricity, Con Edison also supplies the gas system and steam, and several gas and steam pipes also were interrupted.

Resiliency in Energy Infrastructure Systems[11]

Elie Chebli also explained that internal power distribution at the site was owned by the Port Authority of New York & New Jersey. Power distribution had an N-2 design, which means that two 138-kV lines or two power transformers or two 13-kV feeders may be de-energized and the remaining 138-kV feeders, power transformers, and 13-kV feeder are still able to carry the energy load to customers. Five networks totaling approximately 450 MW were lost. The first estimate for repairing the first 138-kV feeder was 14 days. Engineers located places to excavate for splicing the 138-kV cables. The cables then needed to be frozen before being spliced. The time estimate for repairing the feeder was changed because splicers worked around the clock in shifts. Only six days after the collapse, the first 138-kV feeder was repaired and two hours later another was repaired. As a result of the system's design and rapid repair work, there was relatively little service interruption. The performance of the electric power systems during the disaster provides researchers with valuable lessons about system design and resilience that will be useful when addressing other urban disasters.

Telecommunications

Research Issues and Characteristics of the Telecommunications Sector after the World Trade Center Disaster[12]

The main research issues in telecommunications were the performance of the various systems during the emergency and immediate response to September 11th, the characteristics of service outages and system failures, how they were repaired after the initial response, and the nature of the recovery process. Anthony Townsend noted that wireless communication was needed for the emergency response since emergency workers needed connectivity at Ground Zero. AT&T provided unlimited telecommunications for workers. Temporary cell sites replaced destroyed antennas.

The World Trade Center served as a broadcast platform for the New York metro area. Many antennas were located there and the damage to the sector from the disaster was extensive. The collapse also strained the sector as demand for services skyrocketed. AT&T's long-distance service recorded a historic call volume on September 11th with about 431 million calls. The

AT&T emergency response was similar to that of other events, such as the 1994 Northridge, California earthquake. Wireless carriers were not as prepared as AT&T. There were enormous, ongoing, city-wide cell phone service breakdowns in other cities as well, including Boston and Washington, D.C. There were too many calls and not enough capacity. In addition, there was no system in place to set priorities on the calls.

With regard to Verizon, there were outages in lower Manhattan that resulted from damage to 200,000 voice lines, 3.6 million data circuits, 10 cell sites, and other facilities. Despite this heavy damage, phone service was restored. Verizon is rebuilding 140 West Street.

The response of corporations and their effect on the telecommunications sector is also the subject of research, as companies dispersed and there was an exodus to other locations such as midtown Manhattan, Jersey City, Stamford, and Westchester. The resulting competition helped the telecommunications sector through the profusion of carrier hotels, which still tend to be clustered.

Online services performed well because the internet is based on distributed, product-switched networks. The same happened in Serbia during the U.S. bombing of that country's communications systems—the internet survived very well.

Water Distribution and Environmental Services

Performance of the Water Distribution System after the Collapse of the World Trade Center[13]

One of the main concerns in the performance of the water sector after the collapse of the World Trade Center towers was the provision of water for firefighting activities. Diana Chapin explained to the workshop participants that the collapse caused nine water mains surrounding the site to break, which resulted in a loss of water pressure to such a degree that it was insufficient to meet the needs of the Fire Department. However, by about 10:30 p.m. the New York City Department of Environmental Protection, the city's municipal water supplier, was able to restore water pressure to 40-50 psi without having to supplement the system with Hudson River water pumped from the fire boat. Interagency communications immediately after the collapse were problematic. However, communications were reestablished within a short period via face-to-face communications at the command post and use of the agency's standard 800 MHz radio system.

The deeper sewer system sustained much less damage than did the water distribution system, with only one sewer collapsing and another slightly damaged by a 30-foot piece of structural steel. The Department of Environmental Protection sustained very little damage compared to other utilities. The damaged mains are being replaced with concrete-lined ductile iron mains of the same size (12- and 20-inch mains) at a total estimated cost of $14 million.

Watershed Security and Environmental Monitoring of Water Systems[14]

Another major concern in the water sector was enhancing security. According to Diana Chapin, the system is large, and has significant redundancy. Most chemical and biological threats are not effective. Since the September 11th disaster, additional actions were taken both upstate and in New York City. There was no public access to these facilities for security reasons after the attacks. Under the direction of the Department of Environmental Protection, a security action plan was developed by the Bureau of Water Supply and the Department of Environmental Protection's police. There was also consultation with the other law enforcement personnel.

According to David Lipsky, the Department of Environmental Protection has a network to monitor various environmental indicators, including residual chlorine, pH, and others. The department also decided to enhance some chemical and biological monitoring after the attacks. There are many important issues that arise in trying to set up a monitoring system under emergency conditions. For instance, when it is not possible to rely upon air transportation—redundant or local lab support services are needed. It may also be necessary to rely on local expertise to address ancillary risk management issues, such as deciding what type of monitoring programs to implement. This also means that there is a need to address the training and expertise of local teams. In addition, implementation of an effective monitoring system may be difficult if access to internet data is restricted. Another challenge is to provide smaller water companies with adequate and effective direction.

Under such conditions, there is a need for a defensible objectives-based sampling process. For example, should the focus be on risk- or hazard- or exposure-assessment? A strong quality assurance/quality control plan is required for proper data interpretation and comparability. There is also a need for a strong environmental team to make decisions on data, and to identify and qualify labs to support monitoring activities.

Another set of challenges is associated with managing expertise and data. In an urban disaster there may be too many experts offering opinions, which could result in the amplification of risks and the creation of much data but little useful information. In addition, experts may not have ownership of problems or familiarity with the monitoring system. Similarly, there are lots of agencies offering to help and it is necessary to have a coordinated risk management-based approach in order to develop targeted strategies.

September 11th demonstrated a clear need for comprehensive risk management research. It is impossible to prevent all risks but it is possible to effectively manage risk. In order to do this it is necessary to develop defensible risk management programs. This suggests the need for continuing research on early warning monitoring systems. One research issue in this area considers the pros and cons of high-tech vs. low-tech systems. The more high-tech the system is, the more difficult and expensive it will be. It is also necessary to more clearly define risk management objectives.

Interdependencies among Civil Infrastructure Systems[15]

Infrastructure systems are physical components and services that are used or consumed by the public. These infrastructure systems are dependent on each other. For example, if the Metropolitan Transportation Authority is to provide its services, it must have power. There are many examples of interdependencies in New York City among Con Edison, Verizon, the Metropolitan Transportation Authority, and others. Coordinating efforts among these players is sometimes challenging, because private suppliers are not comfortable with sharing proprietary data.

One of the most pressing research activities in the aftermath of September 11th was the collection of perishable information on critical infrastructure interdependencies and supporting inter-organizational coordination of interdependent critical infrastructure systems, as conducted by William Wallace. This allowed for the development of tools to visualize existing and potential interdependencies, to shield proprietary data, and to integrate initial approaches with ongoing efforts. Models of network flows, decision theory, cognitive science, and data visualization were used in these research efforts. Research on the interdependencies of these systems will aid planners and developers in developing strategies to piece back infrastructure systems in a timely fashion after an urban disaster.

Data and Information—
Security Issues vs. Public Access to Data

Information and data are critical in coordinating activities after an urban disaster and in carrying out research to understand what happened and what lessons can be learned. Research in this area includes data collection, data quality and accessibility, information technology firms and their response to disaster recovery, representation of data and information in the media, and disaster assistance in the for-profit sector.

Digital Data Collection Technologies[16]

Research conducted by J. David Frost after the September 11th events used new digital data collection technologies. Traditional data collection is subjective and results in different levels of detail or types of data being gathered. A systemic approach to data collection is very important. Field reconnaissance equipment included palm pilots, digital voice recorders, digital video cameras, and hand-held geographic positioning system (GPS) technology. Laptop computers were used for data analysis. Integrated data acquisition and analysis software allowed for the recording of feature damage, locations, and photographs. New software was also used that allowed for the rapid assimilation of data from multiple users and to query data based on information type and individual features.

Data Accessibility and Quality[17]

There are many issues related to data accessibility and quality. This is one of the greatest challenges to research projects after an urban disaster such as that on September 11th, as noted by Susan L. Cutter. Data can be both public and private, and even confidential. The political climate with data sharing among jurisdictions varies widely. There is a need for spatial data infrastructure before urban disasters. Data need to be stored in more than one location. After the attacks, public data access was discussed in terms of national security vs. the community's right to know. Many researchers consider public access to this data to be critical. For instance, most environmental data is in the public domain, yet access to some of it was limited.

Response and recovery efforts after urban disasters may result in data restrictions that challenge the Emergency Planning and Community Right-to-Know Act (EPCRA), which establishes requirements for federal, state, and local governments; tribes; and industry regarding emergency planning and right-to-know reporting on hazardous and toxic chemicals. The provisions of the act help increase the public's knowledge and access to information on chemicals at individual facilities, their uses, and releases into the environment (U.S. Environmental Protection Agency, 2000). Despite the importance of environmental information, certain internet sources of data were removed from public access after September 11th.

Patterns of Media Coverage of the Attacks on the World Trade Center[18]

In the area of media analysis, a research project undertaken by Christine Rodrigue produced a timeline of events and examined 288 articles on the September 11th disaster from the on-line edition of the *Los Angeles Times*. Ten principal themes dominating the articles emerged in the analysis, including context, diplomacy, investigation, military, and others. In the first three weeks after September 11th, the coverage was focusing on a relatively wide array of topics, especially reactions, military, investigation, and restoration. In the last three weeks, the coverage had begun to narrow to two main themes: the military and investigation.

The ten themes could be grouped further into three main narratives: war, disaster, and crime. According to the researcher, coverage of the World Trade Center and Pentagon attacks shifted into more of a war narrative than a disaster narrative very early on. The military-related categories of military, diplomacy, and mitigation accounted for 31% of the first six weeks of front-page coverage, compared to 36% in the disaster narrative and only 14% in the crime narrative. Stories of the disaster, however, dominated only the first two weeks of the coverage; the war story took over as the dominant narrative by the third week and remained dominant for the rest of the study period.

On the other hand, the context of the attacks was poorly drawn out, garnering only about 1% of the front-screen stories. As in many other disasters, context is subordinated to sensation, in this case, the images of the planes striking the towers, the fear of anthrax, and the furious military response in Afghanistan. The abrupt transition in dominance from the disaster narrative to the war narrative almost certainly did not coincide with the easing of victims' needs and those of the two hard-hit cities.

The recommendations of the project to those non-governmental organizations and victim advocacy groups working in large-scale disasters to keep news coverage focused on the victims and their needs are to try to generate "newsworthy" events, including demonstrations; to cultivate personal relationships with particular reporters; and to consider using the internet to get stories out.

Agency Coordination during Response

The magnitude of the collapse of the World Trade Center resulted in the need for an unprecedented level of agency coordination. Such efforts were led by the New York City Office of Emergency Management. They included interagency coordination, coordinating utilities, communicating risks, coordinating the private sector, managing volunteers, managing emergency efforts at the Federal level, victim information management and victim assistance, and others.

The New York City Office of Emergency Management and Coordination[19]

Michael Berkowitz and Mary Ann Marrocolo explained that the role of the New York City Office of Emergency Management is to serve as incident coordinator. This includes coordinating the integration of emergency and recovery efforts. The collapse of the World Trade Center resulted in about $83 billion of damage to New York City's economy, the loss of 125,000 jobs, and the destruction of 15 million square feet of office space in six buildings.

The Office of Emergency Management was located in building 7, so a new place was needed immediately. The office moved to Pier 92. Over 300 people per shift worked out of Pier 92. Some of the logistics and field coordination activities included coordination and planning of all onsite resources such as urban search and rescue, the U.S. Public Health Service, the New York Fire Department, the New York Police Department, the Red Cross, the Department of Transportation, the Port Authority of New York, the New Jersey Police, and others.

The damage to infrastructure systems was extensive. Con Edison lost a substation but, given redundancy in the power system, was able to continue operating in the area. In the future, this redundancy may need to be dispersed throughout the system. Verizon's building was located at West Street. That was a major switching point for the network and many of the cables were damaged. Such extensive damage also required much interagency coordination.

Another area of work for the Office of Emergency Management was debris removal, which was the costliest and most difficult operation. This consisted of organizing the haul of 1.65 million tons of debris from Ground Zero to five collection sites at the Staten Island landfill, Hamilton, 59th Street in Brooklyn, Pier 6, and Pier 25.

In addition, there were many activities related to health, such as environmental testing, worker and work site safety, and coordination of the five Disaster Medical Assistance Teams operating at the site.

Numerous human services were provided to the public. These included the Disaster Assistance Services Center for families of World Trade Center victims and those displaced by the attacks; the management of volunteers and donations; and respite centers for relief workers. The latter included meals, mental health services, and special events. There were about 12 million meals associated with the site.

Legal issues addressed included spearheading negotiations to obtain Pier 92 for the Office of Emergency Management and coordinating agencies; obtaining the required permits for emergency relief efforts and re-opening the Fresh Kills Landfill; and the declaration of an emergency. In the area of public relations and communications, the Office of Emergency Management coordinated the assembly and dissemination of pertinent information, and the Joint Information Center controlled the flow of information for a consistent message.

In terms of economic recovery, the Office of Emergency Management worked extensively with businesses to coordinate activities. Another activity involved shrinking the restricted zone. There were steady efforts to re-open closed areas of the World Trade Center area to allow vehicles, pedestrians, businesses, and residents back into the vicinity. In addition, there was a continued state of emergency, with the area around the World Trade Center being largely closed and transportation being very difficult. In order to facilitate transportation, the Office of Emergency Management operated buses through Trinity and West streets, and coordinated efforts with the Department of Transportation, the New York Police Department, the Metropolitan Transportation Authority, and other agencies in order to allow downtown access for those affected by transportation limitations.

The main lessons learned from this experience suggest that detailed planning pays off. However, this brings up the emergency management dilemma of managing all hazards vs. being hazard-specific. The best plans must include specific recommendations but also be adaptable and substantive. An example of where these characteristics are desirable is in the area of donations and volunteer management. In addition, the issue of credentials should not be forgotten: controlling access to a disaster site is crucial.

The Role of the Federal Emergency Management Agency in Coordination[20]

The Federal Emergency Management Agency (FEMA), as much as any other agency, was confronted with a new situation, Randolph Langenbach reported. It has been a couple of generations since Americans faced these sorts of threats. People at FEMA were asked to think outside the box. An Urban Hazards Forum was planned, and FEMA was one of the sponsors. Hazard mitigation in a natural disaster is a relatively easy issue, and benefit-cost analysis is required for natural hazards mitigation. In terrorism it is very difficult to do this type of analysis. There is scientific information on how often earthquakes or hurricanes take place. In terrorism this is not easily available. By lowering the risk of terrorism in one place you may well be increasing it in another, so overall benefits are not necessarily obtained. Political scientists need to be involved in discussions about coordination in order to try to understand the fundamental problems behind this type of disaster. It is also important to understand how people outside the United States view this country and what can be done about it. It is also necessary to expand the lessons learned from the September 11th attacks beyond terrorism to fire and explosions in high-rise buildings. Fires can result from a number of causes, not just terrorism. There are many fire safety issues to be addressed through concerted research efforts, including providing better tools for deciding whether rescuers should enter buildings under extreme conditions; and exploring the many technical problems associated with using helicopters to respond to fire in a high-rise building. These are areas of research that should be explored by FEMA.

National Institute of Standards and Technology's Public-private Response Plan to September 11th[21]

S. Shyam Sunder observed that the collapse of the World Trade Center was the worst building disaster in human history, resulting in unprecedented death and destruction. Never before did 400 emergency respondents lose their lives in a single incident. No one was prepared for the dramatic collapse of an occupied building of that magnitude. The disaster had the effect of uniting industry and public sector agencies. There was a multi-agency, public/private program established and partially funded with participation from key private sector organizations. Some of the issues and concerns that emerged as a result of the unexpected collapse of the World Trade Center were changing community expectations of tall buildings and the codes and standards that regulate them. Discussions about the pros and cons of the codes and standards used in 2001 relative to those used in the 1960s are likely to continue. The technical lessons from the collapse of the World Trade Center are relevant to the future of all tall buildings. A key observation is that the buildings performed much better than would be expected under such abnormal loads. There was extreme impact damage to the towers' columns, especially to the exterior columns. The initial National Institute of Standards and Technology response was a code comparison study for the Federal Emergency Management Agency and a 30-day study of repair and rebuilding for the Pentagon.

Interagency Coordination Research[22]

A broad area of research was observing and documenting the inter-organizational responses to the September 11th disaster. About 450 organizations responded to the World Trade Center disaster. The main research issue for John Harrald was understanding how a response to this type of urban disaster is coordinated. He used a number of research tools, including construction of a chronology/timeline, and agency-specific data such as situation reports, media resources, interviews, and other documents.

The findings indicate that the Pentagon response was very effective, and the response system designed for natural disasters is considered effective for terrorist attacks such as the World Trade Center collapse. The long-term implications for research are that a better understanding of multi-organization systems is needed. One of the interesting things about this case is trying to understand and document why things went so well despite the jurisdictional issues associated with trying to coordinate this kind of response.

Public/Private Sector Coordination Efforts[23]

Interorganizational coordination efforts were also complicated by the fact that the private sector played a very important role in the response to the September 11th disaster. Hence, one research issue, examined by Richard T. Weber, focused on the challenges of collaborating across the public and private sectors. Research efforts in this area began about a week and a half after the event. The study revealed that businesses such as Con Edison, security companies, Keyspan, Verizon, and many others were directly affected by the disaster. Some of them helped out directly with the recovery efforts according to their role and area of expertise. For example, AT&T provided 3,000 phones and cell phone lines. Some of these companies' activities were not well coordinated, while others were. A major issue was checking who was moving in and out of the disaster site. Information was scarce and this was a problem for the private sector. Eighteen million square feet of office space were lost, and an effort was made to find new office space to keep these businesses in New York City. Debris removal was another challenge for the public and private sectors. For example, the Federal Emergency Management Agency had to investigate the alleged theft and selling of steel from the World Trade Center. The study also examined changes made to evacuation plans after the 1993 World Trade Center bombing and how these changes affected the number of lives lost in 2001.

Interagency Coordination during Response in Other Locales[24]

Response efforts on September 11th were not confined to New York City, Washington, D.C., and Pennsylvania. In Oregon, research about response efforts to the disaster centered on the evaluation of interagency communications. Robert Parker reported that surveys were carried out and followed up with focus groups. The responses obtained suggest that agencies within that state had a clear understanding of their roles. But they also suggested that there is overreliance on standard communications. There is no extensive evidence of back-up communications systems. Out of 22 agencies surveyed, about 10 play an active role in response. After September 11th, most of the agencies thought communication between them was effective. The results of such research indicate that more frequent exercises and training on existing regulations would probably improve interagency communication. Research in this area will be broadened to address how current systems are working between response activities and mitigation initiatives. It is important to understand how current communications systems can be strengthened and how redundant communications can be used in post-disaster situations.

Coordination for Flight 93 in Pennsylvania[25]

David Hoover conducted research on coordination of the response efforts in Pennsylvania, beginning within 48 hours of Flight 93's crash. The personnel involved in response efforts were interviewed. On-site observations, triangulation, and focused interviews were also used, and a number of local documents examined. This analysis showed that emergency plans and interagency exercises make a difference in the ease of operation during disaster response and recovery. The presence of agencies (such as the FBI and the Disaster Mortuary Response Team (D-MORT)) that were not familiar with the local plans and procedures caused a few problems that smoothed out as personnel continued to work together and became familiar with the reasons behind the activities. The 13-county agreement in western Pennsylvania for emergency response was essential during recovery as it enabled the final sweep of the debris field once the FBI had completed its investigation and released the site.

Impacts and Ramifications of September 11th for Federal Emergency Management [26]

Understanding the impacts and ramifications of September 11th on federal emergency management was another research focus, this one undertaken by Claire B. Rubin. Secondary sources of information, and after-action reports were used, as well as the conceptual framework of the Disaster Time Line. Legislation, regulation, executive orders, response plans, and organizational changes were analyzed. The main finding was that it is not possible to overstate the dramatic changes in political culture, attitudes, and philosophy of the federal government with regard to emergency management resulting from September 11th. There was a major sea change for the Bush Administration and for the government at every level, and notable changes in the willingness to address terrorism. Between September 11, 2001 and the end of 2001, several U.S. General Accounting Office reports, three new laws, three executive orders, and two national security directives were issued.

Victim Information Management and Victim Assistance[27]

David Simpson and Steven Stehr found that the management of disaster victims included a search and rescue phase, the retrieval of bodies and human remains, victim identification, and the disposition of bodies. A research question in this area was how the specific characteristics of the World Trade Center collapse altered the task of victim management. This event had several unusual characteristics. For example, the nature of the site (simultaneously a crime scene and a mass grave) pitted victim retrieval against debris removal. The nature and timing of the attack (against iconic symbols of capitalism) led to fear of other attacks. The scope and intensity of the destruction was another defining characteristic.

Additional research areas were the organization, coordination, monitoring, and evaluation of victim assistance programs, the definition of victims, the provision of resources and access to programs, and the facilitation of paperwork. Initial recommendations that emerged from this area of work was the need to have a victim identification and information clearinghouse on the internet or in another medium, and to have a centralized donation and distribution monitoring system.

Response and Resilience in the World Trade Center Attack[28]

Kathleen Tierney described her project, which focused on the multi-organizational response that was undertaken by New York City agencies after the attack on the World Trade Center. Data for the study were collected through direct observation at the city's Pier 92 Emergency Operations Center (EOC) and at other sites at which decisions were made and response activities carried out after the attacks, including command posts, staging areas, and the Ground Zero site. Preliminary findings indicate that New York City responded in a resilient fashion to a near-catastrophic disaster that far exceeded the scope of prior planning. The city was creative in its use of resources to offset the loss of its state-of-the art EOC, which had been located at 7 World Trade Center. That structure had to be evacuated at the height of the emergency and later collapsed due to fire on the afternoon of September 11th.

The reconstituted EOC at Pier 92, which became the coordinating center for the emergency response, was significantly larger than the EOC the city had lost. The newly-improvised facility, with nearly 250 computer workstations, was able to accommodate representatives from hundreds of organizations and provided a wide array of support functions, including functions not envisioned in earlier planning. At the same time, the spatial arrangements at Pier 92 mirrored those of the original EOC, which facilitated interorganizational interactions and provided needed continuity for the overall emergency response effort.

Behavioral Responses to September 11th

A number of research projects focused on the response of different groups of people to the September 11th disaster. Besides transportation users, whose behavioral changes were described above, other research examined the response of the general population, and of specific groups such as volunteers, advocacy organizations, students, and faith-based groups.

Responses to the World Trade Center Attacks[29]

One research team worked for four and a half days in the field two weeks after the disaster, interviewing about 40 people, including police, fire fighters, military/national guard, religious leaders, Ground Zero workers, New York City teachers, and New York City Board of Education members. The last two groups were the main focus of the study. As explained by researcher Paul O'Brien, the project used a theoretical model of risk communication to test the hypothesis that, given the nature and magnitude of the event, certain differences could be expected from previous literature on response to disasters.

The early results suggested that this event had a massive impact on the community and the nation, that the threat was ongoing, and that the long-term national impact will be substantial in politics and other fields. Among students, the long-term impacts are also substantial: a major increase was expected in the dropout rate in New York City's high schools. To anticipate events such as this, emergency plans need to be revised. New scripts for public officials are also needed. For instance, it is important to address the question of whether the country is at war during such an emergency or whether people should carry on with their lives.

Community Response to Attacks—Spontaneous Volunteers[30]

One of the most visible behavioral responses in the New York City community after the attacks of September 11th was volunteering. A project described by Seana Lowe and Jenna Peck studied the volunteer community by interviewing 23 people, 20 spontaneous volunteers and three professional volunteers. They were asked how they heard of the attacks, what their feelings were, and how they knew what to do. The responses suggest that people were motivated by a compulsion to react. A commonly cited barrier to volunteering was the overwhelming number of volunteers. There was a sense of increased solidarity. Overall, people were extremely impressed by the emergency response.

Advocacy Organization Response to a Suddenly Imposed National Disaster[31]

One lesson confirmed by the September 11th disaster was that in the aftermath of this type of event groups that do not usually work together begin to do so, according to an investigation by Bob Edwards and Patrick Gillham. This was also observed in Mexico City after the most recent earthquake and in North Carolina after the floods. After September 11th a number of advocacy organizations responded by working together, which presents an interesting research area in behavioral science and organization theory.

Ethnic Issues on University Campuses after an Act of Terrorism: Arab and Muslim Student Response[32]

Another research area in the field of behavioral responses to urban disaster is the response of specific ethnic and religious groups. Researcher Lori Peek conducted interviews with Arab and Muslim students in a number of New York City schools: New York University, Baruch College, Brooklyn College, City College of New York, Hunter College, Columbia University, and Queens College. These interviews were completed in late September and early October, 2001. Responses varied by gender and levels of ethnic and religious identification. One of the outcomes observed after this event was that there was a "racialization" of Muslims.

In December 2001, the researcher met with students in Colorado and with additional students in New York City. The total number of students interviewed was 126. The ongoing changes in behavior observed were more dramatic during the second round of interviews. Additional work with a more diverse sample population was considered necessary, including interviews with Arabs and Muslims of different ages and different education levels.

Psychological Reactions among Students after September 11th[33]

David Sattler examined the psychological reactions to September 11th of over 1,280 college students in New York, South Carolina, Colorado, and Washington, three weeks after the attacks. Between half and three-quarters of the students interviewed expressed concerns about future attacks such as "afraid family member might lose his/her life or be seriously injured in future terrorist attack;" "afraid I might lose my life or be seriously injured in future terrorist attack;" or "I have a family member or friend who will be at increased risk due to U.S. response to the event." About one-tenth of the students reported difficulty sleeping, avoiding things that remind them of the attacks, anxiety, nightmares, emotional numbness, feeling irritable, and difficulty remembering important things about the situation. More students in the East reported these symptoms than in the West. On the other hand, almost three-quarters of the students reported feeling patriotic, setting new priorities about what is important in their lives, having new respect for people in their communities, appreciation of each day, feeling closer to family members, and discovering that they are stronger than they thought they were. A study of the Arab and Muslim population conducted by Lori Peek showed different results, indicating that they are not feeling resilient or patriotic.

Faith Community Responses to New York City Terrorist Attacks[34]

There is little research describing the response and recovery activities of faith-based organizations after disasters. Jeannette Sutton conducted research through which she interviewed members of several New York City interfaith groups, Presbyterian congregations, the Church World Service, and faith-based groups affiliated with the American Red Cross, in an attempt to understand the ways in which those organizations rose to new challenges after the disaster.

Future Research Issues and Cross-cutting Themes

At the conclusion of the workshop, a group discussion identified the research themes listed below. These themes are a starting point for further discussion about research needs identified as a result of the September 11th disaster. A message board was set up on the ICIS website (http://www.nyu.edu/icis/Recovery) called "Online Discussion Forum" for the purpose of posting additional suggestions.[35]

Social and Human Behavioral Dimensions

Resiliency, Vulnerability and Criticality of Human Systems

The events of September 11th provided an important context for developing a better understanding of sources of psychological, social, and economic vulnerability, and for exploring ways in which human systems achieve resilience in the face of major disasters. Topics for future research are

- Methods, models, etc. to assess both vulnerability and resiliency of social, political, and economic systems across different units of analysis (e.g., individuals, organizations, institutions in both the public and private sector, as well as non-governmental organizations), geographic scales, and phases of emergency management (i.e., preparedness, response, recovery, and mitigation);
- Assessment of direct, (psychological, social, economic), indirect, and ripple effects resulting from September 11th; and
- Risk factors affecting both impacts and outcomes.

Relationships and Connections between Human and Physical Systems

Research is needed to identify ways in which the built environment and human and organizational behavior interact to either amplify or reduce vulnerability. Topics for study include

- Models, methods, and data focusing on the interface between human and physical (engineered) systems, and in particular ways in which these systems can be better integrated. Examples include building designs and emergency plans to enhance life safety by protecting building occupants and facilitating emergency egress; and
- Risk communication, pre-event planning, and post-event response management to protect lives and property and encourage appropriate self-protective behavior.

Institutional Arrangements

Additional research is needed to address institutional, multi-organizational, and organizational dimensions of pre-event mitigation and planning and post-event response and recovery. Research focusing on the following areas is needed:

- Capability and adaptability of institutions (e.g., governmental and private-sector organizations and entities responsible for infrastructure maintenance) to deal with vulnerability both before and after a disaster;
- Interorganizational and intergovernmental relations, including dynamics of multi-agency decision making and challenges associated with horizontal (among organizations) and vertical (among different governmental levels) integration in major crises;
- Communications and information sharing among individuals, groups, and organizations, especially with respect to the various phases of the emergency management cycle (e.g., preparedness, response, recovery, and mitigation); and
- Social, political, legal, administrative, and other factors that influence institutional behavior and response in large-scale and near-catastrophic events.

Decision Making and Risk

Additional research is needed to improve decision making at different units of analysis. Topics for further research include

- Models and approaches to characterize tradeoffs and decision processes employed by individuals, organizations, and institutions across the emergency management cycle; and
- Decision processes at various levels of analysis across the entire hazards/emergency management field and those that provide linkages among the various stages of the cycle.

Structures and Physical Systems

Analytical Models/Simulation of Performance

This capability has been developed in other areas and can be applied to structures. Data from the World Trade Center collapse is needed to validate such models and simulations. The design and operation should be considered under normal and extreme events. Data from other buildings and cases should also be included.

Analytical Models/Simulation of Building Systems

This area refers to the electrical and mechanical aspects of buildings. Examples include temperature, air flow, and other aspects. Data from the World Trade Center disaster is needed to validate these models and simulations. Design and operation under normal and extreme events should be included.

Analytical Models/Simulation of Emergency Management and Human Response

Such tools can be used in planning and execution. Data from the World Trade Center disaster should be used to validate these models and simulations.

Analytical Models of Information Flows and Information Sharing

This research topic consists of looking at what was done in terms of data sharing and what could be done better in the future. An area of research within this topic is the availability and incentives for sharing information, and being able to demonstrate the consequences of lack of sharing. How access to information can be preserved while respecting security needs is a key challenge.

Debris Field and Collateral Damage

This research area addresses questions related to where the collapsed pieces are likely to go and what the structure of the collapsed material is likely to be. The area of analysis includes both the surface and subsurface, and also includes infrastructure.

Structure of Collapsed Buildings

This refers to three areas: safety and removal; prediction of void spaces; and strategies for search and rescue.

Environmental Consequences

This area includes, but is not limited to airborne/plume models; waterborne and land-based pollution; evolution of source over time; and model validation for the World Trade Center and other crises in urban terrains.

Intelligent Buildings and Bridges

This research addresses the role of advanced technologies on intelligent structures/buildings and their future performance goals.

Distributed Networks

Given New York City's unique energy network, an important research question is what a similar disaster would do in a setting with a different energy network configuration. This area also refers to strategies for resilient networks and complex adaptive systems, such as energy, communications, water, and others. The World Trade Center disaster and other cases can be used to understand what worked and why.

Tools for Making Risk-informed Decisions

This includes databases of networks, models and processes. The main research question is how models of structures, networks, and processes can be integrated into risk models and risk management.

Fragility Curves for Organization Collapse

This area of research refers to the application of models from physical systems to organizations. An example could be how organizations perform under different levels of stress.

Interdependencies among Infrastructure Systems

At the time of the September 11th attacks, systems that were functionally and structurally interdependent may have magnified the impact on any given system. Research on infrastructure interdependencies needs to address the ramifications on other systems of one system's failure.

Cost/Consequence Models

Issues related to costs and benefits should be considered for normal and extreme events, as well as for response efforts.

Cross-Cutting Issues

Many research needs span both engineering and the social sciences. These common areas include the need for

- Improved theories, models, methods, and analytical tools, including tools that are capable of integrating data both spatially and temporally;
- Strategies to ensure maximum data availability, access, and sharing;
- Research focusing on documenting and analyzing both successes and failures in engineered and human systems (e.g., robust and redundant structures and systems, successful organizational coping and adaptation in crises);
- Research to better understand similarities and dissimilarities among varied disaster agents— natural, technological, and terrorism-related disasters; and
- Studies that address the needs of a wide range of users and target audiences (e.g., organizations charged with responsibility for managing response, recovery and reconstruction activities).

Notes

1. Based on a presentation at the workshop by Anthony Townsend, Lecturer, Robert F. Wagner Graduate School of Public Service and Associate Research Scientist, Taub Urban Research Center, New York University.

2. Based on notes from presentations at the workshop by Abolhassan Astaneh-Asl, Professor of Civil and Environmental Engineering at the University of California at Berkeley; W. Gene Corley, American Society of Civil Engineers; and Michael Bruneau, Professor of Civil, Structural and Environmental Engineering and Deputy Director of the Multidisciplinary Center for Earthquake Engineering at the University of Buffalo, State University of New York (summary not reviewed by presenters).

3. Based on a presentation by Frederick W. Mowrer, Associate Professor, Department of Fire Protection Engineering at the University of Maryland.

4. Based on a presentation by W. Gene Corley, American Society of Civil Engineers.

5. Based on a presentation by Thomas O'Rourke, Thomas R. Briggs Professor of Engineering, Cornell University.

6. Based on a presentation by Cruz Russell, Director of Policy and Planning, Port Authority of New York & New Jersey.

7. Based on presentations by Joe Englot, Chief Structural Engineer, Port Authority of New York & New Jersey; and Joseph N. Siano, Vice President/Program Executive, System Expansion, Capital Program Management, New York City Transit.

8. Based on a presentation by David Woloch, Chief of Staff, New York Department of Transportation.

9. Based on a presentation by José Holguín-Veras, Professor of Civil Engineering at Rensselaer Polytechnic Institute.

10. Based on a presentation by Elie Chebli, Manager of Network Design, Consolidated Edison.

11. Based on a presentation by Elie Chebli, Manager of Network Design, Consolidated Edison.

12. Based on a presentation by Anthony Townsend, Lecturer, Robert F. Wagner Graduate School of Public Service, New York University.

13. Based on a presentation by Diana Chapin, New York City Department of Environmental Protection.

14. Based on presentations by Diana Chapin, New York City Department of Environmental Protection, and David Lipsky, New York City Department of Environmental Protection.

15. Based on a presentation by William Wallace, Professor and Research Director, Center for Infrastructure and Transportation Studies, Rensselaer Polytechnic Institute.

16. Based on a presentation by J. David Frost, Professor, School of Civil and Environmental Engineering and Director of the Regional Engineering Program, Georgia Institute of Technology.

17. Based on a presentation by Susan L. Cutter, Carolina Distinguished Professor of Geography and Director of the Hazard Research Lab, University of South Carolina.

18. Based on a presentation by Christine Rodrigue, Professor and Chair, Department of Geography, California State University.

19. Based on a presentation by Michael Berkowitz and Mary Ann Marrocolo, New York City Office of Emergency Management.

20. Based on notes from a presentation by Randolph Langenbach, Federal Emergency Management Agency (summary not reviewed by presenter).

21. Based on a presentation by S. Shyam Sunder, National Institute of Standards and Technology.

22. Based on notes from a presentation by John Harrald, Director of the Institute for Crisis, Disaster, and Risk Management and Professor of Engineering Management in the School of Engineering and Applied Science at George Washington University (summary not reviewed by presenter).

23. Based on notes from a presentation by Richard T. Weber, Emergency Administration and Planning Adjunct Faculty and Professional Development Coordinator at the Center for Public Management in the Development of Public Administration at the University of North Texas (summary not reviewed by presenter).

24. Based on a presentation by Robert Parker, Managing Director of the Community Service Center at the University of Oregon.

25. Based on a presentation by David H. Hoover, Professor of Emergency Management and Fire Protection Technology at the University of Akron.

26. Based on a presentation by Claire B. Rubin, independent consultant at Claire B. Rubin & Associates.

27. Based on a presentation by Dave Simpson, faculty member in the Department of Urban and Public Affairs and Associate Director of the Center for Hazards Research and Policy Department at the University of Louisville; and Steven Stehr, Associate Professor in the Department of Political Science and Criminal Justice at Washington State University.

28. Based on a presentation by Kathleen Tierney, Professor of Sociology and Director of the Disaster Research Center at the University of Delaware.

29. Based on a presentation by Paul W. O'Brien, faculty member of the Department of Sociology and Criminal Justice of California State University.

30. Based on a presentation by Seana Lowe, Program Director of the International and National Voluntary Service Training Program at the University of Colorado at Boulder; and Jenna Peck, Research Assistant at the Natural Hazards Research and Applications Information Center at the University of Colorado at Boulder.

31. Based on a presentation by Bob Edwards, Associate Professor and Graduate Director of the Department of Sociology at East Carolina University; and Patrick Gillham, doctoral candidate in the Department of Sociology at the University of Colorado at Boulder.

32. Based on a presentation by Lori Peek, doctoral candidate in the Department of Sociology, University of Colorado at Boulder.

33. Based on a presentation by David Sattler, Professor of Psychology, Western Washington University.

34. Based on a presentation by Jeannette Sutton, doctoral candidate at the Department of Sociology, University of Colorado at Boulder.

35. The moderators of the two breakout sessions were Priscilla Nelson, Director, Division of Civil and Mechanical Systems at the National Science Foundation and Rae Zimmerman, Director of the Institute for Civil Infrastructure Systems at New York University.

References

Extreme Events Mitigation Task Force. 2002. "A Briefing on the World Trade Center Attacks." *Fire Protection Engineering* (13): 46–52.

U.S. Environmental Protection Agency. 2000. "The Emergency Planning and Community Right-to-Know Act Factsheet." EPA 550-F-00-004. Washington, D.C. EPA. http://yosemite.epa.gov/oswer/ceppoweb.nsf/7b422e07538ad30585256bde007a7bf7/85256b9f0072443385256c2c0057ce03/$FILE/epcra.pdf.

Acknowledgments

"Learning from Urban Disasters: National Science Foundation Response and Opportunities for Future Research Workshop," was funded by the National Science Foundation under grant no. CMS-0203371. NSF's support is gratefully acknowledged.

Correspondence should be addressed to Rae Zimmerman, Professor of Planning and Public Administration and Director, Institute for Civil Infrastructure Systems, New York University, Wagner Graduate School of Public Service; e-mail: rae.zimmerman@nyu.edu.

Part 2

Issues in Disaster Response

The Crash of United Flight 93 in Shanksville, Pennsylvania

Nancy K. Grant
Center for Emergency Management and Homeland Security Policy Research
The University of Akron

David H. Hoover
Center for Emergency Management and Homeland Security Policy Research
The University of Akron

Anne-Marie Scarisbrick-Hauser
Center for Emergency Management and Homeland Security Policy Research
The University of Akron

Stacy L. Muffet
Center for Emergency Management and Homeland Security Policy Research
The University of Akron

Introduction

The crash site of the fourth airplane hijacked during the terrorist attacks of September 11, 2001, was very different from the other three. Somerset County is a rural area of southwestern Pennsylvania known as a good place to ski and hunt. The fire departments are predominantly volunteer and the Emergency Medical Services a combination of paid and volunteer staff. Due to the scarcity of personnel, the different departments frequently assist one another on mutual aid calls. Likewise, the Pennsylvania State Police frequently provides law enforcement assistance and fire and rescue scenes, especially in connection with vehicular accidents on the state turnpike.

Although the events in Shanksville were unique in that they were part of the first terrorist attack on American soil, they were also similar to other emergency responses to natural and technological disasters, especially when considering the response from the perspective of the Integrated Emergency Management System (IEMS). The response to the crash site was multi-jurisdictional, multi-agency, and intergovernmental in character. Using this perspective, the researchers initially approached the events in Shanksville with the concept of building upon previous research opportunities in small communities, specifically the collapse of the Lonz Winery Balcony on Middle Bass Island and the explosion of the steam engine tractor in Medina, Ohio, both of which were mass casualty incidents occurring in small communities involving multiple jurisdictional local response as well as intergovernmental response. Other than the fact that the United Flight 93 crash was a multiple fatality rather than mass casualty incident, much of the response and even recovery activity were relevant to the questions at hand. Initial exploratory research from the first two case studies led to the development of the research questions to be examined in the response to the Flight 93 crash.

Research Questions

The initial research questions built upon findings from previous case studies.

Question 1. To what extent was the response to a disaster in a small town/rural area involving responders from multiple emergency response jurisdictional entities enhanced by serious exercising of existing emergency response plans?

Question 2. To what extent was the response to a disaster in a small town/rural area involving responders from multiple emergency response jurisdictional entities enhanced by personal knowledge of and trust in fellow emergency responders, especially those in charge?

These two items had emerged from each of the previous case studies and the research team wanted to continue investigation to see if there was enough evidence for theory development. Given the fact that this was a terrorist incident while the others, also classified as technological disasters, were accidents, there is an added element to the Shanksville disaster site control. Since the cause was terrorism, the crash site was a crime scene and the final jurisdiction of the scene rested with the Federal Bureau of Investigation (FBI). This led to the third question, designed to determine if the impact of the exercises and trust of fellow responders is mitigated by control of the site by external (i.e., federal/FBI) authorities.

Question 3. Did the presence of external control of the site (by the FBI) mitigate the impact of the exercises and trust of local responders?

A fourth question of the initial research emerged as the researchers observed the outpouring of volunteer assistance across America. This portion of the study focused more on the community response than the emergency response operations at the scene. It was based on the work of Quarantelli and Dynes (1976) dealing with behavior within a convergence crisis.

Question 4. To what degree did the behavior of volunteers conform to theoretical expectation according to the convergence theory of volunteers in a consensus crisis?

Together these questions focused the gathering and analysis of information relating to the crash of Flight 93 in rural southwestern Pennsylvania.

Methodology

Given that this project is an exploratory and interpretive, qualitative methods were used. Qualitative research has gained acceptance as a valuable policy analysis/development and management practice analysis tool. The qualitative research design facilitates the emergence of questions for further research as well as building foundations or providing support for theories during and after data collection (Jacob, 1988; Kruse, 1997). Thus, the approach is appropriate in attempting to identify the influence of prior professional interaction via training and exercise and the development and impact of trust in fellow responders in inter-organizational and multi-jurisdictional response to and recovery from a disaster. Especially appropriate was the emphasis on the importance of conducting research in a natural setting with a focus on understanding participants' perspectives (Lincoln and Guba, 1985; Wilson and Mosher, 1996). The interpretive or qualitative paradigm emphasized the local, individually based constructions of meaning that enabled greater understanding of the context of the emergency response (Kruse, 1997; Lincoln and Guba, 1994; Merriam, 1988).

One of the benefits of using qualitative methods in this project was that the qualitative interpretive approach moved the research from the objective to the perspective view. Individuals had perspective views of the disaster scene and Kruse recognized the need to understand the multiple views of the same phenomenon taken from different "lenses" relating to the position of the individual in order to understand a complicated and changing system or scene (Kruse, 1997). Thus, for example, on the site of a disaster due to terrorism, the FBI viewed it as a crime scene, medical responders viewed it as a mass casualty scene, law enforcement viewed it as a security challenge, business viewed it as an economic challenge for continuity, waste management entities viewed it as a major debris challenge, and emergency management viewed it as a constantly changing puzzle that they were trying to put together and hold in place. Each of these views provided one piece of the whole reality of the situation. In order to understand emergency management at a disaster scene, one needed input from each of the perspectives—thus the need to conduct numerous in-depth interviews with the various responders to disaster sites.

Thus, the qualitative research paradigm was appropriate for the exploratory and interpretive nature of the research questions of this study. Merriam has stated that qualitative methods enable the researcher to "understand humans as they engage in action and interaction with the contexts of situations and settings" (1988, p. 29). Kruse also pointed out that most qualitative researchers prefer data that includes quotes from interviews, participant observations, and information that describes the research setting as well as the people in the setting, precisely the primary approach taken (1997).

The researchers used a three-part data collection method. Initially, the plan was to arrive at the scene fairly early, two to three days after the incident, and conduct interviews with key personnel if possible. Upon arrival Friday morning, however, the situation was such that the methodology was amended to practice primarily observation and hold informal conversations with response personnel on the scene rather than to conduct the planned focused interviews. Thus, the first methodological approach involved observation of the scene and of interaction between and among response personnel. Several pages of field notes were gathered and compared at a later date.

The second method of data collection was the focused interview technique. Due to the tense nature of the scene, the investigators returned to the area to interview response and recovery personnel on two separate occasions. In addition, they had the opportunity to speak with responders to Flight 93 at two different conferences on emergency management. Given that these responders also presented their experiences, this proved to be an additional source of information. The researchers conducted eight full-scale focused interviews, held 11 different identifiable informal conversation interviews, and heard six different formal presentations.

The focused interview technique has been in use for at least 50 years and involves the construction of an interview schedule that is used for each of the subjects in the study. The questions were carefully designed to elicit responses focusing on the research question. The primary question could be enhanced through the use of secondary "prompt" questions that aid in getting the subjects to talk and in keeping them on the topic at hand (Yin, 1984; Merriam, 1988). Keeping the conversation focused ensured that the researcher could gather a rich body of information on which to perform content analysis (Miles and Huberman, 1984; Yin, 1984). This approach had the advantage of permitting the investigator to direct the interview while allowing the respondent freedom to answer in an open, unstructured fashion (Merton and Kendall, 1946). Interviews are best used to find out what kinds of things are happening rather than to "determine the frequency of predetermined things." The personal contact between the researchers and the subject was the principal advantage of the interview as it also led to the following advantages:

- The researchers could re-contact the respondents as necessary.
- The higher response rate of this technique was important due to small sample size.
- Ambiguities, questions, or concerns of the respondent could be resolved.
- Interviews were more flexible and could employ more complex lines of questioning than is generally possible with surveys.
- The interviewer could ask sensitive questions during the interview when he/she sensed the respondent was at ease and more likely to answer.

The weakness of focused interviews identified by Nathan (1986)—that they are "inadequate to do without other supporting evidence" (p. 109)—was mitigated by the use of triangulation using secondary data from reports and newspapers. The strength of the use of focus interviews was that it was very good for building emerging understandings of relatively new phenomenon, which was the focus of the research questions.

The informal conversations occurred when it was not possible to schedule the interaction in advance and when it was not possible to take detailed notes or record the conversation at the time. In these instances, the researchers would make notes individually as soon as possible and then compare individual notes to compile a complete record of the information gathered. Generally these conversations provided background information, perceptions, and sometimes offered points of clarification. They were not as in-depth or detailed as the other verbal data sources, however.

The third method used was content analysis of newspaper coverage of the response and recovery operation. The particular benefit to this component of data collection was that, while national news sources were utilized, the primary source was local coverage of events, which contained more details than the national news stories. This was in part because of the competition with the events at the World Trade Center in New York City, but also because of the personal aquaintance of many local responders by the members of the local media. These stories were talking about local residents as well as about others who were there to lend a hand.

Thus, this study incorporated triangulation using a combination of secondary empirical data/information analysis and primary qualitative analysis utilizing the focused interview.

Background of Flight 93

On September 11, 2001, 44 passengers and seven crewmembers left Newark, New Jersey, aboard United Airlines Flight 93 bound for San Francisco. The aircraft, a Boeing 757, was scheduled to depart at 8:01 a.m., but was delayed 40 minutes. As the plane headed west, it was passed to the control tower at Cleveland, Ohio, for monitoring through the Midwest portion of its journey. When the controllers tried to contact the plane, asking the pilot to verify altitudes, they received no response. Minutes later the plane made a hairpin turn, heading towards Washington, D.C. The air controllers reported hearing a thickly accented voice saying, "This is your captain. There is a bomb on board. We are returning to the airport" (Breslau, 2001, pp. 3–4).

According to Major Lyle Szupinka of the Pennsylvania State Police, air traffic controllers in the Cleveland control tower taped conversations from the plane's open microphone in the cockpit. There was screaming and then a voice said, "Get out of here. We don't want to die." The traffic controllers saw the plane execute a U-turn, turn southeast, and begin to descend in altitude. There were sounds of a continuing struggle and then a heavily accented voice said, "Your pilot has chosen to die. We have a bomb. Please stay in your seats" (Szupinka, 2002). These two brief reports, with their minor differences, were illustrative of the many questions surrounding the details of the last moments of Flight 93.

Passengers made telephone calls to their families and then, knowing what had happened in New York City, the passengers decided to fight back. With the famous battle cry of "Let's roll!" passengers attacked the hijackers and in the ensuing struggle, the plane plunged to the ground at 10:08 a.m.. There were no survivors. United Flight 93 had ended, but the emergency response and recovery had just begun.

The content of telephone calls between passengers and loved ones painted a picture of tense moments, honor, and love of county and fellow humans. These calls became part of our nation's history. For the responders to the crash scene, however, it was another telephone call that first alerted them to the terrible tragedy that was about to take place and put in motion the emergency response to the crash site.

Initiation of Response

Dispatch operators at the Westmoreland County Emergency Operations Center received a 911 call at about 9:58 a.m.. Glenn Cramer, a 911 supervisor, was quoted as saying, "We got the call about 9:58 this morning from a male passenger stating that he was locked in the bathroom of Flight 93 traveling from Newark to San Francisco, and they were being hijacked. We confirmed that with him several times and we asked him to repeat what he said. He was very distraught. He said he believed the plane was going down. He did hear some sort of an explosion and saw white smoke coming from the plane, but he didn't know where. And then we lost contact with him" (Silver, 2001, p. 2).

The report of the sound of a muffled explosion led to some of the speculation that there was a bomb or that the plane had been shot down. However, in cross-referencing other details that emerged, Hoover and Grant discovered references to the probable use of makeshift weapons in the passenger attack on the hijackers. Mentioned as a probable weapon was a portable fire extinguisher which, when discharged, emits an unusual explosive-type sound caused by the sudden release of the compressed nitrogen in the cylinder, followed by a discharge of dry chemical extinguishing agent that has the appearance of white smoke. Grant and Hoover believe that the caller locked in the restroom heard the sudden discharge of the fire extinguisher and that the white smoke he reported was in fact the chemical extinguishing agent (Hoover and Grant, 2002a).

The 911 dispatch center immediately contacted the Pennsylvania State Police Barracks at Greensburg to report the call. Meanwhile, the Federal Aviation Administration had contacted the Pennsylvania State Police and the Johnstown airport stating that there was a commercial carrier over Cambria County and the low altitude led them to believe that they were attempting a landing at Johnstown. State troopers were dispatched at once and were on their way to the airport when reports of the 911 call and of the crash were received. They were then redirected to the crash site (Szupinka, 2002).

Question 1: Impact of Emergency Response Plan Exercises

The emergency management preparedness and planning phase requires participation by multiple agencies and organizations and delineates intergovernmental relationships that are in existence and others that will be enacted in case of a declared disaster. Initial research concerning this planning phase indicates the need for cooperative participation by all responders (Waugh and Hy, 1995; Schneider, 1995). The role of exercises or drills in achieving this cooperation has been addressed somewhat, but specific research on their impact on disaster response is scarce.

Campbell (1999) and Payne (1999) point out that exercises have enhanced the response to a disaster and emergency, but provide little direct link between the exercise of the plan and the execution of the plan in terms of differences and performance. Auf der Hiede (1989) likewise points out the fundamental use of preparedness, including conducting exercises to test the plan, to enhance the actual response. Tierney et al. (2001) document the need for exercising as an integral part of the preparedness process and offer some examples on how this has improved the effectiveness of response. The examples on how this has improved the effectiveness of response. The researchers at the scene of the Flight 93 crash wanted to understand how the plans and exercises themselves affected the response. Earlier research by the investigators identified how deviations from the plan and prior exercises caused problems at the scene and how activities performed in accordance with previously exercised protocol went smoothly. The investigators focused on whether the same was true with the response to the crash of Flight 93. This was the perceived unique aspect of the research. A good illustration of this phenomenon was the events surrounding the challenge of feeding the responders during the response and recovery phase.

The two agencies primarily responsible for mass feeding were the American Red Cross and the Salvation Army. Both agencies had participated in mass casualty drills and had plans in place for responding to a disaster. However, due to the extraordinary nature of the events surrounding the crash, each had to deviate from its protocol. In addition, the local fire department, which did not have mass feeding as a responsibility, attempted to assist in ways that did not conform with the response plan. As a result, several challenges emerged during the response and recovery.

The Salvation Army has its disaster services specialty response located in Pittsburgh. While this seems to be right next door, it must be remembered that they were also contending with as many as 20–50 planes being diverted to the Pittsburgh airport, with as many as 100 people per plane stranded there. Given that the airport was closed, which meant all of the food vendors were shut down, if the passengers were indeed stranded, it was going to be a major challenge to provide food and shelter. Thus, the initial response to the Shanksville crash scene came from local Salvation Army personnel from Somerset, Fayette, Westmoreland, and Cambria counties.

In the initial response, a volunteer for the Salvation Army, who was in fact the communications officer for disaster response, was working at AAA as a dispatcher. He received an activation call from the Pittsburgh office and his employer let him go immediately. He went to the Emergency Operations Center and served there for about 14 hours along with another officer who arrived a couple of hours later. Thus, the Salvation Army operation at the site was staffed by a combination of volunteers and disaster response specialists. Due to the nature of the activity, it was clarified that the disaster response specialists were in charge and that the procedures for disaster response services superceded other Army policies (Myers, 2001). An interesting contribution by the Salvation Army was the distribution of 300 bibles to recovery workers at the crash site. In addition, five worship services were conducted there.

The philosophy of coordinating rather than doing everything is held by support organizations such as the Salvation Army and Red Cross as well as emergency management. For example, after the first 36–48 hours of providing food to the responders on site, the Salvation Army implemented the newly negotiated agreements with local providers, such as Camp Allegheny and three local commercial restaurants. These entities prepared the meals that were then distributed by the Salvation Army. Once these agreements, similar to those of the Red Cross, were in place, the two volunteer organizations coordinated the distribution so that neither organization was overwhelmed by the responsibility. Unfortunately, even with the agreement, there was some difficulty with duplication and mis-coordination.

The challenge in providing mass feeding in long-term disaster recovery operations is primarily one of quality. Getting food is seldom the difficulty because of the generous outpouring of support from individual citizens as well as local businesses. However, the challenge is to maintain the quality of the food to ensure that no one suffers illness due to consumption of contaminated food. Unfortunately, while individuals may prepare food properly, they do not have the facilities to store and transport the food under safe conditions, thus causing deterioration before it is consumed by the response and recovery personnel. The situation in Shanksville was compounded by the tremendous generosity of the local citizens in Shanksville and the members of the Shanksville Volunteer Fire Company. The statement of Fire Chief Terry Schaffer best illustrates the commitment of the Fire Company, "This may be a little town but it is big enough for whatever happens. Whoever and whatever needs us, we're there for them" (Schaffer, 2002). This philosophy led to an outpouring of donations of time and effort as well as food. For example, the local Boy Scouts used mounds of donated apples to make apple cider that could be more easily distributed and consumed. In fact, local generosity was so great, that after three or four days they were overwhelmed with so much product coming into the fire station that they actually had tractor trailer loads of "stuff" inside the station and their fire apparatus was outside. They worked with the Salvation Army to sort supplies and to donate what could not be readily used in feeding the recovery teams to the local food bank. The fire department continued to do some feeding throughout the recovery operations, primarily serving first the perimeter personnel and then staffing an open-table food operation at the Fire Station, which could be taken advantage of by anyone. Upon termination of on-site operations, Chief Schaeffer estimated that they still had food for 10 days (Myers, 2001; Schaefer, 2002).

Due to the fact that there were three agencies providing the same service to the scene, a coordination meeting was called by the Pennsylvania Emergency Management Agency to solidify communications and to identify coordinated feeding activities in order to eliminate duplication and simultaneously ensure comprehensive provision of quality meals to all on-site personnel on a consistent basis. This is illustrative of the key coordination and facilitation role of the emergency management personnel. This was necessitated because of the deviation from the plans and exercises performed in the past that had very defined roles of support agency responsibilities.

The remote character of the crash site led to unique food preparation and distribution challenges. The Salvation Army established contracts with a private caterer and with Camp Allegheny to set up a week's menu at a time. The Eat 'n' Park restaurant chain was their primary caterer, utilizing both their local and Johnstown restaurants to keep either from being overtaxed. One of the more unique partners in meeting this challenge was Camp Allegheny, a Methodist Church Camp located approximately one mile from the site. According to Bob Myers of the Salvation Army, Camp Allegheny would clean all dirty meal containers and return them for the next meal as well as providing housing for Salvation Army personnel as needed (Myers, 2001). Camp Allegheny is one of only about 15 camps nationally with a completed Emergency Response Plan. The Somerset EMA had worked with the camp in writing the plan and the camp is part of the county response plan as well as an identified resource in case of disaster. Camp Allegheny housed from 12 to 25 personnel each night and helped transport meals and Red Cross workers to the site. Camp maintenance personnel made a run to the site each mealtime to provide assistance. They would carry ice in garbage bags along with soda and sandwiches as necessary to keep things cool. According to the Reverend Duane Slade, Camp Director, the primary reason that they were able to work well with response was that (1) they had the plan in place, (2) they reviewed the plan annually with their staff, and (3) they knew the emergency management personnel from Somerset County as well as other local responders on the scene (Slade, 2001).

In terms of the research questions, the mass feeding programs demonstrated that, indeed, those who had participated in this type of response before were better prepared and able to perform the required tasks. This is especially true due to the long-term nature of the operation. The fire department and community members were trying to assist and did the best they could. However, their primary role is not mass feeding and they were not able to maintain the quality and consistency necessary, especially in distribution. The extreme nature of the event required operations beyond the scope of past exercises and resulted in discrepancies that had to be resolved through negotiation.

Question 2: Personal Knowledge of and Trust in Fellow Emergency Responders

Hightower and Coutu (1996) noted that when organizations have interacted and coordinated with each other in prior situations, they have fewer problems interacting during a disaster response. Recognition of people's professional skills and abilities built trust, while drill exercises offered participants the opportunity to demonstrate competencies (Hightower and Coutu, 1996). The reverse has been shown to hold true as well. Lack of information about the activities of other responders has resulted in failure to respond to some disaster demands and to duplication of effort (Caluse et al., 1996). Lindell and Whitney also referenced the impact of trust among planning committee members during the planning as well as response activities as being a major factor in the degree of success (1995).

Indeed, at the Shanksville crash site, the initial local responders had worked with each other frequently in the past, both during mass casualty drills and during actual emergency responses, although these were on a much smaller scale. The personal trust extended to agency trust. For example, given prior experience with individual Pennsylvania State Police Troopers, local fire and EMS personnel readily accepted all state troopers as competent and professional. Likewise, Salvation Army and Red Cross personnel accepted the adjacent county emergency mangers that arrived to assist Somerset County emergency management.

Smooth operation and cooperation was secured initially with those organizations with whom the Salvation Army and Red Cross had responded in exercises and in smaller incidents in the past, such as the County Emergency Management Agency and local fire departments and EMS units. Additionally, the fact that the initial volunteers were from the area and knew many of the responders facilitated the early set-up of operations at the command site.

Question 3: External Site Control and its Impact on Local Responders

Background of Scene Control

The fact that the crash was due to a terrorist attack placed the scene under the direction of the FBI, whose primary duty was to collect and preserve evidence. The FBI worked closely with other federal agencies such as the National Transportation Safety Board; the Bureau of Alcohol, Tobacco, and Firearms; and even the Department of Health and Human Services. Agent Wells Morrison was assigned the task of establishing the on-site command post for the FBI, which brought in over 150 agents and support personnel from ten offices. The crash site of Flight 93 was different from the other terrorist incidents on September 11th in a couple of very important ways. It was by far the most productive in terms of collecting evidence from a crime scene, since there were not tons of structural debris to sift through looking for evidence. The aircraft slammed into the ground at a high rate of speed, causing severe fragmentation and scattering pieces of the plane and its contents that had evidentiary value across an open field, and through a section of trees and across a lake beyond. However, the site was quickly and fairly easily containable. Evidence gathered at the scene was flown to headquarters in Washington, D.C., once a day for analysis. According to Morrison, the first significant piece of evidence was found the first night by a Pennsylvania State Police Trooper assigned to security, who found the wallet and passport of one of the hijackers (Morrison, 2002).

Impact on Interaction and Trust

The designation of the crash site as a federal crime scene made the response event somewhat unique for the local and state responders. The coordinated response and recovery activities among the various local and state agencies proceeded smoothly, according to the plans that were in place. The initial perception of the FBI by several of the local responders was that the local coordination was working well and the FBI was just doing its job without really working with the locals. Federal interagency cooperation appeared to be coordinated; however, there were initial rough spots between the federal and local responders.

An example of how trust had to be better learned on the scene when there was no prior experience with it is that while the FBI was very receptive and supportive of the services of the Salvation Army, they did not initially recognize the need to have a canteen deliver food to the personnel working at the recovery point of the crash site. Their concern was especially strong before the discovery of the flight recorder, which has high evidentiary value. Once the need and the benefit of hot food "and a smile" was explained in terms of human performance and reduction of errors at the scene, and they had the opportunity to observe and obtain feedback about the truth of this position, the FBI became even more supportive of the modes of providing service as well as of the service itself.

Throughout all these events, the Somerset County Emergency Management Director, Richard Lohr, with support from the Western Regional Office of the Pennsylvania Emergency Management Agency, particularly Thomas Hauger, and surrounding county EMAs, did his job facilitating, supporting, and coordinating resources and supplies. The EMA on scene was the clearinghouse to gather and distribute necessary resources. This facilitated quick access since suppliers had been identified in the response plans in advance. It also eliminated duplication as multiple requests for the same items were checked to see whether they were duplicates of one request sent in by different agencies or really individual needs. Donations management was an important issue during this recovery operation as many local and regional citizens and organizations wanted to do their part. The Somerset EMA managed the donations to maximize their utilization and eliminate waste. Initially the coordination role of the EMA was not appreciated by the FBI and other federal agencies. However, as the recovery operations continued, their value and efficiency became clear. By coordinating resources, EMA took the burden off the federal personnel who could then focus on their tasks.

The Somerset EMA processed 205 resource requests during the recovery operation. It took the position of providing logistical support for the responding agencies on the scene and coordinating services among agencies to ensure there was no duplication or wasted resources. For example, at one point two requests for BFI recycling containers came to the EMA resource area within half an hour, one for 150 containers and one for 200 containers. EMA immediately went to work to secure the containers, but also tracked down the two requesting entities to determine whether they were duplicates or if the recovery operation really needed 350 containers at that time. Having resource requests centrally processed eliminated duplication and facilitated quick acquisition. Somerset EMA already had purchase agreements in place with numerous local suppliers. Therefore, they knew who had what products available from their comprehensive emergency response plan resource book and simply had to contact people who were already familiar with the EMA director personally to activate contracts and get resources in quickly.

Sometimes the need to coordinate with EMA was not clear. For example, one morning at about 10 a.m. a call came in from the decontamination area to report that no one had emptied the Port-a-Johns the night before. Director Lohr checked and determined, both from the company servicing the Port-a-Johns and from a supervisor on the night shift, that the Port-a-Johns had, in fact, been emptied the night before. The problem was that the FBI had increased the number of personnel at the recovery site by over 100, which severely taxed the capacity of portable toilets placed at the decontamination site. In order to meet the increased demand, EMA immediately negotiated increasing the number of Port-a-Johns.

Again, those local responders who knew the EMA personnel and had participated in local exercises were immediately receptive of their role and recognized the existing units. Had EMA been informed the day before the additional personnel were assigned, they would have simply increased the number of units on site. This illustrated the fact that, while agreeing in principle to the function of the EMA, some of the state and federal personnel were not willing to be proactively inclusive. However, their value became apparent rather quickly and trust began to build. By the end of the operation, when the FBI released the scene, they were extremely complimentary of the resource management and truly recognized the value of the system in place. In fact, the FBI stated that there was nothing they asked for that they didn't or couldn't get in triplicate and the National Transportation Safety Board later stated that this was one of the smoothest-run recovery operation scenes in which it had ever participated. Basically the IEMS plan of having the experts perform their areas of responsibility with central coordination and facilitation through the support of EMA worked at the scene of the crash of Flight 93.

While local responders had worked with federal agencies such as the Federal Emergency Management Agency, the National Transportation Safety Board, and the Small Business Administration in the past, they had not worked with federal law enforcement agencies such as the FBI and the Bureau of Alcohol, Tobacco, and Firearms. Likewise, while the FBI had worked with local and state law enforcement agencies, they were not accustomed to working with fire, EMS, emergency management, and non-profit organizations under these conditions. Thus, this became a learning situation.

Building Trust and Confidence on Scene

Perceptions presented during the first 36 hours changed significantly during follow-up interviews held with local responders six weeks later. Basically, as the FBI began to understand the various roles of the other response agencies and why they did things the way they did, their attitudes and positions shifted. This was a new role for the FBI and one that may reoccur with more frequency in the future. The FBI demonstrated the ability to learn new methods of intergovernmental operation and cooperation.

Likewise, the local agencies and response and recovery personnel gained an understanding of some of the reasons why the FBI approached recovery matters as it did. They understood the principle that this was a crime scene and that the major focus of the FBI was to recover evidence, but it was only as the operation proceeded that they began to understand more fully what that meant in terms of specific operation protocol and preferences. Thus, this was a learning situation in both directions. Supervisors of the local operations recognized the benefits and contributions made by the FBI, which enhanced the local response and recovery efforts. Examples of these included the completion of the memorandum of understanding with the Department of Health and Human Services, which enabled the deployment of the Disaster Mortuary Response Team (D-MORT) and the assumption of many of the expenditures associated with the recovery operation. This latter was extremely important since no federal disaster relief monies were available because this site never received a Presidential declaration of disaster. In this instance, the FBI demonstrated an understanding of the situation of the local jurisdictions and offered support that they otherwise would not have received.

Summary for Questions 1, 2, and 3

The exercise and training among the emergency responders before the September 11th disaster did make a difference in the response in Shanksville, Pennsylvania. Preparing for disaster did affect the response and recovery operations at the disaster scene. In answer to the second research question, working together and exercising response plans did build trust. There were stronger bonds of trust forged between and among responders who had worked together before than among those who had not had the opportunity to work together. Thus, given that the presence of the some of the federal agencies was not standard for natural hazard disasters, there was unfamiliarity and, therefore, discomfort associated with the initial coordination with these agencies. However, trust was built during the longer term of the response and recovery operations and was solidified among those who held it at the onset.

Question 4: Convergence of Volunteers in a Consensus Crisis

A fourth element of the research project examined volunteers and their behavior within the context of a "convergence crisis," as outlined in the work of Quarantelli and Dynes (1976). Specifically, this question addresses the degree to which the Shanksville event matches the theoretical expectation according to the convergence theory of volunteers in a consensus crisis. Wenger developed a system of classifying events from an emergency to a crisis and then social crisis level depending upon the adequacy of the rules of existing societal rules and daily operating procedures (1978).

A review of research activities indicates that there have been relatively few systematic studies of many of the forms of emergent crowd behavior, particularly those associated with immediate post-impact disasters (Wenger and James, 1994, p. 230). Most research to date confirms the premise that convergence on the site is typically massive and widespread during the early post-impact stages (Barton, 1969; Dynes, 1970; Drabek, 1986; Drabek, 1996). However, there is still a need to conduct research on identifying the social and organizational drivers that are necessary to enhance interagency and multi-level, national, regional, state, and local cooperation and interaction. More importantly, there is a need for those involved in emergency management education and training to build a greater understanding of how volunteer emergency response personnel and non-voluntary disaster site personnel navigate the conflict and bureaucracy and find ways to cooperate with each other in the emergent post-impact disaster period.

It is important to note that, before September 11th, the Flight 93 plane crash, like many other plane crashes, would have been classified as an "emergency" from an organizational and community level (Wenger, 1978).

Indeed, the Shanksville area and surrounding agencies were in a better position to respond to a plane crash as a result of enhanced preparedness and response training after the 1994 crash of USAir Flight 427 in Hopewell Township just northwest of Pittsburgh (Lohr, 2001; Myers, 2001; and Hauger, 2001). However, given that the cause of the crash was criminal terrorist activity, the classification of the Flight 93 event quickly progressed from an "emergency" to a "crisis" as a result of the overwhelming convergence of recovery and investigative personnel. This and the public perception of a national crisis served to classify this disaster as a "social crisis."

A social crisis is identified when the nature of the event affects the social fabric of the entire nation. In the case of September 11th, the operation of the entire American society was affected, the regular ritual of routine activities of society stopped—schools were evacuated, workplaces were closed and evacuated across the country, the financial stock exchanges ceased trading for days, and national sports events were cancelled. Over the next two weeks, the overall mood of society turned away from the daily routine, people were glued to their television sets and television executives took the extraordinary step of suspending advertisements in deference to the national mood of mourning (Alexander, 1985).

Convergence

The concept of convergence has been discussed in both collective behavior and disaster texts since the 1950s (Fritz and Mathewson, 1957). Convergence refers to a mass movement of personnel and informational and material resources into a disaster site in the hours and days after the event. The characteristics, motives, and behaviors of volunteers in the emergent, post-impact period has long engaged the interest of disaster scholars (Fritz and Mathewson, 1957; Zurcher, 1968; Barton, 1969; Mileti et al., 1975; Dynes and Quarantelli, 1980; Adams, 1983; Wenger et al., 1980; Drabek et al., 1981, 1996; Holland, 1989; Stallings, 1989; Mileti and O'Brien, 1991; Perry et al., 1983; Hull and Wenger, 1992; Wenger and James, 1994). As a result of these many investigations, a number of typologies, schemas, and classification trees outlining the many different types of volunteers who have been observed converging on post-impact disaster sites have been well documented.

Quarantelli and Dynes (1976) have identified disaster settings as excellent examples of "consensus crisis reactions." As outlined in their findings, the characteristics that distinguish a **consensus** event include an agreement among participants regarding the goals and objectives for the work that needs to be done and pertinent action steps. Observable characteristics include a decrease or leveling of social distinctions; an emphasis on the needs of the present timeline; the emergence of altruistic norms in both individuals and the community; the recognition of the presence of an external threat to the system; the expansion of the citizenship role and increased identification with the community; and the presence of immediate and non-immediate ambiguous problems leading to a cooperative action to solve issues created by the crisis (Dynes and Tierney, 1994, p. 32).

The next section examines the degree to which these characteristics of volunteer behavior in a consensus crisis, as outlined by Quarantelli and Dynes, are observable using preliminary data from the Shanksville study.

Who is a Volunteer?

Several types of volunteers converged on the disaster site. Many, if not most, of the volunteers present at Shanksville were activated using emergency response procedures. These volunteers did not arrive haphazardly or unannounced.

Up to 74 different agencies were involved in some way in the recovery efforts (O'Brien, 2001a, 2001b). Up to 1,118 people were working on the site during the first 13 days of the recovery/crime scene investigation. This number included federal, state, and local paid employees. This number also included paid volunteers, activated private citizens who become paid volunteers, unpaid volunteers, and volunteers involved in support functions who converged on the scene. Other volunteers may converge on a scene through their proximity to the event; these people may live next door to the site, may have been driving by at time of impact, or may possess local information that is required in the first few post-impact hours of response. For example, Shanksville community members traveled to the crash perimeter from the time of impact through the night to dawn, bringing food to the state troopers guarding the site. Community volunteers began the process of establishing a number of memorial tribute areas—gestures noted by grateful bereaved family members (Rock, 2001).

The activities that occurred during the response to Shanksville supported the various characteristics associated with convergence.

Broad Agreement on Goals and Objectives

As mentioned, personnel, material, and informational convergence were expected at the scene at Shanksville on September 11th. However, once the lack of survivors was confirmed, the goal for site management changed from one of rescue to recovery as it quickly became obvious that the Shanksville site had the potential to yield the most evidence regarding the perpetrators of the day's activities. The entire area was quickly designated a crime scene with federal and state oversight. Volunteers were activated in this instance on the basis of specialized technical skills, e.g., D-MORT and support workers such as the Red Cross, Salvation Army, and mental health professionals. Counselors were also activated to provide therapy for the emergency workers. By Friday, September 14, 2001, more than 400 volunteer personnel had gathered at the crash scene to assist with the recovery process. It was estimated that up to 150 D-MORT volunteers were activated and assigned tasks at different locations around the Shanksville area to assist with locating and identifying human remains within 48 hours of the impact.

It has been noted on more than one occasion that responses to disasters actually serve to create social order or organization as opposed to managing disorganization (Dynes, 1970.) Many of the emergency responders on

September 11th had also responded to the 1994 Hopewell Township jetliner crash. Time and effort had been dedicated in the intervening years to preparing a comprehensive intercommunity response team to respond to similar occurrences. Looking at the Shanksville event, there appeared to be a broad understanding of the tasks necessary to achieve the goals and objectives of the criminal investigation and the recovery effort within the local community. However, there were unavoidable conflicts in the coordination of tasks between jurisdictional and volunteer agencies, particularly when new working relationships had to be established between disparate groups with emergent roles, tasks, and objectives.

Despite some private observations to the contrary, the FBI credited the cooperation among a variety of state, federal, and local agencies as well as assistance from disaster relief volunteers and local businesses in accelerating the work at the site. As William Crowley, FBI special agent, commented, "Virtually anything we requested, we got in triplicate." The Somerset Coroner, Wallace Miller, also commented that "the joint effort brought a decent end to a horrible circumstance" (Gibb, 2001a).

Decrease in Social Distinctions

The perceived social impact of the event combined with an increased propensity towards altruistic activities was likely to result in a leveling of social distinctions during the emergency period and thereby to facilitate participation by broader segments of the community. Community volunteers in collaboration with the united disaster support group and the Red Cross worked quickly with every level of the community to produce a community memorial service in Somerset on September 14, 2001. The researchers who observed this ceremony noted that the entire community was united in an expression of grief for the family survivors and that there was also a need to express gratitude for saving their community from further loss of life. Community volunteers, irrespective of occupation or status, donated their time and resources to respond to multiple requests during this period. As School Superintendent Gary Singel said, "We thought we lived in a place that was almost untouchable" (Gibb, 2001b).

Emphasis on Present Orientation

The investigative and recovery tasks at the Shanksville site continued on a 24-hour basis. In fact, in the initial weeks, the focus on the work at hand was so intense that responders forgot what day it was. Several were surprised when the Salvation Army held a worship service and they realized it was Sunday morning. Each response agency worked hard to establish a realistic work schedule for its personnel, ensuring breaks and rest times that would keep them alert while on the job.

The Presence of Altruistic Norms

From the time that volunteers and paid personnel arrived at the disaster site, a message was quickly communicated that the site was to be secured and preserved mindful of the 44 lives that had been lost in the area. Response to the initial crash was a disappointment for many of the volunteer emergency response personnel. Keith Custer, a member of the Shanksville Volunteer Fire Company, said, "I thought we'd get there and there would be a big plane on fire and victims . . . but you get there and there are just little pieces. Now, I don't want to see anything like that again, seeing that massive loss of life" (Gibb, 2001b).

With the assistance of experienced volunteers, the community resources worked to make sure that family members of the victims who traveled to the site were protected from public and media scrutiny in a local resort area. The community continually expressed a commitment to maintaining the privacy of the family members, and the sanctity of the crash site.

Everyone from school children to public officials identified the passengers as personal heroes and tried to contribute visible honors such as signs and ribbons as well as cards and notes of thanks to responders and families of victims. The donations of food and contributions toward the funding of a permanent memorial were examples of attempts of community members to give of themselves.

After the departure of the FBI and the return of control of the scene to the authority of the Somerset County Coroner, more than 300 volunteers responded to his request to conduct a final sweep of the crash site to secure remaining airplane fragments and human remains. These volunteers were primarily from the Southwestern Emergency Response Group 13, supported by 50 members of the State Funeral Directors Association (Miller, 2001).

Expansion of Citizen Role

Shanksville residents expressed their patriotism and collective grief in the days after the disaster. Judi Baeckel led a drive to erect a memorial to the victims in her front yard on Bridge Street because the residents wanted to make sure that Flight 93 was not forgotten in all the media focus on New York and Washington, D.C. (Schaeffer, 2002). Memorials, displays of American flags, and yellow roses were erected in and around the crash site and Somerset county. Community members traveled the crash perimeter from nightfall to dawn bringing food to the state troopers guarding the site. Community members began the process of establishing a number of memorial tribute areas—gestures noted by grateful bereaved family members. The Somerset community prepared an early public memorial service, held on the courthouse steps, for the families of the victims who arrived in the area hours after the crash.

Increased Identification with Community

This characteristic was best displayed by evidence of cooperative community action to solve problems created by the crisis and alteration of community priorities and values. This change in community perspective was expected to produce a heightened emphasis on mutual support and participation with a potential reduction in the normal workloads to permit people to participate in nontraditional emergent volunteer behavior (Marx and McAdam, 1994).

The residents of the small town of Shanksville (estimated population of 245) reported that, two weeks after the event, they felt more close-knit as a community with a greater sense of pride than before (O'Brien, 2001b). The Shanksville Fire Chief expressed his gratitude to the community and indicated that his group had received nothing but praise from the FBI and the Pennsylvania State Police on how the volunteers had handled themselves, not only as a fire company, but as a community.

From its perspective, the FBI might have felt a closer connection to the community at the end of its investigation, although the creation of a community bond would not have been in line with its larger criminal investigative mandate. According to the *Daily American*, the FBI estimated that up to 1,500 personnel, including its own staff and volunteers, had been deployed to the site by the time the job was completed 15 days later (O'Brien, 2001a). Originally a five-week timeline had been set, but the work was done in 13 days. The FBI credited the cooperation among state, federal, and local agencies, as well as assistance from disaster relief volunteers and local businesses, for the accelerated work at the site.

The Somerset County Coroner expressed his appreciation for the participation and contributions of D-MORT, and other groups such as emotional support specialists, security, x-ray technicians, photographers, fingerprinting, DNA expertise from the FBI, as well as assistance from local firefighters and emergency personnel.

To summarize, the Shanksville event fulfilled the criteria necessary to confirm the convergence crisis process as outlined by Quarantelli and Dynes.

Recommendations for Further Research

Further research should be conducted in the immediate post-disaster period in small communities with multi-jurisdictional response and recovery operations. This exploratory research has demonstrated a definite pattern of interaction and comfort level of working together among responders who have prepared for major disasters together and who have responded to more contained emergencies in cooperative efforts in the past. This familiarity of professional capability as well as understanding of what to expect leads to trust that responsibilities will be carried out. The nature of the inter-organizational interaction and decision making appeared to have a pattern that develops during the course of the response and recovery process. Some of the pattern appeared to emerge as the response and recovery progressed, regardless of the length of time involved. Others appeared to be time dependent, emerging only after three or four days of recovery operations.

The initial patterns of interaction and decision making, both formal and informal, indicated the influence of previous preparedness activities such as planning, training, and exercising, and demonstrated different patterns between and among those who participated together in preparedness (and emergency response) activities and those who did not. These interactions need to be better documented and analyzed. More in-depth, on-scene research and analysis along with a larger scale systematic long-term interview plan, would facilitate clear documentation of the inter-organizational, multi-jurisdictional interactions and the associated influences of preparedness activities.

Learning and applying learning in emergency response improves response operations. The responders learned the importance of interagency communication through events such as the Port-a-Johns scenario. Agencies learned to listen to other agencies in their areas of expertise, such as the Salvation Army's explaining why hot meals were important. Future research addressing the mechanism, process, and speed of learning on disaster scenes would offer insight on how to speed up the learning process in order to improve response operations.

Policy Recommendations

The most obvious policy recommendation growing out of this research is to maintain if not increase the support for preparedness activities, especially those involving multi-jurisdictional and multi-agency participation and interaction. Another recommended policy shift would be to train agency and department supervisors and federal agency leaders and managers, who may be in charge of operations at a scene, in the practice of the IEMS and in the role of local emergency managers. Although the theory of the IEMS might have been understood, the familiarity with the operation of the system at the local level and roles and expertise of the various agencies did not appear to be clear to many participants. Exercises involving some of these individuals along with local emergency mangers and small-town responders could prove beneficial.

References

Adams, D.S. 1983. "Selection, Socialization and Types of Members in the American Red Cross Chapter: A Typology of Voluntary Organization Members." *Journal of Voluntary Action Research* 12: 31–45.

Alexander, Jeffrey C. 1985. "Three Models of Culture/Society: The Watergate Crisis in the U.S." *Sociological Theory* 2: 290–314.

Auf der Heide, Erik. 1989. *Disaster Response: Principles of Preparation and Coordination*. St. Louis, MO: C.B. Mosby Co.

Barton, A. 1969. *Communities in Disaster: A Sociological analysis of Collective Stress Situations*. New York: Doubleday & Company.

Breslau, Karen. 2001. "The Final Moments of United Flight 93." MSNBC Newsweek Web Exclusive. September 22. http://www.msnbc.com/news/632626.asp?cp1=1.

Campbell, Timothy R.S. 1999. "Exercises Bring Competency to Both Daily Emergencies and Large Disasters." *Voice* 28(3): 9–11.

Drabek, T.E., H.L. Tamminga, T.S. Kilijanek, and C.R. Adams. 1981. *Managing Multiorganizational Emergency Responses: Emergent Search and Rescue Networks in National Disasters and Remote Area Settings*. Environment and Behavior Monograph No. 33. Boulder, CO: Institute of Behavioral Science, University of Colorado.

Drabek, T.E., 1996. *Disaster Evacuation Behavior: Tourists and Other Transients.* Environment and Behavior Monograph No. 58. Boulder, CO: Institute of Behavioral Science, University of Colorado.

Drabek, T.E. 1986. *Human Responses to Disaster: An Inventory of Sociological Findings.* New York: Springer-Verlag.

Dynes, R.R. 1970. *Organized Behavior in Disaster*. Lexington, MA: Heath Lexington Books.

Dynes, R.R., and E.L. Quarantelli. 1980. "Helping Behavior in Large Scale Disasters." Pp. 339–354 in D.H. Smith, J. Macauley, and Associates, eds., *Participation in Social and Political Activities*. San Francisco, CA: Jossey Bass.

Dynes, R.R. and K.J. Tierney. 1994. *Disasters, Collective Behavior and Social Organization*. Newark, DE: University of Delaware Press.

Fritz, C.E and J.A. Mathewson. 1957. *Convergence Behavior in Disasters*. National Research Council Disaster Study Number 9. Washington. DC: National Academy of Sciences.

Gibb, Tom. 2001a. "FBI ends Site Work, Says No Bomb Used."*Post-Gazette*. (September 25). http://www.post-gazette.com/headlines/20010925scene0925p2.asp.
Gibb, Tom. 2001b. "It Dropped out of the Clouds."*Post-Gazette*. (September 13). http://www.post-gazette.com/headlines/20010913scene0913p2.asp.

Hauger, Thomas. 2001. Western Region Pennsylvania Emergency Management Agency. Personal interview on November 11.

Hightower, Henry and Michel Coutu. 1996. "Coordinating Emergency Management: A Canadian Example." Pp. 69–98 in Richard T. Sylves and William Waugh, Jr., eds., *Disaster Management in the U. S. and Canada.* : Springfield, IL: Charles C. Thomas Books.

Holland, C.J. 1989. "Effective Utilization of Victim Volunteers in the Emergency Response." Pp. 321–324 in *Proceedings of the International Workshop on Earthquake Injury Epidemiology for Mitigation and Response.* Baltimore, MD: John Hopkins University.

Hoover, David H. and Nancy K. Grant. 2002a. "United Flight 93." Presentation to Detection and Suppression Systems I, class and open lecture series at the University of Akron, March 12 and 14.

Hull, R.B., and D.E. Wenger. 1992. *Relationship among Voluntary Evacuation and Place Attachments: Charleston after Hurricane Hugo*. College Station, TX.: Hazard Reduction and Recovery Center, Texas A&M University.

Jacob, E. 1988. "Clarifying Qualitative Research: A Focus on Traditions." *Educational Researcher* 17(1): 16–19, 22–24.

Kruse, Sharon. 1997. "Reflective Activity in Practice: Vignettes of Teachers' Deliberative Work." *Journal of Research and Development in Education* X (10): 23–30.

Lincoln, Y. and E. Guba. 1985. *Naturalistic Inquiry*. Beverly Hills, CA: Sage.

Lincoln, Y. and E. Guba. 1994. *Naturalistic Inquiry*. Beverly Hills, CA: Sage.

Lindell, Michael K. and David J. Whitney. 1995. "Effects of Organizational Environment, Internal Structure, and Team Climate on the Effectiveness of Local Emergency Planning Committees." *Risk Analysis* 15(4): pp. 439–447.

Lohr, Richard. 2001. Emergency Management Director of Somerset County, Pennsylvania. Personal communication on November 11.

Marx, G.T. and D. McAdam. 1994. Collective Behavior and Social Movements. Englewood Cliffs, NJ: Prentice Hall.

Merriam, S. 1988. *Case Study Research in Education: A Qualitative Approach.* San Francisco, CA: Jossey-Bass.

Merton, R. K and P.L. Kendall. 1946. *The Focused Interview: A Manual of Problems and Procedures*. Glenco, IL: Free Press.

Miles, M.B. and A.M. Huberman. 1984. *Qualitative Data Analysis: A Source of New Methods*. Beverly Hills, CA: Sage.

Mileti, D.S. and P. O'Brien. 1991. *Public Response to the Loma Prieta Earthquake Emergency and Aftershock Warnings: Findings and Lessons*. Fort Collins, CO: Hazards Assessment Laboratory, Colorado State University.

Mileti, D.S., T.E. Drabek, and J.E. Haas. 1975. *Human Systems in Extreme Environments: A Sociological Perspective*. Environment and Behavior Monograph No. 2. Boulder, CO: Institute of Behavioral Science, University of Colorado.

Miller, Wallace. 2001. Somerset County Coroner. Personal interview on November 5.

Morrison, Wells. 2002. "How to Handle a Disaster in Your Own Backyard." Presentation at the Pennsylvania State Association of Boroughs 91st Annual Conference and Exhibition. Seven Springs Mountain Resort, Charleroi, Pennsylvania, June 11.

Myers, Robert. 2001. Salvation Army. Personal interview on November 4.

Nathan, H. 1986. *Critical Choices in Interview: Conduct, Use and Research Role.* Berkeley, CA: Institute of Government Studies.

O'Brien, Mike. 2001a. "More than 70 Agencies involved in Flight 93 Crash Investigation." *Daily American* (September 21). http://www.dailyamerican.com/disaster.html.

O'Brien, Mike. 2001b. "Tragedy Creates Unity in Shanksville." *Daily American* (September 27). http://www.dailyamerican.com/disaster.html.

Payne, Christopher. 1999 "Contingency Plan Exercises." *Disaster Prevention and Management* 8(2): 111–117.

Perry, J., R. Hawkins, and D.M. Neal. 1983. "Giving and Receiving Aid." *International Journal of Mass Emergencies and Disasters* 1:171–88.

Quarantelli, E.L. and R.R. Dynes. 1976. "Community Conflict: Its Absence and Its Presence in National Disasters." *Mass Emergencies* 1:139– 52.

Quarantelli, E.L. and R.R. Dynes. 1977. "Response to Social Crisis and Disaster." *Annual Review of Sociology* 3:23–49.

Rock, Vicki. 2001. "Emotional Support given to help Area Cope with Tragedy." *Daily American* (September 21). http://www.dailyamerican.com/disaster.html.

Schaeffer, Terry. 2002. "How to Handle a Disaster in Your Own Backyard." Presentation at the Pennsylvania State Association of Boroughs 91st Annual Conference and Exhibition, Seven Springs Mountain Resort, Charleroi, Pennsylvania, June 11.

Schneider, Saundra K. 1995. *Flirting with Disaster: Public Management in Crisis Situations*. Armonk, NY. M.E. Sharpe.

Silver, Jonathan D. 2001. "Day of Terror: Outside Tiny Shanksville, a Fourth Deadly Stroke." *Pittsburgh Post Gazette*, September 12, as presented in American Memorials, http://www.americanmemorials.com/memorial/tribute.asp?idMemorial=1320&idContributor=7 .

Slade, Reverend Duane. 2001. Director, Camp Allegheny, Somerset County, Pennsylvania. Personal interview on November 5.

Stallings, R.H. 1989. "Volunteerism Inside Complex Organizations: Off Duty Hospital Personnel in a Disaster. *Nonprofit and Voluntary Sector Quarterly* 18:133–45.

Szupinka, Major Lyle. 2002. "How to Handle a Disaster in Your Own Backyard." Presentation to the Pennsylvania State Association of Boroughs 91st Annual Conference and Exhibition, Seven Springs Mountain Resort, Charleroi, Pennsylvania, June 9–12.

Tierney, Kathleen J, Michael K. Lindell and Ronald W. Perry. 2001. *Facing the Unexpected: Disaster Preparedness and Response in the United States*. Washington, D.C.: Joseph Henry Press.

Waugh, William L., Jr. and Ronald John Hy. 1995. *The Hyatt skywalk Disaster and other Lessons in the Regulation of Buildings*. Working Paper No. 91. Boulder, CO: Natural Hazards Research and Applications Information Center, University of Colorado.

Wenger, D.E. 1978. "Community Response to Disaster: Functional and Structural Alterations." Pp. 17–49 in E.L. Quarantelli, ed., *Disasters: Theory and Research.* London: Sage.

Wenger, Dennis E., Charles E. Faupel, and Thomas F. James. 1980. *Disaster Beliefs and Emergency Planning*. Newark, DE: University of Delaware.

Wenger, D.E. and T.F. James. 1994. "The Convergence of Volunteers in a Consensus Crisis: The Case of the 1985 Mexican Earthquake." Pp. 229–243 in R.R. Dynes and K.J. Tierney, eds., *Disaster, Collective Behavior and Social Organization*. Newark, DE: University of Delaware Press.

Wilson, J.M. and D. Mosher, D. 1996. *The Prototype of the Virtual Classroom: Computers in Education*, 7th ed. Guilford, CT: Duskin Publishing Group/Brown and Benchmark Publishers.

Yin, R.K. 1984. *Case Study Research: Design and Methods*, 2nd edition. Thousand Oaks, CA: Sage.

Zurcher, L.A. 1968. "Socio-Psychological Functions of Ephemeral Roles: A Disaster Work Crew." *Human Organizations* 27:281–97.

Acknowledgments

Funding for this research was supported by a quick response grant from the Natural Hazards Research and Applications Information Center, University of Colorado.

Correspondence should be directed to Nancy Grant, Center for Emergency Management and Homeland Security Policy Research, The University of Akron, Akron, OH 44325-7904; e-mail: ngrant@uakron.edu.

Victim Management and Identification after the World Trade Center Collapse

David M. Simpson
Center for Hazards Research and Policy Development
University of Louisville

Steven Stehr
Department of Political Science and Criminal Justice
Washington State University

Introduction

This paper focuses on the social and organizational processes at work in the management of the mass casualties and victim identification after the collapse of the World Trade Center towers in New York City on September 11, 2001, with special emphasis on how the unusual circumstances of this disaster affected victim recovery and identification processes. This aspect of disaster research has not been widely studied because in most instances these processes are managed in a relatively routine manner. Owing to the nature of the destruction in New York City and the disruption of normal social and administrative systems, a new set of mechanisms was developed to undertake post-event victim location and identification.

The collapse of the World Trade Center was unusual in many ways, particularly in the area of victim identification and management. The examination of characteristics of this case, in what we are calling a "mega-casualty incident,"[1] has focused attention on the manner in which this event differed from other disasters.

This research builds on an initial quick response grant supported by National Science Foundation, which enabled the authors to visit New York and Ground Zero 14 days after the event to observe the disaster response and to conduct semi-structured interviews with personnel in several agencies involved with victim management. We were able to conduct semi-structured interviews with representatives of the Greater New York Hospital Association, which established and coordinated the official patient locator system on its website; several area hospitals; and (because we are using a broad definition of "victim") the Deputy Executive Director for New York City's Center for Animal Care and Control. The research team also observed activities at the Family Assistance Center at Pier 94 (the central location to obtain disaster assistance), the City Disaster Command Center and Emergency Operations Center at Pier 92, several New York City Fire Department locations, and other disaster-related sites. We also met with Professor Rae Zimmerman, New York University, who was designated by the National Science Foundation as the research facilitator for the Quick Response Program. Additional data were collected through extensive analysis of published reports in the *New York Times* and other publications.

One of the gripping scenes that found media attention in the first three days of the aftermath was the difficulty many experienced in trying to find information about their family members and loved ones. Media clips showed people going from hospital to hospital, desperately seeking information. It was clear to most who were watching those clips that the system was not very effective in providing information to those who needed it most. These scenes led to the initial quick response grant to examine victim identification issues. That research has since been broadened to also explore other issues, such as victim assistance and the collection and distribution of aid.

The Research Questions

The questions initially posed were direct and straightforward—how did the victim identification process work (or not work) in this event? What were the circumstances that made the process different in this case, and how might the process be changed in the future? To explore these questions the authors used several sources of information: a small set of interviews, on-site observation, newspaper and other current accounts, as well as a comparative search of how these issues have been handled in parallel disaster situations in the literature.

Unusual Aspects of the Disaster in New York City

Four aspects of the collapse of the World Trade Center are important when examining how the victim recovery and identification processes were undertaken in New York City: (1) the nature of the disaster scene, (2) the delayed collapse of the towers, (3) the unique characteristics of the disaster itself, and (4) the relatively large loss of life. Each of these observations highlights the extent to which this event deviated from what we might call a "normal" disaster.

First, and most importantly, the response activities at Ground Zero were shaped by the fact that the scene was simultaneously considered a disaster area, a crime scene, and—it was soon realized—a mass grave. Among other things, this meant that the routinization of recovery activities that typically takes place soon after a disaster, was instead spread out over a much longer period as new processes were established. Sifting debris for evidence, human remains, and personal effects took considerable time. It also resulted in conflict and confusion between and among different official response agencies, non-governmental organizations, and families and friends of victims as they struggled over competing needs and priorities. For example, while the mayor's office considered a rapid cleanup of debris a high priority (the mayor was quoted as saying that he would like the site cleared by the time he left office on December 31, 2001), others were concerned that debris removal was being done at the expense of searching for bodies and remains. One example of this conflict was the well-publicized scuffle that took place between New York City firefighters and police in early November after the mayor's office announced that the number of people searching for human remains would be cut back so that more heavy equipment could be used at the site.

Second, the fact that the World Trade Center remained standing for a short period of time after the plane crashes allowed official responders—primarily firefighters from the New York City Fire Department—to mobilize in advance of the main devastation of the subsequent collapse of the buildings. Tragically, this caused many official response personnel to become victims themselves. Thus, while many disasters are of the "hit and run" variety (i.e., the disaster agent has a well-defined time frame for its impact), this event was actually two disasters: the initial impact of the airliners, followed by the collapse of the towers. There have been no similar examples of large losses of response personnel that are affected by a secondary disaster event. More research will be needed in this area.

Third, a combination of factors, including the cause of the disaster (a surprise terrorist attack), the scope of the physical destruction, the nature of the targets (the World Trade Center towers were widely considered to be symbols of American economic might), the ongoing threat of further attacks, and the fact that events were televised live to a horrified nation, placed considerable pressures on government officials to act swiftly to identify and quantify the missing and the dead. Since it was impossible to know who was at the World Trade Center on that morning, unlike in airline crashes where a flight manifest is available, and it was not known who among those at the scene were injured and taken to hospitals and who were killed, the official numbers of missing and dead released in the first six weeks varied considerably from day to day (sec Table 1). The effort to quantify number of victims was also complicated by the fact that several news organizations, such as the Associated Press, *USA Today*, and the *New York Times*, among others, began calculating and reporting their own "unofficial" numbers.

Finally, until September 11th, a mass casualty disaster in contemporary America usually involved at most 200 or so victims. The disaster in New York not only produced many more victims, but the destructive forces unleashed were among the worst of U.S. disasters experienced in our lifetimes. The forces at work (identified in more detail below) meant that for the first time in the modern U.S. experience, there would likely be hundreds of victims who cannot be positively identified. Prior experiences from airline crashes, or the Oklahoma City bombing, involved numbers in the hundreds, while the numbers in this event were of another magnitude.

Victim Management Activities

Immediately after large-scale disasters, several activities related to the management of victims, including both the injured and the dead, typically take place. These activities can be divided into four broad, sometimes overlapping, phases: search and rescue; recovery of bodies and human remains; identification of victims; and the disposition of bodies (Blanshan, 1977; Blanshan and Quarentelli, 1981). Although the manner in which these four activities unfold depends upon such factors as the scope of the disaster, the number and location of victims, and the availability of adequate resources, equipment, and response personnel, two generalizations about the management of victims seem warranted. First, the time frame for locating, identifying, and handling victims is generally fairly short. Time is obviously an important factor in saving the lives of those who are injured and, in many cultures, retrieving human remains quickly so that they may be returned to the families is a high priority (Hershiser and Quarentelli, 1979).

The second generalization is that victim management activities are, by and large, a role undertaken by official response personnel such as law enforcement, firefighters, emergency medical personnel, and trained search and rescue teams. Neither of these generalizations explains very well what occurred after the September 11th disaster in New York City. Owing to the nature of this particular disaster, the identification of victims and the retrieval of human remains took place over several months. Indeed, some of these activities are still taking place more than one year after the event.

The mechanisms for generating and disseminating information about the identity of victims also were different in this case when compared to most disasters. In the immediate aftermath of the collapse of the World Trade Center towers, official administrative channels with primary responsibility for disaster response were overwhelmed (or in some cases, destroyed in the collapse). In the short term, a loosely coordinated network formed around the efforts of family, friends, co-workers, and non-governmental organizations, as well as official responders, and included the use of relatively new technologies such as patient locator sites on the World Wide Web to help identify the location of missing persons. In addition, family members produced and distributed homemade posters and flyers—most including pictures of the missing and the location where they were last seen—that were attached to the walls outside of hospitals and other publicly accessible locations.

Over time, as hope gave way to resignation that those still missing were most likely fatalities, another set of processes involving DNA testing and other forensic techniques was established with the intent of identifying the remains of as many victims as possible. This task has proven to be daunting. The force of the initial explosion and subsequent collapse, coupled with the searing heat of the fires fed by massive amounts of jet fuel, has made locating missing persons and identifying the human remains extremely difficult.

Counting the Missing

The images of the towers collapsing were horrific, but it was impossible to know how great was the loss of life. The media pressed for numbers, and immediate estimates were in the thousands, some as high as 10,000. These reports were based solely on estimates of the working population of the two towers, the shopping mall below grade, and the collapse of and damage to the other buildings in the complex.

As more information became available, the numbers came down but still fluctuated. Two days after the event the number was 4,947. Two weeks later it had climbed to 6,566. Table 1 shows the frequent fluctuation of "official" missing as reported by the *New York Times*. This table, and the time frame indicated, illustrate the variance in the number of victims, and show how it can change rapidly as officials try to gain more accurate data.

With over 90% of the death certificates issued, the number has stabilized at around 2,825 (Lipton, 2002). Estimates of the number of people in the buildings when the planes struck indicate that approximately 95% of the twin tower occupants were able to evacuate (Kugler, 2001).

The fluctuating "missing" count can be explained by several conditions. First, the media typically press for the numbers of dead and injured so that they can report a "count." Second, the manager of the official list was the Internal Affairs Bureau of the New York Police Department, which took all reports from all sources (embassy lists, company rosters, and all missing persons reports) and compiled them into a single list. Accepting all missing persons reports at face value no doubt inflated the list. The Police Department dedicated resources to cull the list and eliminate redundancies, but it took time to investigate each report. More than 200 officers were assigned to investigate the missing persons claims. With respect to the official list, Deputy Police Commissioner Antenen stated, "Our goal is not to be fast; it is to be accurate" (Lipton, 2001, p. B-1). Third, some initial assumptions were made about the number of potential casualties of visitors to the buildings who did not normally work there. A later analysis of the death certificates shows that 98% of the victims were at work when the collapse took place (Lipton, 2002).

Table 1. Fatalities and casualties after the collapse of the World Trade Center (selected dates).

Date	Total Missing Persons	Confirmed Dead	Missing & Dead	Bodies Identified
Sept 13	4,763	184	4,947	34
Sept 17	4,957	190	5,147	39
Sept 20	6,333	233	6,566	n.a.
Sept 24	6,453	261	6,714	188
Sept 27	5,960	305	6,265	238
Oct 1	5,219	314	5,533	255
Oct 3	5,219	363	5,582	289
Oct 6	4,974	380	5,354	321
Oct 9	4,815	422	5,237	370
Oct 12	4,715	442	5,157	385
Oct 15	4,688	453	5,141	398
Oct 18	4,404	456	4,860	404
Oct 21	4,313	461	4,774	411
Oct 24	4,129	478	4,607	425
Oct 27	3,958	506	4,464	454

Source: *New York Times*
Note: The number of confirmed dead includes 157 passengers aboard the two airliners.

Waiting for the Missing

Four minutes after the first plane hit, the Greater New York Hospital Association (representing 40 area hospitals) had initiated its emergency call-down system to alert area trauma centers to prepare for incoming casualties. This is a normal and practiced procedure used for multiple casualty incidents. Hundreds of living victims (the walking wounded) were treated throughout the area, the so-called "first wave" of casualties. The first wave did not overload the system. The anticipated second wave, one that would normally include those more seriously injured and could possibly overwhelm available resources, never came.

Looking for the Missing

The chaos of evacuating thousands of people, and the high number of those unaccounted for, as well as the real-time coverage of the event, meant that everyone with a connection to a loved one in those buildings wanted information. Phone systems were down, an entire area code serviced by equipment on the towers was gone. Two systems evolved to deal with the trial of locating information about victims, one high tech, and the other low tech.

The high-tech solution went on-line 72 hours after the incident. The Greater New York Hospital Association set up a web page called the "Patient Locator Service." Anyone with access to the internet could enter a name and if that person was being treated at an area hospital, the hospital name would be returned. Though not highly publicized, the site received over 700,000 "hits" in the next three days, and in a week's time had received 1.2 million hits (Figure 1). Several web sites also hosted "virtual" missing posters, in which a scanned photo and basic information about where that person was last seen was available on-line. These sites were hosted by CNN and other news carriers, and some were privately hosted. Relatives or friends could upload their information directly to the sites.

The low-tech approach was something that is more commonly seen with kidnappings or abductions: a missing persons poster was created that had a picture of the person; a list of distinguishing clothes, features, or other

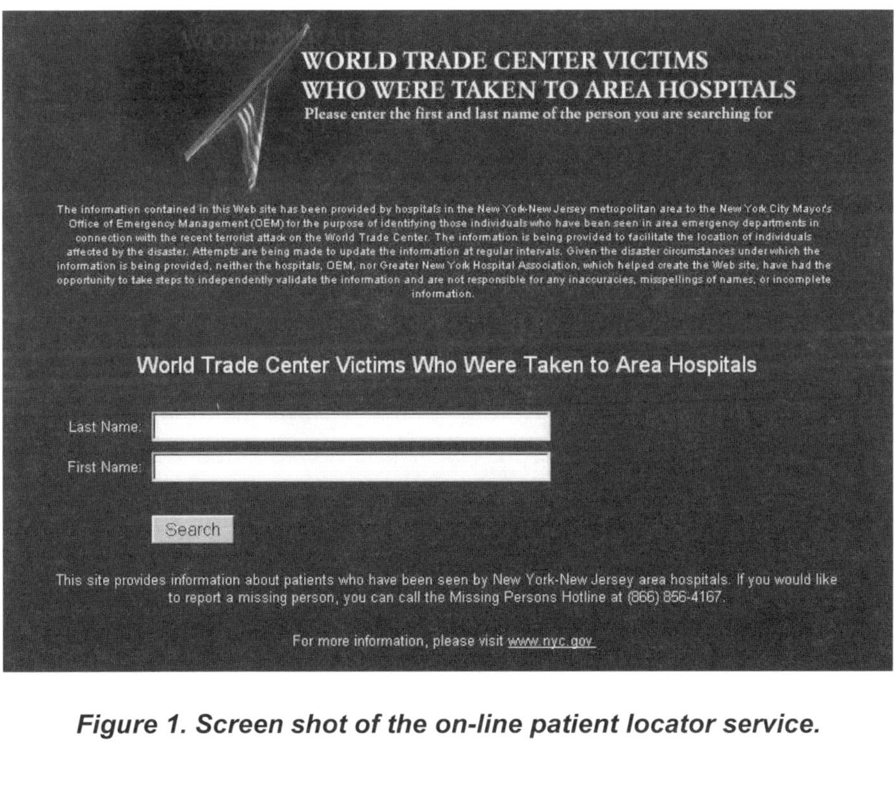

Figure 1. Screen shot of the on-line patient locator service.

identifying information; where he or she worked; and similar data. Vendors would offer assistance with scanning and in some instances making free copies for those wishing to make flyers.

The flyers were posted at hospitals, subway stops, and on construction barriers at the assistance centers. Outside of Bellevue Hospital, a construction partition outside of the main entrance quickly became covered with flyers. As hope faded, the posters took on a second function—that of a memorial remembrance. Soon the partition became a place of mourning and became known as the "Wall of Prayers" (Figures 2 and 3).

Identifying the Missing

After two weeks it was clear that reducing the number of missing would only come about through forensic identification of the human remains found in the 1.8 million tons of debris. This proved to be a daunting task, and one that has yet to be completed over a year after the event. The crush of tons of debris meant that very few bodies were found to be "whole." In fact, only 293

Figure 2. One view of the "Wall of Prayers."
[photo by Stephen Stehr, 2001]

Figure 3. Second view of the "Wall of Prayers."
[photo by Stephen Stehr, 2001]

"nearly" whole bodies were recovered. The only means left of reliably identifying the missing would be through DNA testing of the remains.

The process of recovering the remains took place in several steps. While excavation was taking place at Ground Zero, on-site "spotters" would call a halt to backhoe operations if suspected remains were seen. Just before the debris left the site, it was given a second look to locate remains. The debris was then transported to the specially designated landfill where examiners on either side of a conveyor belt examined the debris for personal effects and any potential remains.

Once an "official sample" was identified at any of these locations, it was transported to the morgue where a forensic anthropologist (Medical Examiner's Office, Department of Forensic Biology) would determine whether the sample was actually human. Because of several restaurants in the complex, there were samples of chicken and hot dogs initially identified as human remains. If determined to be human, the sample then went to the temporary morgue set up by the Disaster Mortuary Response Team.

Once in the Disaster Mortuary Response Team system, the sample would be matched to one of three sources to help identify an existing tissue sample (most reliable in terms of match); a DNA match to the victim's possessions (hairbrush, toothbrush, etc); or a DNA match to a sample from a relative.

Over 20,000 individual remains have been recovered. Identification has been difficult for a number of reasons. For example, some samples are not testable because they are too pulverized. Other samples are too burned to yield good results. Because untrained forensic personnel were involved in the chain of custody, there were complications from inaccurate labeling, missing data, or similar mistakes.

The first round of DNA testing found useable results on only half of the samples. Typical DNA testing uses 400 base pairs, and requires a certain quantity of sample. Because of the poor quality of the samples, and in some cases the small amount of material available for testing, the 400-pair test produced results in approximately 10,000 out of 20,000 samples. The testing procedure has moved to more complex technology. A more advanced testing procedure can get results using a smaller sample size, and obtains results using just 100 base pairs. The procedure is more complex, more expensive, and takes more time. For those samples that can be tested, the 100-pair test is the next step in the process.

The Medical Examiner's Office has identified 1,102 of the missing and anticipates that the number eventually will reach 2,000, but that this will take up to two years after the event. It is anticipated that 800 of the missing may not be identifiable for whatever reason. The Medical Examiner's Office intends to store the samples indefinitely, with the hope that as technology advances the remaining samples can be identified (Emling, 2002).

Conclusion

The events of September 11th and the initial observations of the response efforts have demonstrated that a number of new elements must now be included in disaster preparedness and response plans. First, city emergency planners must be prepared to think and plan for what were previously unthinkable events involving thousands of victims. While the Office of Emergency Management in New York City had planned for the possibility of a small plane hitting a high rise in the city, it had not considered the possibility that two wide-body airliners would be flown into the twin towers.

Second, emergency planners also need to anticipate and create logistical mechanisms for dealing with large numbers of casualties and fatalities. As the September 11th disaster demonstrated, major population centers should be prepared to produce and distribute accurate information to victims' families through web sites and patient locator systems that can be created in advance and activated immediately after catastrophic events. As we have seen, victim identification and patient locator services are vital to the interests of the victim's loved ones and, if utilized effectively, may reduce the burden on public officials. However, while technology can greatly assist in victim identification processes, the public must be aware that these mechanisms cannot completely solve the information-based problems associated with large numbers of victims.

Third, planning and emergency drills and simulations need to address the possibility of larger-scale incidents and how they might be dealt with in terms of information management and communication. Responder education and training that explicitly plans for mass casualty events should become part of all emergency planning.

Future Research

Finally, there are several recommendations that can be made regarding the need for future research. First, there needs to be additional research into how to manage large numbers of victims, from patient information to the process of DNA collection and matching. Second, there are a variety of ways in which the World Wide Web could be utilized in the immediate aftermath of a large-scale disaster, but little research has been performed to determine how best to utilize this as a resource. Lastly, there would be a benefit to additional research that explores how the definition of victim and victim assistance has changed, and how that then affects policy development and policymaking.

Notes

1. We are adapting a term used by the New York Office of the Medical Examiner in which they referred to this incident as a "mega-fatality incident" in a presentation at the 2002 National Disaster Medical System Conference, April 2002. (Office of the Chief Medical Examiner, 2002).

References

Blanshan, S. 1977. "Disaster Body Handling." *Mass Emergencies* 2:249–258.

Blanshan, S. and E.L. Quarantelli. 1981. "From Dead Body to Person: The Handling of Fatal Mass Casualties in Disaster." *Victimology* 6:275–287.

Emling, Shelley. 2002. "Waiting for Closure after Sept. 11." *Austin American Statesman* (August 4): A-7.

Hershiser, M.R. and E.L. Quarantelli. 1979. "The Handling of Dead in a Disaster." Pp. 132–144 in R.A. Kalish, ed., *Death and Dying: Views From Many Cultures*. Farmingdale: Baywood Publishing Co..

Kugler, Sara. 2001 "Despite Horrific Casualties, Evacuation Plans Saved Thousands on Sept. 11." *The Associated Press, Domestic News Section* (December 8): p. 1.

Lipton, Eric. 2001. "Numbers Vary in Tallies of the Victims." *The New York Times* (October 25): B-1.

Lipton, Eric. 2002. "A Nation Challenged: The Toll; in Cold Numbers, a Census of the Sept. 11 Victims." *The New York Times* (April 19): A-14.

Office of the Chief Medical Examiner, New York City. 2002. "Living the Worst Case Scenario: WTC2001, AA FLT 587, Office of the Chief Medical Examiner (OCME) and Disaster Mortuary Teams (DMORT)." Presentation at the National Disaster Medical System Conference, Atlanta, Georgia, April 13–17.

Acknowledgments

The authors wish to thank the Natural Hazards Research Applications and Information Center at the University of Colorado–Boulder; Jacquelyn Monday; and the efforts of three anonymous reviewers. Initial funding for this research was provided in part by a quick response grant from the National Science Foundation (Grant No. CMS-0080977).

Correspondence should be directed to David M. Simpson, Director, Center for Hazards Research and Policy Development, and Assistant Professor, Department of Urban and Public Affairs, University of Louisville, 426 West Bloom St., Louisville, KY 40208; e-mail: dave.simpson@louisville.edu or to Steven Stehr, Associate Professor and Chair, Department of Political Science and Criminal Justice, Washington State University, Pullman, WA 99164-4880; e-mail: stehr@mail.wsu.edu.

Creativity in Emergency Response to the World Trade Center Disaster

James Kendra
Disaster Research Center
University of Delaware
and
Department of Public Administration
University of North Texas

Tricia Wachtendorf
Disaster Research Center
University of Delaware

Introduction

Creativity is an important element of successful disaster response. While advance planning and preparedness serve as the backbone of disaster response efforts, creativity enhances the ability to adapt to the demands imposed upon individuals and organizations during crises and bolsters capacities to improvise in new physical and social environments. The focus of this paper is on creativity not merely as a talent or characteristic, but as a process undertaken by the organizations and emerging collectives that respond to new or changing situations. Disaster researchers and practitioners need to think about creativity much more explicitly, not just as something that "is needed" or "that happens." An enormous body of academic and popular literature focuses on creativity in business settings, where it is expected that managers will need to meet changing competitive or economic environments. Borrowing from the literature on entrepreneurial creativity, we apply Amabile's (1997) framework for categorizing creativity in private sector firms to the activities of responders to the September 11, 2001, disaster at the World Trade Center. We first introduce Amabile's framework and briefly discuss other relevant perspectives on creativity. We choose Amabile's conceptualization of entrepreneurial creativity because the dimensions of creativity that she identifies are analogous to the broad classes of creative activities that we observed, and that, though treated differently, have been relevant to other researchers (e.g., Ross, 1978; Forrest, 1978). In particular, we focus on the development of new emergency response activities or systems; new kinds of products; new or unusual sources of supplies and expertise; and new constituencies for disaster-related services.

At the same time, Amabile's definition of creativity is expansive enough to include solutions that are novel to *those involved*: the solution doesn't have to be one that has never been seen before. Similarly, Ross (1978, p. 216) also considered innovations that were "new to the organizations in question." Creativity within the context of their experience is what is important when looking at how people respond to new events, not whether they have the fortunate inspiration to create something that never existed anywhere.

We relate creativity to such well-recognized features of disaster response as emergence and improvisation. Emergence, as presented in the disaster literature, is fundamentally a creative sequence of actions (see Forrest, 1978; Stallings and Quarantelli, 1985). It is the development of new relationships, new processes, or new ways of getting materials, often while simultaneously locating new constituencies. Emergent groups are creative organizations: they form to be creative. Yet thus far disaster research has not looked at them as such, and this has slowed making connections to other bodies of literature that address creativity more directly. Improvisation certainly requires creativity, understood as the skill, quality, or capacity for generating new ideas. Entrepreneurial creativity, as understood by Amabile and adapted by us, and improvisation as used in the disaster literature (Kreps and Bosworth, 1993; Kreps et al., 1994; Webb, 1998; Mendonca, 2001), are interactive, and may exist as subroutines of each other in different spheres of activity. Multiple improvisations may yield something entirely new, even if that was not the explicit goal at the outset, which may require smaller-scale improvisations in the process of implementing a novel solution.

After a discussion of some of the contradictions of creativity, we describe a number of instances of creative action by both established and emergent organizations. We also identify certain challenges created by the exercise of creativity and suggest preliminary strategies that might be useful in mitigating them. We conclude by suggesting that there are important differences between organizations usually studied in the research on creativity and those that are typically involved in disasters.

Creativity

The literature on creativity is vast, spread among the arts, psychology, business and management, and philosophy. In a summary, Clemen (1996, p. 188) describes creativity as

> . . . new alternatives with elements that achieve fundamental objectives in ways previously unseen. Thus, a creative alternative has both elements of novelty and effectiveness, where effectiveness is thought of in terms of satisfying objectives of a decision maker, a group of individuals, or even the diverse objectives held by different stakeholders in a negotiation.

He observes that "[a]ll definitions include some aspect of novelty. But there is also an element of effectiveness that must be met" (1996, p. 188). In looking at **entrepreneurial creativity**, Amabile (1997, p. 18) defines creativity "as the production of novel and appropriate solutions to open-ended problems in any domain of human activity; we have defined innovation as the implementation of those novel, appropriate ideas." In other words, this conception of creativity involves both success as well as newness: it is both positive and adaptive.

Amabile (1997, p. 20) further specifies four dimensions of entrepreneurial creativity: "(a) the products or services themselves, (b) identifying a market for the products or services, (c) ways of producing or delivering the products or services, or (d) ways of obtaining resources to produce or deliver the products or services." These dimensions, though derived with respect to business enterprises, provide a useful way of conceptualizing the kinds of creativity that responders often exhibit in disaster situations. In addition, they allow us to make connections with other well-developed literatures on creativity that will both enrich our understanding of creativity in disasters and, through further research, help us to use disaster experiences to advance understanding of creativity more generally.

Similarly, Woodman et al. (1993, p. 293) have defined **organizational creativity** as "the creation of a valuable, useful new product, service, idea, procedure, or process by individuals working together in a complex social system," which they further characterize as a common understanding of creativity "placed within an organizational context." Amabile (1997, p. 20) is also careful to distinguish between what is and what is not entrepreneurial creativity:

> It is not limited to the establishment of new businesses, because it can be found when new enterprises are established within existing businesses. Moreover, it is not necessarily present in the creation of any new business; some significant degree of novelty must be involved, at some stage of the process. . . . Entrepreneurial creativity is not present in many of the incremental product or service improvements within established systems or paradigms, unless some significant novelty is required. . . . Moreover, even when a truly novel product or service idea is present, or when there is a novel insight about a market opportunity, entrepreneurial creativity does not exist unless the ideas are implemented in the creation of a new business or enterprise.

An important difference between the disaster and the business environment is, of course, the overall urgency of the creative decisions to be made; nevertheless, the types of creativity are analogous. Some or all of them are apparent in various instances of creative action that occurred in New York City. Comfort (1999, p. 29) argues that creativity is also strongly related to the capacity for "sensemaking" that Weick (1993) describes: the ability to comprehend aspects of the environment and to make decisions. She draws on Luhmann's (1989) conception of "autopoiesis," noting that it is "a powerful, driving force for creative self-expression . . . in individuals that, if extended to social groups and organizations through articulated communications processes, serves as a vital source of creativity, renewal, and regeneration in social systems undergoing change." Comfort (1999, p. 59) observes that "[a]utopoiesis necessarily involves interaction with the environment."

Woodman et al. (1993, p. 294) draw upon Woodman and Schoenfeldt's (1989, 1990) interactionist model of "creativity [as] the complex product of a person's behavior in a given situation." They further argue (1993, p. 296) that "group creativity is a function of individual creative behavior 'inputs.'" Creativity thus is a kind of human-environment interaction, by which we mean part of an evolving ecology of local demands, existing resources, individual skills, and organizational dynamics. What emerges from these different research approaches is a view of collective creative action rooted in gathering environmental information, considering the implications of that information with respect to ambient challenges, and then generating, identifying, and selecting actions that are anticipated to meet those challenges.

Creativity and Emergence

These concepts of creativity correspond with those of emergence. In his analysis of emergent groups, Forrest (1978, p.110), relied upon similar concepts of environmental interaction: "[S]ocial systems interpret cues from the environment so that necessary adaptations or corrections can be made to achieve systematic objectives." Much of the adaptive response that occurred in New York took place through processes of emergence, a phenomenon that is well established in the literature on disasters. Emergence is, at its heart, the development of processes that did not exist before. Entrepreneurial creativity is the foundation of emergence, as an example drawn from that literature demonstrates. Forrest (1972, pp. 30–33) described the evolution of a group of volunteers who assisted in registering evacuees and providing support services to firefighters during a 1970 southern California fire. Civil authorities were burdened by proliferating demands and, unable to cope with the challenges of evacuee processing, "turned to the local community for assistance." An officer of a women's club and her husband volunteered to help. They soon organized a system for tracking donations, offers of assistance, and assigning tasks to incoming volunteers. A friend was recruited to divide supervisory responsibilities and to help manage the storage and disbursement of supplies, the establishment of a medical treatment area, and other relevant tasks. When firefighters needing rest began to arrive, the nature of the operations shifted to meet those needs: furnishing cots and providing personal-care products, food, and laundry service:

> Five basic activities crystallized: general support and assistance, providing food and clean clothing for firefighters; a medical department staffed by two nurses who administered first aid; a stockroom to receive, record and dispense all material resources; a food and coffee relay system, which operated between CDHQ and the firefighters; and lastly, a communications department which received, sent and recorded all incoming and outgoing telephone calls.
>
> (Forrest, 1972, pp. 32–33)

As this example shows, emergence is a significant coping response in times of crisis, augmenting the capacity of established organizations to meet shifting demands. Emergence is also rooted in various kinds of creativity. In this paradigmatic example, none of the participants was really engaged in any particularly new activity. If one considers their tasks solely as segmented entities, making lists, answering phones, unloading a truck, and stacking boxes are, in fact, everyday tasks, familiar to everyone. But they were organized into a **production process** that had not existed before, either in that place or for that group, and that involved new kinds of relationships.

Creativity and Improvisation

A disaster is an event that is defined, in large part, by the improvisational aspects of the response (Tierney, 2002). Since disasters disrupt the patterns of what can be governed or absorbed by routine procedures, an event that does not demand the exercise of improvisation does not, by definition, constitute a disaster. Indeed, Kreps and Bosworth (1993) argue that the pioneers of disaster research intended that the field would place a theoretically based focus on organizational stability and change in the crisis context. This research is well represented by a large body of literature examining emergent groups (Stallings and Quarantelli, 1985); organizations that form new or altered organizational structures and perform non-routine tasks in a disaster (Dynes, 1970); organizational adaptation in disaster (Stallings, 1970); improvisation in organizational domains, human and material resources, tasks, and activities (Kreps et al., 1994); role improvisation (Webb, 1998); and enhancing improvisation through decision support tools (Mendonca et al., 2001).

We distinguish between our uses of creativity and improvisation in two ways. First, within the literature on disasters and emergency response, improvisation refers to unplanned-for activities that take place after a crisis. In contrast, creativity, including in the specialized entrepreneurial sense, is important across the entire spectrum of emergency work, from planning, implementing, and responding, to post-event learning. Second, there is a more theoretically significant distinction to be made between creativity and improvisation. Improvisation refers to the aligning of individual organizational components which, acting separately, nevertheless maintain a mutually shared vision of desirable outcomes. For example, Weick, Mendonca, and others use jazz as a context in which to examine the organizational aspects of improvisation. Weick's introductory essay in the 1998 special issue of *Organization Science*, which focused on jazz improvisation, draws on a number of perspectives on improvisation, in particular those that liken it to conversation (Weick, 1998, p. 548). Weick also argues:

> Considered as a noun, an improvisation is a transformation of some original model. Considered as a verb, improvisation is composing in real time that begins with embellishments of a simple model, but increasingly feeds on these embellishments themselves to move farther from the original melody and closer to a new composition. Whether treated as a noun or a verb, improvisation is guided activity whose guidance comes from elapsed patterns discovered retrospectively.
>
> (Weick, 1998, pp. 546–547)

At the same time, creativity is an aspect of improvisation. The catastrophic collapse of the World Trade Center after the September 11th terrorist attacks and the magnitude of the impact on New York City necessitated a wide range of improvised activities (Wachtendorf and Kendra, 2002) which themselves were, in turn, creative processes in the entrepreneurial sense elaborated by Amabile (1997). Hundreds of thousands of people were evacuated by boat from lower Manhattan; telephone communication was, in large part, temporarily disabled in parts of the city due to the destruction of telephone lines and cellular phone towers; the city's Emergency Operations Center (EOC) at 7 World Trade Center was evacuated and eventually collapsed, necessitating the establishment of interim and then semi-permanent EOC facilities; and the damage to the World Trade Center area necessitated complex site management, security, safety, and clean-up processes (while response and recovery activities overlapped) in ways that had not previously been implemented by any of the organizations involved. Organizations and individuals improvised, some more successfully than others, to meet the demands generated by these and other emerging challenges, often with very new and innovative results.

Planning and creativity work in concert to produce effective improvisation. The new social arrangements that emerge after a disaster and in response to an evolving crisis situation cannot be divorced from previously existing arrangements (Kreps and Bosworth, 1993). Prior preparedness increases the ability to improvise (Kreps, 1991). This planning forms the basis for decision making in emergent environments, and informs decisions by anticipating possible challenges or pitfalls that could come as a consequence of improvised activities. Planning often provides some element of stability—whether of organizational structure, role, task responsibility, resources, or the physical environment—when other elements are in flux or demand unplanned-for action. At the same time, the very need for improvised action points to the inability of plans to take adequately into account one or more specific demands—sometimes quite understandably so, since it is not practical or feasible to fully plan for every possible scenario. Existing social arrangements are always subject to change (Kreps and Bosworth, 1993), particularly when coupled with the ambiguity and confusion that often accompanies large-scale disasters (Webb et al., 1999). For these reasons, creativity emerges as an instrumental contributor to successful improvisation.

The Contradiction of Creativity

Along with researchers, emergency management practitioners appreciate the creative aspects of their work. Creativity is a trait or characteristic often strongly associated with emergency managers and is often cited as a prime job-related skill, as the following passages indicate:

> The Texas Emergency Manager (TEM©) certification is an indicator of experience, hard work, continuing education, dedication to integrity, and creativity.
>
> (Emergency Management Association of Texas, 2002)

> "A disaster is any event that overwhelms your ability to respond," [Judi Van Swieten] says. "You have to be prepared for the worst and work from there, often changing the plan as you progress. Flexibility, adaptability and creativity—those words guide my career."
>
> (Thomson, 2002)

One publication by The International Emergency Technical Rescue Institute notes that

> [t]he future belongs to those who can recognize the needs of an emergency situation and respond with speed, accuracy, creativity, innovation and calm leadership.
>
> (*USARAA News*, 1999, p. 1)

Creativity is an important quality for disaster managers even outside the environment of a disaster: it is important during hazard identification, plan development, and communication and outreach to the public, processes that often have strong entrepreneurial aspects as well. A disaster plan may have to be developed and "sold" to elected officials or corporate officers, for example. Yet even though creativity and flexibility are regarded as important qualities of emergency managers, and people involved closely with emergency response recognize that emergencies demand these qualities, exercising creativity during a response is, paradoxically, often regarded as dysfunctional for emergency personnel. It appears as an indication of failure to plan properly ahead of time. This is because emergency management plans, apart from their function as guides to action, serve rhetorical or political purposes (Clarke, 1999). Clarke argues that they are meant to attest to the competence of emergency planners to foresee events. Moreover, plans fulfill the symbolic function of converting the uncertainty that surrounds hazards or accidents into the kind of certainty that can then be managed. Sometimes the planning process can be stretched beyond credulity; at that point plans become "fantasy documents" (Clarke, 1999) that accept as possible that which is improbable. In other words, planning is such an important activity that plans must be written for situations in which the event will almost certainly differ from what is anticipated, and the anticipated response will be based on preconditions that are likely to be radically altered.

Given the emphasis on plans, even those that are impossible to execute, it is not surprising that departing from them can seem to be evidence of a failure. Disasters, however, break the rules that guide the ordinary conduct of business and government, at least for a period of time. Disasters create new environments that must be explored, assessed, and comprehended. They change the physical and social landscape, and therefore disasters require a period of exploration, learning, and the development of new approaches.

Method

The findings presented in this paper are based on qualitative, inductive analyses of data gathered during exploratory fieldwork commencing within two days after the terrorist attack on the World Trade Center and continuing for two months thereafter. Researchers from the Disaster Research Center conducted over 750 collective hours of systematic field observations. In particular, we closely observed key planning meetings at highly secured facilities, including New York City's EOC, incident command posts, and the federal Disaster Field Office. We also spent extensive periods observing operations at volunteer, supply, and food staging areas, the "Ground Zero" area, family assistance centers that were established for victims' families, and respite centers that were established for rescue workers. Additionally, we observed activities at major security checkpoints in lower Manhattan and at locations central to the emergency response. In the course of our fieldwork, we generated a large volume of notes providing a rich description of observations and experiences, took over 500 photographs, and sketched and collected floor plans of various facilities to track the spatial and organizational changes over time.

In addition to direct observation in New York City, we collected a wide array of documents produced by local, state, and federal agencies as well as by individuals and organizations with less formal ties to response efforts. These documents included but were not limited to internal and public reports, requests for information or resources, informational handouts, internal memos, schedules, meeting minutes and agendas, maps, and internal directives.

To supplement observational and documentary data sources, the Disaster Research Center also compiled an extensive electronic database of articles and web-based information. Newspaper articles from major New York City papers were collected for six months after the attack. Articles from major periodicals, selected articles from newspapers from around the world, and information from the many government, charity, community-based, individual, and private internet sites that emerged after the disaster event were included in this database. The diverse subject matter was later coded according to relevance to the response and early recovery as well as to primary emergency response functions. The functional categories that formed the basis of the coding scheme were informed by the literature on disasters and based in large part on the activities observed during the fieldwork component of the research. The use of multiple methods and data sources—direct observation, informal interviews, reports and other documents produced internally by New York City responding agencies, documents produced by victims of the disaster and informal supporters of the official response, newspaper accounts, and internet-based data—allowed us to triangulate the data, comparing the information collected from one source with other sources as a means to check for accuracy and validity of the data (Denzin, 1998).

The discussions below are based on direct observation, informal discussion with participants and, where noted, secondary sources. They illustrate that response activities involved combinations of the dimensions of creativity identified by Amabile. We analyze (a) new products or services that responding agencies provided or used; (b) situations in which responders identified a particular constituency for products or services; (c) creativity in producing or delivering response-oriented products or services; and (d) the acquisition of resources for the disaster response. We emphasize that, in our use of this entrepreneurial model, we are not suggesting that responders were acting like business entrepreneurs. Rather, we use the model in a more strict analytical sense because of its usefulness in conceptualizing the different manifestations of creativity and in characterizing an operational environment in which new ideas, strategies, and methods came to fruition under extreme conditions.

Creativity in New York City after the Attacks

Mapping and GIS

The emergency response in New York City after the attacks was created on virtually a daily basis as needs were identified, solutions considered, and actions implemented. In other work (Kendra and Wachtendorf, 2003) we describe how the New York City EOC was reconstituted after the destruction of its very advanced facility at 7 World Trade Center. After moving to a succession of intermediate facilities and making use of a mobile communications van, the Office of Emergency Management (OEM) finally moved to Pier 92, a cruise-ship pier that had been scheduled to be used for a bioterrorism exercise on September 12. OEM re-constructed the EOC within this space, bringing in or facilitating the delivery of computers, fax machines, printers, desks, chairs, and even carpet. Emergency managers, in many respects, faced a new operational environment consisting of many more agencies than previously dealt with and in a cityscape that was fundamentally altered, both by the destruction itself and by road closures, detours, and facilities that were put to new, unusual uses. A hotel and a university student center became respite areas for rescue workers, for example. Stated most generally, emergency managers had to explore and reclaim an altered environment. They had to develop a new "map" of a response that had not been previously envisioned and identify the important locations for staging and coordinating response activities, which themselves were changeable as the response evolved.

The term "mapping" can be used in a literal and not merely metaphorical sense for these activities. One noteworthy example of creativity was the development of a geographic information system (GIS) and map-distribution function, amounting virtually to a cartography factory. Ground Zero, itself a new term for a transformed area, was an entirely altered landscape, difficult even for New York residents to orient themselves with respect to the familiar features of the area. Command posts, respite centers, warehouses, and washdown stations were among the needed facilities for which space had to be found, locations mapped, and maps made available to responders. Apart from the reconstitution of the EOC as a whole, development of the mapping capability within the EOC shows creativity (see *ArcNews*, 2002 for an extensive narrative. See Thomas et al., 2002 for a discussion of use of GIS and other technologies) that exemplified all of Amabile's (1997) creative dimensions involving product and process. The original EOC at 7 World Trade Center had GIS equipment, but the scope of the operation at Pier 92 was much larger, providing a variety of map products for different users. Furthermore, a number of personnel from different organizations worked there: students and professors from local colleges; information technology and other specialists from New York City; and representatives from ESRI (the ArcInfo vendor) among others.

This operation evolved in sophistication over time, so that eventually there was even a customer service desk, at which a responder could place a request for a map and pick it up later. This activity and the burgeoning organization that supported it were not pre-planned; rather, they emerged in answer to developing needs. This capability involved not only bringing in people and hardware and software from a variety of sources, but also a process of learning, by spatial analysts and emergency managers, what spatial information was required for the response and what was possible to produce given the available information.

Forrest (1978, pp. 120–122) observed that "viable" groups must secure their legitimacy with respect to other groups, such as by advertising their existence in some way, and that they must further successfully process resources, information, and "demands" (requests for action). Interacting with the environment requires the establishment of a "boundary position" or "position of entry." Our observations of the GIS function are certainly congruent with Forrest's observation: some genuine marketing was involved—a process of reaching out to new users of spatial information—as GIS specialists displayed their products and kept potential map users informed about what might be available. The customer service desk/person was a boundary-spanning position, both organizationally and spatially: the desk was located at the entrance to the GIS area. At the same time, demonstrating the importance of entrepreneurial creativity to the formation of emergent groups, this function was supplemented by a deliberate creative undertaking: the map requests became so numerous that a tracking system was required.

> ESRI staff in Redlands worked with the New York City team to develop an online map request system that entered map requests into a database and provided a prioritized queue to the mapping staff. "We trained folks on how to use it and then wrote up a quick user manual," says [Mike] Tait. "It made it much easier to track the status of a map request with all the specifics, including contact information, right there."
>
> (*ArcNews*, 2002, p. 6)

The activities related to mapping and spatial analysis illustrate all four types of entrepreneurial creativity suggested by Amabile (1997). New products were produced in the form of maps and the online request-tracking system. The overall organization and the customer service system constituted new processes (in turn supported by the online tracking system). New resources were seen in the GIS and remote sensing expertise brought in from public and private sources. Finally, complex and overlapping markets emerged, consisting of the various participants in the response who requested and supplied maps and other spatially referenced information. GIS specialists created new relationships to supplement those that already existed, and they used technologies in ways that had not been envisioned before September 11th (Tierney, 2002).

Waterborne Evacuation

The waterborne evacuation of lower Manhattan immediately after the attack provides an example of creativity along two dimensions identified by Amabile (1997): "ways of producing or delivering the products or services," and "ways of obtaining resources." Here the service is the evacuation itself, and the resources are the people and materials that contributed to it. It is also an example of emergence, in which responders departed from their normal and even their disaster-related roles and in which many responders took part on an unplanned basis. An evacuation of that magnitude was not planned; one Coast Guard officer referred to it as an "ad hoc" event, while another described it as an extension of the agency's existing catastrophic search and rescue plan (which had been designed for the thousands of people who might be, for example, involved in a ferry accident). The Coast Guard is now working to "memorialize" the procedures that emerged on September 11th, which demonstrated the importance of multiple, sequential improvisations to overall entrepreneurial creativity. Available vessels arrived to assist and were assigned by Coast Guard officers working aboard the Sandy Hook Pilots' pilot boat and then aboard a cutter (Sherwood and Schoenlank, 2001). According to Coast Guard officials, approximately 500,000 people left Manhattan by boat, whether by tour boat, military vessel, passenger ferry, or private craft. In another instance of people using existing skills and capabilities to perform new tasks the pilot boat *New York* fueled fire trucks and other vehicles (Sherwood and Schoenlank, 2001), which, for the firefighters, was a new way of acquiring resources. Refueling a truck was a new service provided by the pilots, with the firefighters constituting a new market for it. The waterborne operation was a creative exercise in which people rose to the occasion with all sorts of vessels, and it is also an instance, especially initially, of the kind of self-organization that is important in complex adaptive systems (Comfort, 1999). As one account noted:

> "We moved about 30,000 people on our six boats," says Peter Cavrell, senior vice president of sales and marketing for Circle Line. "It wasn't any kind of coordinated effort. We just started doing it." Continues Cavrell, "In its own

> small way, Circle Line is a symbol of New York. We just wanted to do our part."
>
> (Snyder, 2001)

Not every instance of creativity in New York City involved creating a new product or item; for example, some of the creativity involved "the means for creating or delivering the product—the identification of new market opportunities, or the organization and the systems that are established for bringing the product to market" (Amabile, 1997, p. 18). Although the Coast Guard and other vessels respond from time to time to smaller-scale emergencies, there had never been an evacuation of Manhattan by water; creativity was key to how they developed this response. In addition to developing new "products" or new systems, creativity can take the form of altered procedures, i.e., doing or not doing something that would be done ordinarily. With respect to the seagoing evacuation of Manhattan, Coast Guard inspectors at the point of embarkation were authorized to use their discretion to permit vessels to exceed their certificated passenger capacities. The Coast Guard example is just one of many where process was adjusted with respect to ambient conditions and authority devolved to personnel closer to the scene for greater flexibility. Working closely with Department of Health officials, New York State Department of Environmental Protection officers also relaxed the issuance of citations to truckers hauling debris from Ground Zero without tarpaulins, recognizing that it was impossible, because of the risk of fire, for them to comply with the regulations requiring that their cargoes be covered.

In their discussion of high reliability organizations, Weick et al. (1999, p. 103) noted that "[w]hat is distinctive about HROs is that they loosen the designation of who is the important decision maker in order to allow decision making to migrate along with the problem." It wasn't, however, merely the loosening of regulations that was significant in the waterborne evacuation; it was the capability of the inspectors to apply their experience and judgment. Weick et al. also stress the importance of circumventing hierarchy when greater expertise is located at lower levels. Their focus is on detection of error in complex, high-intensity operational environments: aircraft carriers or air-traffic control, for example. In Manhattan on September 11th, the challenge was avoiding two kinds of error: overloading boats, or not permitting as many people to evacuate as might be able to. In this instance the force of the Coast Guard hierarchy was crystallized in the Certificate of Inspection, which regulates the number of passengers and crew a vessel can carry. The simple act of empowering inspectors to act at their discretion placed expertise in deciding how many people was a safe number where and when it was needed.

Credentialing

Procedures that developed around security and credentialing constitute an additional instance of "creative ways of obtaining resources to produce or deliver products or services." Not only was the September 11th incident a high-impact disaster that produced numerous casualties, it was also a complex emergency with added ambiguous dimensions such as the ongoing terrorist threat, the criminal investigation, an ongoing process of remains recovery and identification that persisted more than six months after the attack, and a very dangerous collapse site situated within close range of an extremely densely populated urban area. Early in the response, it became clear that controlling access to various affected sites would prove a significant challenge. In addition, the standard OEM visitor badges had been lost in the destruction of the original EOC and, even if available, these badges would have been entirely inadequate for the hundreds of people who passed through the reconstituted EOC on a daily basis or who required access to other secured zones and facilities throughout the area.

One of the ways this complex emergency was dealt with was through the development of a credential system. This system, in the form it took after September 11th, was not a previously existing process. While based on other credentialing procedures, it evolved over the course of the response. Initially, government-issued identification sufficed for entry into the EOC. Beginning on September 15, and continuing over a few weeks, OEM developed a series of badges and progressed through several phases. At first, OEM used a relatively simplistic credentialing system where anyone given "clearance" received a blue and yellow badge featuring the OEM insignia. This computer-printed badge was essentially a piece of paper placed in a name-tag holder, could be easily duplicated, and had no identifying information. Eventually, OEM issued plastic badges with a white background and the label "WTC 2001." These badges displayed a digital color image of the individual, the person's title and organizational affiliation, and a variety of codes indicating particular areas to which the person could have access. At the same time the more sophisticated WTC 2001 badges were developed and distributed, temporary badges were developed for contractors and volunteers who needed short-term access to specific areas. These badges also evolved over time. The process involved in obtaining badges was at times very time-consuming for some individuals. Although it was important for the city to restrict the number of people with access, the city also had a real and legitimate need to move along with critical assessment and recovery tasks, including the inspection and repair of surrounding buildings. Some of the contracted workers used in these aspects of the response employed creativity to obtain resources—in this case, the resource was access badges—in order to deliver their response services and meet their responsibilities in an expedited fashion.

Due to safety and security considerations, supervisors of construction workers were only allowed a certain number of contractor badges. At the same time, demands were placed on the supervisors to carry out their responsibilities in an expedited manner. The number of badges allocated to them occasionally fell short of the number of contractors needed to undertake or promptly complete these tasks. The supervisor would then contend with a certain competing tensions that needed to be resolved. On the one hand, the contract workers needed to do a task and on the other hand they did not have the resources—access badges—that would allow them to complete the task. This tension resulted in some supervisors engaging in creative strategies in order to achieve their ultimate response goal.

This scenario recounted by one supervisor of contract workers illustrates their employment of creativity. The supervisor received approximately 20 badges needed for access to complete the inspection or repair of a building. More workers were needed, however, to finish the task at hand. As a solution, 20 workers would go in, one worker would take their badges, and then this worker would give separate groups of 19 workers the same badges for access to the building. Temporary badges for contract workers did not have identifying information, but instead expired after a certain time period. Supervisors retained control over the badges and a contract worker could not enter or exit that building or area without a badge. Still this solution enabled responders in charge of inspection and repair to "make do" with the badges they were allocated by implementing a creative approach for accessing resources needed to achieve their ultimate goal.

The credentialing system represents an instance of creativity of process. Emergency managers classified sensitive areas, such as Ground Zero, and ascertained who required access. They instituted a system for issuing and tracking badges, and they improved the system over time. Others within this system, such as the contractors, viewed these badges as resources and undertook creative means to acquire them in order to do their work.

Challenges of Creativity

Although creativity is accepted by researchers and practitioners as significant in managing emergencies, and although feats of creativity were significant in New York City's response on September 11th, exercises of creativity during the pressure of a response to an emergency may give rise to future complications. We can anticipate that, the greater the magnitude, scope, and/or duration of a disaster, the greater or more frequent the complications might be. Plans promise coherence in a dynamic situation, and the ability to comprehend and respond to a disaster as a total unit. They attempt to bring many possible contingencies within the ambit of predictability. Response strategies that involve creativity, however, approach disasters as more disaggregated entities, comprised of micro-events that require separate management. Detailed plans developed in advance of an emergency are intended to provide coherence and predictability to the response; a plan with which everyone is familiar should be a source for re-establishing an orderly, predictable response in the uncertain and dynamic post-event environment.

The prime difficulty with the exercise of creativity is that, by necessity, it occurs outside of a framework of control. Sometimes individuals exercise creativity; other times groups or organizations do so. Creativity is a function of inspiration and artistry (Kendra and Wachtendorf, 2003). It does not emerge on schedule, and as a consequence creative and innovating steps can occur out of sequence with other actions being undertaken by responding organizations and groups. Creativity can introduce a random and unpredictable element into the response milieu. One person's or group's creative insight can become another's challenge, and creative activities also become a new part of the operational environment about which people must learn and to which they must adjust, precisely at the time when they prefer stability and predictability. The spatial analysis, waterborne evacuation, and credentialing activities described above had successful outcomes but also had challenges associated with them. The examples outlined below highlight other types of creative activity, but here we discuss in more detail the challenges that can accompany creative action in the disaster context.

One example of challenges associated with creativity was tension that developed within the formal disaster response organizations regarding the nature and scope of creative efforts, in particular over what timeframe to consider emergent needs. The time horizon is an important consideration when planning courses of action; some officials have jobs that compel them to look at different spans of time when contemplating actions. Creativity within the response milieu developed as an iterative process among various officials and, as in any work setting, there were clashes over the direction of the creative endeavor.

One of the needs identified early in the response was washing down debris and vehicles, especially trucks and heavy equipment that would be leaving the Ground Zero area. Much of the debris was dangerously hot after having been extracted from the rubble pile (hot enough in some instances to ignite the tarpaulins on the trucks), and in addition, the dust and ash posed a health hazard. Emergency managers needed new washdown equipment and procedures to deal with the hazard, i.e., new products and new processes. Officials from the New York City Department of Health (DOH) and the New York City Department of Design and Construction (DDC) quarreled over whether it was better to have washdown apparatus in place as quickly as possible, or whether some time should be taken to design a more carefully engineered structure that would be heated ("winterized"), in anticipation of the cold weather that would arrive in December. The official from DDC argued that building winterized facilities required a "substantial planning process," and that responders should plan how to develop that plan. The official from DOH was perplexed by what he saw as unreasonable delay in meeting immediately pressing needs as opposed to problems that could develop a couple of months later. Ultimately the sophisticated equipment was built, custom-designed for this application.

The quarrel over the washdown apparatus illustrates an aspect of conflict that is apparent in other settings. The argument between DOH and DDC reveals a "core-overlay" structure, a term used by Burgess and Burgess (1995, p. 107) to characterize the dynamics of environmental controversies. In that pattern, a fundamental moral or philosophical "core" issue is "overlain" by "confused interests . . . disagreements over technical facts, [and] questions of procedural fairness." Although the core issue (development versus conservation) is the real source of the conflict, it is manifested as a sequence of overlying disputes that propel the controversy. Burgess and Burgess (1995) distinguish between conflict, the fundamental divisive issue, and dispute, the overlying opposing contentions. Their goal in a controversy is to identify the core-overlay structure and to treat the overlying disputes. The core, representing well-developed moral views, may be intractable, though some progress may be possible; then the strategy is to mitigate the rancorousness. The core-overlay structure can be seen in the washdown argument: the core issues were the fundamental organizational missions of DDC and DOH and the respective professional imperatives sensed by the participants, made more complex by the different time horizons for planning. The overlying issues were clustered around operational challenges or linguistic details, such as what the word "shall" meant in regulations involving transport of potentially hazardous materials.

Another instance shows the challenges of adopting new methods, tools, or procedures. The introduction of emergency management software, although not precisely an example of entrepreneurial creativity, had analogous effects on the EOC organization since it was new to many of those involved. Before September 11th, OEM had decided to adopt E-Team, a web-based application that allows for tracking of resource requests and deliveries. The decision had only recently been made, but OEM decided to make use of the software in this emergency and to institutionalize new organizational routines, even though most of the agency representatives staffing the EOC had little or no experience with it. OEM brought in E-Team personnel, as well as other emergency management specialists familiar with its use, to install the software at Pier 92. Because few workers in the EOC had any experience with E-Team, it was necessary to run training sessions to acquaint people with its use. This introduction of a new product and a new way of delivering services associated with the distribution of resources was a highly effective tool to reduce duplicate allocation of resources and for accounting for resource requests and fulfillment. Still, the timing of its introduction drew some criticism from those responders who were required to use the program; although OEM was familiar with the software, it was a new product to most of those who worked with it and required a period of learning and adaptation. One logistics officer said that the middle of an emergency was a bad time to bring in new software. Yet this is also an example of the importance of the timescale over which creativity operates; the early introduction of E-Team, a new process for the EOC organization, allowed it to be used during nearly the entire course of the response. When American Airlines Flight 587 crashed in November shortly after leaving John F. Kennedy airport, New York City EOC staff were experienced with E-Team and able to use it to manage the response.

Creativity is not the sole province of official emergency responders. Just as important as the creativity exhibited by emergency managers in the official response structure is that exhibited by the convergers. As noted earlier, the subtext of emergence is creativity: while people may not always be creating something that has never been seen before, the essence of creativity is that the actions undertaken are new to them. As Amabile (1997, p. 18) observed, "[N]ovelty may appear in the means for creating or delivering the product . . ." and in reaching new markets, not just in creating something new. Emergent groups and convergers often display considerable imagination and ingenuity in meeting their objectives. In many instances it was a matter of adapting their existing talents to the new post-event environment (for example, boat operators). We encountered, for example, bicycle couriers who delivered food along the secured perimeter when they weren't permitted to help in other ways (Kendra and Wachtendorf, forthcoming). Some of the volunteers exhibited skills that were quite entrepreneurial, not in a business or financial sense, but there was a kind of volunteer "market" in place; many people were competing for an opportunity to help, not in a direct sense but certainly implicitly. Those with particular skills sought to identify, or to create, markets for them. We observed chiropractors who, by skillfully allying themselves with Red Cross workers, gained access to the staging sites surrounding Ground Zero and eventually the EOC and worked on a stack of pallets with a pad thrown across. The imagination and resourcefulness of such well-meaning volunteers, to say nothing of the creativity shown by exploiters and the disaster opportunists who also converged, were sometimes an irritant to emergency managers. Convergers can often be a source of additional assistance to emergency managers, bringing skills that may not exist when and where they are required, but they can also present challenges, since they are another potentially uncontrollable element in the response milieu whose appearance can create complications for security and site safety.

Conclusions and Implications for Planning

New York City's OEM had conducted many drills and exercises that addressed responses to different kinds of emergency events. Included at these drills were representatives from a broad range of local departments and agencies. When responding to the September 11th disaster, these agencies essentially recreated their ongoing and planned relationships on a daily basis, accounting for changes in the social and physical context but also using sets of skills and capabilities that were developed in earlier training and practice. At the same time, other individuals and organizations (that had not been involved in any of the city's exercises) played important roles in the response. These individuals and organizations, however, were able to draw upon their experiences, informational resources, and existing networks and to augment those established resources with creative ideas. For all of these groups, the requirement in this disaster was to deploy these skills and capabilities in new ways that were adapted to the emerging situation. Although creativity is generally regarded as emerging from flashes of inspiration or insight, it is also founded on broadly applicable abilities. Bruner, for example (1983, p. 183, cited in Weick, 1993), argues that creativity is "figuring out how to use what you already know in order to go beyond what you currently think."

In other work (Kendra and Wachtendorf, 2003) we considered the tension between anticipation and resilience, especially as articulated by Wildavsky (1991), who argued that the likelihood of experiencing events that could not be planned for was such that a strategy of developing resilience to stressors would be better than trying to anticipate and plan for every type of event. Since it is not possible to anticipate everything, such an effort would lead to failures in many cases. In our view, however, anticipation and resilience are not in opposition. Instead, the sought-after quality of resilience can be achieved only by the prior fostering of sets of capabilities that can be applied in a variety of disaster situations. Indeed, we argue that the World Trade Center disaster response shows that creativity is such a significant feature of response to an extreme event that planning and training should move explicitly toward enhancing creativity and the resultant improvisation at all levels of responding organizations. Given that creativity undergirds improvisation, and is an important dimension of resilience (Weick, 1993), such a widely recognized and vital component of emergency response should not be left for emergency managers to acquire by chance, nor should it rely on emergency managers fortuitously bringing these skills to the job or developing them on their own.

With this objective in mind, Mendonca (2001) is building a decision support system with a training mode that features improvisation, and he notes that there are other techniques that can be used within organizations to promote creativity, such as brainstorming. Clemen (1996) summarizes some methods that are used in corporate settings to develop creativity skills; these might be applied in the emergency management field as well. He first distinguishes between "fluent" and "flexible" thinking. "Fluency is the ability to come up with many new ideas quickly. Flexibility . . . stimulates variety among these new ideas" (Clemen, 1996, p. 203). Relevant exercises that Clemen mentions include thinking of new uses for familiar objects, using "idea checklists," and using or generating lists of questions such as Osborn's "Idea-Spurring Questions" (Clemen, 1996, p. 204, citing Osborn, 1963).* Emergency managers should investigate other techniques that might be useful in their particular circumstances. They should also consider aspects of their work environment that can be conducive to creativity. Given the environmental-interaction aspect of creativity, a well-designed EOC that facilitates information exchange (both with personnel in the field and among those staffing the EOC) is essential (see Perry, 1991).

In our discussion, we have considered creativity in a disaster response setting by using concepts developed to describe entrepreneurial creativity in business settings. Future research should examine whether or not the same organizational factors that impede or facilitate creativity in business settings have an impact in the disaster response environment, but it seems that, at a minimum, emergency managers should try to identify and mitigate the features inside and outside their organizations that might suppress or impede creativity, such as deleterious reward structures and other maladaptive motivational influences (see Amabile, 1997 and Woodman et al., 1993 for a discussion of some of these barriers). Emergency managers should also consider the meta-organization that forms to deal with disaster. It is not just the agencies that are usually thought of as the "emergency management" agencies that respond to community-wide disaster. Such a response will include agencies that had never worked together before, and between whom there may be vast differences in organizational culture. The longer a crisis lasts, the more tension there is likely be among officials whose jobs and whose professional imperatives involve different timeframes for action. That tension may work against the development of adaptive, creative solutions. The exchange concerning the washdown stations occurred several weeks after the initial impact when these tensions had increasingly begun to develop in the New York City response environment, and it was extremely contentious. We don't suggest that emergency managers turn themselves into mediators or alternative dispute resolution specialists. Nevertheless, it might be useful to develop an alertness for the core-overlay structure, especially in response

* For example: "*Substitute?* Who else instead? What else instead? Other ingredient? Other material? Other process? Other power? Other place? Other approach? Other tone of voice?"

activities that will be carried out over a period of time and that will require the participation of multiple agencies, including agencies not normally involved in disaster response and for whom the entire environment is new. Such alertness may help an emergency manager forestall potentially divisive quarrels.

We have noted that creative action, though both a sign of and a requirement for intelligent responses to disaster, can create other challenges, inconveniences, or discord within and between organizations. These pitfalls are not really those of creativity; rather, they are the inevitable consequence of fundamental inabilities to foresee every contingency. These inabilities will be considerable when confronting surprise disasters (Mitchell, 1996) such as the World Trade Center disaster. People have to act creatively in such situations, by definition. The challenge is in mitigating any drawbacks. Comfort (1999) holds out hope that communications and information technology can help organizations align their actions more coherently. E-Team, with its capacity for displaying information simultaneously to many users, is such a technology, and our observations suggest it was useful at keeping emergency managers "on the same [web] page."

But as Quarantelli (1982) pointed out, what is communicated is as important as the effectiveness of the medium. The medium is not always the message. In order to have an awareness of what must be communicated, emergency managers must have at least a working knowledge of how creative processes unfold in organizations. We argue that the entrepreneurial creativity model provides a useful way of conceptualizing the creative processes involved in emergency response. We suggest, however, that this research need not stop with the application of existing theory; instead, we see a number of possible research directions that might be taken to advance theories of creativity more generally. For example, the rapid tempo of disaster response may provide an opportunity to examine a creative process from inception to fruition in compressed time, which may throw into relief the most critical aspects.

This paper attempts to connect phenomena observed in disasters with those observed in other settings. In so doing, we hope to begin a more thorough exploration of creativity in disasters than has previously existed, while at the same time setting the stage for transfer of findings to wider literatures. Much of the literature on creativity is concerned with organizational aspects that foster or diminish creativity, such as reward systems and the critical reception of new ideas. This literature works within an entirely different temporal spectrum from that considered in this research, although even in disasters some decision making is carried out over weeks and months. This, of course, was particularly so over the protracted World Trade Center response. We believe that this line of research can lead to advances in the understanding of organizational creativity more generally, such as by developing a better understanding of the temporal scale in the exercise of creativity.

Furthermore, research on creativity as it affects the emerging self-organization of the group would add a significant dimension to understanding of creativity—for example, how it can affect the structure and composition of a group. Looking at the creative aspect of emergence, our concern in this paper has been with individuals and organizations that do not routinely work together on the activities under examination. Most of the research on creativity focuses on organizations in which the participants have some familiarity with each other and with the dynamics of their organizations. The motivational factors that are examined in such research can only be a factor when people have an expectation of a particular unfolding of events and distribution of rewards within their organizations. But emergent groups are often composed of strangers, or people who have not worked with each other before in cooperative/collaborative activities. Interactions within the group are developing simultaneously with the creative process of meeting disaster-related needs. In these situations, creativity is exercised at the same time that group participants negotiate their evolving relationships with the group and with newcomers. An examination of the exercise of creativity with respect to the emerging division of labor, self-identification of appropriate skills and talents (what Forrest (1978, p. 116) refers to as "usable human attributes"), and group-level validation of individual participation can lead to an enhanced understanding of creativity as an agent or catalyst during group formation and change. Similar research might also illuminate aspects of interorganizational coordination, particularly among organizations that ordinarily have little contact with each other.

The World Trade Center disaster plainly showed the significance of creativity in disaster response; many instances of creativity were featured in the news media and may have been valuable in fostering within the public an appreciation for the unplanned aspect of some disaster response activities. We noted earlier a fundamental contradiction: that creativity is important but simultaneously perceived as an indication of failure, if not by emergency managers then by their constituencies, such as the public they serve and the elected and appointed officials to whom they answer. We suggest that it is important to recast creativity, not as a dysfunctional feature, but as a highly necessary and adaptive response. Examining well-established post-event behaviors, such as emergence, as acts of simultaneous, self-reinforcing individual and collective creativity highlights both the importance of improved understanding of creativity and the need to relax the tension, noted earlier, that often surrounds unplanned-for yet highly adaptive creative solutions. Some treatment of creativity in the research literature on disasters might give practicing emergency managers more leverage to feature creativity in their planning and response activities.

References

Amabile, T.M. 1997. "Entrepreneurial Creativity through Motivational Synergy." *Journal of Creative Behavior* 31(1): 18–26.

ArcNews. 2002. "Managing Chaos Relies on Fast, Accurate Information: New York City—Creating a Disaster Management GIS on the Fly." *ArcNews* 23(4): 6–7.

Burgess, G. and H. Burgess. 1995. "Beyond the Limits: Dispute Resolution of Intractable Environmental Conflicts." Pp. 101–119 in J.W. Blackburn and W.M. Bruce, eds., *Mediating Environmental Conflicts: Theory and Practice*. Westport, CT.: Quorum.

Bruner, J., 1983. *In Search of Mind*. New York: Harper.

Clarke, L. 1999. *Mission Improbable*. Chicago, IL: University of Chicago Press.

Clemen, Robert T. 1996. *Making Hard Decisions: An Introduction to Decision Analysis*. 2nd edition. Pacific Grove, CA.: Duxbury Press.

Comfort, Louise K. 1999. *Shared Risk: Complex Systems In Seismic Response*. Pittsburgh, PA: Pergamon.

Denzin, N.K. 1998. *The Research Act: A Theoretical Introduction to Sociological Methods*. 3rd edition. Englewood Cliffs, NJ: Prentice Hall.

Dynes, Russell R. 1970. "Organizational Involvement and Changes in Community Structure in Disaster." *American Behavioral Scientist* 13: 430–439.

Emergency Management Association of Texas. 2002. "Texas Emergency Manager© Certification Program." http://www.emat-tx.org/summary.doc.

Forrest, T.R. 1972. *Structural Differentiation in Emergent Groups*. Ph.D. dissertation. Columbus, OH: The Ohio State University.

Forrest, T.R. 1978. "Group Emergence in Disasters." Pp. 105–125 in E.L. Quarantelli, ed., *Disasters: Theory and Research*. London: Sage.

Kendra, J.M. and T. Wachtendorf. 2003. "Elements of Resilience after the World Trade Center Disaster: Reconstituting New York City's Emergency Operations Center." *Disasters* 27(1): 37–53.

Kendra, J.M. and T. Wachtendorf. Forthcoming. "Reconsidering Convergence and Converger Legitimacy in Response to the World Trade Center Disaster." *Research in Social Problems and Public Policy*.

Kreps, G.A. and S.L. Bosworth. 1993. "Disaster, Organizing, and Role Enactment: A Structural Approach." *American Journal of Sociology* 428–463.

Kreps, G.A. and S.L. Bowsworth with J.A. Mooney, S.T. Russell, and K.A. Myers. 1994. *Organizing, Role Enactment, and Disaster: A Structural Theory*. Newark, DE: University of Delaware Press.

Kreps, G. 1991. "Organizing for Emergency Management." Pp. 30–54 in T.E. Drabek and G.J. Hoetmer, eds., *Emergency Management: Principles and Practice for Local Government.* Washington, D.C.: International City Management Association.

Luhmann, N. 1989. *Ecological Communication*. Chicago, IL: University of Chicago Press.

Mendonca, David. 2001. "Decision Support for Improvisation During Emergency Response Operations." *International Journal of Emergency Management* 1(1): 30–39.

Mitchell, J.K. 1996. "Improving Community Responses to Industrial Disasters." Pp.10–40 in J.K. Mitchell, ed., *The Long Road to Recovery: Community Responses to Industrial Disasters*. New York: United Nations University Press.

Osborn, A.F. 1963. *Applied Imagination.* 3rd edition. New York: Scribner.

Perry, R.W. 1991. "Managing Disaster Response Operations." Pp. 201–223 in T.W. Drabek and G.J. Hoetmer, eds., *Emergency Management: Principles and Practice for Local Government*. Washington, D.C.: International City Management Association.

Quarantelli, E.L. 1982. "Human Resources and Organizational Behaviors in Community Disasters and Their Relationship to Planning." Preliminary Paper #76. Newark, DE: Disaster Research Center, University of Delaware.

Ross, G.A. 1978. "Organizational Innovation in Disaster Settings." Pp. 215–232 in E.L. Quarantelli, ed., *Disasters: Theory and Research*. London: Sage.

Sherwood, B. and R. Schoenlank. 2001. "Sandy Hook Pilots on Station at Ground Zero: Report of Sandy Hook Pilots." *Response to World Trade Center Disaster Prepared for the NY/NJ Pilot Commissions*. http://www.americanpilots.org/Sandy%20Hook-WTC.htm.

Snyder, J. 2001. "Ferries to the Rescue after World Trade Center Terror Attack." http://www.marinelog.com/DOCS/PRINT/mmiocfer1.html

Stallings, Robert A. 1970. "Hospital Adaptations to Disaster: Flow Models of Intensive Technologies." *Human Organizations* 29: 294–302.

Stallings, Robert and E.L. Quarantelli. 1985. "Emergent Citizen Groups and Emergency Management." *Public Administration Review* 45: 93–100.

Thomas, D.S.K., S.L. Cutter, M. Hodgson, M. Gutekunst, and S. Jones. 2002. *Use of Spatial Data and Geographic Technologies in Response to the September 11 Terrorist Attack*. Quick Response Research Report #153. Boulder, CO: Natural Hazards Research and Applications Information Center, University of Colorado. http://www.colorado.edu/hazards/qr/qr153/qr153.html

Thomson, H. 2002. "Grab and Go Guru: When it Comes to Planning for a Disaster, Judi Van Swieten is UBC's Ace." http://www.publicaffairs.ubc.ca/reports/02jan10/02jan10pro.html.

Tierney, K. 2002. "Lessons Learned from Research on Group and Organizational Response to Disasters." Paper presented at "Countering Terrorism: Lessons Learned from Natural and Technological Disasters," National Academy of Sciences, February 28—March 1.

USARAA News. 1999. "The International Emergency Technical Rescue Institute Press Release" *USARAA News* 1(14). http://www.esb.act.gov.au/usaraa/pdf/14.pdf.

Wachtendorf, T. and J. Kendra. 2002. "Rising From the Ashes: Organizational Adaptation Following the September 11th, 2001 World Trade Center Attack." Presentation to the Research Committee on Disasters at the 15th International Sociological Association World Congress of Sociology, Brisbane, Australia.

Webb, Gary R., Michael Beverly, Megan McMichael, James Noon, and Tabitha Patterson. 1999. "Role Improvising Under Conditions of Uncertainty: A Classification of Types." Preliminary Paper #289. Newark, DE: Disaster Research Center, University of Delaware.

Webb, Gary R. 1998. "Role Enactment in Disaster: Reconciling Structuralist and Interactionist Conceptions of Role." Dissertation #32. Newark, DE: Disaster Research Center, University of Delaware.

Weick, K.E. 1993. "The Collapse of Sensemaking in Organizations: The Mann Gulch Disaster. *Administrative Science Quarterly* 628–652.

Weick, K.E. 1998. "Introductory Essay: Improvisation as a Mindset for Organizational Analysis." *Organization Science* 9(5): 543–555.

Weick, K.E., K.M. Sutcliffe, and D. Obstfeld. 1999. "Organizing for High Reliability: Processes of Collective Mindfulness." *Research in Organizational Behavior* 21: 81–123.

Wildavsky, A. 1991. *Searching for Safety*. New Brunswick, NJ: Transaction Books.

Woodman, R.W. and L.F. Schoenfeldt. 1989. "Individual Differences in Creativity: An Interactionist Perspective." Pp. 77–92 in J.A. Glover, R.R. Ronning, and C.R. Reynolds, eds., *Handbook of Creativity*. New York: Plenum Press.

Woodman, R.W. and L.F. Schoenfeldt. 1990. "An Interactionist Model of Creative Behavior." *Journal of Creative Behavior* 24: 279–290.

Woodman, Richard W., John E. Sawyer, and Ricky W Griffin. 1993. "Toward a Theory of Organizational Creativity." *The Academy of Management Review* 18(2): 293–321.

Acknowledgments

An earlier version of this paper was presented as a paper at the 9th Annual Conference of The International Emergency Management Society, Waterloo, Canada, May 14-17, 2002. This research is supported by grants from the National Science Foundation (supplemental to grant EEC-9701471), the Multidisciplinary Center for Earthquake Engineering Research, and the Public Entity Risk Institute (2001-70). We gratefully acknowledge Kathleen Tierney, Russell Dynes, and the anonymous reviewers for their many valuable comments and suggestions on drafts of this paper. The views expressed here are solely those of the authors.

Correspondence should be directed to James Kendra, Disaster Research Center, University of Delaware, Newark, DE 19716; e-mail: jmkendra@UDel.edu or to Tricia Wachtendorf, Disaster Research Center, University of Delaware, Newark, DE 19716: e-mail: twachten@UDel.edu.

Use of Spatial Data and Geographic Technologies in Response to the September 11th Terrorist Attack on the World Trade Center

Deborah S.K. Thomas
Department of Geography and Environmental Sciences
University of Colorado–Denver

Susan L. Cutter
Department of Geography
University of South Carolina

Michael Hodgson
Department of Geography
University of South Carolina

Mike Gutekunst
Department of Geography
University of South Carolina

Steven Jones
Department of Geography
University of South Carolina

Introduction

The emergency management community is keenly aware of the potential for mapping technologies (geographic information systems (GIS), remote sensing, and global positioning systems (GPS)) in support of emergency response operations (Mileti, 1999, Cutter, 2001). Despite this awareness, systematic knowledge about, and experience with, the use of geo-technologies in this capacity is somewhat limited because of the relatively recent development of this application area. Documenting how geographic (mapping) technologies were used in the aftermath of disasters is an important process so that they can become more refined tools for effectively supporting emergency management decision making.

The September 11th terrorist attack on the World Trade Center provided an opportunity to evaluate the use of geographic technologies in response to a catastrophic event. In the days and months after September 11th, maps and imagery of the sites and surrounding areas appeared in newspapers and on television, depicting the extent of the damage and conveying the level of emergency response operations. The emergence of the maps and related stories in the media clearly indicated that response efforts were employing a wide variety of geographic technologies in a decision-support capacity. The glimpse of map products led to a set of questions:

1. What geo-technologies were utilized?
2. Were they effectively able to be used?
3. How were they supporting response efforts?

In an attempt to better understand these questions, a research team conducted field work in New York City shortly after September 11th in order to identify broad issues associated with implementing geographic technologies in support of emergency response. This information can inform communities beyond New York City that may wish to integrate GIS and related geo-technologies into emergency response plans in order to improve emergency preparedness and response.

Geo-technologies in Support of Emergency Management

Geographic technologies contribute to all phases of the emergency management cycle (Figure 1), especially hazard and vulnerability assessments (Hodgson and Cutter, 2001; Cutter et. al, 2000; Federal Emergency Management Agency, 1997; Carrara and Guzzetti, 1996; Hodgson and Palm, 1992). Increasingly, GIS, remote sensing, and GPS are being used for hazard mitigation as well as response (Showalter, 2001). Many people find maps (a common output from a GIS) accessible for understanding information because of their visual nature (Monmonier, 1997). Consequently, a GIS can be the perfect medium for establishing dialog among stakeholders involved with mitigation, or to guide first responders in rescue operations. Geographic technologies are not just about visualization, however. The most powerful aspect of GIS is the capability to analyze and display risk in conjunction with human systems. For example, a planner could identify areas unsuitable for development based on hazard risk. Or, first responders might be able to quickly identify where vulnerable populations live, such as homebound elderly, for more effective evacuation practices.

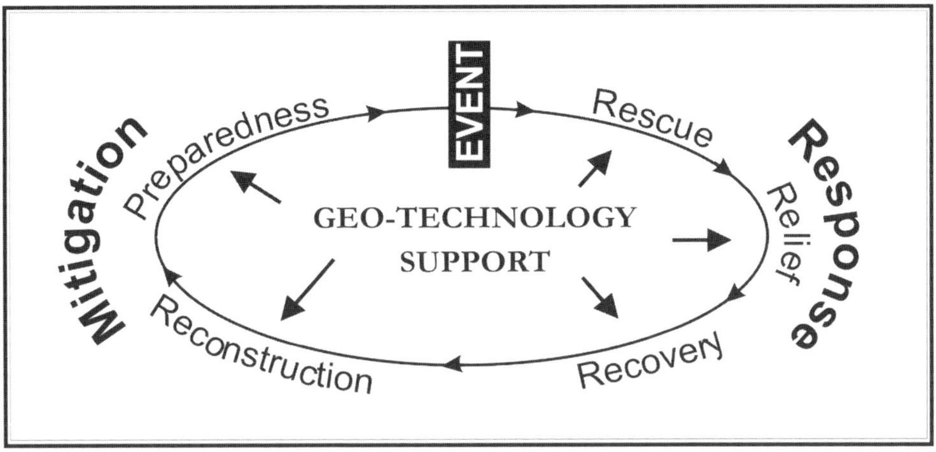

Figure 1. Geo-technologies and the emergency management cycle.

Geo-technologies allow users to create maps and to combine various spatial (geographic) data resources, making them ideal tools to aid in hazard analysis and disaster management (Radke et al., 2000). This has come to mean the development of a wide variety of hazard-based GIS applications, including, but not limited to, hazard detection (Ambrosia et al., 1998; McKean et al., 1991; Kerle and Oppenheimer, 2002), identifying vulnerabilities (Cutter et al., 2001), determining critical needs in the aftermath of disasters (U.S. Geological Survey, 1999), developing evacuation routes (Cova and Church, 1997), damage assessment mapping (Hodgson and Cutter, 2001), and risk perception and communication (Hodgson and Palm, 1992). In essence, GIS supports the decision-making process throughout the emergency management cycle by providing people with a tool for assessing and analyzing the geographic nature of any one or all of its components. In New York City, geo-technologies were implemented for precisely this reason.

Methodology

To facilitate our understanding of the immediate post-event application of geographic technologies, this evaluation concentrated on the first 21 days after September 11th, primarily the rescue and early relief phases, representing "real-time" application. Data were gathered through an interviewing process, as well as by tracking the use of maps in the *New York Times*. While the interviews were the primary source for understanding the broad use of geo-technologies, communication to the general public in the media using maps is also a crucial contribution in support of the emergency management cycle.

Fieldwork conducted on October 8–10, 2001, in New York City provided an impression of mapping activities and was used to identify key people involved in the geo-technological response efforts. Interviews were not conducted at this point since response efforts were still underway and people's attention focused on the events at hand. The site visit did provide an invaluable opportunity to witness the GIS system in place at the Emergency Operations Center (EOC), giving a context of the mapping process for the interviews that followed.

Telephone interviews were conducted throughout November after the site visit, starting with those connections first established in New York City. These people, in turn, identified others to contact. Through this snowball sampling approach, 21 formal and informal interviews were completed. Although this is not the total number of people involved with mapping, these respondents did represent all of the main mapping groups (various levels of government and private sector). Since the focus of the study was on geo-technology implementation in real-time settings, not an end-user assessment, end users were not interviewed, only those involved with providing geo-technology support.

A structured questionnaire regarding the implementation of geographic technologies by agencies and organizations guided the interviews (see Table 1). First, contact and organization information was collected, followed by 11 questions that garnered information about the types of geo-technologies used, how they were used, and who was involved in mapping activities. Recognizing that geographic technologies are data-driven, several of the questions addressed this specifically. Generally, the questions were open-ended; the interview was not conducted in the form of a survey.

Common themes emerged from the interviews; similar issues were often identified by the respondents, although with a slightly different perspectives. The following discussion draws on these, as well as supporting literature on hazards, GIS, and remote sensing. The review includes all facets of geographic technologies, including data, personnel, software integration, hardware infrastructure, and organizational arrangements in the rescue and relief and preparedness stages of the emergency management cycle.

Mapping Efforts in New York City

Maps were created to support local and federal response efforts in the days and months following September 11th. Multiple types of geo-technologies were utilized, including GIS, remote sensing technologies (such as LIDAR (light detection and ranging), thermal radar, and orthophotography), and air monitoring/modeling, as well as GPS-based technologies. In addition, numerous maps depicted various aspects of the aftermath to the public in the *New York Times* and other media outlets.

Table 1. Eleven broad questions used to guide interviews.

Which types of geo-technologies were used? For those that were used, rate the effectiveness on a scale of 1–5 (5 being excellent).
When were these used (what point in the emergency management cycle)?
What types of geographic data were used?
What data were not available?
Did you experience any difficulty obtaining data?
Of the data that were used, were there problems?
Were expectations met by agencies providing data?
What were some of the pitfalls in the technologies used?
What were some of the successes, and which use has the most potential in the future for emergency management?
From your perspective, what was the most important lesson learned?
Do you have any additional comments?

GIS Response

There were three primary mapping endeavors in New York City directly supporting emergency response efforts. The Urban Search and Rescue teams supported by the National Incidence Management Team (including Federal Emergency Management Agency (FEMA) personnel) provided micro-scale mapping focused largely on the World Trade Center site itself and maps with a focus on national issues. The Phoenix Group out of the New York City Fire Department used GIS and remote sensing, again focusing primarily on the 16-acre site, for search and rescue efforts. At the EOC on Pier 92, the Director of Citywide GIS oversaw mapping activities, particularly those supporting local response efforts. The maps produced here were generally at a more macro, city-wide scale, although they also included many site-specific maps. Many people and groups supported mapping at the EOC, including GIS specialists from agencies throughout New York City's local government, vendors (notably Environmental Systems R), volunteer mappers, and local universities (for example, the Department of Geography at Hunter College).

Using broad ways in which geo-technologies are used for hazard applications as a guide, Table 2 provides some examples of the types of maps created during the response phase in New York City. The general categories on the left are not necessarily discrete categories, nor are the New York City examples on the right a comprehensive listing of all geo-technology uses.

Normally, the technology application categories are relevant for most types of hazards, including natural and human-induced. However, the nature of the terrorist attacks limited the use of these technologies for mapping the event itself and for conveying evacuation routes. In the face of a different hazard, a hurricane or flood for example, these uses would certainly be more extensive.

Table 2. Uses of geo-technology in New York City after September 11th.

General Geo-technology Hazard Applications	NYC Map Examples
Event mapping *Prediction and warning* *Monitoring event*	Showing routes of airplanes
Response coordination/resource allocation	Displaying deployment of rescue workers Showing search and rescue grid of the World Trade Center site
Damage assessment	Mapping damaged buildings to establish extent of impacted area; to convey which buildings were re-habitable
Environmental monitoring	Monitoring of air quality, asbestos, and particulate matter
Risk assessment	Assessing debris piles and temperature hot spots on World Trade Center site Examining location of underground storage tanks
Risk communication *Public* *Emergency workers*	Illustrating extent of smoke plumes Showing debris and fire hazards on World Trade Center site
Relief & resource locations	Depicting where people could go for support services
Identification of vulnerable populations	Showing evacuated areas
Coordination and monitoring of cleanup	Planning clean-up efforts and portraying progress
Lifeline status	Illustrating utility service provision and status of electric, water, and telephone
Evacuation efforts/status of transportation routes	Portraying the changed subway network enabling people to plan alternate routes to work

The EOC processed over 1,500 map requests in the few months after September 11th. In essence, geo-technologies were used to re-map the changed geography of Manhattan. This included the creation of lower Manhattan base maps with affected buildings, as well as search and rescue grids, utility outages, and the altered nature of the transportation system. These maps were used not only to document the impacts of the hazard and identify affected people and places, but also to aid in resource allocation for rescue worker deployment and getting affected people to the proper services.

Remote Sensing Efforts

The New York Office for Technology (OFT) coordinated the remote sensing activities and the production of derived products. Numerous remote sensing data collection efforts were planned and successfully implemented by public/commercial groups. Importantly, several remote sensing data collection activities took place much earlier than September 11th. These data found a new use because of the event.

Remotely sensed data were used at the World Trade Center site for several hazard-related purposes. The public most commonly saw the graphic images of the building destruction. Images were collected from low-altitude aircraft (both fixed-wing and helicopters) and through commercial satellites. Imagery from commercial satellites has relatively low spatial resolution (i.e., about 1 meter x 1 meter) and therefore offered somewhat limited use. Several companies and agencies collected vertical aerial mapping photography over the disaster site (Figure 2). These images became the most current "map" of Ground Zero as previously mapped features had been obliterated. Because of rescue operations, and later cleanup efforts, the "map" changed daily.

Historic remotely sensed imagery can be used to document the myriad set of landscape features around a hazard event. For example, Earthdata, an imaging company, had collected aerial photography of the World Trade Center area in July of 2000. Analysis of the archival and new imagery became useful for damage assessment as they documented construction materials as they were transported away from the site.

Airborne LIDAR data over the World Trade Center were used to map the surface elevations each day and to analyze the day-by-day changes of the debris pile. Ground control points for the LIDAR-derived surface models were collected by National Oceanic and Atmospheric Administration staff. In part, these data were used to estimate the volume of debris. It was also anticipated that the spatial changes in volume would reveal unexpected shifts in the pile and thus identify risks to the response personnel on the ground.

Thermal imagery was also collected on the same overflights of the World Trade Center site as the LIDAR data. Essentially these data became a map of the absolute temperature of the surface each day (at the moment of imagery

Figure 2. World Trade Center site after September 11th. (photography collected by the National Oceanic and Atmospheric Administration)

collection). Expected uses of thermal data included documenting the location/spread of continuous and new fires within the debris pile. Again, identifying risks to response personnel was the intended use, but was not ultimately particularly useful because of processing issues.

Maps for Communicating to the Public

In terms of communicating to the public, a content analysis of maps in the *New York Times* is revealing because the newspaper was a major source of information about the September 11th disaster. Maps were common additions to the stories, and in several instances, the focus of the story itself described how geographic technologies were aiding response efforts. The newspaper was one of the primary ways of reaching the public with information about the event. In a practical sense, the maps gave local residents the means for finding relief resources, returning home, or getting back to a daily routine.

During the first month after September 11th, approximately 84 maps or mapping stories about the World Trade Center attack appeared in the *New York Times* (Figure 3). Coverage peaked in the first week after the event and diminished as the newspaper's focus shifted to Afghanistan beginning on September 19. The initial map coverage reflected the broad uses for emergency response such as damage assessments, alternative transportation routes, service provision status, risks in the environment, and relief resources. During the first week all 35 maps appeared in Section A of the *New York Times*. Thirty-eight were printed in Section B in the following three weeks, with five in Section C. The quality and number of these maps only highlights how mapping may be used for communicating to the public in a future event.

Response GIS Realities

The post-event response experiences in New York City support the premise that geo-technologies can and do support response efforts. While they contain a wealth of valuable information, real-time or near-real-time geo-technology efforts are resource intensive and require significant advance planning to perform most effectively. Pre-impact planning serves multiple purposes, putting infrastructure in place for response as well as identifying and implementing mitigation measures to minimize hazard impacts.

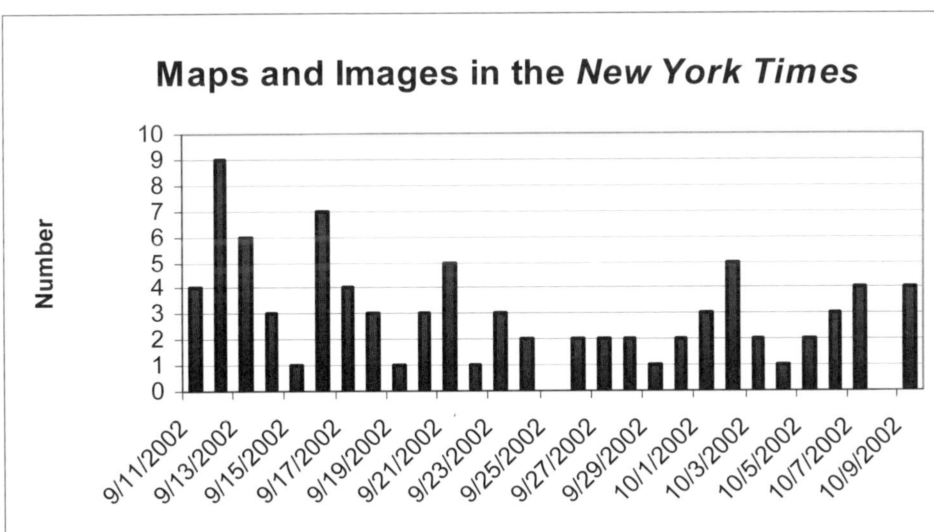

Figure 3. Trends in newspaper mapping coverage.

Organizational Plans for Geo-technologies

Geographic technologies do not just take root overnight within an organization. Creating an effective mapping system requires substantial planning, effort, time, and most importantly, money. It often has modest beginnings within an organization, evolving and becoming increasingly integrated into the decision-making process over time (Chrisman, 1997). The challenge at the moment is that preparedness for terrorism or bioterrorism is at the forefront of the American consciousness and there is an impending need to have better mapping capabilities supporting the entire emergency management cycle, but particularly response. Facing this challenge requires immediate attention to implementing geo-technologies and integrating them into the emergency management decision-making process.

Planning the flow of information through the organizational structure and explicitly defining how geo-technologies fit into this plan is vital to their successful use in a post-event situation. In the face of any disaster, having a coordinated GIS in place beforehand is clearly the ideal situation.

Although a single city-wide GIS did not exist before September 11th, many New York City agencies within the local government already had established formal and informal relationships and coordinated efforts were underway. These efforts and arrangements became the basis for building a response GIS in the post-event period, which was coordinated by the Director of Citywide GIS. In addition, federal, state, and local mapping efforts also required coordination. Several of those interviewed noted that jurisdictions ideally would not want to depend on informal relationships, having to develop information flow processes, or creating data on the fly. This translates to the need for technology plans (including organization and personnel) within the response plan, as well as designing a mechanism for integrating data sets. In this way, the duplication of effort and resources among agencies will be minimized and the most effective tools made available within the necessary time frame.

Post-September 11th meetings to identify lessons learned were held between some participants associated with remotely sensed data collections. Most parties agreed that having one agency—the New York Office for Technology (OFT)—coordinating the remote sensing collection streamlined the mission planning process for both federal and private partners. The OFT became the focal point (or "go-between") for user data requests and for coordinating collections. Unlike the contractual problems observed in historic natural hazard events, the relationship between the State of New York and federal and private partners was quickly established, enabling almost immediate collection of photography, LIDAR, and thermal imagery of the New York site. Finally, it should be noted that the remote sensing collections required numerous participants for each overflight, ground control, and subsequent analysis. Collections of remotely sensed imagery after a natural hazard or technological hazard event are often problematic because of the physical environment as well as the political, contractual, and legal hurdles.

In conjunction with the vast array of geographic technologies deployed in New York City and the unprecedented local coordination of efforts, there were still some challenges in managing geo-technologies. While people's first priority was, no doubt, to aid in the search and rescue, relief, and recovery efforts, there was evidence of some competition among vendors and contractors to demonstrate their capabilities. This may have resulted in uncoordinated efforts or the duplication of GIS and/or remotely sensed data collection. More importantly, different endeavors may have complicated risk communication by conveying dissimilar messages. Outside the vendor arena, even some of the efforts among levels of government may not have been as efficient as possible. Having detailed plans in place for geo-technologies and for the flow of spatial information limits the potential for this.

Alternative Plans

Among the most significant lifeline disruptions in New York City was the destruction of the city's EOC, the nerve center for coordinating response and recovery. As the EOC was recreated, the GIS system also had to be reconstructed, including the spatial information. Mapping efforts cannot depend on internet availability, accessible mapping experts, or even one location for data, software, and hardware. Experiences in New York City clearly point to the need for alternative and flexible plans for geo-technology capabilities. As with all response efforts during an event of this magnitude, mobilization and coordination were challenging in the initial few days, especially because the original EOC had been destroyed. In this case, mapping efforts initially depended on local efforts in a make-shift environment. Support staff was not immediately available since air travel was prohibited and additional experts from anywhere outside of New York City were unable to gain access swiftly. Resources of all types were stretched, and this was no less true of mapping. Computers, people, software, and data were hastily mobilized and the use of geographic technologies expanded in the days after September 11th.

Data Accessibility and Quality

Having a spatial data infrastructure in place before any event is vital to ensure successful mapping during rescue, relief, and recovery (National Research Council, 1999). A uniform spatial data infrastructure is an absolute necessity in emergency response, especially when the applications occur in real time or near real time. With data in place that integrate into a single platform and have the appropriate spatial and temporal resolution, the foundation is set for utilizing geographic technologies to their fullest during response efforts.

New York City had many elements of a city-wide spatial data infrastructure in place before September 11th. Most importantly, base layers, such as parcel information and street centerlines, had already been created and were commonly used. The structure, however, was distributed throughout many agencies, with different agencies housing data in sometimes disparate systems. So while many New York City agencies were utilizing GIS extensively, it was not an entirely integrated GIS system containing all of the necessary data elements (both spatial and attribute). For example, some of the sub-terrain features, such as subway lines or underground storage tanks, were not housed in a GIS, but instead were in CAD (computed aided design) system, some other database system, or even existed only as paper maps. Although incomplete at the time of the World Trade Center attack, the city was in the process of creating uniform (and unique) building identification codes that integrated those used by various agencies. Consequently, while not seamlessly integrated, once re-established, the EOC did have the spatial data to produce many of the requested maps. In addition, data were added from the field and some data sets updated daily.

The quality of available GIS data varies extensively across departments, jurisdictions, and communities in the United States. In addition, many desirable data sets simply do not exist. For example, if we are interested in vulnerable populations, there is little information on homeless people, undocumented workers, or even the day-versus-night populations of urban centers and/or buildings. These data gaps must be filled for effective response using GIS. Another important point is that data collection is not a finite process. Instead, it is ongoing either because of new data needs or the maintenance and updating of data already collected.

Another challenge is the creation of integrated data sets across multiple jurisdictions. Although the databases were maintained by various agencies in the case of New York City, at least they were not faced with trying to integrate spatial data from multiple jurisdictions. In most other metropolitan areas faced with this different scenario, spatial data requirements would require data sharing between communities. Unfortunately, few places across the United States maintain regional GIS databases.

Data sharing agreements must be in place before any event to ease the transition to real-time, response-based GIS. During emergency response, privately held data, such as utilities, as well as classified data will likely be needed. A mechanism for obtaining this type of data should already be negotiated. This may translate to having data stored in a secure environment or obtaining it from a secure site. One point is clear after the destruction of the EOC in building 7 of the World Trade Center—these data should be stored in multiple locations in addition to at the main center. The reality of forming integrated, accessible data sets can be quite difficult for political or economic reasons, but these events only highlight the potential value.

Given the unique threat that terrorism presents, many agencies took data off-line in the aftermath of September 11th. While not directly part of the New York City local response, the broader implications for data sharing are immense. Policies of publicly available data are being re-evaluated out of concern that this same information can fall into the wrong hands, prompting public debate on data access and its corollary, privacy issues. While there are philosophical issues surrounding this debate, in a practical sense it has very real implications for the GIS and emergency management communities. Making data on hazardous threats more difficult to obtain could hinder preparedness or mitigation, and could create even greater vulnerabilities.

Need for Technical Expertise

Even though New York City experiences heightened the awareness of how mapping products can support emergency response, implementing the suite of geo-technologies is no small task. The challenge of integrating geo-spatial data, platforms, and software into the response efforts is technically daunting for most emergency managers, even at a pre-impact stage. Real-time or near-real-time applications in the aftermath of a disaster are even more complicated. Considering all the potential technologies available, such as the variety actually used in New York City, advanced knowledge is required up front (pre-positioned) to know where the assets are, to know what types of activities they can support during response efforts, and to understand how to process the information with the software. The mechanisms for obtaining and processing data from various sensors and sources should be in place before any event and links to technical support established, including universities; private firms; and federal, state, and local government agencies.

Geo-technologies as Decision-support Tools

The goal of implementing geo-technologies is to improve emergency workers' abilities to do their jobs by giving them useful tools. If these technologies fail on this account, then they have not effectively supported response efforts. There is a distinct need to ensure that products, including models and maps, meet the needs of end users and that appropriate tools make it into the hands of the right users. Many of the maps and visualizations created in New York City were used in support of risk communication. This is a key element of the emergency management cycle, creating useful tools for improved decision making in the face of a disaster. The informational needs are not the same for all groups, however, nor are GIS and map-reading skills. Emergency responders' information requirements differ from those of managers, or even the public.

In the case of New York City, many maps were requested, which suggests response teams and managers were using these products. This research was unable to establish details on how maps were actually used by end users, since they were not interviewed. Several respondents stated that maps were widely used, while others expressed a lack of use, even indicating that the maps went directly in the trash. A number also suggested that vendors were using this as an opportunity to showcase their software capabilities, rather than truly supporting the needs of response efforts. For example, remotely sensed imagery was used to monitor the debris pile, but apparently was not a major source of information for the emergency response or cleanup personnel at the site itself. An analysis of map requests (unavailable for this research) with follow-up interviews in New York City would provide some insight into user needs in that situation. In creating GIS implementation plans for support of emergency response, user need assessments are essential so that appropriate and usable maps will be created.

Conclusion

This study provided an overview of how geo-technologies were used in the aftermath of the September 11th disaster in New York City, as well as supplying some practical considerations for other communities when incorporating GIS into emergency management plans. We are still far from understanding the full potential for these technologies during response efforts, however. There is clearly a need for a true post-audit of the experiences in New York City to assess the full range of successes and shortcomings associated with mapping technologies. This would include an assessment of how people actually used the maps, as well as identifying the ways in which they were incorporated into the decision-making process.

The level of recognition of what and how geo-technologies can contribute to response and decision-making efforts most certainly increased in the wake of September 11th. The original EOC in New York City had one computer terminal devoted to mapping. The makeshift EOC on Pier 92 had an entire section with over 20 computers, a server, and a whole staff dedicated to creating maps. Because of this demonstration in this national disaster of how maps can aid in the rescue, relief, and recovery efforts, the role of mapping in any future New York City EOC and potentially other locations, will certainly be reassessed. The number of maps included in the *New York Times* over this period also points to an increased role of mapping. In fact, the public and emergency responders and managers may now expect high-quality informative maps in any future events after having had a glimpse of mapping products. The full range of ways that mapping can support emergency management is only beginning to be realized, and the extensive use of mapping in New York City further emphasizes the need to explore the effective integration of these geo-technologies into the emergency management cycle for all hazards.

REFERENCES

Ambrosia, V.G., S.W. Buechel, J.A. Brass, J.R. Peterson, R.H. Davies, R.J. Kane, and S. Spain. 1998. "An Integration of Remote Sensing, GIS, and Information Distribution for Wildfire Detection and Management." *Photogrammetric Engineering and Remote Sensing* 64(10): 977–985.

Carrara, A., and F. Guzzetti, eds.. 1996. *Geographical Information Systems in Assessing Natural Hazards*. Dordrecht: Kluwer Publishers.

Chrisman, N. 1997. *Exploring Geographic Information Systems*. New York: Wiley.

Cova, T. J. and R. L. Church. 1997. "Modeling Community Evacuation Vulnerability Using GIS." *International Journal of Geographical Information Science* 11:763–784.

Cutter, S.L., ed.. 2001. *American Hazardscapes: The Regionalization of Hazards and Disasters*. Washington, D.C.: Joseph Henry Press.

Cutter, S.L., J.T. Mitchell, and M.S. Scott. 2000. "Revealing the Vulnerability of People and Places: A Case Study of Georgetown County, South Carolina." *Annals of the Association of American Geographers* 90(4): 713–737.

Federal Emergency Management Agency. 1997. *Multi Hazard Identification and Risk Assessment: A Cornerstone of the National Mitigation Strategy*. Washington D.C.: U.S. Government Printing Office.

Hodgson, M.E. and R.I. Palm. 1992. "Attitude and Response to Earthquake Hazards: A GIS Design for Analyzing Risk Assessment." *GeoInfo Systems* 2(7):40–51.

Hodgson, M.E. and S.L. Cutter. 2001. "Mapping and the Spatial Analysis of Hazardscapes," Pp. 37–60 in Cutter, S.L., ed., *American Hazardscapes: The Regionalization of Environmental Risks and Hazards*. Washington D.C.: Joseph Henry Press..

Kerle, N. and C. Oppenheimer. 2002. "Satellite Remote Sensing as a Tool in Lahar Disaster Management. *Disasters* 26(2): 140–160.

McKean, J., S. Buechel, and L. Gaydos, 1991. "Remote Sensing and Landslide Hazard Assessment." *Photogrammetric Engineering and Remote Sensing* 57 (9): 1185–1193.

Mileti, D.S. 1999. *Disasters by Design: A Reassessment of Natural Hazards in the United States*. Washington D.C.: Joseph Henry Press.

Monmonier, M. 1997. *Cartographies of Danger: Mapping Hazards in America.* Chicago: University of Chicago Press.

National Research Council, Board on Natural Disasters, Commission on Geosciences, Environment, and Resources. 1999. *Reducing Natural Disasters through Better Information*. Washington D.C.: National Academy Press.

Radke, J., T. Cova, M. F. Sheridan, A. Troy, L. Mu, and R. Johnson. 2000. "Application Challenges for Geographic Information Science: Implications for Research, Education, and Policy for Emergency Preparedness and Response." *URISA Journal* 12(2): 15–30.

Showalter, P. S. 2001. "Remote Sensing's Use in Disaster Research: a Review." *Disaster Prevention and Management* 10(1): 21–29.

U.S. Geological Survey. 1999. *Digital Atlas of Central America.* Washington D.C.: Center for Integration of Natural Disaster Information.

Acknowledgments

This research was supported by a quick response grant from the Natural Hazards Research and Applications Information Center, University of Colorado. The authors are grateful for the comments of three anonymous reviewers, and also thank the respondents to the interviews for their time and insight.

Correspondence should be directed to Deborah Thomas, Department of Geography and Environmental Sciences, University of Colorado at Denver, P.O. Box 173364, CB 172, Denver, CO 80217; e-mail: deborah.thomas@cudenver.edu.

Part 3

The Engineered Environment—Buildings and Infrastructure

Managing Disruptions to Critical Interdependent Infrastructures in the Context of the 2001 World Trade Center Attack

William A. Wallace
Department of Decision Sciences and Engineering Systems
Rensselaer Polytechnic Institute

David Mendonça
Department of Information Systems
New Jersey Institute of Technology

Earl E. Lee
Department of Decision Sciences and Engineering Systems
Rensselaer Polytechnic Institute

John E. Mitchell
Department of Mathematical Sciences
Rensselaer Polytechnic Institute

Joe H. Chow
Department of Electrical, Computer and Systems Engineering
Rensselaer Polytechnic Institute

Overview

Critical infrastructure systems provide services that are essential to both the economy and well-being of nations and their citizens. As documented in a recent report to the U.S. Congress (U.S. General Accounting Office, 2001), it is of vital importance that these services not be degraded, whether by willful acts such as terrorism or by natural or random events such as earthquakes, design flaws, or human error. Yet infrastructure systems and the organizations that manage them are now recognized as components of highly coupled systems that increasingly rely on one another in order to deliver key services. In addition, as complex, interconnected systems, they are vulnerable to disruptive events that propagate from system to system.

The September 11, 2001, attack on the World Trade Center in New York City illustrates the importance of understanding relationships among infrastructure systems and of managing these relationships in order to ensure continuance of necessary services after disruptive events. This research is intended to improve understanding of and support for the management of critical infrastructure interdependencies after large-scale, disruptive disasters. As discussed more fully below, infrastructure interdependencies occur when, due to either geographical proximity or shared operations, an impact on one infrastructure system is also an impact on one or more other infrastructure systems. The particular focus of this work is on developing techniques that can be used either to mitigate against or respond to events that have the capability of impacting interdependent infrastructure systems. The approach taken is to model salient elements of interdependent critical infrastructure systems and to provide decision makers with means of manipulating this model for purposes of mitigation or response. Models can provide powerful means of understanding (Wallace, 1994), monitoring, and controlling large-scale infrastructure systems (Beroggi and Wallace, 1998). The need for powerful but parsimonious models is particularly acute as modeled infrastructures increase in complexity, as when a number of infrastructures are interdependent.

The first objective of the present research—improving understanding—involves identifying, classifying, and describing incidents of critical infrastructure interdependence related to the World Trade Center attack. Three steps involved in accomplishing this objective are discussed. First, a coding methodology is presented for identifying instances of infrastructure disruptions, particularly related to interdependence that became evident after the World Trade Center attack. Second, results of the application of the methodology to newspaper articles for the period September 12, 2001, to December 12, 2001, are presented. Third, a number of potentially rich cases identified in step two and involving electric power and telecommunications infrastructures were developed and used in formulating and assessing the mathematical models. One such case serves as the basis for an exercise in the application of a mathematical model developed to contribute to the second objective, described next.

The second objective—improving support for the management of infrastructure interdependencies—involves development of analytic techniques embedded in computer-based tools. Such computer-based decision support is intended to assist decision makers in reducing expected loss of service due to disruption and in restoring service more quickly if loss of service actually occurs. Definitions of infrastructure interdependence and related concepts are refined in order to allow development of a mathematical representation of infrastructure systems and their interdependencies. Next, such a mathematical representation is presented. The representation permits the development and use of algorithms for searching and locating solutions to problems associated with disruptions to interdependent critical infrastructures. A model of interdependent infrastructure systems operating under normal operating conditions is next developed. The model is intended for use in determining whether services provided by impacted infrastructures can be continued without undertaking extensive restoration operations. A second model is then presented to support decision making when the restoration of services is required. An example based upon restoration activities following the World Trade Center attack illustrates the application of this second model.

This paper proceeds as follows. The section on Strategies for Emergency Management places the problem of management of impacted critical infrastructures in the context of emergency response. The next section, Incident Identification and Classification, summarizes newspaper reports on post-World Trade Center impacts to critical infrastructure. Formal descriptions of infrastructure systems and their interdependency relationships are given in the section after that (Modeling Infrastructure Systems), followed by a discussion of how a network flow approach may be taken in modeling interdependent infrastructures. Models for supporting decision makers during response and restoration activities are presented in the section after that, and with their mathematical representations presented in the two appendices. A demonstration of the models' use is presented in the section entitled Model Demonstration. The paper concludes with a discussion of ongoing research and suggestions for further work.

Strategies for Emergency Management

The focus of the current research is on assisting emergency managers in responding to degradations in service that arise following events that impact infrastructure interdependencies. In these situations, infrastructure and emergency managers are faced with identifying, assessing, and mitigating the effects of the impact in order to restore necessary services. The implementation of strategies to achieve these goals requires responding personnel to marshal and apply available resources in a timely manner. Table 1 provides a construct that has been found useful in emergency management (Wallace, 1990) for discussing and analyzing a range of emergency management strategies. Mitigation and preparedness strategies are designed to reduce the impact from threats *before* disaster occurs, either by

Table 1. Emergency management strategies.

Strategy	Examples	When is Impact Reduced?
Mitigation	Building codes, insurance programs	Before occurrence, reducing the consequences, partially or in total
Preparedness	Warning systems, inventories of food and medical supplies	Before occurrence
Response	Rescue teams, fire fighting	After occurrence, as impact is being felt
Recovery	Disaster relief funds, rebuilding assistance	After occurrence, when full impact has been felt

reducing the impact of the disruption caused by the threat or by providing advance warning in order to lessen an event's impact. Response and recovery strategies are, on the other hand, designed to reduce the impact from threats *after* disaster occurs. As an example, response teams attempt to reduce impact by containing the effects of disruptions; disbursement of disaster relief funds is intended to lessen the burden on affected individuals and organizations.

Elimination or reduction of threats, whether they are human or technological, random or willful, is a key consideration in managing critical infrastructures. Assuming that for some threats it is possible to reduce, but not eliminate their probability of occurrence, additional management strategies are required to deal with such consequences as loss or degradation in service. The present research addresses pre-event strategies of mitigation and preparedness by increasing understanding of how organizations respond to disruptive events and by designing model-based tools for supporting organizational response to these events. Post-event response activities can be supported by identifying feasible alternatives for providing service and by assisting in developing new approaches to service restoration.

Incident Identification and Classification

Identification of instances of infrastructure dependence and interdependence is intended to support model-building (as described in subsequent sections) and expansion of knowledge of how organizations respond to infrastructure disruptions. Instances of disruption to critical infrastructures in the borough of Manhattan are here summarized by drawing upon reports published in the *New York Times* Metro edition for the period September 12, 2001, to December 12, 2001. This time period closely approximates the length of the response phase to the World Trade Center attack.

Independent coders were provided with hard copy of all the above issues of the *New York Times*, along with definitions of critical infrastructures and with instructions on how to identify interdependency relationships among them. Eight infrastructures, as defined in the President's Commission on Critical Infrastructure Protection (1997), are included in the analysis: emergency services; transportation; information and communications; electric power; banking and finance; gas and oil production, storage, and transportation; water supply systems; and government. Additional definitions given to the coders were as follows:

An infrastructure is coded as **dependent** on one or more other infrastructures if any one of the following three conditions holds:

- **input**: the infrastructure requires as input one or more services from another infrastructure in order to provide some other service;
- **shared**: some physical components and/or activities of the infrastructure used in providing the service are shared with one or more other infrastructures;
- **exclusive-or**: either the infrastructure or some other infrastructure (but not both) can be in use during provision of the service.

Two or more infrastructures are coded as interdependent[1] if the following condition is true:

- **interdependent**: two or more infrastructures' physical components or activities are co-located within a prescribed geographical region.

Note that a disrupted infrastructure might not be involved in a dependent or interdependent relationship with another infrastructure. Thus, the final category:

- **none**: indicates that the infrastructure was not involved in a dependent or interdependent relationship with another infrastructure.

Results of the coding are summarized in Tables 2 and 3. Table 2 shows the number of disruptions for each of the eight infrastructure systems, regardless of whether or not a particular system was dependent or interdependent on one or more other systems. A total of 244 disruptions was

Table 2. Reported disruptions to critical infrastructure systems.

Infrastructure	Count
Emergency services	26
Transportation	44
Information and communications	29
Government services	43
Electric power	15
Oil and gas production and storage	2
Banking and finance	66
Water supply	19
Total	*244*

Table 3. Interdependence and dependence relationships.

Relationship	Count	Median
Input	18	2
Shared	1	1
Exclusive-or	2	2
Interdependent (co-located)	30	2
None	155	1

reported during the 90-day period. As shown in Table 3, 51 instances of interdependence or dependence were reported. The median number of infrastructure systems involved in a particular type of relationship is also given.

Based on these data, a number of potentially rich cases were identified. A number of organizations involved in these cases were contacted and asked to participate in a study to investigate interorganizational aspects of the management of these disruptions. Representatives of two public service providers agreed to be interviewed for this research to assist in choosing and developing a number of cases.

Modeling Infrastructure Systems

Definitions

Previous work provides definitions of the concept of critical infrastructure interdependence (President's Commission on Critical Infrastructure Protection, 1997; Little, 2002; Rinaldi et al., 2001), as well as discussions of key terms. Further refinement and generalization of these definitions are undertaken here before proceeding with the development of a mathematical representation.

An **infrastructure** is defined as a linked set of physical components with associated activities. **Physical components** are the built part of an infrastructure; **activities** are tasks necessary to operate physical components of the infrastructure. An **intersection** is the area where two or more physical components meet or are joined. An intersection circumscribes the activities and physical components necessary to manage the connection between the joined physical components. As an example, the intersection of two roadways may have one or more physical components (e.g., a traffic signal) and activities (e.g., manipulation of the signal via sensors embedded in the roadway). All intersections in a given infrastructure must have a physical component.

A **service** is something made available by the infrastructure for use or consumption. A service may be used by people or by other infrastructures; it is provided in order to meet a real or perceived need. An infrastructure can provide one or more services. **Material** is any physical entity or "substance or substances out of which a thing is or can be made" (Pickett, 2000, p. 837). Examples include electrons, people, product, and electromagnetic signals. Provision of a service requires activities such as movement, collection, transformation, or storage of material. Activities may be initiated at one or many locations and may be terminated at one or many locations. Assuming that traversal of a connection between two intersections requires a set of activities from beginning to end, management activities are necessary when provision of the service requires traversal of more than one intersection.

A **disruption** in an infrastructure is said to occur when one or more of the physical components or one or more of the activities needed to operate a physical component cannot function at prescribed levels. Disruption may or may not result in service degradation. **Service degradation** is said to occur when the service itself cannot be provided at its prescribed level.

An infrastructure is said to be **dependent** on one or more other infrastructures if any one of the following three conditions holds:

- **Input**: the infrastructure requires as input one or more services from another infrastructure in order to provide some other service.

- **Shared**: some physical components and/or activities of the infrastructure used in providing the service are shared.
- **Exclusive-or**: only one of two or more services can be provided by an infrastructure. (Note that a disturbance in an infrastructure that is dependent on another by virtue of its inability to operate if the other infrastructure is operating will affect just its own provision of service.)

A collection of infrastructures (denoted *I*) is said to be **mutually dependent** if the following condition holds:

- **Mutually dependent**: at least one of the operations of any infrastructure in *I* is dependent upon each of the other infrastructures in *I*. An example of mutual dependence involving two infrastructures occurs when an output of infrastructure A is an input to infrastructure B, and an output of infrastructure B is an input to infrastructure A.

Two or more infrastructures are said to be **co-located** if the following condition holds:

- **Co-located**: any of their physical components or activities are situated within a prescribed geographical region.

Collectively, these five conditions—input, shared, exclusive-or, mutual dependence, and co-location—will be denoted types of **interdependence**, since all imply that an impact on one infrastructure system is also an impact on one or more other infrastructure systems.

Examples of Infrastructure Interdependence

The following examples are intended to illustrate the above concepts of infrastructure interdependence.

Input Dependence

At 10:20 a.m. on September 11, 2001, after the collapse of the first World Trade Center tower, transit authorities decided to suspend all subway service, issuing an order to send all trains to their yards or to secure them in the tunnels. Around 10:15 several subway lines were left without alternating current, which supplies power to the trains through the third rails, and without direct current, which runs the signals. Officials did not know exactly how the power was disrupted. But because of the power loss, the closing of all stations in lower Manhattan and the possibility of further explosions or collapses, "the general consensus was that the best thing to do was discharge all passengers and secure the trains temporarily," a transit authority spokesperson said (Kennedy, 2001, p. A8).

Shared Dependence

New York Waterway put all 24 of its boats into service, some to work as floating ambulances from piers in lower Manhattan and others to go to Hoboken, Hunts Point in Queens, and the Brooklyn Army Terminal (Kennedy, 2001).

Exclusive-or Dependence

After the collapse of the World Trade Center towers, financial services could not be provided because employees could not use the streets and sidewalks to travel to work. While there was pressure on financial firms to open, the firms worried that a partial opening might further damage investor confidence. They therefore pressed for a return to full operations. Some areas in New York's Financial District are narrow and congested even under normal circumstances; debris and vehicle and human traffic made them more so after the collapse. Moving people through lower Manhattan presented a challenge to the transit system (Berenson, 2001).

Mutual Dependence

This type of interdependency was not included in the review of *New York Times* reports. This type of interdependency would be said to occur when a power plant uses coal that is shipped by trains that require power from the plant in order to operate.

Co-location

There were numerous examples of both power lines and fiber optic cables being located in the same manhole, thus creating the possibility that the organizations responsible for these infrastructures would have to coordinate efforts at the manholes. A second example occurred at bridge and tunnel entrances which also served as the locations for security checks (*New York Times* Editorial Staff, 2001).

Modeling Infrastructure Systems as Networks with Flows

Infrastructure systems involve material flow, signals, water, commodities, people, and the like. A highway system, a power grid, a telephone network, or an airline network all provide the physical structure and associated activities to support the movement of material through the system (Berge, 1962).

A common feature in models of infrastructure systems is their essential dependence on a geometrical figure called a graph, a figure in a two-dimensional plane consisting of lines and points. In modeling the physical components of infrastructure systems, both lines and points represent components of the system. Points are also called nodes or vertices, while lines are referred to as arcs, links, branches, or edges. A network is a graph where a direction is specified for every line, meaning that a line begins at one point and ends at another. In addition, lines in networks typically represent the movement of some material, whereas lines in graphs represent connections with the possibility of a direction. The physical structure of an infrastructure system can be represented by a graph; however, when the intention is to model an infrastructure providing a service, the graph must represent not only the activities necessary to generate, transfer, and terminate a service but also the service itself. Such a graph representation becomes a network with flows (Frank and Frisch, 1971), with points as nodes and lines as arcs, all with specified values representing characteristics of the infrastructure system being modeled. Physical components of an infrastructure are then modeled as a network, with nodes representing such components as communities, highway intersections, railroad yards, power generators, phone switching systems, and water reservoirs. In general, these are points where flow originates, is relayed, or terminates. The arcs of a graph can represent such elements as roads, railroad tracks, transmission lines, airline routes, and water pipes. In general, these are the channels through which material moves.

While the graph representations of different infrastructure systems may appear similar, the characterization of the nodes and arcs may be quite different. Telephone networks are characterized by parameters such as cost per unit length, capacity of a wire, and number of wires; a power grid will be characterized by parameters such as resistance, or capacitance. In addition, parameters that represent limitations and capabilities of nodes—sources of material, transshipment points, and destinations, and channels for material flow—can be incorporated into the graph model as numbers on the nodes and arcs. For power systems, a node representing power generation might have values for maximum power output, reliability of a generator, and cost per kilowatt-hour. An arc might have values representing capacity, reliability, and cost.

A network representation is very useful for modeling systems that involve connectivity. Indeed, in the models discussed below, the services provided by infrastructures are modeled as network flows. Given a system and its network, the question of determining if a particular material can be moved from one location to another can be understood as determining if there exists a path between two nodes. Since infrastructure interdependencies are connections, and the objective is to see if, after a disruption, material can be moved from a source to a particular destination (i.e., can be used to provide a service), a network representation is appropriate. In addition, to determine if a given level of service can be provided to a particular destination from a particular source, material flow over an arc can be represented by values on the nodes and arcs. The objective here is to find the maximum flow between the specified locations and determine if it satisfies the desired level of service.

Interdependent infrastructures are here viewed as networks, with material corresponding to flows and with services corresponding to a desired level of these flows. For ease of representation, each network (i.e., each infrastructure) is defined as a collection of nodes and arcs with commodities (i.e., material) flowing from node to node along paths in the network. Activities, physical components, and intersections are considered to be contained within a node. Similarly, management activities are not considered in traversal of an arc; they are contained within the arc itself. For each commodity, each node will either be a supply node, a demand node, or a transshipment node. Arcs may, of course, have limited capacities (Ahuja et al., 1993). Infrastructure systems operate in an environment subject to disruptions—natural, human-caused, or willful acts. Given a system having interdependent infrastructures, the analysis must determine likely system degradation following an event. Based upon performance criteria, an infrastructure system can then be designed to minimize possible service degradation. In addition, once a disruption occurs, alternative ways of restoring service can be determined.

The following section employs a network with flows representation in the formulation of models to aid emergency managers in response and, if necessary, restoration of service.

Decision Support for Response and Restoration

The focus of the present research is on assisting organizations responsible for responding to events that disrupt services provided by infrastructures they manage. Managers in these organizations are responsible for developing response and restoration strategies and proposing them for review by stakeholders or regulators both within and external to their organization. Once a strategy has been determined, it is implemented by field personnel. This section describes models intended to be embedded in computer-based systems to support strategy development, both before and after the occurrence of disruptive events. In the context of management of interdependent infrastructures, these strategies are likely to be directed towards restoration of key services. Figure 1 presents stages of decision making for emergency response and restoration actions using the models.

Normal Operations

Under normal conditions, all the demands of all the infrastructures are met. Since interdependency considerations come into consideration only when there are unmet demands, each infrastructure may be considered to be operating independently and analyzed independently. It is assumed that, before the disruption, the system operates at a minimum cost-optimal

Normal Operations

Network flow model of infrastructure systems is used to describe conditions of normal operation.

Disruption Occurs

Impact Assessment Stage: Impacts of disruption on physical components of infrastructure systems are identified (includes identifying disabled nodes and arcs, reduced supplies, and increase or reduction in demand).

Response Stage: Run normal operations model to assess feasibility of infrastructure systems to meet demands. If not feasible, identify resources available for restoration, including estimates of time and resources needed for alternative possibilities for restoration. Prioritize unmet demands.

Restoration Stage: Enter post-disruption network configuration into restoration model, including disabled supply, transshipment and demand nodes, disabled arcs between nodes, reduced supply, reduced or increased demand, and reduced arc capacity. Identify interdependencies and enter into model. Formulate alternative restoration possibilities as a network flow model and include in restoration model. Run model and present optimal solutions to management.

Figure 1. Process of decision support for response and restoration.

solution, denoted **normal operations**. This solution can be found by solving the normal **operations model** for each infrastructure system separately. This network flow model is described in detail and represented mathematically in Appendix A.

The normal operations model consists of an objective to be optimized and constraint requirements representing flow conservation and structural requirements. The objective is to minimize the cost of operation of each of the infrastructure networks, while satisfying demand. Therefore, constraining the solution is the requirement that the flow out of the supply nodes must be less than the available supply and the flow into the demand nodes must meet the required demand. Since transshipment nodes have neither supply nor demand, the flow into a transshipment node must equal the flow out. Also, we cannot exceed the capacity of transshipment nodes. Similarly, the flow on any arc cannot exceed its respective capacity. Last, the flow on the arcs must be nonnegative, with a value of zero denoting that there is no flow on that arc. Structural requirements model the network's configuration and identify whether or not arcs and nodes are in operation.

The normal operations model is intended for use by infrastructure managers during routine operations. However, it is envisioned that the model is also a component of a decision support system to be used by a state or local emergency operations center for emergency response and restoration.

Impact Assessment Stage

When an incident occurs that has the potential to cause a major disruption in service, initial activities include assessing (1) its likely impact on physical components of infrastructure systems, (2) the potential loss of service, (3) its impact on the safety of humans, and (4) its effect on the security of sensitive systems in the natural and built environments. Reductions either in capabilities of supply and transshipment nodes or capacities of the arcs between the nodes need to be identified. Assessment of new demands must also be made, since post-event conditions can result not only in decreases but also in increases to demand.

Response Stage

The impact assessment may reveal that the desired or normal demand levels cannot be met. However, once the desired demand levels are ascertained and prioritized, it may be possible to satisfy revised demands using functioning supply points that, before the event, had been operating at less than full capacity. Absent such a situation, the normal operations model must be modified by changing supply, transshipment, and demand nodes and flow conservation constraints. In addition, connections between nodes must be revised—either eliminating arcs or reducing capacity over an arc—to reflect damage caused by the incident, i.e., changing the structural constraints. The model is then run, taking into account the revised and prioritized demands, to determine the set of feasible solutions. If at least one feasible solution is found for each infrastructure system, response to the incident can proceed. If not, alternatives for restoring service must be developed. The infeasible solution provided by the modified normal operations model for the infrastructures that could not provide a service, i.e., meet the revised demands, may be used to identify unmet demands.

Restoration Stage

Once the need for developing ways of restoring service has been determined, physical and personnel resources available for implementing restoration strategies must be identified. Time may be one of the factors considered in selecting a restoration strategy and may indeed be critical. Many infrastructure organizations have database systems with visualization capabilities that enable rapid determination of resource availability and location. A **restoration model** for decision support in selecting restoration strategies is described in this section; Appendix B provides more detail, including the mathematical representation.

As previously noted, if all the revised demands for all the infrastructure services can be met without restoration, each infrastructure system can be considered to be operating independently. However, when unmet demand for any infrastructure service is found, interdependencies among infrastructure systems must be considered and incorporated in any model that seeks to support the restoration decision-making process. Also, this process must involve those emergency managers who have the authority to make decisions concerning prioritization of unmet demand among systems.

The material in the remainder of this section will describe how the various interdependencies defined above are represented in the restoration model. Each interdependency will be described with a textual description of how it is modeled. Then the overall model will be discussed.

Input

Input dependency is modeled as follows. Consider a telephone switching station. The switching station itself is a transshipment node within the telecommunications network. However, this same switching station from the perspective of the electrical network is seen as a demand node since it needs an adequate source of electricity to operate. From the perspective of the electrical network, the switching station is therefore a dependent component. More formally, denote the demand node for the switching station in the electrical network to be node j. If there is an adequate flow of electric power into node j, the switching station can function. If power is not available at this level, then the switching station fails. A binary variable, y, is used in this case to represent the two states of the switching station. If adequate power is available at j, then $y = 1$; if not, then $y = 0$. The phone switching station also has some maximum capacity u within the telecommunications network. Consider the station's capacity to be the product of the binary variable y and the rated capacity w. When adequate power is available the station can operate to its capacity w (since $y = 1$). On the other hand, if adequate power is not available then the capacity of the station is 0. This binary variable y serves as a virtual connector between the two systems. Its value is set by the conditions existing in one system, and affects the operating characteristics of a second system. Events affecting the power network that have an effect on node *j* in turn impact the model of the telecommunications network. The effect on each system can be analyzed in a similar manner.

Shared

Methods similar to those used in multi-commodity flow problems (Rardin, 1998) may be used to model systems with the shared dependency. The use of one or more shared components by all systems is constrained by a limit on maximum flow. In the context of the example of ferry service given earlier, regardless of whether or not the ferry is used for transit services or medical services, it cannot exceed its maximum capacity. In the context of the telecommunications and power systems and the World Trade Center restoration, it could have been advantageous to route the shunts used to restore phone and power through the same temporary enclosures. This would reduce time and cost since only one enclosure would need to be built.

Exclusive-or

When multiple infrastructures share a component, for example, an arc, but the component can only be used by one infrastructure at a time, exclusive-or dependence occurs. In the example given previously, streets (i.e., shared components) could not be used by both the emergency response personnel and financial district workers. Considering power and telecommunications, it can also be the case that a power and a telecommunications shunt cannot be routed in close proximity to each other. This would be the case with a T1 line, which cannot be too close to a power line due to radio frequency interference considerations. Emergency managers would establish the priority for which lines could be run, i.e., which services are more critical, and others desiring a similar path would have to be re-routed. Exclusive-or dependencies are modeled by determining which service has the highest priority, making an appropriate change to the objective function, and selecting additional constraints to restrict flow to one commodity or the other.

Mutual Dependence

A collection of infrastructures is said to be mutually dependent if at least one of the operations of any infrastructure system is dependent upon any other infrastructure system and at least one of the operations of this other infrastructure system is dependent upon the first infrastructure system. As previously noted, no cases of mutual dependence in infrastructure systems were identified in news reports from the *New York Times*. However, consider two mutually dependent systems, a natural gas system pump and a gas-fired electric power generator. From the perspective of the natural gas system, the pump is a transshipment node and the generator is a demand node. From the perspective of the electrical network, the generator is a supply node and the pump is a demand node. The generator needs gas to produce electricity; the pump needs electric power to deliver gas through the system to the generator.

In this case, two variables are used: y_{gas}^{power}, the connection of the electricity to the pump, and y_{power}^{gas}, the connection of the gas to the generator. Failure of one component causes its corresponding binary variable to be set to zero, thus reducing the capacity of the other component to zero. In other words, if the pump were to fail, supply of gas to the generator would be inadequate and then y_{power}^{gas} would be set to zero. When $y_{power}^{gas} = 0$, the capacity of the generator is now zero (since its capacity is the product of y_{power}^{gas} and its capacity u). Since the generator is a supply node, all flows on the arcs (i.e., the power lines) leaving the generator would now be zero, by flow conservation.

Alternately, if there is a lack of power at the pump demand node in the electrical generating network, its binary variable y_{gas}^{power} is set to zero and the capacity of the pump reduced to zero. To correct this situation, either an alternate source of gas must be found for the generator or an alternate source of power must be found for the pump.

Co-location

Co-location occurs when activities or physical components of two or more infrastructures are situated within a prescribed geographical region. An event that impacts the entire region, as occurred in the World Trade Center attack, impacts all infrastructures in that region. The implication for any restoration model is that capacities of the supply, demand, and flow capacity nodes and arcs in the model must be revised based on their location with respect to the impact of the disruptions. This determination is made during the assessment stage of the decision process. Once necessary adjustments have been made, the normal operations model (see above) can be run to determine any feasible paths through the networks. If no feasible paths exist, the restoration model discussed in the remainder of this section is run to assist in determining where alternative nodes and arcs can be constructed.

Summary of Restoration Stage

In the restoration phase, alternatives for restoring services are considered. The objective is to find the alternative that meets unmet demand at a minimum cost. However, different demands for the same service as well as demands for different services from the same source will likely emerge and must therefore be reconciled and prioritized. The objective function of the restoration model must be able to incorporate different priorities in addition to modeling interdependencies.

Once all the interdependencies have been identified, two sets of demand nodes can be defined among all the infrastructure systems. One set would be those nodes that do not affect nodes or arcs in any other infrastructure. We will call these nodes independent. The remaining demand nodes would be associated with a connection to some other infrastructure. These would be the dependent nodes.

In the restoration model, we must also meet the flow conservation constraints. That is, the flow from supply nodes, through transshipment nodes, and to demand nodes must balance to satisfy the demand while not exceeding the supply. The node capacity constraint must be modified to represent the fact that the total flow into the node would be less than or equal to its revised capacity, *w*, multiplied by the connector variable, *y*. Constraints are included in this restoration model to shift the connector variable from 1 (operating) to 0 (failed) when the required demand is not met at a dependent node. As in the normal operations model, arc flow is limited to its capacities. Supply and demand constraints as well as structural constraints representing the reconfigured network are also in the model.

Note that some interdependent infrastructure system failures may result in reducing capacity to some value other than zero. For example, loss of supervisory control systems in a subway system may result in operators' exercising greater care and slowing trains, so that the post-disruption capacity may be one-third of normal. In this case, the connector variable *y* would shift between 1 and 1/3. The exact effect of each disruption must be evaluated.

In summary, the restoration stage consists of first identifying each of the interdependencies among infrastructure systems, modeling them, and incorporating them into the restoration model. Emergency managers must revise and prioritize the supplies and demands in order to have the restoration model provide support to those responsible for restoring services.

Model Demonstration

The following excerpts from a news report on September 14, 2001 (Pristin, 2001, p. A12), illustrate impacts of the World Trade Center attack on electrical and telecommunications infrastructures. The article states that "[i]n Lower Manhattan, about 300 Con Edison workers are trying to restore service to about 12,000 commercial and residential buildings without electricity," according to a spokesman for the utility. The previous day, electrical, gas and steam service "were normal throughout the city, except for the areas affected by the loss of two substations that were knocked out when 7 World Trade Center caught fire and collapsed." Of the approximately 500,000 phone lines south of 14th Street, at least 200,000 remained out of service on September 14 (though "most of those lines served locations that are either not in use or no longer exist").

Five Verizon switching centers—one of which is on West Street near the location of 7 World Trade Center—serve these 500,000 lines. The loss of power to the West Street switching center affected the 200,000 phone lines below 14th Street and also about three million private data lines for corporate customers. About 20% of these data lines that serve the New York Stock Exchange were among those affected. "Even more than the West Street office," the same article notes, "the New York Stock Exchange depends on a Verizon switching center on Broad Street that handles about 80 percent of the exchange's data lines." The center on Broad Street was not physically damaged by blasts from the attack but did lose power shortly afterward. As of September 14, diesel generators were continuing to provide power, and power had been restored to the site on September 12.

The foregoing news report illustrates both input and mutual dependence and serves as the basis for the example used to exercise the modeling tools presented earlier in this section. Data associated with the World Trade Center attack (e.g., locations of equipment and personnel, generating capabilities, capacities of feeder lines and shunts, power demands) is, of course, sensitive and has not been used. Rather, the example takes the incident as a starting point and is supplemented with additional, simulated events (e.g., an additional instance of mutual dependence, a system failure) and with information on aspects of operations provided by Consolidated Edison and Verizon personnel and by other domain experts.

To illustrate the mutual dependence of telecommunications and power, damage to a Controlled Environment Vault (CEV) in the phone system is simulated. The result is a loss of phone service. With this failure in the phone system, the supervisory control and data acquisition (SCADA) system for the power company becomes unable to notify supervisors of the impending failure of another component, which therefore also fails. Figure 2 includes a depiction of a section of these systems after impact assessment. (In Figure 2, CO refers to the Verizon Central Offices where cables from the CEVs enter and switching of calls occurs.) The power system considers only the distribution portion. The distribution system starts at the high voltage supply feed to a 13kv substation. The lines from the feeder to the 120/208v transformers and the demand nodes will be the customers.

Illustrative Example

Referring to Figure 2, the demonstration example begins with the failure of Transformer *A*, which is in the Pearl Street feeder. This failure results in loss of power to Metropolitan Transit Authority (MTA) facilities and systems and to one CEV and residential and commercial customers. The failed CEV had been carrying SCADA lines in addition to residential and business service. This SCADA system had been monitoring the power feeder and associated

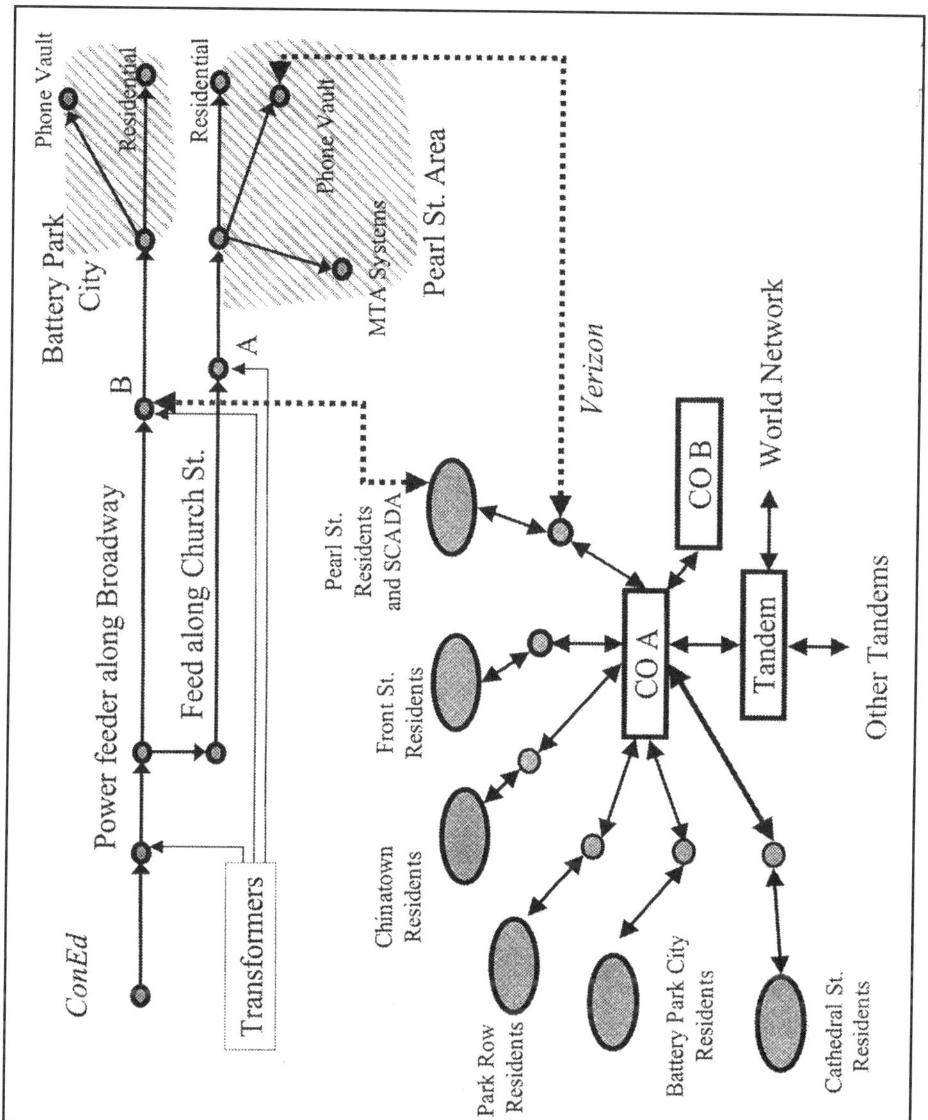

Figure 2. Interdependent telecommunications and power infrastructures.

components along Broadway. Consequently, failure of the SCADA results in an inability to monitor these power distribution components. A high temperature condition at Transformer *B* goes unreported and the transformer subsequently fails. This failure results in loss of power to another CEV and another residential/commercial area. Another CEV failure leads to loss of phone service in a second area.

After assessing the impact and making the required modifications, the normal operations model is run and determines that no feasible solutions exist for power and telecommunication. The model also identifies unmet demands. The next step in the decision process is to examine the solution generated by the model, identify unmet demands, and obtain priorities from managers on restoration. During the response stage, available resources are identified. If the capacity at the substations were to exceed the required load for the demands, the restoration strategy would be to develop a routing plan for feeder lines and inform managers of the demand level and the routing plan. However, if these substations do not have the power available to satisfy the demands, then restoration will require the use of temporary generators.

In this example, it is assumed that managers place priority on restoring power to the two CEVs (requiring five and six units each) and the MTA systems (requiring 10 units). Managers conduct a resource assessment and determine that the three closest substations to the affected areas each have five units of power capacity available. However, these 15 units are not sufficient for the three loads that are candidates for restoration. There are also two portable diesel generators, each having capacity of four units, which can be moved into the area. Managers identify four possible sites for the generators. Each site can accommodate up to two generators.

Figure 3 depicts the decision situation facing the emergency managers. A question is what combinations of transformers (substations) and temporary generators should be selected for presentation to stakeholders (e.g., New York City Office of Emergency Management). Temporary feeder lines or shunts have to be routed from a supply site (e.g., substation or generators), to a demand site that requires resources (e.g., cable, crews). In addition, one or more sites to house the generators must be selected, based on consideration both of the proximity of the generator site to the demand site, and of the resources required to relocate the generators. In this example, there are 33 possible ways to satisfy the unmet demand. Actual implementation, including routing through city streets, must be planned but is not considered here.

The restoration model is designed to support selection of a restoration strategy. The constraints have been specified in terms of available resources and unmet demand; the objective function that drives the search algorithm needs to be specified. Based on discussions with domain experts, reasonable cost estimates have been developed and are depicted in Table 4.

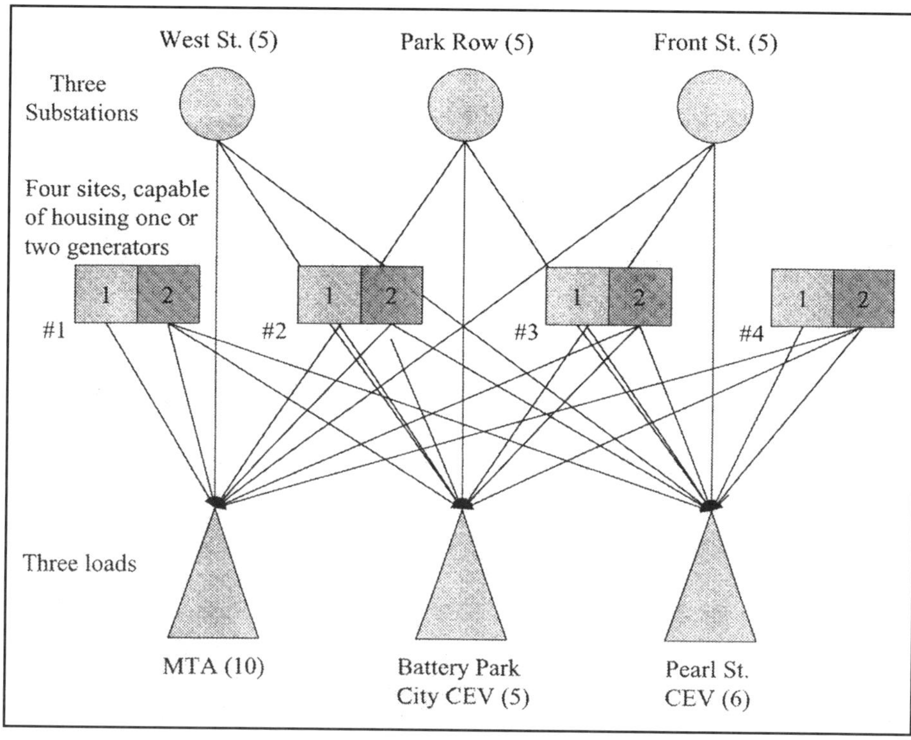

Figure 3. Decision situation.

Since the temporary generators could be located closer to the demand sites than the distance to the transformer substations, shunts from the generators could be installed at lower cost than those from substations. However, transportation and installation of generators do incur costs. With these values, the restoration model can again be run to provide the best alternative restoration strategies, as discussed next.

The decision situation facing the emergency managers is, in essence, to construct a new network utilizing power from the substations from unaffected sections of the power grid. This restoration model does not require the y variables in its formulation. The specific objective function for this example is to minimize cost of operation of the shunts and the generators. The restoration model is formulated as follows. Each shunt has fixed cost, k, and power cost, c, which is a function of generator or transformer use at the substations. A unique k is determined for each shunt and includes the cost of generator transport and setup as appropriate. The cost of power is set at 1 for power

Table 4. Estimated generator and shunt costs.

Site	Location, Qty.	Demand Sites (shunt cost + generator cost)		
		Metropolitan Transit Auth.	Battery Park City CEV	Pearl Street CEV
1	substation, 1	30+0	40+0	50 +0
2	substation, 1	40+0	30+0	40 +0
3	substation, 1	50+0	40+0	30 +0
4	generator site 1, 1 gen.	25+10	35+10	55 +10
5	generator site 2, 1 gen.	35+10	25+10	45+10
6	generator site 3, 1 gen.	45+10	30+10	35+10
7	generator site 4, 1 gen.	55+10	40+10	25+10
8	generator site 1, 2 gen.	25+20	35+20	55+20
9	generator site 2, 2 gen.	35+20	25+20	45+20
10	generator site 3, 2 gen.	45+20	30+20	35+20
11	generator site 4, 2 gen.	55+20	40+20	25+20

coming from the distribution grid and to 1.5 for power from generators. The cost to operate each shunt is therefore $c_e^i x_e^i + k_e^i$. Using binary variables r to indicate whether or not a shunt is installed, set $r_e^i = 1$ when $x_e^i > 0$. The objective is to minimize the total cost of installing and operating all shunts, as follows:

$$\text{minimize} \quad \sum_{e \in E^i} c_e^i x_e^i + k_e^i r_e^i$$

$$\text{subject to} \quad \sum_{e \in \delta^+(j)} x_e^i - \sum_{e \in \delta^-(j)} x_e^i \geq b_j^i \qquad \text{for } j \in V^i \quad \text{with } b_j^i < 0$$

$$\sum_{e \in \delta^+(j)} x_e^i - \sum_{e \in \delta^-(j)} x_e^i = b_j^i \qquad \text{for } j \in V^i \quad \text{with } b_j^i \geq 0$$

$$x_e^i \leq 100 r_e^i$$

$$x_e^i \geq 0$$

$$r_e^i \quad \text{binary.}$$

In order to meet the constraint of only two diesel generators, additional binary variables w and t are introduced. The variable t is assigned a value of one if there is flow from a one-diesel generator site. The variable w is assigned a value of one if there is flow from a two-generator site. Therefore, there are four *t* variables corresponding to the four sites having one diesel generator each, and four *w* variables for the four sites if they have two diesel generators. The number of diesel sites is controlled by constraining the sum of the *t* variables to be less than or equal to two, and the sum of the *w* variables to less than or equal to one. Additionally, *w* is constrained to zero if *t* is greater than zero, and *t* is constrained to zero if *w* is greater than zero.

For this example, solution of the restoration model produces two solutions with minimum costs, as shown in Figures 4 and 5. Solution 1 has two parts: (1) to run shunts from the West Street and Park Row substations to the MTA facilities, and from the Front Street substation to the Battery Park City CEV and (2) to locate two diesel generators at site 4 to power the Pearl Street CEV. Solution 2 also has two parts: (1) to run shunts from the West Street and Front Street substations to the MTA facilities, and from Park Row substation to the Battery Park City CEV and (2) to locate two diesel generators at site 4 to power the Pearl Street CEV.

Given these solutions, managers can now decide which solution to pursue. However, if neither is found to be acceptable, it may be possible to incorporate other costs into the objective function and to develop another set of solutions for review. If a restoration strategy is proposed, the additional cost of using that strategy versus one of the minimum cost strategies can be determined. Note that estimates of these costs result from evaluation by managers of available resources and of the time and effort required to deploy these resources.

Conclusions and Suggestions for Further Research

The nature of the September 11th attack on the World Trade Center—its scale, scope, and type of target—demonstrates the need for better understanding of the interdependencies among critical infrastructure systems. This research employs a systems approach to addressing this need by modeling interdependent infrastructures as systems of systems. The approach allows for optimization of restoration strategies and is a step towards integration of models of infrastructure interdependence with decision support systems. In the future, models such as those presented here could be used to identify opportunities for reducing vulnerabilities, developing countermeasures to mitigate impact of disruptions, and guiding actions for response and recovery.

Summary of Work to Date

Initial investigations of infrastructure interdependence as reported in the *New York Times* produced a starting set of incidents, some of which have been and continue to be explored with the appropriate organizations. In order to

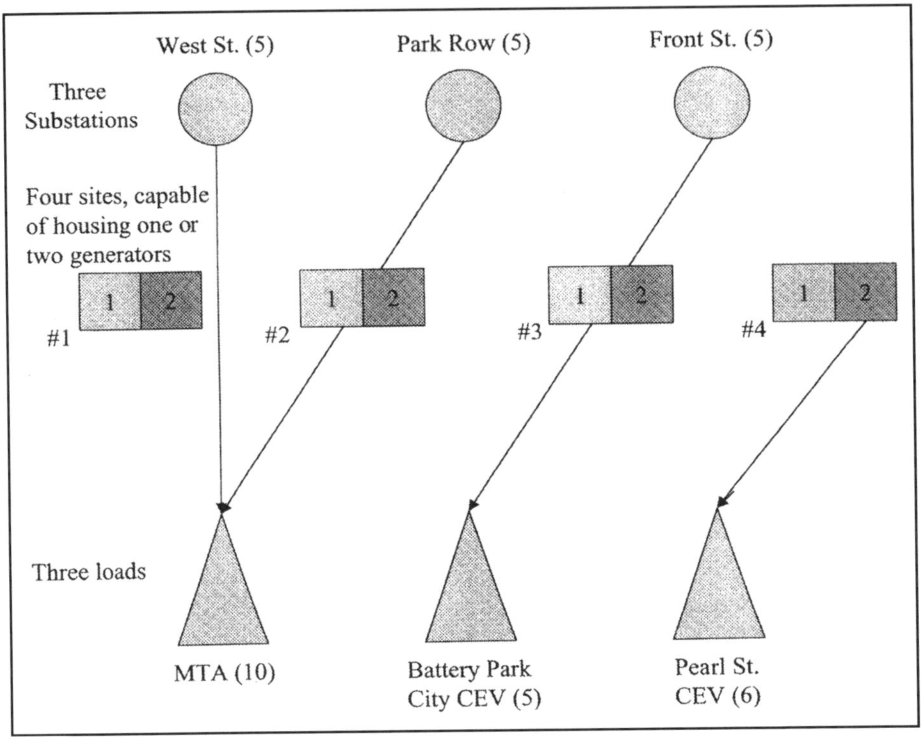

Figure 4. Solution 1 to decision situation.

conduct modeling in later stages of the work, definitions of interdependence among and between critical infrastructure systems were operationalized. A mathematical representation of the physical components of an infrastructure system and the services it provides was developed. A decision support process was proposed to assist infrastructure managers in responding to disruptive incidents that involve infrastructure interdependencies. Two models, one for more normal operations and one for restoration of services, were proposed as components of a decision support system. An illustrative example centered on events following the World Trade Center attack was presented to demonstrate how the models could be used in response and restoration.

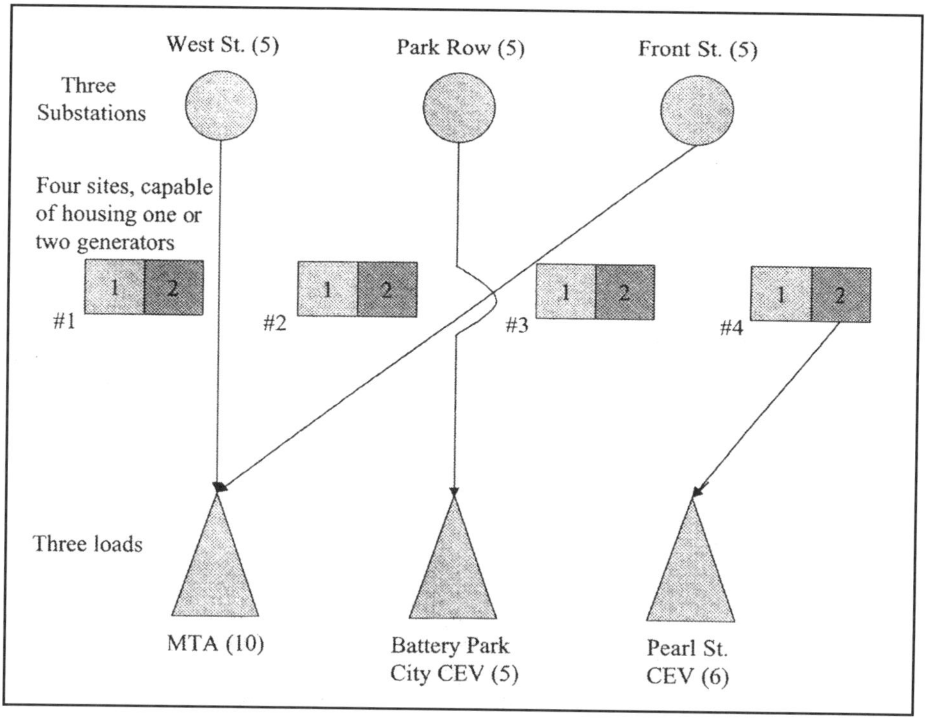

Figure 5. Solution 2 to decision situation.

Ongoing and Future Work

Ongoing work is being conducted on improving understanding of decision-making processes in the management of infrastructure interdependencies. Based on data from *New York Times* reports, a number of potentially rich cases of infrastructure dependence and interdependence have been identified. Organizations involved in these cases are participating in a study to investigate decision making in the management of these disruptions. Of particular interest are non-routine cases, since these provide an opportunity to examine organizational flexibility and improvisation (Kreps, 1991). Additionally, since one concern for many organizations is the proprietary nature of relevant data, cases in which ad hoc or temporary solutions were employed are also being investigated. The Critical Decision Method (Klein et al., 1989; Hoffman et al., 1998; Flanagan, 1954) of knowledge elicitation is well-suited to these situations. The Critical Decision Method (CDM) can be used for uncovering information about how individuals responded to critical situations. It is intended to uncover critical decisions and their content, particularly for non-routine decision making. It has proven useful for guiding training, identifying lessons learned, and developing decision support tools (Klein et al., 1989). Additional information is being provided through an examination of other materials such as activity logs, maps, and after-action reports. Approximately five cases involving emergency services, electric power, and telecommunications are currently being investigated, with a number of participants being interviewed for each case. Results of these investigations will continue to inform the construction of models to support the management of infrastructure interdependencies.

Additional modeling efforts are focusing on addressing the various types of interdependence, again drawing upon case studies. Efforts are also underway to incorporate considerations of time into the models, since some effects of a disruption in service take time to develop. For example, a generator may be able to produce additional power to cover a shortfall only for a limited amount of time. These and other time-varying consequences of disruption should appear in the restoration model. The result is in a **time-expanded network** (Ahuja et al., 1993). Longer-term work may involve the use of models as an aid in exploring vulnerability of systems, particularly during planning and design of infrastructure systems. One approach is to develop event scenarios, evaluate network performance in the scenarios, and re-design as necessary.

Finally, it should be noted that visual models capitalize on a fundamental, native expertise of humans: the capability to solve complex problems by reasoning with graphical representations. Visual models can offer advantages over purely lexical models by increasing interpretability and reducing cognitive load, thus enabling decision makers to devote additional cognitive resources to problem solving (Larkin and Simon, 1987). Indeed, informal observation on the results of interviews conducted for this research suggests that visualization tools such as geographic information systems had widespread use during the response to the World Trade Center attack. Future work should contribute to capabilities for visualizing both the assumptions and implications of models of infrastructure interdependence.

Notes

1. As discussed below, the term "co-located" is later used for this definition. The term "interdependent" is later redefined.

References

Ahuja, R.K., T.L. Magnanti, and J.B. Orlin. 1993. *Network Flows*. Englewood Cliffs, NJ: Prentice Hall.

Berenson, A. 2001."A Full Reopening of Stock Trading Is Set for Monday." *New York Times* (September 14): C1.

Berge, C. 1962. *The Theory of Graphs and its Applications*. New York: John Wiley & Sons.

Beroggi, G.E.G. and W.A. Wallace. 1998. *Operational Risk Management: The Integration of Decision, Communications, and Multimedia Technologies*. Amsterdam: Kluwer.

Flanagan, J.C. 1954. "The Critical Incident Technique." *Psychological Bulletin* 51: 327–358.

Frank, H. and I.T. Frisch. 1971. *Communication, Transmission and Transportation Networks*. Reading, MA: Addison-Wesley Publishing Company.

Hoffman, R.R., B. Crandall. and N. Shadbolt. 1998. "Use of the Critical Decision Method to Elicit Expert Knowledge: A Case Study in the Methodology of Cognitive Task Analysis." *Human Factors* 40(2): 254–276.

Kennedy, R. 2001. "With City Transit Shut Down, New Yorkers Take to Eerily Empty Streets." *New York Times* (September 12): A8.

Klein, G., R. Calderwood, and D. MacGregor. 1989. "Critical Decision Method for Eliciting Knowledge." *IEEE Transactions on Systems, Man and Cybernetics* 19: 462–472.

Kreps, G.A. 1991. "Organizing for Emergency Management." Pp. 30–54 in T.E. Drabek and G.J. Hoetmer, eds., *Emergency Management: Principles and Practice for Local Governments*. Washington, D.C.: International City Management Association.

Larkin, J. and H.A. Simon. 1987. "Why a Diagram is (Sometimes) Worth Ten Thousand Words." *Cognitive Science* 11(1): 65–99.

Little, R. 2002. "Controlling Cascading Failure: Understanding the Vulnerabilities of Interconnected Infrastructures." *Journal of Urban Technology* 9(1): 109–123.

New York Times Editorial Staff. 2001. "A Sacrifice for Commuters." *New York Times* (September 27): A20.

Pickett, J.P. 2000. *The American Heritage® Dictionary of the English Language*. 4th edition. Boston: Houghton Mifflin Co.

President's Commission on Critical Infrastructure Protection. 1997. *Critical Foundations: Protecting America's Infrastructures: The Report of the President's Commission on Critical Infrastructure Protection*. Washington, D.C.: U.S. Government Printing Office.

Pristin, T. 2001."Phone Service Improving, but Many Still Lack Power." *New York Times* (September 12): A12.

Rardin, R.L. 1998. *Optimization in Operations Research*. Upper Saddle River, NJ: Prentice Hall, Inc.

Rinaldi, S.M., J.P. Peerenboom, and T.K. Kelly. 2001. "Identifying, Understanding, and Analyzing Critical Infrastructure Interdependencies." *IEEE Control Systems Magazine* (December): 11–25.

U.S. General Accounting Office. 2001. *Critical Infrastructure Protection*. GAO-01-323. Washington, D.C.: U.S. General Accounting Office.

Wallace, W.A. 1990. "On Managing Disasters: The Use of Decision-aid Technologies." Pp. 261–279 in A. Kirby, ed., *Nothing to Fear: An Examination of Risks and Hazards in American Society*. Tucson, AZ: University of Arizona Press.

Wallace, W.A., ed. 1994. *Ethics in Modeling*. Oxford: Pergamon Press.

Acknowledgments

This research has been funded by National Science Foundation Supplemental Grant for Exploratory Research number CMS-0139306. We thank personnel at Consolidated Edison, Verizon, and the Jersey City (New Jersey) Fire Department for their participation and continuing cooperation, and Dr. Rae Zimmerman of the Institute for Civil Infrastructure Systems for her coordination efforts.

Correspondence should be directed to William A. Wallace, Rensselaer Polytechnic University; e-mail: wallaw@rpi.edu.

Appendix A

Normal Operations

As discussed in the section on Modeling Infrastructure Systems, interdependent infrastructures are modeled as networks having services represented as flows in networks. Let I denote the set of infrastructures. Infrastructure $i \in I$ has nodes V^i and directed arcs E^i. Associated with each node $j \in V^i$ is a scalar b^i_j representing its supply or demand. If node $j \in V^i$ is a demand node then $b^i_j < 0$; if it is a supply point then $b^i_j > 0$; and if it is a transshipment node then $b^i_j = 0$. If $j \in V^i$ is a supply node then b^i_j equals the maximum possible amount that could be produced at that node. A non-negative vector of variables, x^i, represents the flow on each arc of the infrastructure. Associated with each arc in E^i are non-negative costs c^i and capacities u^i, where $0 \le x^i \le u^i$ for each element in x^i. The objective is to find the minimum cost feasible network flow under normal operating conditions.

Arcs are represented using either the endpoints of the arc or the index of the arc. For a node $l \in V^i$ for some infrastructure $i \in I$, let $\delta^+(l)$ denote the set of arcs in E^i that enter node l and let $\delta^-(l)$ denote the set of arcs in E^i that leave node l. Define $\delta(l) := \delta^+(l) \cup \delta^-(l)$, the set of all arcs incident to node l. Without loss of generality, assume that every supply node has no incoming arcs (i.e., $\delta^+(l) = 0$ if $b_l^i > 0$) and that demand nodes have no outgoing arcs, (i.e., $\delta^-(l) = 0$ if $b_l^i < 0$). A transshipment node l may have a limited capacity, w_l^i, modeled by placing an upper bound on total flow across the arcs $\delta^+(l)$. Included in the model are flow conservation constraints (i) that for supply nodes ensure that total flow out of the node is no greater than the available supply, (ii) that for demand nodes ensure that demand is met, and (iii) that for transshipment nodes ensure that flow into the node equals flow out of the node.

Under normal conditions, all the demands of all the infrastructures are met. Since interdependency considerations come into consideration only when there are unmet demands, each infrastructure may be considered to be operating independently and analyzed independently. It is assumed that, prior to disruption, the system operates at a minimum cost feasible solution, denoted the normal operations solution. This solution can be found by solving the model for each infrastructure separately. The model prior to disruption is the solution to the following minimum cost network flow problem for each infrastructure $i \in I$, where the total flow into node j is given by $\sum_{e \in \delta^+(j)} x_e^i$ and the total flow out of the node is given by $\sum_{e \in \delta^-(j)} x_e^i$, as follows:

$$\text{minimize} \qquad \sum_{e \in E^i} c_e^i x_e^i$$

$$
\begin{aligned}
\text{subject to}\quad & \textstyle\sum_{e\in\delta^-(j)} x_e^i \le b_j^i && \text{for } j\in V^i \text{ with } b_j^i > 0\\
& \textstyle\sum_{e\in\delta^+(j)} x_e^i = -b_j^i && \text{for } j\in V^i \text{ with } b_j^i < 0\\
& \textstyle\sum_{e\in\delta^+(j)} x_e^i - \sum_{e\in\delta^-(j)} x_e^i = 0 && \text{for } j\in V^i \text{ with } b_j^i = 0\\
& \textstyle\sum_{e\in\delta^+(j)} x_e^i \le w_j^i && \text{for } j\in V^i \text{ with } b_j^i = 0\\
& x^i \le u^i && \text{for } i\in E^i\\
& x^i \ge 0 && \text{for } i\in E^i
\end{aligned}
$$

Appendix B

Restoration Model

In the restoration phase, alternatives for restoring services are considered. Different demands for the same service as well as demands for different services from the same source will likely emerge and must therefore be reconciled and prioritized.

Define the set $V^{i,+} \subseteq V^i$ to be the nodes $j \in V^i$ with $b^i_j > 0$ (supply nodes); sets $V^{i,=} \subseteq V^i$ (transshipment nodes) and $V^{i,-} \subseteq V^i$ (demand nodes) are defined similarly. Let $D(i,i_1) \subseteq V^{i,-}$ be the set of nodes in i that some other infrastructure i_1 depends upon (parent nodes) and let $D^i := \cup_{i_1 \in I, i_1 \neq i} D(i,i_1)$. This subset of nodes is the interdependent nodes. The remaining nodes in $V^{i,-}$ will be referred to as the independent nodes. The binary variable $y^{i,l}_{i_1,j}$ is the connection between node l in infrastructure i (where it is a demand node) and node j in infrastructure i_1, where it may be either a supply, demand or transshipment node and is only defined for $l \in D(i,i_1)$ (this connector variable concept was discussed in section entitled Restoration Stage). Let $C(i_1,i) \subseteq V^{i_1}$ be the set of nodes in i_1 that depend on some other infrastructure i (child nodes) and let $C^{i_1} := \cup_{i \in I, i \neq i_1} C(i_1,i)$. Without loss of generality, all nodes have been disaggregated to the point where, given infrastructures i, i_1, and l in $D(i,i_1)$, there is a unique node j in $C(i_1,i)$ such that $y^{i,l}_{i_1,j}$ is defined, and given infrastructures i, i_1, and node j in $C(i_1,i)$, there is a unique node l in $D(i,i_1)$, such that $y^{i,l}_{i_1,j}$ is defined.

The objective function of the restoration model must be able to incorporate different priorities in addition to modeling interdependencies. On independent nodes, the available supply may meet the required demand, but there may be some shortfall. The slack variable s^i_j represents the shortfall in meeting demands at independent nodes. In the model, there is no consideration for partial slack at the interdependent nodes. Because these interdependent nodes control the operation of nodes in other infrastructure systems, they either must be fully operational or they are in a failed condition. There is no benefit to partially meeting the requirement. After the response phase, when the operator realizes there are unmet demands across one or more systems, one choice for the objective function is to minimize the total shortfall (slack) plus the unmet interdependent demands, as follows:

$$\text{minimize} \quad \sum_{i\in I}\sum_{j\in V^{i,-}\setminus D^i} k^i_j s^i_j + \sum_{i\in I}\sum_{l\in D^i}\sum_{i_1\neq i} b^i_l (1-y^{i,l}_{i_1,j})$$

$$\text{subject to} \quad \sum\nolimits_{e\in\delta^-(j)} x^i_e \le b^i_j \qquad \forall j\in V^{i,+}, \forall i\in I$$

$$\sum_{e\in\delta^+(j)} x^i_e - \sum_{e\in\delta^-(j)} x^i_e = b^i_j \qquad \forall j\in V^{i,=}, \forall i\in I$$

$$s_j^i + \sum_{e\in\delta^+(j)} x_e^i = -b_j^i \qquad \forall j \in V^{i,-}, \forall i \in I$$

$$\sum_{e\in\delta^+(j)} x_e^i \le w_j^i \qquad \forall j \in V^{i,=}, \forall i \in I$$

$$\sum_{e\in\delta^-(j)} x_e^{i_1} \le b_j^{i_1} y_{i_1,j}^{i,l} \qquad \forall j \in C(i_1, i\ ) \text{ with } b_j^{i_1} > 0, \forall i, i_1 \in I, i \ne i_1$$

$$s_j^{i_1} + \sum_{e\in\delta^+(j)} x_e^{i_1} = -b_j^{i_1} y_{i_1,j}^{i,l} \qquad \forall j \in C(i_1, i\ ) \text{ with } b_j^{i_1} < 0, \forall i, i_1 \in I, i \ne i_1$$

$$\sum_{e\in\delta^+(j)} x_e^{i_1} \le w_j^{i_1} y_{i_1,j}^{i,l} \qquad \forall j \in C(i_1, i\ ) \text{ with } b_j^{i_1} = 0, \forall i, i_1 \in I, i \ne i_1$$

$$s_l^i \le (1 - y_{i_1,j}^{i,l}) b_l^i \qquad \forall l \in D(i, i_1), \forall i, i_1 \in I, i \ne i_1$$

$$x_e^i \le u_e^i \qquad \forall e \in E^i, \forall i \in I$$

$$x^i \ge 0 \qquad \forall i \in I$$

$$y_{i_1,j}^{i,l} \qquad \text{binary}, \ \forall l \in D(i, i_1), \forall i, i_1 \in I, i \ne i_1$$

$$s_j^i \ge 0 \qquad \forall j \in V^i \text{ with } b_j^i < 0, \forall i \in I$$

Investigation of a Damaged High-rise Building near Ground Zero

Gordon Warn
Department of Civil, Structural & Environmental Engineering
State University of New York at Buffalo

Jeffrey Berman
Department of Civil, Structural & Environmental Engineering
State University of New York at Buffalo

Andrew Whittaker
Department of Civil, Structural & Environmental Engineering
State University of New York at Buffalo

Michel Bruneau
Multidisciplinary Center for Earthquake Engineering Research
State University of New York at Buffalo

Introduction

Within two weeks of the September 11, 2001, attacks on the World Trade Center towers in New York City, the Multidisciplinary Center for Earthquake Engineering Research (MCEER) dispatched a research team to the site to collect perishable data related to the collapse of the two 110-story towers and collateral damage to buildings and infrastructure surrounding the World Trade Center complex. Two visits to Ground Zero were undertaken, the first on September 21 (see Bruneau et al. (2002)) and the second on September 23. The visit on September 23 involved a walk-through of one high-rise building that was badly damaged by large pieces of debris that were ejected from World Trade Center tower 2 as it collapsed.

The objectives of the work described in this paper were two-fold: (1) to collect information about the structural and nonstructural damage suffered by the building at 130 Liberty Plaza due to the collapse of tower 2, and (2) to investigate whether simple numerical tools used in the structural analysis of buildings can be used to help explain the observed structural behavior. The first objective was achieved by the September 23 walk-through of the building. Summary information on the building damage is presented in the reconnaissance section. The second objective was addressed using linear and nonlinear analysis tools that are used by earthquake engineers. Berman et al. (2002) and Warn et al. (forthcoming) present detailed information on the September 23 reconnaissance and the subsequent studies.

The building studied is a 39-story office building at 130 Liberty Street in lower Manhattan. It was located to the immediate south of tower 2. Figure 1 shows the building's location with respect to the Trade Center complex.

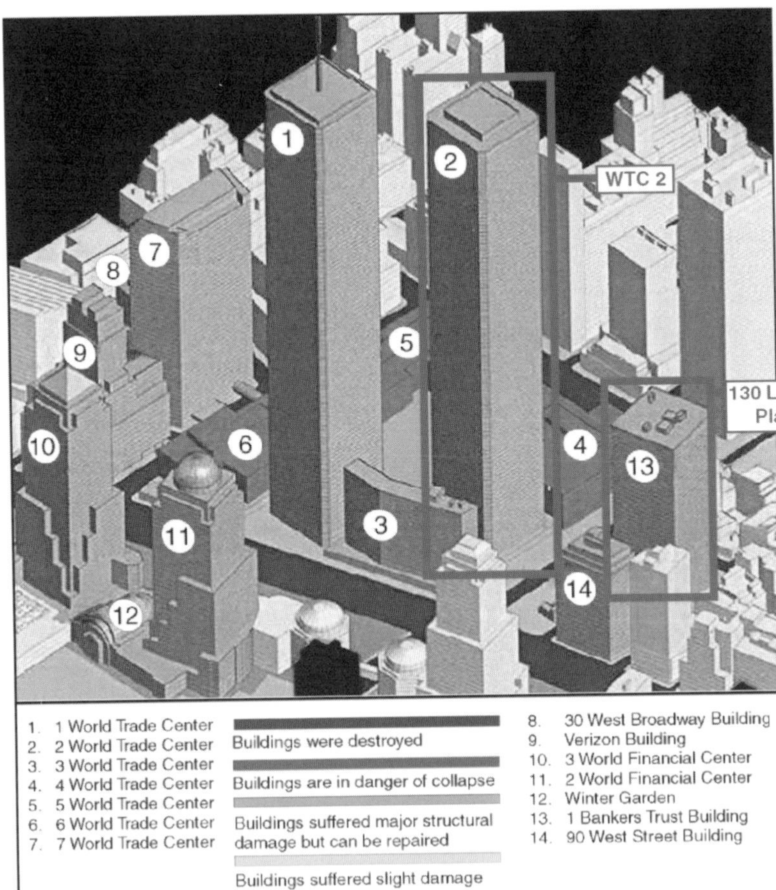

Figure 1. Location of subject building with respect to the World Trade Center complex.

The 130 Liberty Street building was designed in the early 1970s by the structural engineering firm of James Ruderman LLP. Structural drawings were not available to the reconnaissance team from the building owner but information on the framing system was gathered during the reconnaissance. The typical gravity framing system is composed of a non-composite unreinforced concrete slab on metal decking spanning approximately 8 feet 8 inches to steel beams that span 26 feet to steel girders that frame into steel wide-flange columns. The concrete slab is 2.5 inches thick atop a 20-gauge metal deck. The lateral framing system consists of a three-dimensional steel moment-resisting space frame (that is, all beam-to-column connections are rigid connections) and a steel braced core. Figure 2 shows a part plan of a typical floor; the member sizes indicated on the figure were established by members of the reconnaissance team after independent analysis of the frame as described in the building analysis section. The grid marks (A through H and 1 through 8) were selected by the reconnaissance team to aid in the identification and interpretation of damage.

The façade of the building consists of windows and a lightweight cladding system. A sketch of the northern façade badly damaged by falling debris is presented in Figure 3. The elliptical line in the figure identifies the zone of damage to the façade of the building, where TL denotes "tenant level."

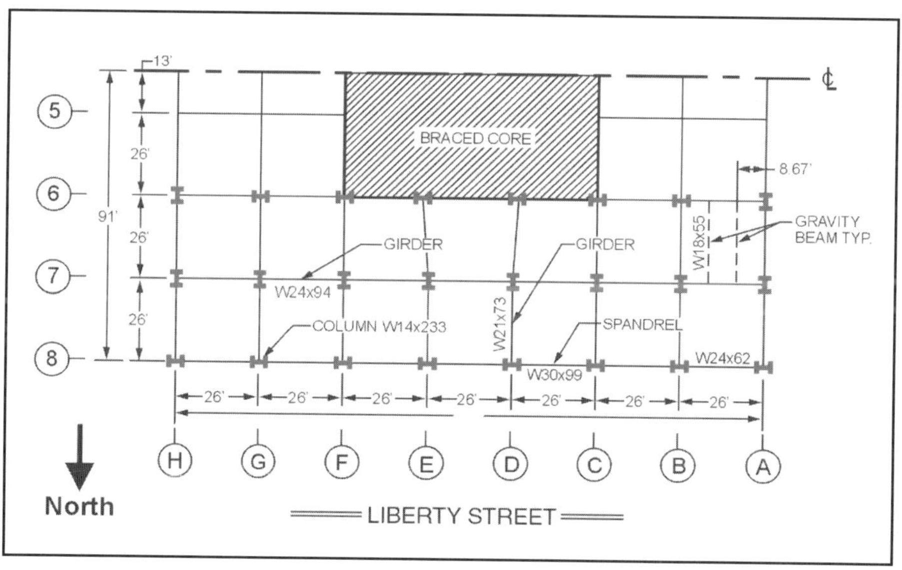

Figure 2. Part plan view of a typical floor with estimated section sizes at tenant level 25.

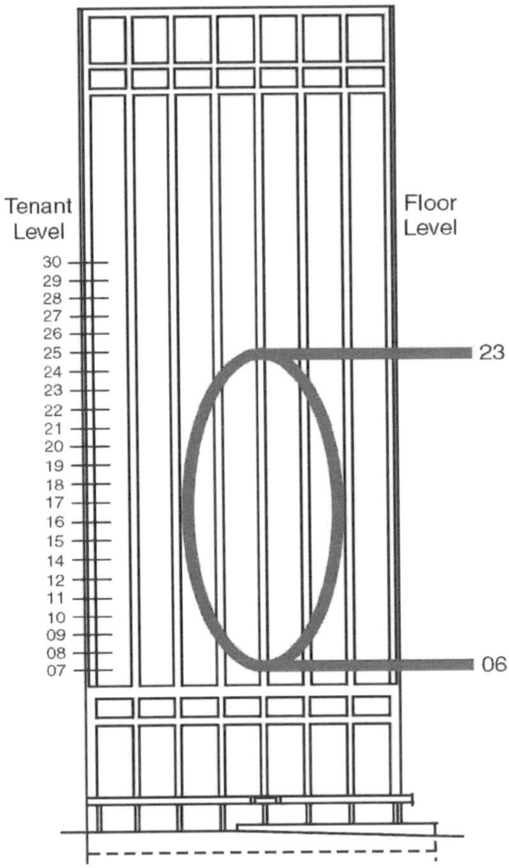

Figure 3. Sketch of northern façade showing the zone of damage.

Reconnaissance

Exterior Reconnaissance of September 21, 2001

The visit of September 21 involved an exterior inspection of the building from ground level as part of a broader survey reported elsewhere (Bruneau et al., 2002). Figure 4 presents four photographs of the building taken on September 21, 2001. Part a. is a view of the northern façade showing the large gash in the exterior wall caused by falling debris that was ejected from tower 2 as it collapsed. One large piece of the façade of tower 2 can be seen in this figure at approximately the 9th floor level. It is highly likely that this three-story-high piece of debris caused much of the damage that is evident above the 9th floor. Clearly seen in this figure is the loss of a column on line D (see Figure 2) between tenant level 7 and tenant level 25.

Part b of Figure 4 shows damage to the northern and western faces of the building. Windows were broken on both of these faces of the building but the damage on the western face was confined to a one-bay width between lines 7 and 8. The cladding to corner column A8 was lost between the ground and tenant level 5 and between tenant level 9 and 12.

Parts c and d of Figure 4 show damage to the lower levels of the building on its north face. Burning debris ignited small fires in the lower levels of the building but these fires appear to have been confined to the perimeter of the building. Most of the windows in the lower levels, especially below the mechanical plant room at the 5th floor, were broken but it is not known how many of these windows were broken deliberately by the rescue teams to eliminate falling glazing hazards to the rescue workers operating to

Figure 4a. Damage to the 130 Liberty Plaza building from the exterior—view of northern façade.

Figure 4b. Damage to the 130 Liberty Plaza building from the exterior—view of northern and western facades.

Figure 4c. **Damage to the 130 Liberty Plaza building from the exterior—view of lower 10 stories of northern façade.**

Figure 4d. **Damage to the 130 Liberty Plaza building from the exterior—damage to northern façade at the entry level.**

the immediate south of tower 2. Part d shows the debris pile just north of the building, 10 days after the attacks. Much of the debris was from tower 2.

Figure 5 presents information on the moment-resisting connections used in the building. Part a of the figure shows fire damage to the lower framing, a cover-plated moment connection of the beam to the column, and the connections used to join the steel floor beams to the perimeter girders. No spray-on fireproofing is evident on the side or bottom of the perimeter beam although it may have been knocked off upon impact. Façade construction details can be clearly seen in this figure. Part b of the figure shows a typical beam-column moment connection on the western face of the building. Cover plates are clearly visible in this connection and damaged (or incomplete) fireproofing can be seen on the exterior face (web) of the perimeter beam.

Interior Reconnaissance of September 23, 2001

One member of the MCEER reconnaissance team returned to Ground Zero on September 23, 2002, to accompany an expert structural engineer from LZA/Thornton Tomasetti on a detailed inspection of the interior of the building. That inspection involved climbing directly from the entry level to the roof via the two stairwells located in the core of the building and returning to the entry level after a floor-by-floor inspection of the building, in which all quadrants of the building were inspected at each floor but with attention being focused on that portion of every floor to the north of grid 6.

Figure 5a. Exposed steel moment framing in the first story of the building—first story of northern face.

Figure 5b. Exposed steel moment framing in the first story of the building—first story of the western face.

Both structural and non-structural components in the building suffered significant damage but all of the non-dust- and non-mildew-related damage above the sixth floor appeared to be substantially confined to a two-bay-by-two-bay zone contained by lines 6 and 8 and C and E (see Figure 2 for details). Non-structural damage did extend a short distance to the west of line C and the east of line E.

Figure 6 presents photographs taken at the roof level of the building. Part a of the figure is a photograph looking east across the roof. Silver U-shaped aluminum panels from tower 2 can be seen in the figure together with life vests from American Airlines (AA) Flight 11 that struck the north face of tower 1 (approximately 120 meters from 130 Liberty Plaza). Part b of the figure is a photograph looking west across the roof. Aluminum panels and AA life vests are also visible in this figure. Parts c and d of the figure are close-up photographs of an AA life vest and a seat from the AA Flight 11.

No structural damage was observed at the roof level of the building and there was no evidence that the roof floor slab had been punctured by falling debris from tower 2. This observation was confirmed by the walk-though of the plant-room spaces immediately below the roof. Apart from broken glass in the stairwells from the roof-level skylights, the structure and mechanical plant in the 38th story were undamaged by the collapse of tower 2.

Figure 6a. Debris at the roof level of 130 Liberty Plaza—photograph looking east.

Figure 6b. Debris at the roof level of 130 Liberty Plaza—photograph looking west.

Figure 6c. Debris at the roof level of 130 Liberty Plaza—American Airlines life vest.

Figure 6d. Debris at the roof level of 130 Liberty Plaza—seat from American Airlines plane.

Figure 7a. Office space damage above tenant level 29 — penetration of tower 2 fascia piece.

Figure 7b. Office space damage above tenant level 29— typical damage in office space.

Damage to the building above tenant level 29 was modest (relative to the damage below tenant level 25) and was limited to broken glass (caused by debris ejected from tower 2 as it collapsed). Figure 7 shows sample non-structural damage above tenant level 29. Part a of the figure shows damage caused by a section of fascia that was ejected from tower 2 and penetrated a window on the northern face of the building. Part b shows debris in a corner office at grid A8. The angle section lying atop the overturned chair in this figure penetrated the north-facing window of the building. (The hard hat and flashlight in the photograph belonged to the photographer.)

The first major structural damage was observed at tenant level 29 and was due to the impact of one single-story-tall structural steel column from tower 2. (As described in Bruneau et al. (2002), this column was one of three in a typical spandrel module, so the two steel beams linking the three columns were destroyed before the single column hit the 130 Liberty Plaza building.) Figure 8a is a sketch of a typical three-column module. Figure 8b shows the upper end of the column above Level 29. Figure 8c is a photograph of the underside of the column and the damage it caused to the ceiling system. It is noteworthy that the unreinforced slab on metal deck arrested the fall of this substantial steel member.

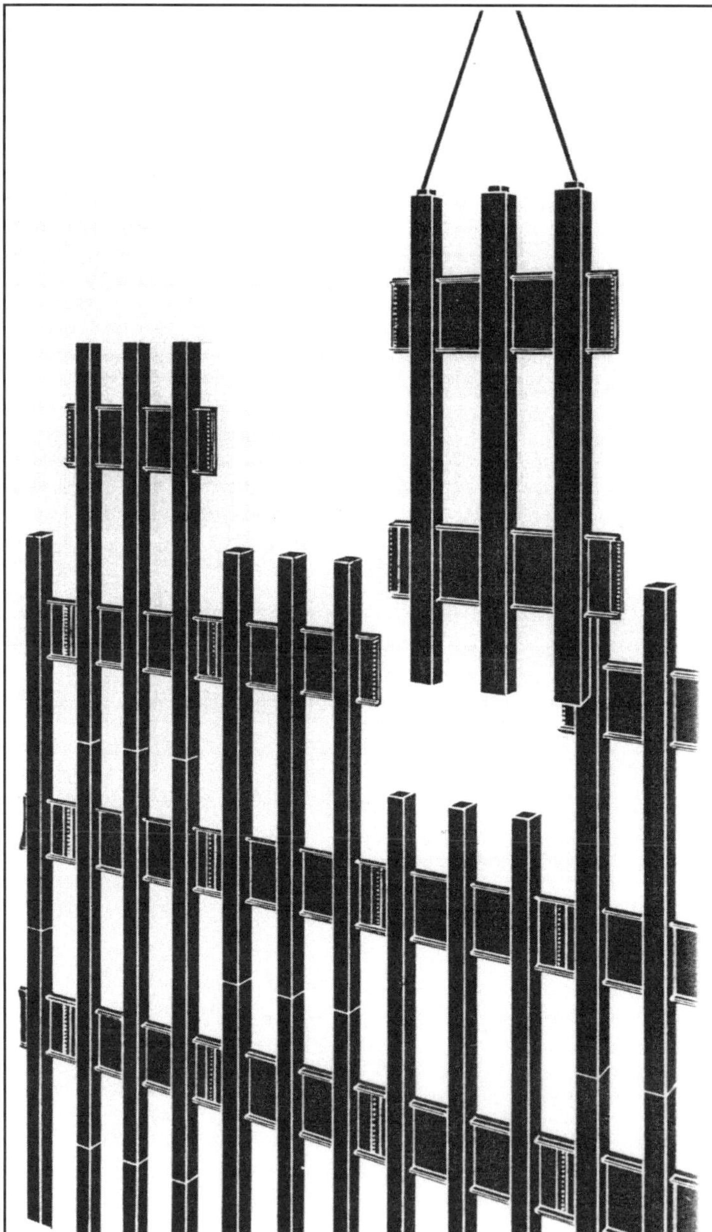

Figure 8a. Damage due to column-missile ejected from tower 2—three-column module used in the construction of tower 2 (from Hart et al., 1978).

Figure 8b. Damage due to column-missile ejected from tower 2—upper end of column-missile above level 29.

Figure 8c. Damage due to column-missile ejected from tower 2—Lower end of column-missile below level 29.

There was little evidence of damage to or distress in the building frame in the stories immediately above the zone of structural damage. There was no sign of excessive deflection in the framing above tenant level 25 as evinced by the total lack of deflection-induced damage in brittle components such as glazing and plasterboard partitions.

The degree of damage to the structural framing increased substantially below tenant level 26. Figure 9a shows a north-south spanning floor beam at the underside of tenant level 23 that has lost its support on line 8 due to the failure (loss) of the spandrel beam between lines D and E. Note the clean separation of the metal deck from the beam (made possible by the lack of studs joining the beam to the slab). Part b of the figure shows the underside of the floor slab at tenant level 22. At this level both north-south spanning floor beams were lost; the locations of these beams are marked by dark stripes on the underside of the metal decking. Of importance to the integrity of the floor system is the fact that the tenant level 22 slab at this location was able to span three times the distance assumed in design, namely, 26 feet, albeit with large deflections in the slab system. Part c of the figure shows a view of the slab of part b looking north-east. Part d of the figure is a view looking northeast of the underside of the slab at tenant level 21: a photograph taken from

Figure 9a. Interior damage at tenant levels 22 and 21—underside of level 23 slab looking northwest.

Figure 9b. Interior damage at tenant levels 22 and 21—underside of level 22 slab looking northwest.

Figure 9c. Interior damage at tenant levels 22 and 21—underside of level 22 slab looking northeast.

Figure 9d. Interior damage at tenant levels 22 and 21—underside of level 21 slab looking northeast.

approximately the same location as Figure 9c but one floor lower in the building. Fracture of the metal deck floor system at the line of the girder spanning north-south on line E can be seen in part d of the figure together with gross distortion of two steel floor beams spanning north-south to the (lost) spandrel.

Figure 10 shows the remnants of a moment-resisting connection as photographed from tenant level 19 in the building. The fractured flange connections and the welded web tab are clearly visible. Part b of the figure is a photograph taken from the same location as the photograph of part a but looking towards the north-south spanning girder on line C. The World Financial Center and Winter Garden can be seen in the background of this photograph. Figure 10c is a view of two stories of framing along line D. Note the distortion in the column flanges at the level of the beam-to-column connection and that the metal decking fractured cleanly along a butt (noncontinuous) joint atop the girder on line D. Large deflections in the floor framing are not seen in this figure. Part d of the figure shows the fractured column on line D at approximately three feet above tenant level 18, immediately below the bottom of the framing shown in part c of the figure. The rubble pile at the base of tower 2 can be seen in the bottom of the Figure 10d.

Figures 11a and 11b are photographs of the fractured column on line D taken from tenant levels 18 and 17, respectively. The façade of tower 2 can be seen in the background of part a of the figure and World Trade Center 4 can be seen in the background of part b of the figure.

Figure 12 presents two photographs taken from outside the building looking toward the damage zone described in the previous pages. The box superimposed on Figure 12a shows the column of Figure 11. The fracture evident in Figure 11a is located at the top of the box. The distorted but intact two-story section of column seen in Figure 11b is located in the center of the box. In Figure 12a, the upper dashed line is at tenant level 25 and the lower dashed line is at tenant level 7. The dashed line in Figure 12b corresponds to the lower dashed line of Figure 12a. The large piece of debris that likely caused most of the damage above tenant level 10 is seen in Figure 12b precariously attached to the façade of the building.

The damage to the structural and non-structural components between tenant levels 16 and 10 was most severe. The near-total destruction of the ceiling, mechanical, and plumbing systems in the areas surrounding the collapsed structural framing prevented the reconnaissance team from gaining access to the perimeter of the damage zone. As seen in Figures 12a and 12b, the zone of extreme damage expanded below tenant level 16 to the two-bay-by-two-bay zone bounded by lines C and E and 6 and 8.

Figure 10a. Interior damage photographs from tenant level 19—failed moment connection at grid D8.

Figure 10b. Interior damage photographs from tenant level 19—floor framing on line 8 between lines C and D.

Figure 10c. Interior damage photographs from tenant level 19—damaged framing at levels 19 and 20.

Figure 10d. Interior damage photographs from tenant level 19—failed column immediately above level 18.

Figure 11a. Fractured column on line D from tenant levels 18 and 17—view at level 18.

Figure 11b. Fractured column on line D from tenant levels 18 and 17—view at level 17.

Figure 12a. Damage to the northern face of the building between tenant levels 23 and 10— view of damage zone between levels 23 and 10.

Figure 12b. Damage to the northern face of the building between tenant levels 23 and 10—view of damage zone adjacent to level 10.

Figure 13 shows two photographs of damage at tenant level 9. Figure 13a shows the complete destruction of one zone of the floor immediately adjacent to the northern face of the building. Part b of the figure shows the interior face of the section of tower 2 façade seen in Figures 4a and 12b that caused much of the damage to the building.

Structural damage below the mechanical plant rooms that were located below tenant level 9 was modest and no photographs were taken at the levels between tenant level 9 and the entryway to the building. Figure 14 is a photograph taken inside the building looking north towards a standing section of the tower 2 façade.

Building Analysis

The observation that the subject building did not collapse despite the loss of key structural elements and severe damage motivated the research team to analyze the building to understand the cause of the observed behavior. Standard tools for the linear and nonlinear analysis of buildings subjected to earthquake shaking were employed for these studies. Linear analysis was performed to determine demand-to-capacity ratios for an undamaged state as well as three damage states, one of which corresponds to the observed damage. Linear analysis further provided an estimate of the elastic limit of the framing system for each of the damage states considered. Two-dimensional and three-dimensional linear analyses were performed. Small displacement theory was employed for these analyses. Simple plastic analysis was then performed to determine an upper bound to the capacity of the framing system for each of the damage states considered. Both two- and three-dimensional framing systems were considered for simple plastic analysis.

Detailed information on the structural framing system was not available to the research team, although approximate sizes were noted during the building reconnaissance. To facilitate the linear and nonlinear analysis of the building, sizes of the beams and columns in the moment-resisting frame were estimated by analysis of the building frame for gravity and winds loads as described in the following section. All beam-to-column connections were assumed to be moment resisting. The estimated sizes of the W-shape beams and columns were checked against the approximate sizes noted during the building reconnaissance. Because no information was available on the steel braced core, sizes were not estimated for the steel braces.

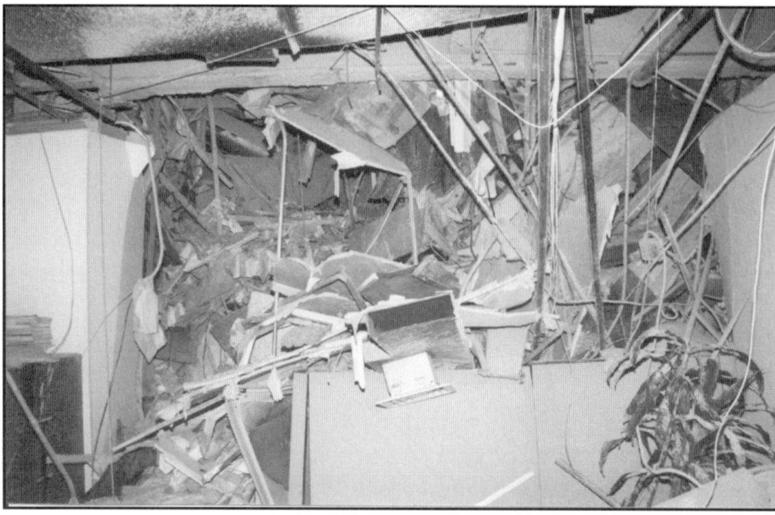

Figure 13a. Building damage at tenant level 9—complete destruction of one zone of the building.

Figure 13b. Building damage at tenant level 9—view of inside face of tower 2 façade element.

Figure 14. View of the tower 2 façade looking northwest from the foyer of the building.

Preliminary Analysis and Design

A preliminary design was undertaken per the 1970 Building Code of the City of New York (City of New York, 1970) to determine beam and column section sizes. Resulting sizes of the beams and columns were used in the analytical studies presented herein. Because actual sizes were unavailable, the reader should interpret the analysis results presented below with care with regards to the performance of the 130 Liberty Plaza building.

Both gravity and wind loads were considered for the preliminary design. Gravity loads were taken to be 50 pounds per square foot (psf) (dead load) and 100 psf (live load), based on measured structural properties and the occupancy of the building. Live loads were reduced for the design of columns. Wind pressures on the face of the building were taken to be 30 psf for elevations above 301 feet, 25 psf for elevations between 101 feet and 300 feet, and 20 psf for elevations below 100 feet. A one-third increase in allowable stress was used for the gravity and wind load combinations per the 1970 building code. Limits on maximum lateral drift under wind loads were not considered.

Only the moment-resisting frame along line 8 (see Figure 2) was considered for preliminary analysis and design. The tributary building width for calculating wind loads for the line 8 frame was taken to be 39 feet: one and one-half bay widths. Design actions were first estimated using simple analysis tools, including the portal method. Steel sections were then sized using the *Manual of Steel Construction— Allowable Stress Design* (American Institute of Steel Construction, 1989).

To facilitate three-dimensional finite element analysis of the building, sections sizes along frame lines 6 and 7, including intermediate perpendicular framing, were determined based on information from the September 23, 2001, reconnaissance visit and gravity load considerations. The resulting column and beam sections at tenant level 25 are shown in Figure 2 for one-half of a typical building floor plan.

Linear Elastic Analysis

Two- and three-dimensional finite element models were constructed using SAP2000 (Computers and Structures, Inc., 2000). First, a two-dimensional model (Figure 15a) was prepared that considered structural framing over all 39 stories along line 8 (see Figure 2). This model was constructed to study the response of a single frame with varying degrees of damage (or damage states). Second, a three-dimensional model was prepared that considered structural framing over all 39 stories along lines 6, 7, and 8, including intermediate perpendicular framing (Figure 16). This model was prepared to better understand the response of the building for the observed damage state and to

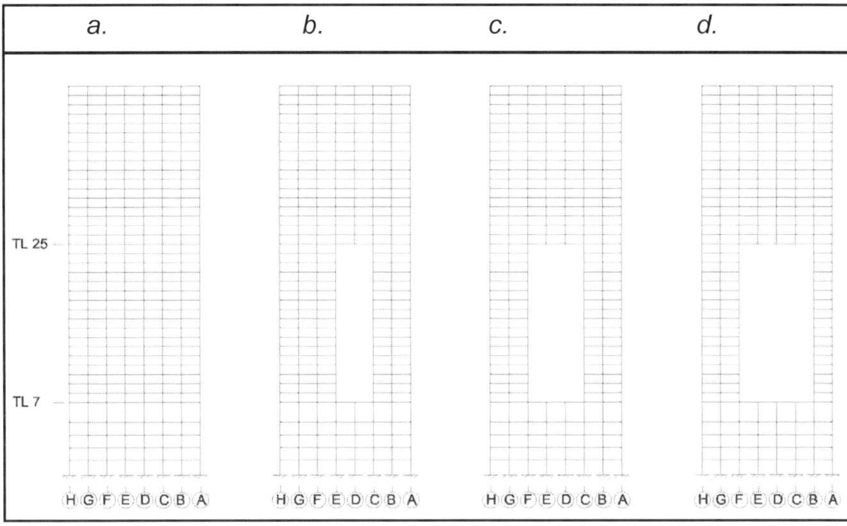

Figure 15. Two-dimensional mathematical models of framing on Line 8: a. ND model; b. DS1 model; c. DS2 model; d. DS3 model.

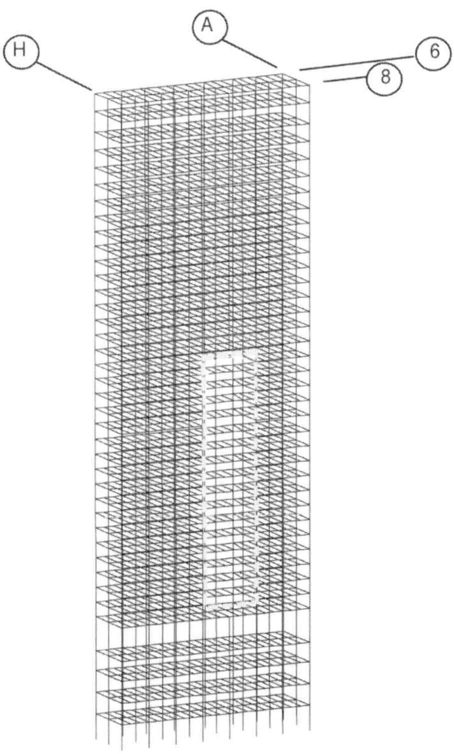

compare the results of two- and three-dimensional analysis. The light dashed line in Figure 16 identifies the zone of observed damage per Figure 15b.

The analyses presented below considered only gravity loads with a dead load and reduced live load of 50 psf. A uniform distributed load of 260 pounds per foot per floor was assumed for the curtain wall loading. Mathematical models were analyzed for the undamaged state, denoted ND, and three damage states, one of which corresponds to the existing damage shown in

Figure 16. Three-dimensional mathematical model of framing on lines 6, 7, and 8 for DS1.

Figure 4a. Each damage state involved the removal of columns on line 8 from tenant level 7 to tenant level 25. The three damage states involved the removal of (1) the column on line D (the observed damage per Figure 15b); (2) the columns on lines D and E; and (3) the columns in lines C, D, and E, denoted DS1, DS2, and DS3, respectively. Maximum member actions under gravity loads were calculated for the undamaged state and the three damage states.

Figures 17 and 18 present some of the results of the two- and three-dimensional analyses, respectively. Shown in these figures are moments normalized by the yield moment for the assumed beam section sizes at tenant level 25—the floor level immediately above the observed damage. The spandrel beam designation (e.g., DE) refers to the grid lines between which the beam spans (e.g., lines D and E). The girder designation (e.g., E87) refers to the grid line along which the girder is aligned (e.g., line E) and the grid lines between which the girder spans (e.g., lines 8 and 7). Such normalized moments represent demand-to-capacity (D/C) ratios for these elements, albeit not exactly, because M_y is used in lieu of ϕM_{nx} and each beam is assumed to be fully braced.

First consider damage state DS1 (the observed damage) and two-dimensional analysis. All D/C ratios are substantially less than 1. The three-dimensional analysis shows similar results. These results provide an explanation for the observed behavior of the framing along line 8 after the impact of debris from tower 2 and the loss of a column on line D, namely, that the moment-resisting framing above tenant level 25 provided an alternate (redundant) path for gravity loads around line D and to the foundation without distress of the structural framing. As observed from DS1, models with increasing levels of damage (i.e., DS2 and DS3) showed that the moment-resisting framing above the damage provided an alternate path for gravity loads.

The mathematical models for damage states DS2 and DS3 were prepared to evaluate the robustness of a building frame with characteristics similar to those of 130 Liberty Plaza, where robustness herein is judged by the ability of the framing system to support gravity loads after the loss of multiple perimeter columns. The results of the two-dimensional analysis of the DS2 model show that the frame on line 8 would have been compromised by the loss of columns on lines D and E unless the moment-resisting connections were ductile (i.e., possessed some degree of inelastic rotation capacity).

Review of the three-dimensional analysis results of Figure 18, however, shows that use of the results of the two-dimensional analysis leads to conservative conclusions and that moment-resisting framing perpendicular to

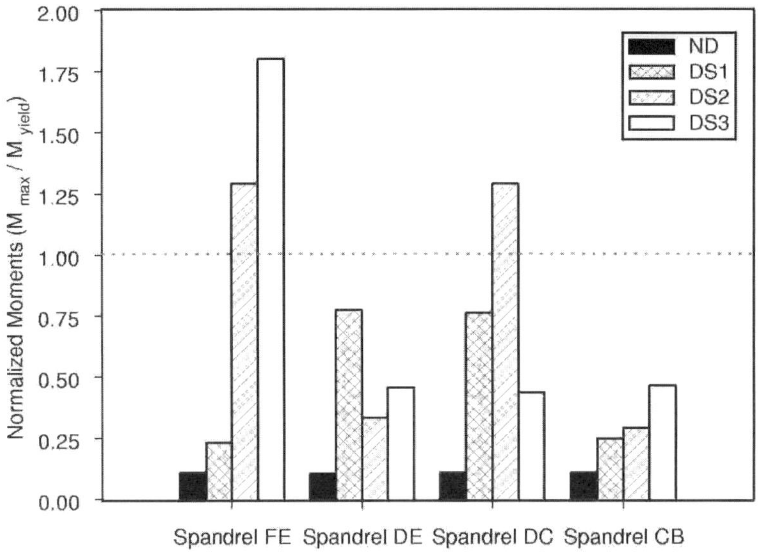

Figure 17. Demand-to-capacity ratios for two-dimensional linear elastic analysis.

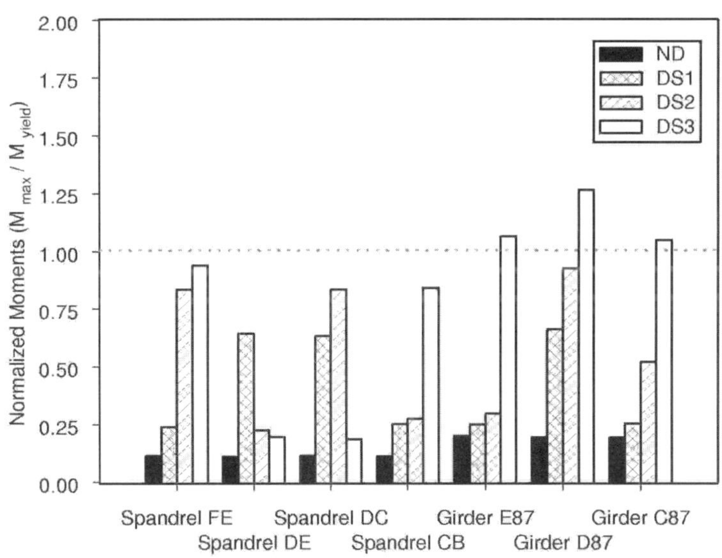

Figure 18. Demand-to-capacity ratios for three-dimensional linear elastic analysis.

line 8 also participated in the redistribution of load around the lost columns on lines D and E. Analyses for both the two- and three-dimensional models of DS3 (the loss of columns on lines C, D, and E) show modest overloads in both cases. For the three-dimensional analysis, the D/C ratios are greatest in the framing perpendicular to line 8, namely, girders C87, D87, and E87.

Demand-to-capacity ratios were calculated for columns on line 8 between tenant levels 24 and 25 (see Figure 15) for both two- and three-dimensional analyses. The column designation (e.g. F8) in Figures 19 and 20 refers to the column between tenant levels 24 and 25 at the intersection of lines F and 8. Shown in Figures 19 and 20 are column D/C ratios. These ratios were calculated using the 1998 edition of the *Manual of Steel Construction—Load and Resistance Factor Design* (American Institute of Steel Construction, 1998) nominal strength equation for members under combined forces (LRFD Eqn. H1-1a), namely,

$$\frac{P_u}{\phi P_n} + \frac{8}{9}\left(\frac{M_{ux}}{\phi M_{nx}} + \frac{M_{uy}}{\phi M_{ny}}\right) \leq 1.0 \tag{1}$$

To facilitate calculation of the D/C ratios and comparison of analysis results, the value of the effective length factor was assumed to be 1.0 for all columns. This value is the largest assuming that side-sway is prevented, which is a reasonable assumption for the moment frame considering the high lateral stiffness of the braced core (significantly greater than that of the moment frame) and the presence of rigid floor diaphragms.

The results of the two-dimensional analyses shown in Figure 19 indicate D/C ratios less than unity for the undamaged state ND and the damage states DS1 and DS2. For damage state DS3, the D/C ratios for columns F8 and B8 exceed unity. Ratios greater than unity can be attributed to an increase in both axial forces and moments due to the removal of columns C8, D8, and E8. This observed increase in bending moment for the two-dimensional analyses ranged from essentially zero for ND to approximately $0.25M_p$ for DS3, where M_p is the plastic moment. Noting that the span between adjacent columns for DS3 is four times the span in the undamaged state ND, an increase in moment demand is expected. The D/C ratios for the three-dimensional model (Figure 20) are less than unity for ND and DS1, DS2, and DS3. Again, the results of the three-dimensional analyses indicate that the use of two-dimensional analysis leads to conservative conclusions and that the presence of perpendicular framing provides additional redundancy and capability for gravity loads to be redistributed to adjacent framing.

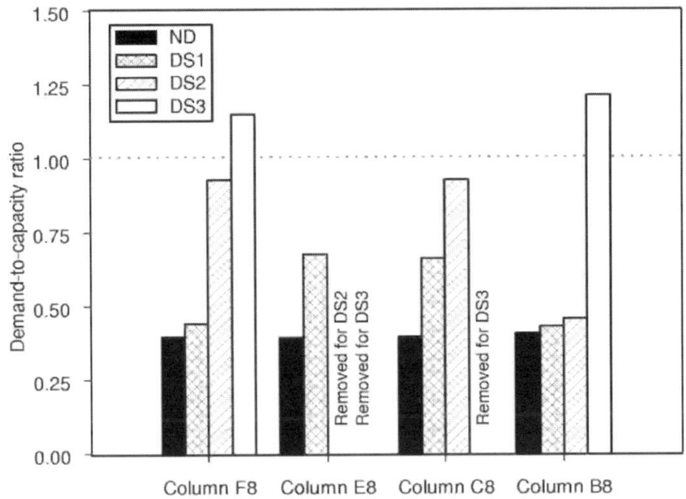

Figure 19. Demand-to-capacity ratios for two-dimensional linear elastic analysis.

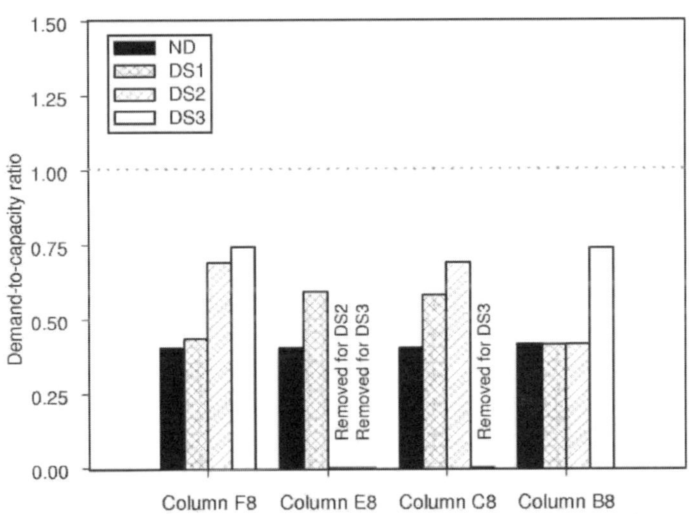

Figure 20. Demand-to-capacity ratios for three-dimensional linear elastic analysis.

Simple Plastic Analysis

Simple two- and three-dimensional plastic analyses were performed to determine an upper bound on the load-carrying capacity of the framing system for each of the three damage states identified previously. In both the two- and three-dimensional analyses, beam plastic moments were calculated assuming a yield stress of 36 ksi (kips per square inch) and the section sizes determined in the preliminary design (see Figure 2). All sections were assumed to be compact. Further, all beam-to-column connections were assumed to have unlimited rotation capacity. For the two-dimensional analyses, vertical panel mechanisms were assumed to form for each damage state. Figure 21 shows the assumed mechanism for DS1. The assumed mechanisms for the three-dimensional analyses were similar, with the addition of hinges forming in the girders of the perpendicular framing where appropriate. Curtain wall loads and beam plastic moments were taken as known quantities, and the maximum corresponding floor load that could be sustained by the framing was determined for each case. Joint mechanisms involving plastic hinges in the column above and below the beams of tenant level 25 and in the beams of adjacent bays were also considered, however, the gravity loads necessary to cause those mechanisms were found to be larger than those required to form the mechanism of Figure 21.

The results of these analyses are presented in Table 1 in terms of both the maximum floor loading (measured in psf) and the maximum floor loading normalized by the likely *maximum* loading at the time of the September 11th attacks, judged by the reconnaissance team to be approximately 100 psf.

These analyses support the results of the elastic analysis described above, namely, that the framing system could have tolerated the loss of two columns without collapse. Note the additional load-carrying capacity that results from consideration of the framing perpendicular to line 8.

However, it must be noted that the cover-plated moment-resisting connections employed in the construction of the building likely have limited plastic rotation capacity as demonstrated by a series of tests conducted after the 1994 Northridge earthquake (Kim et al., 2000). Therefore, the results of the simple plastic analyses must be interpreted with care.

Vertical Nonlinear Static Analysis

Two-dimensional nonlinear static analysis of the framing along line 8 was performed for the three damage states to obtain insight as to the displacements that could be expected at the maximum floor loads predicted by simple plastic analysis. Elastic perfectly plastic moment-rotation relationships were assumed for all beams. Connections were assumed to have infinite rotation capacity. The analyses were run using SAP2000 under displacement control and used the node at frame line D and tenant level 25 as the control node. The loading pattern was a series of equal valued point loads at the locations where the floor beams and perpendicular girders framed into the spandrel elements on line 8.

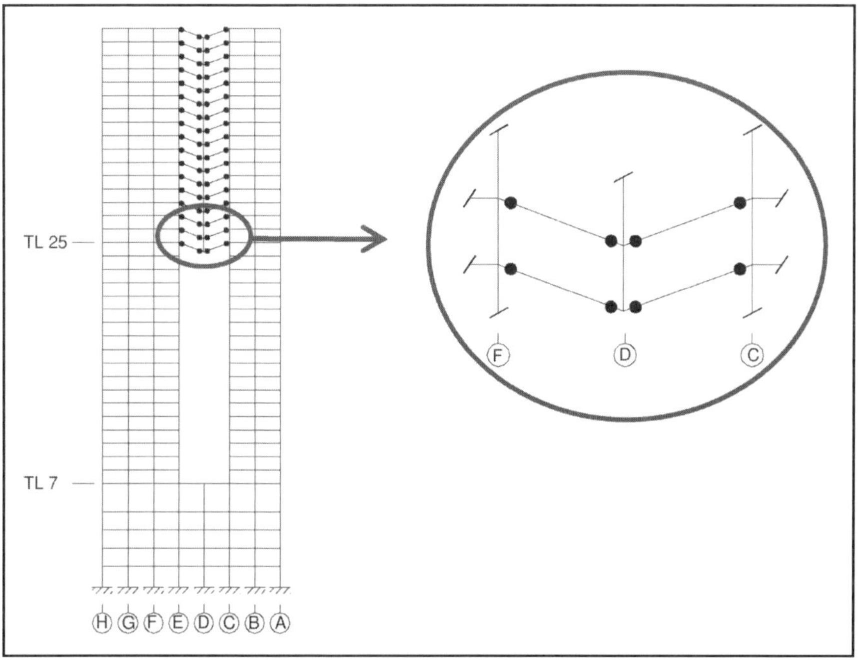

Figure 21. Two-dimensional collapse mechanism for DS1.

Table 1. Plastic analysis results.

	Two-dimensional Analysis		**Three-dimensional Analysis**	
Damage State	Floor Load (psf)	Normalized Load	Floor Load (psf)	Normalized Load
1	293	2.9	308	3.1
2	136	1.4	218	2.2
3	84	0.8	182	1.8

Before running the displacement controlled nonlinear static analysis, the curtain wall load was applied as a single force controlled step to be consistent with simple plastic analysis calculations. Figure 22 shows the observed progression of plastic hinging for DS1. The hinging patterns for DS2 and DS3 were similar to DS1. Hinging started at the beam-to-column connections on lines E and C near tenant level 25, progressed to the connections on frame line D, and then vertically up through the framing on line 8.

Figure 23 shows the resulting relationships between floor load (psf) and the deflection of line D at tenant level 25 for the three damage states, where the floor load is assumed to be present over the entire width of the damaged zone (i.e., between lines C and E for DS1, lines C and F for DS2, and lines B and F for DS3). The initial displacements seen in the figure correspond to the deflection due to the prior application of the curtain wall load. Also shown in this figure is the research team's estimate of the likely maximum floor loading at the time of the September 11th attacks, of 100 psf. As expected, an increase in damage led to increased structural flexibility and reduced the maximum permissible floor load. The results of the nonlinear static analysis and those of the simple plastic analysis are in good agreement for all damage states.

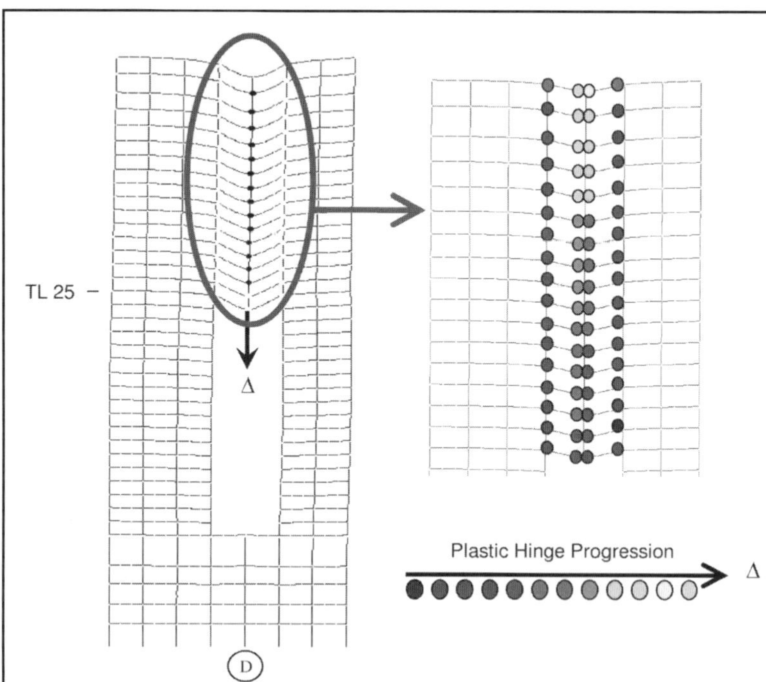

Figure 22. Progression of plastic hinge formation in DS1.

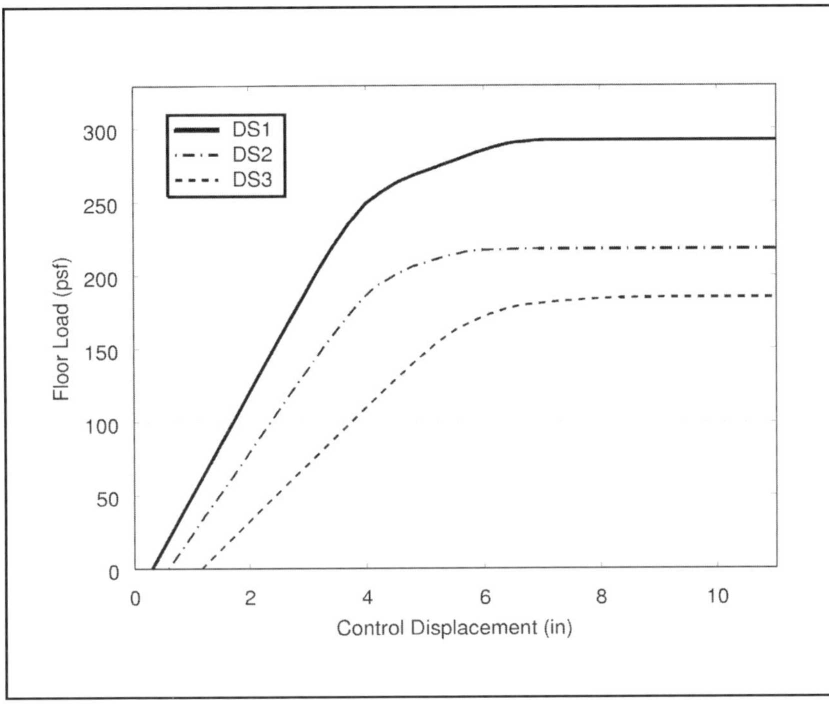

Figure 23. Two-dimensional vertical nonlinear static analysis for DS1, DS2, and DS3.

Summary and Conclusions

The 130 Liberty Street building sustained severe damage from falling debris during the collapse of the World Trade Center tower 2. Reconnaissance efforts on September 21 and 23, 2001, documented the exterior and interior damage to the building. Despite the loss of a perimeter column over a 17-story height, the building did not collapse because the lateral and gravity load resisting systems were highly redundant. The redundant structural systems permitted gravity loads to be redistributed around the badly damaged region, an observation supported by preliminary elastic and plastic analyses of a building frame with characteristics similar to those of the damaged building. Key observations from the work to date are

- Highly redundant gravity and lateral-force-resisting systems are key to the construction of damage-tolerant buildings.

- The use of ductile details (ability to deform well into the inelastic range) will improve the damage tolerance of buildings.

- Simple framing systems such as unreinforced slabs on metal decking can span substantially farther than assumed in design, and such capabilities should be included in the evaluation of buildings for damage tolerance. The addition of inexpensive details (such as continuous slab reinforcement and continuity in the metal decking) could further enhance building performance and prevent partial collapses.

- Simple two- and three-dimensional analysis tools such as those adopted for the work presented in this summary report can be used to judge, in a preliminary sense, the damage tolerance of buildings.

- Two- and three-dimensional analysis tools that are commonly used for the earthquake engineering of buildings showed that the 130 Liberty Street building could sustain the loss of an exterior column over a 17-story height as observed during the post-September 11th reconnaissance work.

References

American Institute of Steel Construction. 1989. *Manual of Steel Construction—Allowable Stress Design*. Chicago, IL: AISC.

American Institute of Steel Construction. 1998. *Manual of Steel Construction —Load and Resistance Factor Design*. Chicago, IL: AISC.

Berman, J., G. Warn, A.S. Whittaker, and M. Bruneau. 2002. *Reconnaissance and Preliminary Assessment of a Damaged Building Near Ground Zero: Engineering and Organizational Issues Related to the World Trade Center Terrorist Attacks*. MCEER-02-SP03. Buffalo, NY: Multidisciplinary Center for Earthquake Engineering Research, State University at New York.

Bruneau, M., A.S. Whittaker, and A.M. Reinhorn. 2002. *Overview of Damage to Buildings Near Ground Zero: Engineering and Organizational Issues Related to the World Trade Center Terrorist Attacks.* MCEER-02-SP02. Buffalo, NY: Multidisciplinary Center for Earthquake Engineering Research, State University at New York.

City of New York. 1970. *Building Code of the City of New York*. New York, NY: Van Nostrand Reinhold Company.

Computers and Structures, Inc. 2000. "SAP2000-Integrated Finite Element Analysis and Design of Structures," Version 7.4. Berkeley, CA: CSI.

Hart, F., W. Henn, and H. Sontag. 1978. *Multistory Buildings in Steel*. English translation, ISBN 0-258-969761. London: Granada Publishing Limited.

Kim, T., A.S. Whittaker, A.S.J. Gilani, V.V. Bertero, and S.M. Takhirov. 2000. "Cover-Plate and Flange-Plate Reinforced Steel Moment Connections." Report No. PEER 2000/07. Berkeley, CA: Pacific Earthquake Engineering Research Center, University of California–Berkeley.

Warn G., J. Berman, A.S. Whittaker, and M. Bruneau. Forthcoming. "Reconnaissance and Preliminary Assessment of a Damaged High-Rise Building Near Ground Zero." *The Structural Design of Tall Buildings.*

Acknowledgments

The authors wish to acknowledge the substantial technical contributions of the expert structural engineering staff at LZA Technology of New York, a division of the Thornton/Tomasetti Group. Special thanks are due to Mr. Edward Sweirtz (Associate Partner, Chicago), Mr. Daniel Cuoco (Managing Principal, New York), and Mr. Emmanuel Velivasakis (Managing Principal, New York).

This work was supported in whole by the Earthquake Engineering Research Centers Program of the National Science Foundation under a supplement to Award Number ECC-9701471 to the Multidisciplinary Center for Earthquake Engineering Research. However, any opinions, conclusions, and recommendations presented in this paper are those of the authors and do not necessarily reflect the views of the sponsors.

Correspondence should be directed to Andrew Whittaker, Associate Professor, Department of Civil, Structural, and Environmental Engineering, University at Buffalo, Amherst, NY 14260, e-mail: awhittak@acsu.buffalo.edu.

Public Infrastructure Service Flexibility for Response and Recovery in the Attacks at the World Trade Center, September 11, 2001

Rae Zimmerman
Institute for Civil Infrastructure Systems
Wagner Graduate School of Public Service
New York University

Introduction

After the terrorist attacks on the World Trade Center in New York City on September 11, 2001, the ability to rapidly restore transportation, power, water, and environmental services to users was absolutely critical, especially to those involved in the immediate search, rescue, and recovery operations. What better way could infrastructure serve its users—both emergency workers and the general public—than to be able to respond quickly in a crisis? The ability to provide these services required a degree of flexibility, often unanticipated and unplanned, that only became apparent as the response efforts unfolded.

The capability of basic infrastructure service providers to respond to public needs for transportation, energy, communication, water, sanitation, and solid waste removal after the September 11th attacks was to a great extent influenced by the flexibility of the initial infrastructure design and management functions to respond to normal system disruptions and to extreme, but not necessarily terrorist-related, events.

The Concept of Flexibility: Definitions and Significance

The ability of physical systems and human services to respond rapidly to user needs is a broad measure of flexibility, and is the focus of this paper. The concept of flexibility as used in this research encompasses both the flexibility of physical configurations and the flexibility of social institutions to bring about changes that enable a system to return to its existing state or to an improved one.

What makes flexibility difficult to define precisely is that it can depend on specific objectives as well as the broad social and environmental contexts in which those objectives play out. Attributes of flexibility as applied to infrastructure services may differ, for example, when the objective is to resist failure or survive shock as opposed to when the objective is for system improvement, such as expansion, in a non-crisis environment.

Flexibility is often related to or associated with other traditional or technical concepts such as resiliency, robustness, redundancy, diversity, adaptability, and interoperability. These are described extensively in the scientific and engineering literature (for a brief summary and references see Zimmerman, 2001). Redundancy is a concept that is often considered synonymous with flexibility, although it is but one component of it. Redundancy has been defined in the context of the physical aspects of communications infrastructure as "having extra capacity available, generally from more than one source" (U.S. General Accounting Office, 2003, p. 92). Diversity, another related term, "involves establishing different physical routes . . . and using different equipment along those routes . . ." (U.S. General Accounting Office, 2003, p. 92).

Flexibility, as used in this paper, is broader than (and encompasses) these other concepts, giving greater emphasis to the socioeconomic, political, organizational, and environmental contexts of infrastructure systems and its explicit interdisciplinary character. In its most general construction, flexible infrastructure supports behavior that does not compromise the goals of the users of the service and includes the ability to change or adapt. An operational definition of the concept of flexibility is provided below in terms of criteria.

Research Objective

The research objective underscores the importance of building in flexibility early in infrastructure decisions. The research hypothesis is:

> If the initial design and operation of an infrastructure service is flexible in dealing with user needs, it will be flexible in reducing the adverse consequences of a crisis.

Methods and Procedures

Zimmerman (2003b) evaluates many specific measures for infrastructure characteristics and behavior with respect to user needs before, during, and after the September 11th disaster at the World Trade Center. These measures are used to test the hypothesis that flexibility supports the performance of infrastructure in crises if it is incorporated as a basic or fundamental characteristic of infrastructure services. Infrastructure as used in this research encompasses transportation, energy, communication, water supply, wastewater treatment, and solid waste management services.

Criteria to Characterize Flexibility

In order to operationalize the concept of flexibility for infrastructure services, several guiding principles or criteria are used to identify and characterize flexibility as it applies to infrastructure services. These criteria provide one basis for designing measures to evaluate how the infrastructure performed on and after September 11th. The criteria below apply primarily to systems facing crisis conditions.

The first criterion is the existence of **alternative routes** to build new capacity or redistribute existing capacity between the sources of production and the intended users (including emergency workers) whose services have been curtailed or otherwise diminished as a consequence of an extreme event. A corollary to this is the existence of **alternative production facilities and locations** as well that permit production to be rerouted or otherwise substituted in an extreme event.

The second criterion is managerial and organizational capability and capacity to **quickly identify, acquire, and manage resources** needed for the safety and security of the population, including knowledge of how those systems operate and the power to control them in a crisis. It includes materials, labor, and supplies for construction as well as to support workers, residents, and business operations. In the management literature these resources are often referred to as "slack resources" (Pfeffer and Salancik, 1978, p. 275), that is, extra resources that can be mobilized when needed. This ability is usually considered an important characteristic of disaster management.

Third is the ability to **transfer information** to users about the state of the system and the alternatives available to them in a way that is simple to use, easy to understand, and can be rapidly disseminated in order to reduce the consequences of a crisis that are specifically related to infrastructure.

Approach

The approach consisted of the following steps to explore the validity of the hypothesis:

- Numerous activities or conditions were identified that involved the use of infrastructure immediately following the attacks and thereafter. The overall research project is incorporating events well beyond this period, however, in order to ascertain the longer-term responses of the infrastructure and to evaluate characteristics other than flexibility.

- Measures were identified or developed for characteristics and behavior of the infrastructure services in conducting those activities, such as the extent to which people used the infrastructure, how long it took to obtain the service, the adequacy of service capacity, and the overall availability of facilities and services.

- The criteria above were then used to characterize the flexibility of the infrastructure services before September 11th and to ascertain the role flexibility played in the provision of infrastructure services during and after the September 11th disaster.

Many information sources were used to identify and analyze events that relied heavily upon infrastructure and how infrastructure was used. These sources included documents and reports, and presentations by public and private owners, operators, and regulators of infrastructure at meetings, conferences, and workshops. The extensive media coverage during and after the September 11th disaster also was a critical resource. The data used span many months following September 11th.

Assumptions

An underlying, operational assumption of the research is that flexibility is a key attribute of and benefit to crisis response capability. It is clear, however, that flexibility does not always lead to desirable outcomes. That is, under some conditions, the same attributes that promote flexibility may also be detrimental. For example, co-location of different systems on the one hand promotes rapid transfers among systems (e.g., shifts from one water line to another) and economies in construction and maintenance, but can also lead to catastrophic failure across all of these systems if they are affected at the same time. The findings of the research are discussed in terms of both the positive and negative aspects of flexibility.

Selected Findings

Numerous success stories during the period immediately following September 11th portray the positive role of flexible infrastructure systems, in particular, in light of flexibility that predated September 11th. Instances are also noted in which systems that were inflexible before September 11th are likely to have contributed to a more adverse impact at the time of the attacks than might have otherwise occurred. Although each infrastructure system is evaluated separately in this section, numerous interrelationships existed among them and the role of these interdependencies in enhancing or inhibiting flexibility on a systemwide scale is a part of the analysis, presented below in the section on Understanding Infrastructure Interdependencies.

Communication Technologies for Infrastructure

Many types of communications technologies are routinely used specifically to support other kinds of infrastructure. The degree of flexibility originally incorporated into the design (including connectivity) and deployment of those technologies influenced the extent to which they could perform extraordinary functions and meet the needs of their users immediately after the disaster. The response of the telecommunications sector occurred at many levels, reflecting the sector's many different technologies and services. The technologies covered here include wireless communications, the internet, conventional phone lines or land lines, and radio, recognizing, of course, that a variety of sub-types of technologies exist within any one of these areas. Only a few of the many types of communications technologies and their flexibility are described below, although the research (Zimmerman, 2003b) undertakes a more extensive analysis of this sector as it relates to the support of other infrastructure.

Wireless Communications

Restoring cell phone communications undoubtedly had a high priority, given the dependence on and prevalence of use of cellular devices for the communication requirements of other forms of infrastructure. This usage is reflected in part by the escalation in the use of cellular devices and the facilities supporting them over the past decade or so (for nationwide trends before September 11th see Zimmerman (2001) and Cellular Telecommunications and Internet Association (2001)). Three measures discussed below for wireless communications reflect how the degree of flexibility built into the original system influenced the extent of impact and the ability to rebound quickly. The measures covered pertain to the planning for and availability of cell towers to support wireless communications, the ability to meet changes in the demand for additional capacity, and the capability to detect and identify users as part of the immediate search and rescue operation.

The first measure is the design of and planning for physical facilities that support wireless communication devices. In spite of the apparent absence of physical connectivity of wireless devices, these devices depend on land-based towers and sites for signals, and the availability of and impact on these facilities affected cellular communications on and after September 11th. In the case of cell sites, the Wireless Emergency Response Team (2001, p. 12) noted that the number of available cell sites in lower Manhattan after the attacks was limited: although only three cell sites were reported destroyed, about 173 sites were impacted (Verizon's initial figures were five sites destroyed and 160 out of service (Condello, 2001)). In the case of cell towers, mobile cell towers (known as cells on wheels) were rapidly brought into the area to restore the lost service, and Verizon, the provider, (Condello, 2001) reported that 55 such facilities were deployed in the World Trade Center area. Verizon was able to identify, acquire, and take control over resources already in place for other purposes throughout the country. Moreover, it acquired the resources to transport the towers into the city.

A second measure is the ability to meet demand for calling capacity. As is well known, the attack dramatically increased cell phone use. The system was not designed to anticipate such a dramatic rise in usage within the city, regionally, or nationally during and immediately after the attacks. One initial response was call blocking. The providers reported that while normally only 4% of calls are blocked, on September 11th, 92% were blocked (Condello, 2001). However, beyond that immediate time period, Verizon, the major cell phone service provider in the area, reported restoring one-quarter of the capacity within hours of the attacks and 100% within a week (Condello, 2001). Alternate facilities using temporary cabling provided the flexibility to restore capacity. However, had the damage been wider in scale, that approach might have been more limited. A consortium of communications providers and city agencies organized by the New York City Department of Information Technology and Telecommunications and the New York City Office of Emergency Management (OEM) under an agreement called the Mutual Aid and Restoration Consortium 2 Agreement was considered to be an important foundation for post-attack responses by the communications industry. The response might have been more time consuming had such an arrangement not already been in existence before the disaster.

The third measure is the ability and availability of technologies to detect wireless facilities in the rubble and to apply this to the search and rescue operation. The Wireless Emergency Response Team, which was set up within hours of the attacks, led this initiative, with 33 organizations participating (Wireless Emergency Response Team, 2001, p. 7), and provided the flexibility for response that drew upon and very much expanded resources and resource networks set up before September 11th. Detection technologies to identify victims by means of the wireless devices they might have had in their possession (assuming the devices were on) were used extensively by Wireless Emergency Response Team sub-teams in the few days following the attacks to locate victims, although the technical obstacles to detecting wireless devices were serious. Various Wireless Emergency Response Team participants used cell registration patterns, radio emissions from wireless devices of persons trapped in the debris, and database look-up services to track potentially trapped victims (Wireless Emergency Response Team, 2001). Wireless firms outside the immediate area were drawn upon to provide services, such as the Telcordia Routing Administration, with offices in New Jersey, Iowa, and Florida, and this geographic decentralization contributed to at least an implicit organizational flexibility in the system overall. These arrangements probably drew upon relationships established over the years. They not only attempted to meet some of the emergency needs of the moment, but according to the Wireless Emergency Response Team report, provided a core institutional resource for future emergencies. Ultimately, in spite of receiving 120 reports of calls from missing persons in the debris, no lives were saved through this massive and unique effort (Wireless Emergency Response Team, 2002, p. 7), although the approaches developed are likely to advance future response capabilities.

The Internet

One analysis of September 11th impacts defines the internet as "a worldwide collection of networks . . . operated by Internet service providers (ISPs) that accommodates a diversity of applications . . . and numerous more specialized functions" (National Research Council, 2002b, p. 1-1). The internet is not only an infrastructure system in its own right for communications and many other related functions, but it also supports and actually controls other infrastructure by being "embedded" within it (National Research Council, 2002a, p. 135). Similar to other forms of communication, internet use is substantial and has been increasing over the years, although it was reported as being stable between 2001 and 2002. Estimates of the percentage of the U.S. population using the internet by 2001 range from about 54% (U.S. Department of Commerce, 2002, p. 1) to about 72.3% and 71.1% in the following 2002 time period (UCLA Center for Communication Policy, 2003, p. 17). The internet is potentially vulnerable not only through its distribution system but also at a few distinct nodes throughout the system. There are in fact numerous points of entry that could provide the basis for disruption through terrorism (National Research Council, 2002a).

The impact of the September 11th attacks on the internet occurred on many levels. First, on a global scale, the overall impact was relatively minor when compared to other breaches and when the New York City role is averaged globally. Second, in numerous, but often isolated local instances (including within the New York area), internet outages produced profound impacts given the level of dependency of users on it and the absence of alternatives. Third, the internet actually provided an important resource for communications when all other forms of communication became less reliable.

Global impacts of the September 11th attacks on the internet were measured in terms of number of users affected and the time period over which the impacts occurred (in particular, how long it took to return to previous levels). One global measure of internet activity is reachability, defined specifically in the context of the September 11th attacks as "the ability to reach (ping) a select number of sites on the internet in the minutes following the collapse of the first tower" (National Research Council, 2002b, p. 2-4). The results of applying this measure of reachability have been interpreted as showing little sustained impact:

- According to a National Research Council report (2002b), a survey based on approximately 105,000 routes indicated that the impact on the reachability of four internet hosts measured before, during, and after the attacks was relatively minimal. Between the hours before the attacks (a benchmark) vs. 10 a.m. on September 11th, the percent reachability declined 1–7% (depending on the host) and by 7 p.m. that evening had rebounded to within 0.5–1% of the benchmark (extrapolated from National Research Council, 2002b, p. 2-14).
- One provider noted that the overall drop in reachability was about 8% over the day of the attacks, and furthermore, "a loss of this magnitude for an extended period of time would generally be considered a serious problem, but its occurrence for a brief period of minutes is less so—and certainly not unprecedented" (National Research Council, 2002b, p. 2-4).
- In the longer term, the reachability of most routes returned to normal "within 15 minutes of the collapse of the South Tower," however 1–2% of the 105,000 routes did not return to normal for 24 hours after September 11th (National Research Council, 2002b, p. 2-4).

Adverse impacts on the internet that were considered more extreme occurred in the form of localized but often profound outages. Not all of these effects were based in and around New York City or even in the United States, and are in part a function of interconnectivity. For example, 74 U.S. and multinational carriers have equipment in New York and 71 countries have direct links to New York City (National Research Council, 2002b, p. 1-1). New York City carrier hotels (defined as "buildings in which carriers lease space in order to link with other carriers located in the same building") are often the only links for transatlantic cables (National Research Council, 2002b, p. 1-2). For example, as the National Research Council report pointed out, domain name system (DNS) servers are key points in the transmission of information, "the DNS server associated with the .za top-level domain was physically located in New York City," and was damaged during the attacks, resulting in much of South Africa being without internet service (2002b, p. 2-6). Similar kinds of specialized linkages resulting in local outages occurred within New York City and the United States as well, disrupting major services. For example, the New York City Department of Environmental

Protection, the major protector of the city's water supply and wastewater treatment systems, did not have internet access for about two weeks after the attacks, since its internet service provider was located in the Verizon building. As a result, they could only use in-house library resources (Lipsky, 2001). Thus, the considerable concentration of connectivity worldwide in New York City and in particular, through the World Trade Center area, reduced the flexibility of those systems to withstand the effects of the attack, although the ability to eventually reroute to other parts of the internet allowed them to recover. The variety of means of accessing the internet, however, is expected to increase with the introduction of "Wifi" or the "802.11 wireless data standard" (Markoff, 2002).

The internet in many cases played a supportive role in rebounding from the attacks. It proved to be a flexible means of communication for some internet users during and after the attacks when other forms of communication were not viable. For example, when some modes of access to the internet were disrupted, other technologies took their place. Wireless internet services provided by, for example, the Blackberry (of the vendor Research in Motion), introduced in the mid-1990s, were an invaluable internet-based communication tool for government officials and rescue workers. Wireless internet traffic was reported to be 60% higher than normal by 10 a.m. on September 11th (National Research Council, 2002b, p. 2-10), which demonstrates the popularity and responsiveness of this technology. IBM distributed several hundred Blackberry units to city and state agencies and the American Red Cross within days of the attack (Woodworth, 2002). The internet enabled the news media to become a major provider of information as the events unfolded. Traffic to CNN.com in terms of page views increased from a typical load of 14 million to 132 million on September 11th and 304 million on September 12 (National Research Council, 2002b, p. 3-5). CNN was able to meet increased demand for news services by redesigning its websites and adding more servers (National Research Council, 2002b).

Conventional Telephone Lines (Land Lines)

An estimated 300,000 telephone lines and 3.6 million high capacity data circuits were inoperable as a result of the September 11th attacks (Guernsey, 2001, p. G6). As in the case of cell phones, the capacity of the conventional phone system was not flexible enough to compensate for the damaged lines and increased demand, and available phone capacity was soon overwhelmed. Many callers wanted to volunteer their services or provide donations. But many people were unable to phone in, for example, to offer their services to support the health system. They ended up using a more basic means of communication—they arrived at the hospitals to convey their messages. Ackermann (2001) reported that at St. Vincent's Hospital, the nearest hospital to the World Trade Center area, "30,000 people were surrounding St.

Vincent's Hospital to help by the evening. When you plan, think about that. People want to be of help. Eight hundred people lined up to donate blood—there was no administration for that. Twelve Metropolitan Transportation Authority buses took them to other sites in New York City where they could donate blood." A similar situation occurred with the Red Cross, which, according to Lowe and Peek (2001), processed 2,000 offers from 15,570 volunteers. The toll-free calling system, upon which some of the other technologies depended—particularly internet communications—was severely limited in its ability to provide service after the attacks because certain providers had their services disrupted (National Research Council, 2002b).

Through the concerted effort of numerous service providers (not limited to land lines), these communication systems were restored through the use of an extensive above-ground network of cables combined with emergency restoration of power prioritized with respect to users. This ultimately enabled the stock exchanges to function in less than a week after the attack. New sites for operations and the ability to acquire equipment to support them enhanced the rate of recovery. For example, "AT&T established a temporary mobile central office by deploying tractor-trailers with necessary equipment to northern New Jersey. AT&T used telecommunications lines in the tunnels to New Jersey to link service in Manhattan to that temporary facility" (U.S. General Accounting Office, 2003, p. 95).

Radio

Radio played a key part in the communications within the fire and police departments. Many accounts not only of September 11th but of other emergencies as well point to the fact that the flexibility of communications between the New York City Fire Department and New York City Police Department were inhibited by profound cultural differences between the two (Dwyer et al., 2002). In some cases, the technology reflected these differences in that radios operated at different frequencies that made communication between the two departments impossible, except at a relatively high level in the organizations. This has been extensively analyzed in terms of both the technological and managerial dimensions of the problem, including how flexibility can be incorporated into the redesign and use of those communications systems.

Electric Power

Electric power became perhaps the critical link to the viability of infrastructure during and after the attacks, and infrastructure users were not always aware of this. For example, for communications infrastructure, electric energy was needed to run cooling facilities or enable pulses of light to be sent or received over fiber optic cable (National Research Council, 2002b). Major intermediate production and distribution facilities for electricity were damaged. Batteries and generators were used as backup power, but this was often limited for a number of reasons. Batteries failed or eventually expired. Temporary or backup generators were difficult to start, had environmental effects related to the use of diesel fuel (Zimmerman, 2003a), or ran out of fuel and getting fuel to them was difficult given the restricted access to lower Manhattan. While production, transmission, and distribution systems for electric power in lower Manhattan were inflexible in being resistant to the immediate impacts of the attacks, they proved to be flexible thereafter, during recovery or restoration. In the days and weeks after September 11th, electric power was restored at least on a temporary basis by rerouting power from other areas, often with temporary cables, backup facilities, and using mobile facilities such as generators.

Production and Transmission

Substation transformers are a key component of the electrical system. In lower Manhattan, substation transformers step down voltage from 138,000 volts for easy consumption, and 14 operating transformers and six spare transformers out of a total of 355 in New York City and Westchester were located in lower Manhattan at two locations: 7 World Trade Center and the South Street Seaport (New York State Urban Development Corporation, 2002, p. 12-10). Half of that capacity was destroyed in 7 World Trade Center (New York State Urban Development Corporation, 2002, p. 1-2). In spite of the initial vulnerability due to the high concentration of substations in the World Trade Center area, the response reflected a considerable amount of flexibility in terms of the ability to identify, obtain, and deploy additional resources. In the immediate short term, mobile generators were moved in similar to the way in which mobile cell towers were moved in to provide a short-term solution for wireless communications. A vendor supplied over 100 of these generators. Since the vendor routinely supplies such portable generators for emergency power, the units were on hand. This reflected one key criterion of flexibility—an organizational and managerial structure with the ability to identify, acquire, and deploy resources quickly at least to provide short-term needs. As portrayed in the environmental assessment for the rebuilding of 7 World Trade Center, over a longer period of time (to the summer of 2002), the spare transformer vaults from the South Street Seaport site were put into operation and power from other locations in Manhattan enabled some of the lower Manhattan load to be serviced but used up much of its local spare capacity. In the long term, a permanent solution that was put forth is the rebuilding of the two substations that were lost, along with the reconstruction of 7 World Trade Center. This is expected to restore pre-September 11th levels as well as provide for a 1% increase in power (New York State Urban Development Corporation, 2002, p. 1-2, 1-3). Thus, the concentration of transformers in the substations in 7 World Trade Center led to a wider initial impact than might have otherwise occurred. However, the recovery was hastened by the existence of spare power at other locations and in the longer term, the locating of a new building where 7 World Trade Center had been has provided a site for new power and backup capability for lower Manhattan. Institutions adapted to the situation as well. The city's land use and environmental review systems expedited the reviews required to speed up construction of the new building at the 7 World Trade Center site to house the new substations (New York State Urban Development Corporation, 2002).

Distribution

"The Consolidated Edison Company, which provides New York City's electric power, has approximately 102,000 km of primary and secondary electric distribution cables over 1700 km^2, making it the most concentrated in the world" (O'Rourke, 2001). Moreover, the energy density at the World Trade Center site was also extraordinarily high. To compensate for the destroyed distribution lines, Con Edison was able to provide an estimated 36 miles of overland distribution lines as a temporary measure in order to connect the area to alternative sources of energy, and the last temporary cable was removed on May 23, 2002 (Armistead, 2002). According to Con Edison, 98% of the 13,000 customers without service after September 11th had their service restored by means of the temporary cables (Armistead, 2002). The ability to draw power from other areas of their system through temporary distribution lines also contributed to the ultimate restoration of power. Thus, while the considerable density of lines in the area increased the damage, the proximity and similar density of the street system enabled temporary overland cabling to restore some of the lost capacity in the short term.

Water

Two water supply functions were critical immediately after the September 11th attacks. One was ensuring that the supply was adequate for the users whose services were disrupted, particularly for emergency responders for fire-fighting. The other was preventing water from creating floods, and potentially massive ones, if the bathtub was breached in a serious way.

Water Usage

As a result of the attack at the World Trade Center site, about eight to ten water mains and related facilities were ruptured. The broken water lines ranged in size from 16-inch to 24-inch mains (Corley, 2002). New York City consumes about 1.4 billion gallons of water a day. Although the amount lost was a small fraction of this daily total, it was concentrated in a very dense area of the city, constituting a substantial part, if not all, of the water supply of that area. Apparently, neither the quality nor quantity of water supplies throughout the rest of the city was affected. Water was restored later that evening since the city had the flexibility to tap alternative water distribution lines (Figure 1).

In general, dense matrices of pipes are more flexible in securing a water system from the effects of breakages than a branched or linear pattern that has fewer interconnections. Zimmerman (1999) compared the New York City and northeastern New Jersey systems in this regard, noting that when water main breakages occur in New York City, which has a matrix system, outages are usually shorter in duration than in northeastern New Jersey where branched, linear systems provide few alternatives other than bringing in bottled water while a single line is repaired.

The first objective for the provision of water immediately after the attacks was to restore water for fire-fighting at the World Trade Center site. The fires ultimately proved to be inaccessible in the short term to most of the traditional fire-fighting capability. However, the flexibility of the system to bring water into the area in alternative ways was reflected in a range of efforts. These included volunteer bucket brigades, distribution systems for fire-fighting operated by the New York City Department of Environmental Protection through its conventional water system, and the New York City Fire Department fire boat system, which can be connected to either the city's system or the Hudson River. In order to restore water in the city's system for firefighting, the city used a guideline of 40 pounds per square inch to restore pressure after the breakages, and eventually attained that level from a starting point of practically zero due to the breakages. This pressure was restored the day of the attacks due to the city's ability to identify, isolate, and seal off leakage through a systematic series of valve shutoffs and the flexibility to draw water from alternative lines to compensate for the loss of other lines. However, the city's fresh water system was neither of sufficient quantity nor close enough to the fires to provide enough water. Another water source, provided by the fire boat system run by the New York City Fire Department pumping water from the Hudson River, was drawn in to put out the fires. Ultimately, the location of some of the fires and their intensity prevented them from being extinguished for many months.

A second objective for the provision of water met by both the New York City Department of Environmental Protection and the New York City Fire Department pertained to water for dust suppression and for construction needs in general.

A third objective was the provision of water to area residents once rehabitation of the area occurred. Here, quality of the supplies as well as quantity was of critical importance. When water supply lines are breached, an influx of contaminants into the pipes is potentially possible. The New York

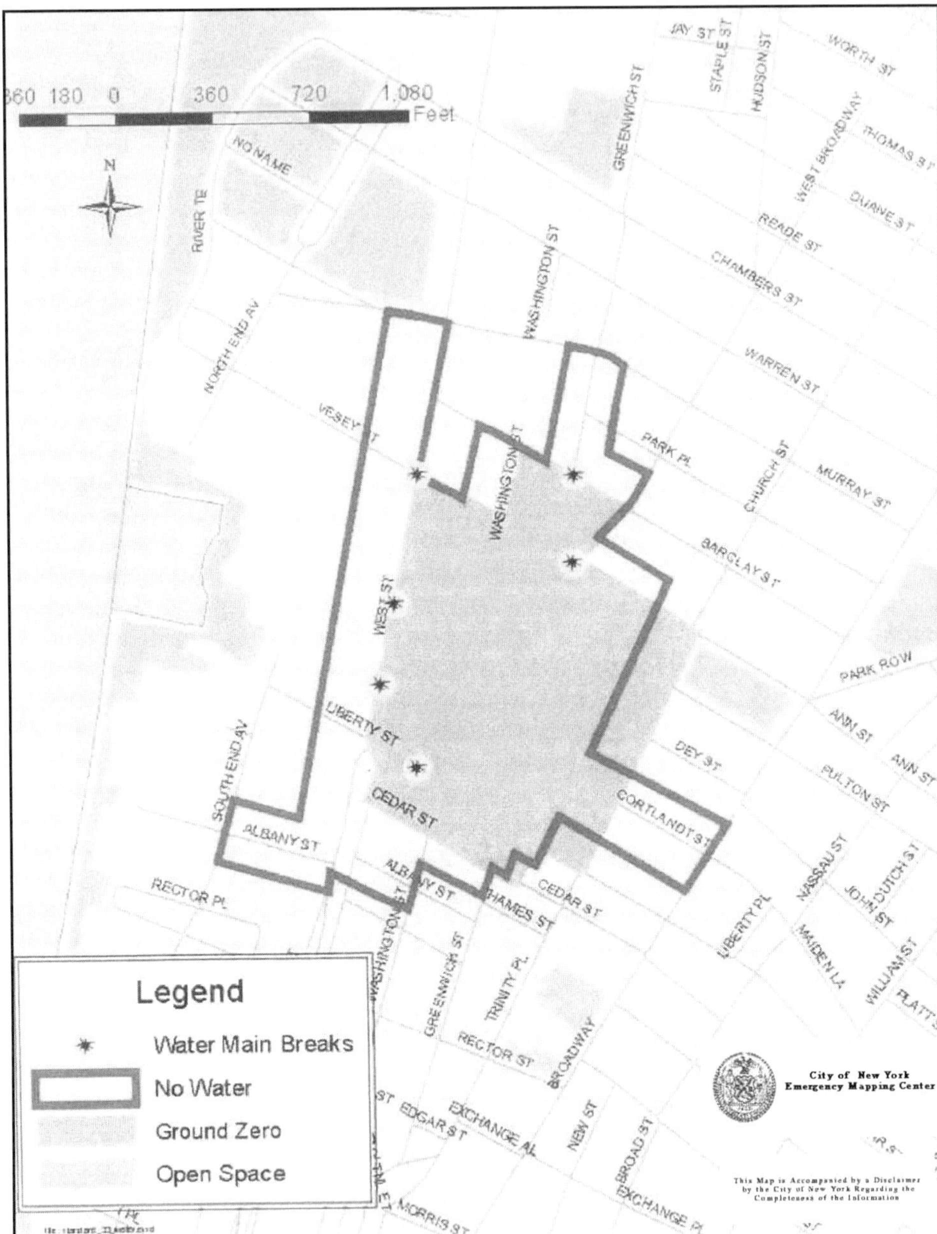

Figure 1. Water outages as of 3:00 p.m., October 7, 2001.
[City of New York, Emergency Mapping Center, Office of Emergency Management]

City Department of Environmental Protection tested the purity of the supplies, and water quality at the time of rehabitation was considered to be meeting water quality standards.

Flood Prevention

A second water supply issue was preventing flooding. One critical flooding problem occurred at the Verizon building at 140 West Street, which was flooded from one of the water main breaks. Another water problem, not directly related to New York City Department of Environmental Protection functions, was the rampant flooding of the Port Authority Trans-Hudson (PATH) tubes from a number of sources. "Flooding of the PATH tunnels was caused by a combination of factors, including rupture of two 1675-mm-diameter pipelines, which circulated cooling water from the Hudson at the World Trade Center Complex, as well as broken water mains, water from fire hoses, and leakage from breaches in the World Trade Center Complex basement walls. Valves were closed to shut off water from the cooling pipelines within a few hours of attack. The northern PATH tunnel was fully flooded at its lowest elevation, with water at station platform level in Exchange Place Station in New Jersey. Water was pumped from the tunnel at a maximum rate of approximately 11,350 liters per minute, and a concrete bulkhead was established at Exchange Place Station to prevent further water ingress on the New Jersey side" (Anonymous, 2002). One of the criteria for flexibility—that of being able to rapidly assemble resources—was met in the sense that the Port Authority had access to the Metropolitan Transportation Authority's pump with its 6,000-gallon-per-minute capacity for its immediate need to prevent flooding, though the more long-term solution was the installation of concrete plugs (Langewiesche, 2002, pp. 70–71). The Metropolitan Transportation Authority maintains the pump for flooding problems from rainstorms, high water table, water main breaks, and other sources of water into its system.

Transportation

New York City is renowned for the variety of ways to travel, and this became a key resource to attain flexibility in the movement of people and goods during and after the attacks. A number of measures reflect the dramatic adaptability of the transportation system. These measures included those that were user-oriented, such as numbers of passengers and trips. Others pertained to the physical system and its ability to expand to meet the travel demands of people by enabling passengers to shift among different modes of transport.

Waterborne Transportation—Ferries

Ferry service provided the flexibility to move people rapidly away from the site on September 11th, and provided ongoing transportation in the months that followed. Ferry service was critical to the movement of New Jersey residents to New York City and back after September 11th. For example, through the end of November, total ferry ridership increased 27.8% with variations of this average for public and private facilities (calculated from Lee, 2001).

According to some people, long-term reliance on ferry service will have to address questions about differences in the safety standards that ferries have to meet relative to other forms of transportation and the effect these differences may have on passenger safety. Such standards cover the areas of fire protection, preventive maintenance, and backup communications systems (Wald, 2002) and the loading and distribution of passengers relative to capacity. Also, as one would expect, uncertain weather conditions will require that alternative means of travel be available.

Transit—Passenger Transportation

New York City and the region are served by an extensive system of long-distance and commuter rail lines and local transit services. These systems, from long-distance rail to local trains and buses adapted their services dramatically in the aftermath of September 11th to accommodate customer needs after an initial system shutdown. It was the initial, pre-September 11th flexibility in the design and management of those systems to allow a considerable amount of rerouting that contributed to such a rapid post-September 11th response.

Long-distance rail service—Long-distance rail service returned during the evening of September 11th after a general shutdown of all service in and out of Manhattan earlier that day. Amtrak, the main long-distance provider, created a critical alternative to air travel when airlines were shut down for many weeks after the attacks. Key long-term issues will be the relative roles of air travel and Amtrak in providing northeast corridor transportation services and the ongoing concerns about the effect that uncertainties in federal subsidies will have on Amtrak's performance. The condition of Amtrak's facilities will undoubtedly play a role in any systematic relative risk analysis of Amtrak vs. other modes of long distance travel. In particular, questions have been raised about the condition of the tunnels through which Amtrak travels near its main terminals, namely New York City's Pennsylvania Station. The federal government announced $78 million to improve the tunnels for fiscal year 2003, which will undoubtedly reduce the uncertainties in this area.

Local transit and commuter rail service—The Port Authority TransHudson (PATH) system, New Jersey Transit, and the Metropolitan

Transportation Authority and its subsidiaries (Metro North and the Long Island Railroad) provide these services. They responded in many ways immediately after the September 11th attacks.

First, the ability to reroute trains quickly while the attacks were occurring was a clear success story. As a result, service was restored that evening. The ability of the Metropolitan Transportation Authority to reroute subway trains around or away from the damaged site, although producing some delays relative to normal service, was a testimony to the inherent flexibility of a multi-route transit system that has existed for close to a century. The ability to reroute trains is built into its multiple track system used for normal operations and anticipation of routine service outages, and this enabled three trains to be converted into shuttle services in the months following September 11th. The ability to rapidly reroute trains as well as the fact that restrictions on surface transportation led many to use transit systems shows up in the pattern of travel. Month-to-month averages on the larger transit systems show a clear pattern of recovery at least in the short term (Table 1).

Table 1. Change in average weekday transit ridership, 2001.

	August—September	September—October
MTA (Subways)	-6.4%	+11.9%
PATH	-23.8%	+10.1%
NJ Transit	-4.6%	+6.7%

[Calculated from Metropolitan Transportation Authority information.]

Second, however, station capacity proved to be inflexible to the changes in ridership. For example, in the PATH system when the northern routes compensated for the loss of the World Trade Center station, station capacity at those northern locations was soon overwhelmed beyond the point of being able to provide comfortable, and to many, safe and adequate service. Ridership at the Ninth Street station increased from 4,100 to 8,900 and at the Christopher Street station increased from 3,600 to 7,400 (Dunlap, 2002, p. B3) in a matter of days. Expansion of station capacity does not come easily, since increasing the number or relocation of stairways does not typically meet with community approval (Dunlap, 2002, p. B3).

Local bus service—Although bus traffic was initially sharply curtailed in the area to reduce street congestion, emergency bus routes were eventually provided to enable emergency workers, other workers, and eventually residents to access the site.

Surface Transportation—Roads, Bridges, and Tunnels

Massive traffic congestion along surface transportation links needed for rapid access to the site by emergency responders typically occurs in disasters, and unlimited roadway access creates the added problem of security. To avoid this, the city, state, and regional authorities adopted two strategies. The first was to shut down roadways in and out of the city as an initial response to the catastrophe. In the months following the attacks, the New York City Department of Transportation had to invoke other roadway shutdowns for a variety of reasons including the repaving of some 37 miles of roadway encompassing about 600 streets (Haberman, 2002, p. 14), relocating utilities, and security. Although the transit system had similar shutdowns, these were lifted much sooner than the road system shutdowns. While the immediate roadway shutdowns provided flexibility for movement of priority vehicles, it obviously limited the full utilization of resources that normally flow in and out of the city, producing an initial impact on the transport of supplies to residential and commercial customers. The second strategy was traffic management, for example, invoking single occupancy vehicle (SOV) lane restrictions.

Thus, many institutional mechanisms for transportation supported flexibility in routing of people immediately after the attacks and for many months afterwards.

Debris Removal

Debris removal was a critical part of the rescue operation as was the recovery of human remains. Much of the emergency site restoration and the search for human remains there could not proceed without debris removal. Over the course of the eight months since the initial attack, debris removal was estimated at 1.56 million tons—equivalent to the total solid waste generated in the city over a month or the municipal solid wastes generated over four months. The ability to undertake such an additional burden was astounding. Of all of the infrastructure activities that took place, debris removal was perhaps the best example of the flexibility of a very diverse set of resources to come together in a crisis to solve the immediate needs of the public.

By May 2002, debris removal was completed more quickly and less expensively than originally expected. In spite of the many demands placed on the process by the need to maintain the structural stability of the area and to allow the maximum opportunity to uncover bodies, debris removal proceeded at a relatively constant pace, with holiday periods being the only exception. The delicate connections among the many different players such as equipment suppliers, construction contractors, and regulators of waste removal to provide rapid debris removal reflected an extraordinary degree of flexibility given the complexity and uncertainties in the debris removal process.

First, the flexibility to obtain large earth-moving equipment from some of the largest construction companies in the country for site excavation invoked a very large network of construction firms that could be tapped quickly and continuously due to a long-standing tradition of heavy construction activity in New York City. At the time the September 11th attacks occurred, some of the largest construction projects in the country were in progress in New York City, such as the Third Water Tunnel. The availability of the construction industry as a resource is reflected in a very dramatic and symbolic act that took place when news of the World Trade Center attacks spread among the workers: many construction workers left their work sites and proceeded immediately to the World Trade Center. Second, the proximity of the World Trade Center site to a waterfront and a tradition of using barges for waste transfer afforded the opportunity to accommodate large barges for debris shipments. The community adjacent to that waterfront was willing to accept the inconveniences at least in the short term. Third (and related to the second condition), was the flexibility afforded by expedited permitting to excavate nearby waterfront areas to allow the rapid removal of material by barge. Fourth was the city's ability to reopen a portion of the Fresh Kills landfill for debris disposal, even though Fresh Kills had received its last barge shipment in March 2001 and its closing had been awaited and negotiated for years. In addition, a steel-recycling network was tapped to receive the steel. This met with considerable opposition from engineers trying to salvage some of the steel for research, for example, to understand its behavior under the very high temperatures to which it was exposed at the World Trade Center site (Astaneh-Asl, 2001). Fifth, JFK airport was adapted to provide storage capacity for some of the larger, intact pieces of debris such as PATH trains that were being salvaged.

Thus, the process of debris removal reflected an ability to tap resources along the entire chain of debris removal activities.

Policy Recommendations

Flexibility as a Measure of Infrastructure Performance in Times of Crisis

Measures of flexibility reflect user and community expectations for infrastructure services that are important inputs into future design and operation for enhanced resiliency and security. Three indicators of or criteria used in this research for flexibility that emerged from the responses to September 11th provide important guidance for policy, and these criteria provide a simple way of organizing and understanding the subtleties of post-disaster events. A summary of the application of these criteria to September 11th infrastructure responses based on the results of the research to date is given below.

Alternative Routes and Alternative Production Locations

In a number of cases, interdependencies among service providers and their support services limited flexibility and led to disruptions that might have otherwise been avoided, yet measures of infrastructure service response after the disaster indicated that public services were able to draw upon alternative means to meet public needs. Mobility in production and distribution units proved to be a key factor in the ability to rebound quickly, as provided, for example, by mobile generators, cell towers, and flexible distribution lines for the temporary restoration of water, energy, and communication. Immediately after the disaster, for example, distribution networks were, in many cases, put in place to provide and connect users to emergency services for power, water, and communication, because service providers could move equipment from one place to another and provide connections quickly. Temporary cables were rapidly deployed to provide these services on a short-term basis, often outside of the city, for example, into New Jersey. Public transportation systems showed extraordinary ability to redirect traffic, and enabled the city's population to keep moving. Production or critical transmission points are also critical to avoiding vulnerability and supporting distribution networks. Greater flexibility in these systems would have enhanced their role in the recovery effort, for example, having less spatially and functionally concentrated electric power substations and telecommunication domain name servers.

Identification, Acquisition, and Management of Resources

Both government and industry utilities were able to quickly identify, acquire, and manage resources for emergency response in the short term. In some cases, the resources were already available in the form of spare parts at nearby locations (e.g., the Con Edison substation transformers). In other cases, the resources had to be brought from very long distances, but the knowledge of the location of those resources and the ability to obtain them was already in place. Utilities were able to obtain temporary generators and cell phone towers (deploying 55 "cells on wheels"), miles of cable and pipe to construct overland utility networks for land lines, wireless and internet services, construction equipment for debris removal, and mobile temporary offices. Most importantly, they were able to tap the human resources needed to put these facilities in operation.

The ability to draw upon organizational resources was critical to obtaining the physical supplies and facilities and identifying those in need. What became clear was that successful organizational arrangements, both informal and formal, were those that had existed before the disaster or at least where linkages among the individuals had existed that could be quickly mobilized. Examples were numerous. In the telecommunications industry, the Mutual Aid and Restoration Consortium agreement among private vendors and city agencies, the Wireless Emergency Response Team, and the financial group of about a half-dozen vendors all had existed or had informal liaisons before September 11th. The GIS mapping unit, which proved valuable to the rescue and recovery effort, was formed from informal professional groups that had professional associations before September 11th. Likewise, the debris removal operations drew resources from the construction industry, whose companies were already very familiar with one another through professional networks.

Information Transfer

Transferability of information for the purpose of reducing human exposure to the attacks, and hence its consequences, was extremely limited by the capacity and compatibility of communication networks in spite of the fact that facilities were made available to support expanded capacity. Information transfer is a vital dimension of emergency services, and more attention is needed in this area.

Understanding Infrastructure Interdependencies as a Key to Infrastructure Flexibility

Flexibility intended to increase the effectiveness of infrastructure performance can backfire, having the opposite outcome, if not approached from a systemwide perspective that takes into account interdependencies within and among infrastructure systems and between infrastructure and social systems.

Infrastructure systems are interconnected both spatially and functionally, and the degree of interconnectedness tends to increase with population density. Interconnectedness has both advantages and disadvantages. The advantages in terms of economies of scale are well known. The disadvantages are that interconnectedness can generate cascading failures not only among infrastructure systems (O'Rourke, 1993) but also between infrastructure and social systems—the failure of one system creates failures in others.

Many of the infrastructure systems described individually above were highly interrelated, and flexibility has to be evaluated systemwide to capture the impact of such interconnectivity. Numerous telecommunications providers, for example, had facilities that were connected both functionally and physically through the World Trade Center site that were disrupted along with facilities and services provided at that location.

Moreover, communications and other infrastructure depended on electric power provided by Con Edison facilities, which was also concentrated in that same location and forced many of the systems to seek alternative routes and draw on alternative resources in order to recover. In the case of electric power, Guernsey (2001, p. G6) points out that "some of those multiple lines travel the same conduits to the same routing centers" and the conduits or routing centers were not redundant enough to withstand the damage. "Redundancy in some instances may only be apparent or limited to a given part of the system only to have a critical link moved to a different point in the system" (Guernsey, 2001, p. G6).

Examples where water systems were connected to other infrastructure and caused failures across different kinds of infrastructure as a result of the September 11th attacks also illustrate the downside of these interconnections. On September 11th, water damaged telecommunication infrastructure at 140 West Street, train lines were flooded by water from various sources, and flooding of electric power systems destroyed telecommunications and computer data systems that were connected to those power systems (Zimmerman, 2001; Guernsey, 2001).

Looking Backwards to Look Forward

One of the reasons for many of the successes in the short term was that many responders used their experiences from previous situations that appeared to at least have similar elements to what happened on September 11th. These past experiences were both positive and negative. Examples of the kinds of previous experiences that may have and certainly could have contributed to the knowledge and experience supporting the responses to and recovery from the September 11th attacks are listed below.

- New York City has 500–600 water main breaks a year, a few involving pipes and volumes of water lost greater than the amount of water lost on September 11th. However negative these experiences seem, they at least provided the city with a protocol for responding to water main breaks that was used on September 11th and the days following.
- Con Edison has had major outages from underground fires and other origins that often have greater consequences than what was experienced on September 11th. Response capabilities from these experiences in part provided Con Edison with the ability to at least respond in the short term.
- The transit system's experience with train failures, including water and electric failures that contribute to train failures, has built considerable redundancy into the system over decades, which paid off in being able to enable the transit system to rapidly rebound in the short term.

- When new information and communication technologies fail, operators typically respond with older, more familiar technologies. The New York City Fire Department did this with respect to radios. Water managers routinely do this when information technology controls fail, and they have to use manual overrides to operate valves and treatment plant controls.

Integrated Decision Tools

Finally, in making the transition from past behavior to action during emergencies and from emergency action to long-term solutions, many tradeoffs have to be made in operationalizing the concept of flexibility and ensuring that it is applied system-wide. Many examples of these tradeoffs were alluded to above. In order to accomplish these tradeoffs, traditional methods for decision-making need to be expanded. For example, the analytical capabilities of various existing techniques need to be integrated. These individual techniques include risk assessment (Haimes, 1998; National Research Council, 1994), decision analysis (Clemen and Reilly, 2001), environmental impact assessment (Gregory et al., 1992), life cycle engineering (O'Rourke, 1993), risk management (Stern and Fineberg, 1996), and the social psychological approaches to risk perception and the application of all of these techniques to the behavior of infrastructure services and their users in extreme events (Slovic, 2000; Bier et al., 1999; Mileti, 1999). First, these individual techniques gain more strength if they are integrated and act in concert with one another. Second, more applications to infrastructure are needed. Some applications to infrastructure exist, for example, in the area of risk assessment of engineered structures (Haimes et al., 1998; Lave and Balvanyos, 1998; Zimmerman and Bier, 2002). Third, and most importantly, these capabilities need to be designed carefully in a way that can be communicated appropriately and be useful to front-line decision-makers.

Longer-term Issues: Security as a High Priority

Security has added a whole new dimension to the provision of infrastructure services, yet it is one that can be made compatible with existing infrastructure. Infrastructure is essential to moving emergency personnel, food, shelter, sanitary facilities, and medical services into a devastated area and moving victims out in the short term. In the long term it is critical to moving debris and other waste out in an environmentally sound manner and getting equipment in for restoration and site. Thus, the need to protect and secure these infrastructure resources is obviously critical.

Estimates after September 11th are in the billions of dollars for securing water and transportation infrastructure alone, and leveraging through flexibility can potentially provide savings. In the area of water supply, the following Congressional testimonies reflect the magnitude of some of these needs:

- $4.7 billion in fiscal year 2003 alone for vulnerability assessments and security systems for the nation's water supply and wastewater treatment systems (Water Environment Federation, 2002, p. 1).
- $267 million in the first year plus $65 million annually will be needed by the Corps of Engineers for security upgrades (Ichniowski, 2001).
- $5 billion for infrastructure work and $100 million for vulnerability assessments will be needed in total for the city of Boston alone, according to the Boston Water and Sewer Authority (Ichniowski, 2001).

A key policy issue is that in light of the very large investments necessary for security, how security can be designed in a way that it meets multiple objectives rather than one. That is, whether or not security needs can be met that also support improved condition and resistance to the adverse effects of natural hazards. Moreover, the use of information technologies is increasing to manage infrastructure and also to support security. Detectors or sensors, for example, will become more common in managing and protecting infrastructure. Accompanying this technology are uncertainties in interpreting the data and matching the data to what we want to know. A final issue is the reliability in the coverage of such devices relative to our expectations.

Initial Observations about Flexibility and Infrastructure Policies in Times of Crisis

The concept of flexibility provides a valuable framework for evaluating and explaining the response and recovery of vital infrastructure services during and after the attacks on the World Trade Center. This concept applies to both physical and social systems involved in meeting the needs of users of infrastructure, and the criteria that define it involve the ability to obtain alternative production systems and distribution routes; the ability to rapidly identify, acquire, and manage resources and transfer information; and the ability to promote the networks to sustain these capabilities. Key policies to improve the adaptability of infrastructure services under crisis conditions would benefit from explicit use of measures of flexibility to enable agencies to manage their services in a crisis and, in particular, to gain access to resources and support connectivity with other functions in order to restore services.

References

Ackermann, M. 2001. Presentation at "Learning from Urban Disasters: National Science Foundation Response and Opportunities for Future Research Workshop," sponsored by the Institute for Civil Infrastructure Systems, New York University, December 12–13.

Anonymous. 2002. Personal communication from an anonymous reviewer(s) of an earlier draft of this paper, December.

Armistead, T. 2002. "Last Emergency Cable Removed in WTC Area." http://www.enr.com.

Astaneh-Asl. 2001. Presentation at "Learning from Urban Disasters: National Science Foundation Response and Opportunities for Future Research Workshop," sponsored by Institute for Civil Infrastructure Systems, New York University, December 12–13.

Bier, V.M., Y.Y. Haimes, J.H. Lambert, N.C. Matalas, and R. Zimmerman. 1999. "Assessing and Managing Risks of Extremes." *Risk Analysis* 19(1): 83–94.

Cellular Telecommunications and Internet Association. 2001. "Semi-Annual Wireless Industry Survey." Washington, D.C.: Cellular Telecommunications and Internet Association.

Condello, K. 2001. "Wireless Industry: Impact and Recovery Efforts Summary Report." Network Reliability & Interoperability Council. Presentation made to the Cellular Telecommunications and Internet Association, Washington, D.C., October 30.

Clemen, R.T., and T. Reilly. 2001. *Making Hard Decisions with Decision Tools*. Pacific Grove, CA: Duxbury.

Corley, G. 2002. *World Trade Center Building Performance Study: Data, Collection, Preliminary Observations, and Recommendations*. New York: Federal Emergency Management Agency Region II.

Dunlap, D.W. 2002. "PATH Project Unloved in the Village." *New York Times* (June 20): B3.

Dwyer, J., K. Flynn, and F. Fessenden. 2002. "9/11 Exposed Deadly Flaws in Rescue Plan." *New York Times* (July 7): A1, A12.

Gregory, R., R.L. Keeney, and D. von Winterfeldt. 1992. "Adapting the Environmental Impact Statement Process to Inform Decision Makers." *Journal of Policy Analysis and Management* 18: 231–245.

Guernsey, L. 2001. "Keeping the Lifelines Open." *New York Times* (September 20): G1, G6.

Haberman, M. 2002. "City Gets Rolling on Downtown Street Fix." *New York Post* (April 2): 14.

Haimes, Y.Y. 1998. *Risk Modeling, Assessment, and Management*. New York: Wiley.

Haimes, Y.Y., N.C. Matalas, J.H. Lambert, B.A. Jackson, and J.F.R. Fellows. 1998. "Reducing Vulnerability of Water Supply Systems to Attack." *Journal of Infrastructure Systems* (December): 164–177.

Ichniowski, T. 2001. "Corps Estimates $267 Million Needed to Upgrade Its Water Facilities' Security." http://www.enr.com.

Langewiesche, W. 2002 "American Ground: Unbuilding the World Trade Center—The Rush to Recover." *Atlantic Monthly* 90 (September): 46–79.

Lave, L.B., and T. Balvanyos. 1998. "Risk Analysis and Management of Dam Safety." *Risk Analysis: An International Journal* 18(4): 455–462.

Lee, D. 2001. "They Don't Just Go to Staten Island Now." *New York Times* (December 2): 14-4.

Lipsky, D. 2001. Presentation at "Learning from Urban Disasters: National Science Foundation Response and Opportunities for Future Research Workshop," sponsored by Institute for Civil Infrastructure Systems, New York University, December 12–13.

Lowe, S. and L. Peek. 2001. Presentation at "Learning from Urban Disasters: National Science Foundation Response and Opportunities for Future Research Workshop," sponsored by the Institute for Civil Infrastructure Systems, New York University, December 12–13.

Markoff, J. 2002. "Talks Weigh Big Project on Wireless Internet Link." *New York Times* (July 16): C1.

Mileti, D. 1999. *Disasters by Design*. Washington, D.C.: Joseph Henry Press.

National Research Council. 1994. *Science and Judgment in Risk Assessment*. Washington, D.C.: National Academy Press.

National Research Council. 2002a. *Making the Nation Safer. The Role of Science and Technology in Countering Terrorism*. Washington, D.C., National Academy Press.

National Research Council. 2002b. *The Internet under Crisis Conditions: Learning from September 11*. Washington, D.C., National Academy Press.

New York State Urban Development Corporation, Empire State Development Corporation. 2002. "7 World Trade Center Reconstruction Project SEQRA Environmental Assessment Form and Supporting Analyses." Prepared by Allee King Rosen & Fleming, Inc. with Eng-Wong, Taub & Associates, Inc. and Sive Page & Riesel, P.C.

O'Rourke, T.D. 1993. "Prospectus for Lifelines and Infrastructure Research." Pp. 37–58 in N.K. Khachaturian, ed., *The Art and Science of Structural Engineering, Proceedings of the Symposium Honoring William J. Hall*. Upper Saddle River, NJ: Prentice Hall.

O'Rourke, T. D. 2001. Presentation at "Learning from Urban Disasters: National Science Foundation Response and Opportunities for Future Research Workshop," sponsored by the Institute for Civil Infrastructure Systems, New York University, December 12–13.

Pfeffer, J. and G.R. Salancik. 1978. *The External Control of Organizations.* New York: Harper and Row.

Slovic, P. 2000. *The Perception of Risk.* London: Earthscan Publications Ltd.

Stern, P.C., and H.V. Fineberg, eds., 1996. *Understanding Risk: Informing Decisions in a Democratic Society.* Washington, D.C.: National Academy Press.

UCLA Center for Communication Policy. 2003. *The UCLA Internet Report, Surveying the Digital Future, Year Three.* Los Angeles, CA: University of California at Los Angeles.

U.S. Department of Commerce, National Telecommunications and Information Administration. 2002. *A Nation Online: How Americans are Expanding Their Use of the Internet.* Washington, D.C.: U.S. Department of Commerce.

U.S. General Accounting Office. 2003. *Potential Terrorist Attacks: Additional Actions needed to Better Prepare Critical Financial Market Participants.* Washington, D.C.: GAO.

Wald, Matthew L. 2002. "Safety Board Cites Flaws and Lapses in Hudson Ferry Fire." *New York Times* (June 12): B1.

Water Environment Federation. 2002. "Water and Wastewater Systems Need $4.7 Billion for Security." *Water Environment Regulation Watch* (May):1.

Wireless Emergency Response Team . 2001. "Final Report for the September 11, 2001 New York City World Trade Center Terrorist Attack." October.

Woodworth, B. 2002. "Response and Recovery Issues," Presentation at the MCEER Workshop: Lessons from the World Trade Center Terrorist Attack, New York, New York, June 26.

Zimmerman, R. 1999. "The New York Area Water Distribution System: Water Main Breaks." In: "Integrated Decisionmaking for Infrastructure Performance." Final Report to the National Science Foundation, grant # 9526057. New York: New York University.

Zimmerman, R. 2001. "Social Implications of Infrastructure Network Interactions." *Journal of Urban Technology* 8(3): 97–119.

Zimmerman, R. 2003a. "Social and Environmental Dimensions of Cutting-Edge Technologies." In R. Hanley, ed., *Moving People, Goods and Information in the 21st Century.* London, UK: Spon Press/Routledge Press (forthcoming).

Zimmerman, R. 2003b. "Urban Infrastructure in a Time of Crisis." Small Grant for Exploratory Research funded by the National Science Foundation, Civil and Mechanical Systems Division, Grant # CMS-0204660. New York: New York University.

Zimmerman, R. and V. Bier, 2002. "Risk Assessment for Extreme Events." Paper presented at Risk Management Strategies in an Uncertain World, Columbia-Wharton/Penn Roundtable held in Palisades, New York, April 12. Http://www.Ideo.columbia.edu/CHRR/Roundtable/Zimmerman_WP.pdf.

Acknowledgments

This work is supported in part by the Institute for Civil Infrastructure Systems (ICIS) at New York University (in partnership with Cornell University, Polytechnic University of New York, and the University of Southern California) that is funded by the National Science Foundation (NSF) under Cooperative Agreement No. CMS-9728805 "The Institute for Civil Infrastructure Systems" and CMS-0204660 "Urban Infrastructure Services in a Time of Crisis: Lessons from September 11th," a Small Grant for Exploratory Research. In addition, two workshops, funded by the NSF to ICIS, also provided important resources for this work. These were CMS-0203371, "Learning from Urban Disasters: National Science Foundation Response and Opportunities for Future Research Workshop," held on December 12–13, 2001, for the purpose of identifying an interdisciplinary research agenda for issues raised by the September 11th attacks, and CMS-0091482, "Bringing Information Technology to Infrastructure: A Workshop to Develop a Research Agenda," held in June 2001 to identify research areas for the use of information technology to improve infrastructure performance. Any opinions, findings, and conclusions or recommendations expressed in this document are those of the author(s) and do not necessarily reflect the views of the National Science Foundation.

Correspondence should be directed to Rae Zimmerman, Professor of Planning and Public Administration and Director, Institute for Civil Infrastructure Systems, Robert F. Wagner Graduate School of Public Service, New York University, 4 Washington Square North, New York, NY 10003; e-mail: rae.zimmerman@nyu.edu; http://www.nyu.edu/icis.

Lessons Learned from the World Trade Center Disaster about Critical Utility Systems

Thomas D. O'Rourke
School of Civil and Environmental Engineering
Cornell University

Arthur J. Lembo
Department of Crop and Soil Sciences
Cornell University

Linda K. Nozick
School of Civil and Environmental Engineering
Cornell University

Introduction

The terrorist attack on the World Trade Center in New York City on September 11, 2001, was unprecedented in terms of damage to and repercussions for civil infrastructure for an event not related to declared military operations. There were 2,830 people killed and 10 major buildings destroyed or subjected to partial collapse. Approximately 2.8 million square meters of commercial office space were removed from service, of which 1.1 million belonged to the World Trade Center complex (Federal Emergency Management Agency, 2002).

The World Trade Center site is located within an intricate complex of infrastructure systems. It is important, therefore, to assess the consequences of such severe damage on the interdependent systems connected to the site.

In this paper, the consequences of the World Trade Center disaster on civil infrastructure systems is explored, and a summary of performance is given for water supply, electrical power, telecommunications, natural gas, and steam energy. The effects of water distribution system damage on fire fighting are addressed. The characteristics of damage and systems performance resulting from the World Trade Center disaster are compared with those of other major incidents in New York City. Preliminary conclusions are drawn with respect to causes of damage, factors affecting the spread and constraint of damage, and factors that contribute to both the resiliency of urban infrastructure and the attendant services necessary to respond effectively to extreme events.

Buildings

The building performance at the World Trade Center complex has been described in considerable detail (Federal Emergency Management Agency, 2002), and only select features of the building response are described herein. On September 11th, two hijacked aircraft were flown into the World Trade Center towers. At 8:46 a.m. the north face of the north tower (WTC 1) was struck, and at 9:03 a.m. the south face of the south tower (WTC 2) was struck. At 9:59 a.m. and 10:29 a.m., WTC 2 and WTC 1, respectively, collapsed.

Figure 1 shows the area of the collapse debris impact as depicted by FEMA (2002). Buildings surrounding the towers are also shown, with special emphasis (dark shading) provided for those that were destroyed as a consequence of being struck and/or ignited by debris from the collapsing towers. The inner circles indicate the approximate radius of inner steel column collapse and other heavy debris. The outer circles indicate the approximate radius of fallen aluminum cladding and other relatively light debris. Of special interest for electric power and telecommunications are WTC 7 and the Verizon Building located along Vesey St. immediately north of the World Trade Center complex.

Figure 2 is a map of the buildings surrounding the complex as published by FEMA (2002). The map shows the damage assessment of surrounding structures based on the inspections of 406 buildings on September 14 and 15 by New York City structural engineers. The building inspection procedures and damage classification were adopted from guidelines developed for rapid assessment of buildings after earthquakes (ATC, 1989).

As noted above, the damage to buildings was of unparalleled proportions for a disaster perpetrated independently of official military operations. Of significance also was the damage sustained by other civil infrastructure systems, which are described briefly in the following sections of this paper.

Water Supply

The New York City water supply is maintained and operated by the New York City Department of Environmental Protection (DEP). About 90% of water provided to New York City is conveyed through City Water Tunnels 1 and 2, which were commissioned in 1917 and 1938, respectively. Since

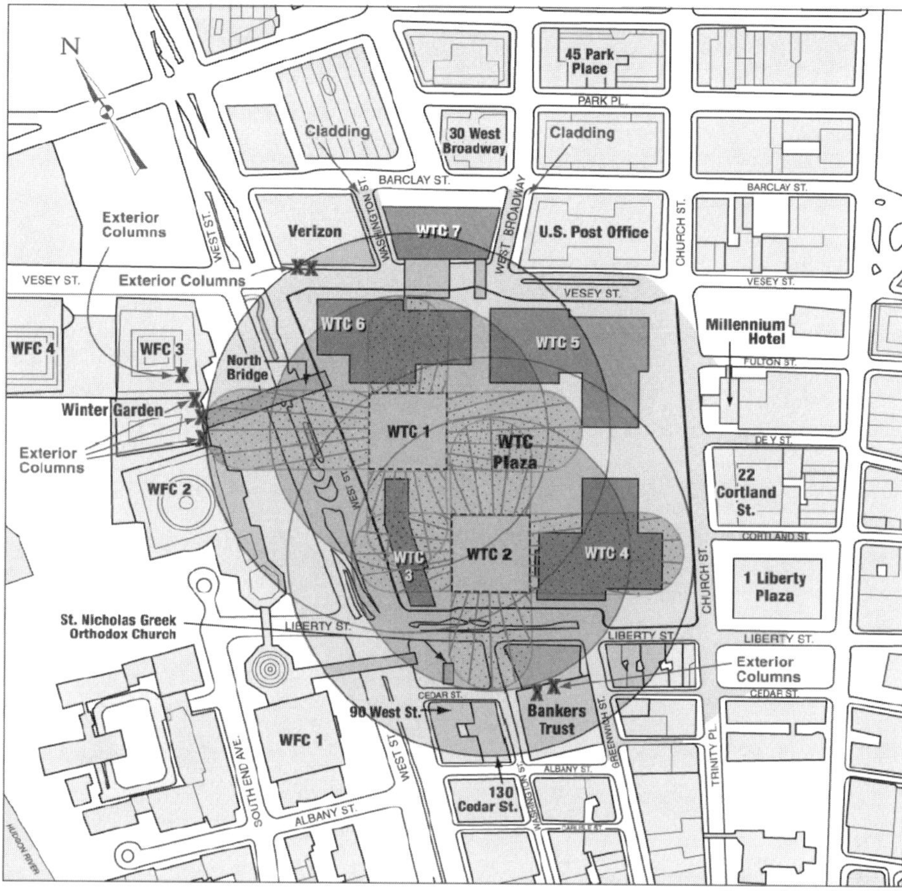

Figure 1. Schematic depiction of areas of collapse debris impact (after FEMA, 2002).

commissioning, neither tunnel has been dewatered for inspection, so their state of repair can only be inferred indirectly. City Tunnel 1 is the backbone of supply for Manhattan. It is about 183–213 meters below street surface and transmits water to the distribution pipeline system through riser shafts. Some valves in the tunnel and shafts have not been used in many years, and no attempt is made to operate them because of concerns about malfunctioning and corresponding impact on the system. A third tunnel, City Tunnel 3, is currently under construction, and will supplement supply and allow for dewatering sections of City Tunnels 1 and 2 for maintenance and repair.

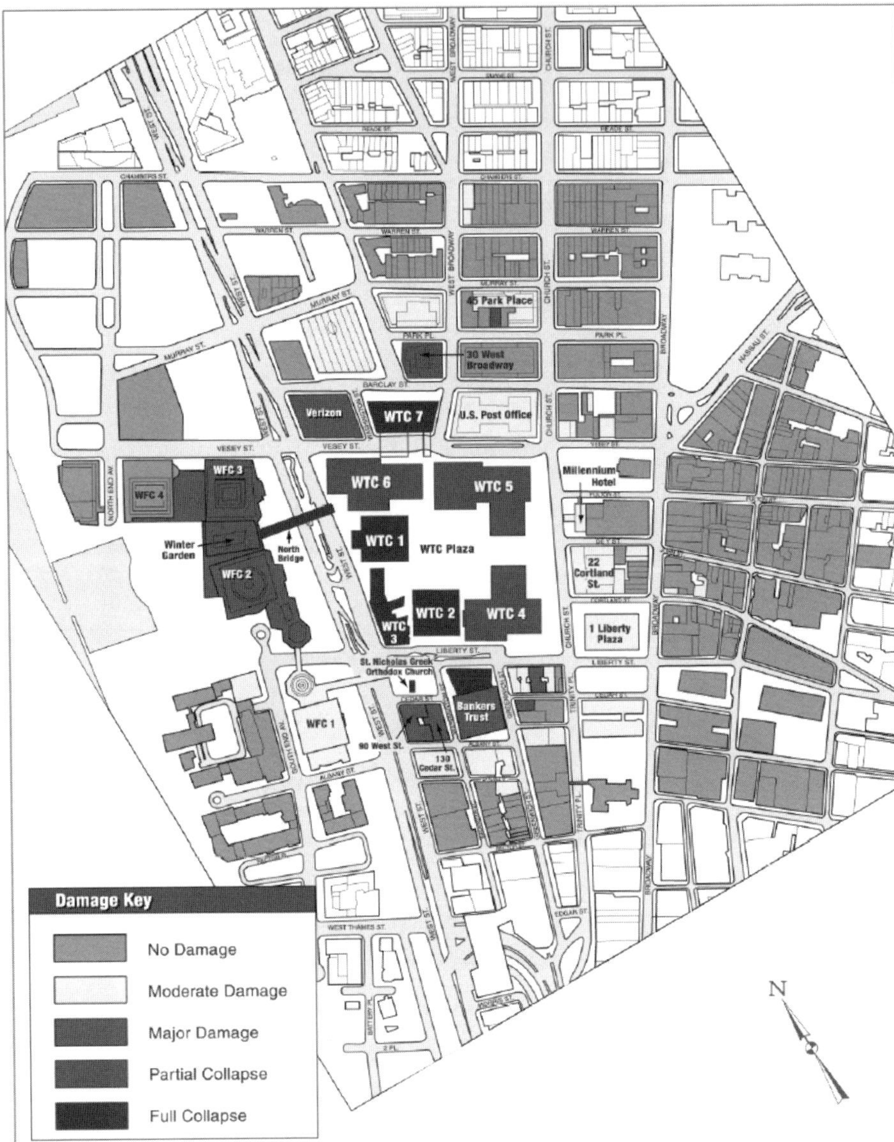

Figure 2. New York City DDC/DoB cooperative building damage assessment map of November 7, 2001 (after FEMA, 2002).

Four shafts in lower Manhattan provide water for the distribution system that serves the World Trade Center area and nearby neighborhoods. The shafts are equipped with regulators to decrease pressure to a level compatible with the capacity of the trunk and distribution pipelines. Continuous flow measurements are performed at each shaft. Because the pressure at each shaft is regulated, the resulting flow measurements provide accurate data on rate of flow at constant pressure.

Collapse of the towers and the subsequent collapse of WTC 7 resulted in about 10 locations of ruptured water mains within the disaster area. These locations do not include the loss of a 406-millimeter-diameter water main that was situated beneath the complex. Damage to 508-millimeter- and 305-millimeter-diameter water mains was caused primarily by direct impact from collapsing structures and falling debris.

On September 11th, maximum total water flow recorded at four shafts in lower Manhattan was 7,140 liters per second compared with 5,610 liters per second on September 10. This resulted in an increase of 1,530 liters per second relative to normal usage. Water from the shaft closest to the World Trade Center site accounted for about half the increased flow relative to normal usage. However, the usage was not normal on September 11th. A close inspection of the DEP flow measurements shows water demand from the four shafts declining to about 4,380 liters per second before the collapse of the towers. After collapse, water flow jumped to 6,790 liters per second, and then rose gradually to a peak of 7,140 liters per second. The initial jump of 2,410 liters per second represents water lost through broken water mains beneath and surrounding the complex. The additional 350 liters per second represents the draw from fire hydrants that were used to fight fires in adjacent buildings.

An independent assessment of water drawn from the distribution system was made by interviewing deputy fire chiefs who were in command on the eastern and western sides of the complex. The chiefs were able to identify hydrants, hoses, and fire engines used at the site from which it was estimated that approximately 280 liters per second were being drawn from the water distribution system. Hence, the estimate obtained from fire department commanders and the rate shown by the flow measurements are in reasonably good agreement.

Water pressures at hydrants adjacent to the World Trade Center complex declined throughout the afternoon. Measurements performed by DEP at 6 p.m. show pressure two to three blocks from the site at a level approximately one-third of that under normal conditions. Fire fighting was impaired by the declining pressures. From about 6 p.m. to 2 a.m., DEP personnel shut gate valves in the pipeline network surrounding the site to isolate damaged pipelines from the rest of the system. Isolating the broken mains restored pressure in the intact system outside the perimeter of closed gate valves.

The area initially isolated was relatively large to provide sufficient coverage and confidence that the then-unknown locations of damage would be included within the isolated zone. The area was bounded on the north by

Murray St., and on the east by Broadway (see Figure 2). The isolation area extended to the western side of the World Financial Center and as far south as Morris Street. The isolation zone was eventually reduced to include only the streets immediately adjacent to the World Trade Center complex.

Fire Fighting

Fires at and adjacent to the World Trade Center complex were fought in essentially two theaters of action. On the eastern side, three tower ladders were deployed to attend to fires at World Trade Center 4, 5, and 7. In the afternoon of September 11th, the decision was made to abandon World Trade Center 7, which collapsed about 5 p.m.

On the western side, fires were fought primarily with water from the Hudson River, which is two to three blocks from the complex. The New York City Fire Department, Marine Division, dispatched four fire boats to the site: *Firefighter* (1,260 liters per second capacity), *John D. McKean* (1,200 liters per second capacity), *Smoke II* (130 liters per second capacity), and *Kevin C. Kane* (410 liters per second capacity). The *McKean* arrived within 15 minutes, just before the second aircraft struck WTC 2. The *Smoke* arrived shortly thereafter. Initially, the *McKean* and *Smoke* were used to ferry wounded and fleeing people to New Jersey City. Between approximately 10 and 11 a.m., the four fire boats took up positions. The *Firefighter* docked on the Hudson River at Vesey Street. The *Smoke* and the *Kane* docked in North Cove, and the *McKean* docked on the Hudson River at Albany Street.

A review of the pumping records, interviews with Marine Division personnel, and inspection of high resolution aerial photographs were used to estimate the amount of water supplied to the World Trade Center complex and surrounding buildings. All fire boats were pumping at high capacity from the Hudson River. However, not all the water was relayed to fire engines. Excess water, unable to flow through hoses and land-based equipment, was discharged into the river without specific measurement. Hence, the fire boat pumping records do not provide a direct measure of water actually used at the site, but an upper-bounds estimate. An alternate estimate was made by obtaining information about the number of hose lines deployed from each boat, and approximating the flow based on hose diameters and distances from boats to fire engines that were relaying the water.

The estimated total flow during September 11th from Marine Division equipment is between 630 and 1,260 liters per second, with the actual value likely to be closer to 630. Hence, fireboats played an important and highly significant role in suppressing fires on and immediately after September 11th. During the critical hours following the collapse of the towers, they provided from two to four times the amount of flow that was pumped from intact water distribution mains near the site.

Electric Power

Almost all the electric power for New York City is supplied by the Consolidated Edison Company of New York, Inc. (Con Edison). Con Edison serves the densest electrical **load pocket** in the world. It supplies over 3.1 million customers in an area of 1,564 square kilometers served by more than 145,000 kilometers of underground distribution cable and 255,000 manholes and service boxes. Within New York City electric power is transmitted by high voltage cables (typically 138-kilovolt), which are encased in butyl-oil-filled steel pipelines. They connect with electric substations where the power is converted to low voltage (27- or 13.8-kilovolt) feeders. Con Edison uses a **distribution network** concept. The networks are grids supplied by the 13.8-kilovolt feeders that cover anywhere from several city blocks to several square kilometers. The networks are mostly underground, beneath the streets and sidewalks. The Con Edison system includes about 55 distribution networks within the city limits, of which 33 are in Manhattan. Each network is independent of its neighboring networks and is fed from multiple distribution feeders. Con Edison has 50% of all the distribution networks in the world, and its engineers are experts who often act as consultants for other distribution networks.

Because each distribution network is virtually independent, local damage usually will not affect the operation of other networks. Moreover, there is redundancy in the number of 27- or 13.8-kilovolt feeders supplying a given network. Even if multiple feeders are lost, there is reserve capacity to sustain power. The basic design of the network system provides for overall high reliability. According to Con Edison representatives, the annualized power outage per customer is 2.6 minutes in Manhattan compared with a U.S. national average of 111 minutes.

Debris falling from the collapse of WTC 1 damaged and ignited WTC 7, which contained two electric substations. Approximately one hour before the collapse of WTC 7, the New York City Fire Department requested that the substation be de-energized. Con Edison had been preparing for this event and power was immediately shutdown to three networks. The World Trade Center complex had lost power earlier in the morning as a direct result of the fires in the twin towers. The substations were totally destroyed when WTC 7 burned and collapsed. This collapse also resulted in damage to five 138-kilovolt cables. The faults generated by the damaged cables triggered switches that disconnected an additional substation at the lower east side of Manhattan. Thus, five local distribution networks lost power as a result of the events. A sixth network, which services the New York Stock Exchange area, lost six

13.8-kilovolt feeders. It continued to function from the remaining ten feeders, which supplied the local distribution network from a substation located north of the World Trade Center complex. The total power loss throughout all affected portions of the system was approximately 450 megawatts.

Restoration of power to the lower west side of Manhattan involved running 58 kilometers of 13-kilovolt cables above ground from areas that had power to the areas that did not have power. This process was completed in eight days. The first network was restored in three days by this method.

Restoration to the lower east side of Manhattan required that the 138-kilovolt cables into the complex be isolated and repaired. Cable repair involved freezing the butyl-oil-filled pipelines with liquid nitrogen, cutting both the pipeline and three cables contained in each pipe, then capping the cables and pipes, filling and pressurizing each pipe with oil, and soaking the cables. The first damaged 138-kilovolt cable was restored by the morning of September 17.

Con Ed crews worked 24 hours a day to complete the restoration work. Con Ed also set up 82 mobile diesel generators, each with a capacity of 1.2 to 1.8 megawatts, throughout lower Manhattan to supply emergency power to parts of Wall Street, Battery Park City, the Merchantile Exchange, and World Financial Center. The combination of emergency generators and repairs to the area's power system were required to be ready for opening of the New York Stock Exchange on Monday, September 17.

Telecommunications

A major telecommunications central office is located immediately north of the World Trade Center complex at 140 West Street, known as the Verizon Building. The structure was built in 1928 and is 32 stories high. Telecommunications equipment was located on the 1st through 9th floors. A cable vault, transformers, generator, and fuel tanks were housed in the five basement levels of the building.

Verizon claims that the telecommunications capacity of the 140 West Street central office was one of the largest in the world and equivalent to that of Austria, Denmark, and Egypt combined. The building housed four digital switches, 500 optical transport systems, 1,500 channel banks, 17,000 optical fiber lines, 4.4 million data circuits, and 90,000 message trunks.

The building was damaged from the collapse of WTC 1 and WTC 7. Damage from the collapse of WTC 7 was especially severe on the eastern side from about the 9th floor down where most of the telecommunications equipment was located.

The worst physical damage to telecommunication facilities occurred in the cable vault and on the 1st, 4th, 7th, and 9th floors. The emergency 911 system failed instantly, but switched to a back-up node at the Verizon central office at Metrotech, Brooklyn. Due to the 911 design, switching occurred immediately, and the new center never missed a call. There were approximately 38,000 calls on the 911 system on September 11th, which is 71% higher than normal volume.

Three class 5ESS switches were located on the 7th floor, and one was located on the 9th floor. Each switch consists of 40 frames and supports 60,000 customer lines. Remarkably, there was very little direct physical damage to the switches. Telecommunication equipment is designed and operated in compliance with the Network Equipment Building System (NEBS) standards (e.g., Verizon Communications, 2001). These standards have been developed with significant input from experience during natural disasters, including the 1989 Loma Prieta and 1994 Northridge earthquakes. In general, the telecommunication equipment remained intact and continued to function at reliable levels even though severe damage was sustained by the structural and building elements surrounding them. Three of the four switches were scheduled for replacement within two years of the disaster because dust had entered the equipment, clogging the filters and impairing reliable operation. Dust also had a deleterious effect on the turbines and cooling systems within the building.

Steel beams and debris, primarily from the WTC 2 collapse, penetrated the cable vault on West Street, severing 70,000 copper pairs and additional fiber optic lines. Water flooded the cable vault. Nearly 41,600 cubic meters of water had to be pumped from the vault during recovery. There was also water damage on the 7th and 9th floors where fire fighters sprayed water on adjacent buildings.

As a result of the damage, Verizon lost 200,000 voice lines, 100,000 private branch exchange (PBX) lines, 4.4 million data circuits, and 11 cell sites. Over 14,000 business and 20,000 residential customers were affected.

Telecommunications traffic was switched from the Verizon Building to other central offices. Of special importance were the Security Industry Data Network (SIDN) and the Security Industry Automation Corporation (SIAC) circuits that are used to execute and confirm block trades on the stock exchange. These circuits were necessary to resume trading and financial activity. The SIDN and SIAC circuits were rerouted from a central office in mid-town Manhattan to by-pass the Verizon Building. This was accomplished by routing traffic through a facility north of the World Trade Center complex to another large central office in lower Manhattan, from which they were connected with the New York Stock Exchange.

AT&T lost two local transport nodes in WTC 2, which burned and collapsed. Two other network switches located in the vicinity of the complex were not accessible because of damage, evacuations, and response activities. The equipment was powered down as the offices were evacuated to provide for easier restoration of service later. Telecommunication equipment showed resilience similar to that in the Verizon Building. AT&T Internet Protocol transport and other equipment in a sub-basement level of WTC 2 continued to work even after the building collapsed.

As a result of the disaster, the two largest AT&T local transport nodes in New York City were destroyed. Six other transport nodes in Manhattan were fully functional, and were rapidly configured to reroute traffic. In all, two switches were lost due to building collapse; two others were temporarily out of service because of lost power. Seven other switches in Manhattan remained in operation.

Two switches were restored by the fourth day after the attack, an optical remote module was upgraded to a full switch within six days, and a third local switch was restored within 18 days. Broadband wireless capacity was established whenever possible. Within 24 days of the attack, 24 38-gigahertz radio systems, including antenna and back-up equipment, were deployed throughout the affected area.

A team of telecommunication companies, brokerages, exchanges, and municipal leaders, in combination with federal authorities, developed a plan to restore services so that the New York Stock Exchange could be reopened on September 17. Reliable telecommunications service could have been supplied on Friday, September 14. However, brokerages and financial banking institutions with offices adjacent to the World Trade Center complex could not have been linked to the New York Stock Exchange on that date because of disrupted telecommunication services. These organizations were potentially at a significant competitive disadvantage. Additional routing of cables and fiber optic lines to connect with these organizations was performed over the weekend to prepare for the opening of the stock exchange.

On September 11th the wireless telecommunication industry experienced a 50% increase in call attempts nationally. In contrast, calls increase nationally by about 30% on Mother's Day. Figure 3 shows a bar chart of wireless traffic demand and blocked call percentages for normal conditions in contrast with September 11th conditions for the northeastern United States, Washington, D.C., and New York City areas averaged over the entire day, and New York City at approximately 11 a.m.. In the 9–11 a.m. period on September 11th, there was an approximate 1,300% increase in wireless call attempts experienced by one carrier before the call-count administrative controls were discontinued to support call processing.

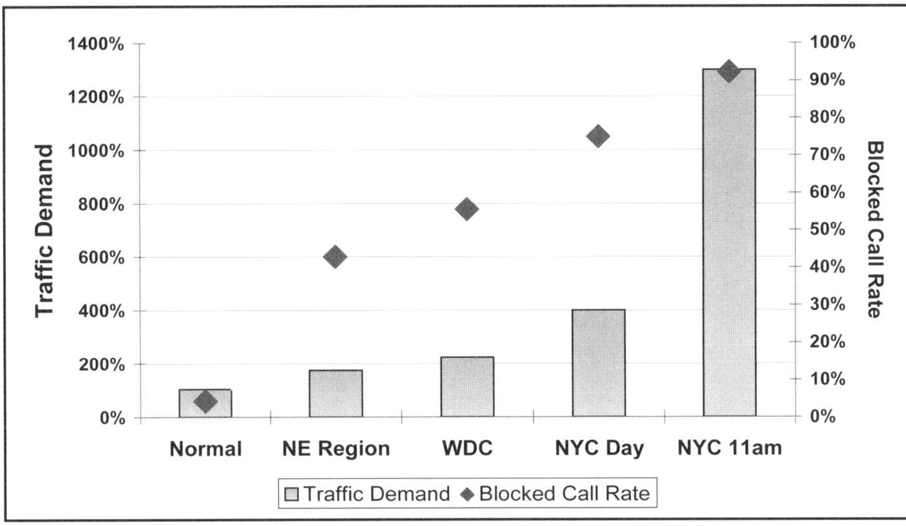

Figure 3. Cellular traffic and blocked call rate on September 11, 2001 (after Condello, 2001).

At 11 a.m., roughly 92% of wireless calls were blocked in the New York City area. Restoration of wireless service was rapid, as shown by the plot of equivalent site capacity near the World Trade Center complex versus time in Figure 4. Equivalent capacity to restore lost wireless sites was provided by restoring sites; deploying portable wireless cells, known as cellulars on wheels; and increasing the radio frequency range allocated to telecommunication providers.

Emergency operations were underway at Verizon, AT&T, and other telecommunication providers immediately after the attack. AT&T, for example, followed a strict and well-rehearsed telephone system protocol in which command-and-control operations were established with multiple technical bridges from upper management to personnel in the field. All telecommunications facilities in the World Trade Center area were secured, and services were restored using the Telecommunications System Priority (TSP) program (National Coordinating Center for Telecommunications, 2002) to assure that critical services received the highest priority. The TSP program is the administrative and operational framework, overseen by the Federal Communications Commission (FCC), for priority restoration and provisioning of telecommunication services in times of crises and events that threaten the U.S. population and national security. Large AT&T trailers containing telecommunications equipment were mobilized and placed in service at a site in New Jersey within 48 hours of the attack.

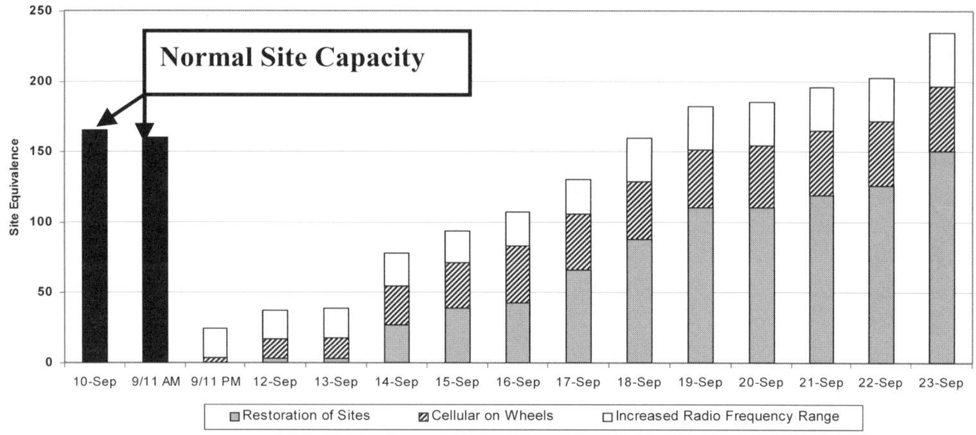

Figure 4. Restoration of cellular telecommunications in the vicinity of the World Trade Center (after Condello, 2001).

Redundant, dispersed facilities and the ability to rapidly by-pass damaged central offices were key aspects of the telecommunications performance. Verizon reports that it was able to process approximately twice the normal daily volume of cell phone calls on September 11th.

Wireless E-mail

Discussions with representatives of the New York City Mayor's Office, FEMA Region II, and the Port Authority of New York and New Jersey confirm that cell phone and landline voice communications were impaired by delays and busy signals after September 11th. To facilitate emergency communication, the IBM Crisis Response Team assisted both the Red Cross and New York City Mayor's Office after the disaster. The team organized a secure information system for the Mayor's Office with linkages to wireless e-mail. The e-mail system employed BlackBerry handheld units operating on the Cingular Mobitex system. IBM distributed about 70 units to the Mayor's Office and approximately 11 units to the New York State Governor's Office.

Cingular lost connectivity to most of its base stations in lower Manhattan because of failed telecommunication interconnections and interruption of electric power. Service, however, was able to continue because the handheld devices simply roamed to other non-affected base stations in New Jersey and Brooklyn that provided overlapping coverage to lower Manhattan.

The e-mail system is based on the transmission of information packets, each of which may travel by different routes and be separated in time intervals that exceed those required for voice messages. Because e-mail is asynchronous, message routing is more flexible and able to accommodate heavy traffic more easily than voice communication, which requires synchronous transmissions. Therefore, e-mail communications experienced less disruption than voice communications.

The wireless e-mail system also has queuing capability. If a message is not sent because of congestion, there is an automatic retry algorithm built into the system so that transmissions initially delayed are sent. In a voice network there is no queuing. The e-mail system is also able to save messages so that retrieval can occur at significant lengths of time after initial transmission.

The use of handheld e-mail devices greatly facilitated communications and contributed to substantial improvements in efficiency. Some advantages of wireless e-mail communication that were identified by handset users include reliability and consistency. Communications are not impeded by background noise, which can be troublesome with voice communication in emergency situations. E-mail communications are in text format so no notes have to be taken. Note taking can be difficult when using voice communication devices in the field. Moreover, e-mail messages are automatically logged and can be archived for future record keeping.

Natural Gas Distribution

New York City is served by a natural gas distribution system that involves over 10,000 kilometers of pipelines. Natural gas distribution in Manhattan is provided by Con Edison through approximately 1,000 kilometers of underground piping.

After the towers collapsed, Con Edison crews were dispatched to the disaster area. The gas company personnel could gain access to only a limited number of curb valves to turn off gas flow because debris, vehicles, and equipment were blocking access. To ensure stoppage of gas flow into areas of greatest damage and because of uncertainties regarding the extent of damage, a relatively large section of the distribution network was isolated. Isolation was performed by cutting and capping mains until an area, which was bounded by Chambers Street, Broadway, Rector Street, and the Hudson River, was disengaged from the remainder of the system. The isolation process required approximately 24 hours.

Leakage on September 11th was confirmed in the near vicinity (within one block) of the complex at only four locations at which steel and cast iron mains were situated. Most of these locations coincide with areas of fallen debris and projectile penetration of street cover.

The gas distribution network was restored by returning service successively to nine different sectors. Sectors at the perimeter of the isolated area were restored first, followed by those closer to Ground Zero. The last sector, which surrounds the complex, was not restored for over a year after the disaster because of ongoing construction and the absence of customers who did not re-occupy their businesses. About 90% of gas service was restored within one week of the disaster. There were about 300 Con Edison personnel engaged in the isolation and restoration efforts, working 24 hours per day.

Steam Energy

New York City is served by a steam energy system consisting of seven steam generating plants and over 160 kilometers of steel transmission and distribution pipelines. The system is operated by Con Edison. The pipelines are typically 254 to 762 millimeters in diameter, and are operated at internal pressures of 1,360 to 2,720 kPa and temperatures of 212 to 246°C. The steam is used for heating, cooling, and processing.

There was damage to several steam pipelines surrounding the World Trade Center complex caused by structural collapse or direct hit by projectiles and falling debris. A 762-millimeter-diameter steam line within the complex was lost. Steam line ruptures in 406- to 610-millimeter-diameter lines were confirmed by video inspection at four locations. On several streets adjacent to the complex, where building damage was especially severe, steam pipelines were cut and capped with no inspection.

Monitoring stations within the distribution network showed a steady drop in steam pressure after the towers collapsed. By 11 a.m. on September 11th, the decision was made to shut down the steam distribution network south of Grand and 7th Avenue in order to preserve service in areas unaffected by damage. The area isolated was chosen conservatively to be sufficiently large to cover locations of damage that were unknown at the time. It was also chosen to expedite the isolation process. In the vicinity of Grand and 7th Avenue, the system configuration was such that only five valves were shut to isolate approximately 30% of the system in lower Manhattan.

It took six days to restore most of the steam service to lower Manhattan. There were several temporary boilers installed and run for several weeks to supply buildings disengaged from the network. Several new mains were installed. The steam system in the immediate vicinity of the complex remained isolated for over a year after the disaster because of construction in damaged buildings and the continued absence of customers who once had occupied the area.

Other Incidents

Two incidents that occurred in New York City before September 11th are used here to show the potential for escalating damage originating from the failure of a single component. There have been similar incidents in other cities. The incident descriptions are taken directly from O'Rourke (1993).

Hellgate Incident

On December 29, 1989, a 750-millimeter-diameter gas transmission line exploded in the Hellgate area of the Bronx, New York. The explosion killed one person and injured six others, who had been working with a backhoe near the pipeline. The explosion and fire severely damaged a recycling plant, shattered windows, forced the evacuation of several nearby buildings, including a regional bus terminal. If confined to the nearby area, this incident would be regarded as serious by any standard. The accident, however, had much greater repercussions.

Heat and fire from the pipeline caused all electric power to be shut off at the adjacent Hellgate Plant. This cut off electricity to about 135,000 customers, including households and entire buildings, and blacked out an area of 5 square kilometers on the upper east side of Manhattan and 7.5 square kilometers in the Bronx. Traffic lights switched off, forcing substantial traffic jams. The loss of electricity suspended service on eight separate subway lines. Although the trains operate on direct current from the third rail, the station lights and track signaling run on utility-supplied electricity. Evacuation of the bus terminal near the fire left 200 buses stranded. As a result, there was severe disruption of bus service on 12 routes in Manhattan and the Bronx.

The accident occurred on Friday afternoon at about 2:00 p.m., just before the start of weekend rush-hour traffic before New Year's Eve. Hundreds of thousands of commuters were left stranded and a similar number of businesses and buildings incapacitated by the blackout, which lasted from three to five hours over major portions of the city.

Garment District Incident

On August 10, 1983, a 300-millimeter-diameter cast iron water main ruptured near the intersection of 38th Street and 7th Avenue. Water from the burst main flooded an underground electric substation, shorting electric circuits and touching off an immense fire. Loss of the substation blacked out approximately 3 square kilometers, including the Garment District and neighboring areas of the city. This involved over 10,000 customers, including Macy's and Gimbel's Department Stores. The blackout affected the telephone company central office, interrupting telecommunication service to tens of thousands of customers until emergency power was switched on. Phones using electric utility power were lost for a considerably longer time.

The accident occurred during Market Week in the Garment District, when most out-of-town buyers come to New York City to order next year's spring clothing lines. Direct and indirect business losses during this critical time have been estimated in the tens of millions of dollars.

Fire in the substation caused heat so intense that firefighters could not approach the blaze directly, but had to attack it with more than 2,000 cubic meters of foam. The blaze flared up an air shaft, igniting the roof of a 25-story building. Burning transformers released polychlorinated biphenyls, thereby generating hazardous emissions. The fire took 16 hours to extinguish.

Electric power was not fully restored for over five days. Distribution substations are designed to operate in isolation from the remainder of the system to reduce the chance of one station's failure overloading neighboring substations. This isolation strategy means, however, that bypassing a given station cannot be accomplished easily. Power was diverted from other substations by the time-consuming process of splicing underground lines, all during a period when many electric utility workers were on strike.

It is clear from these incidents that there was sudden and extensive escalation of damage from an otherwise local event. This process of escalating damage was referred to by O'Rourke (1993) as the cascade effect. Figure 5 shows an event tree for the Garment District incident, which illustrates the cascade effect. The pipeline rupture at first caused local disruption through flooding and undermining in its vicinity. When the electric substation was flooded, causing fire, disruption was transferred to a much larger area by electric power outage. Loss of electricity, in turn, affected business, transportation, and telecommunications. The figure shows three principal tiers, or levels, of disruption generated by the pipeline failure. As the damage cascaded from one level to another, losses were sustained in increasingly larger and complex systems.

Lessons Learned

The World Trade Center disaster was an immense and complex event. The lessons learned, as presented here, should be viewed as preliminary. Work is in progress to collect additional information that will lead to refinements in the databases and further clarification of the issues raised by the disaster. Lessons learned are summarized under the five subheadings that follow.

Damage to Underground Facilities

September 11th damage to underground water, gas, and steam pipelines and telecommunication conduits was caused by direct impact from falling debris or projectile penetration of the ground. There is no direct evidence of damage caused by ground vibrations and secondary effects, such as water hammer. There are records of gas leakage in streets traversed by heavy construction vehicles, but such instances were not exceptional or unanticipated by gas personnel. Virtually all damage of major consequence was the result of direct

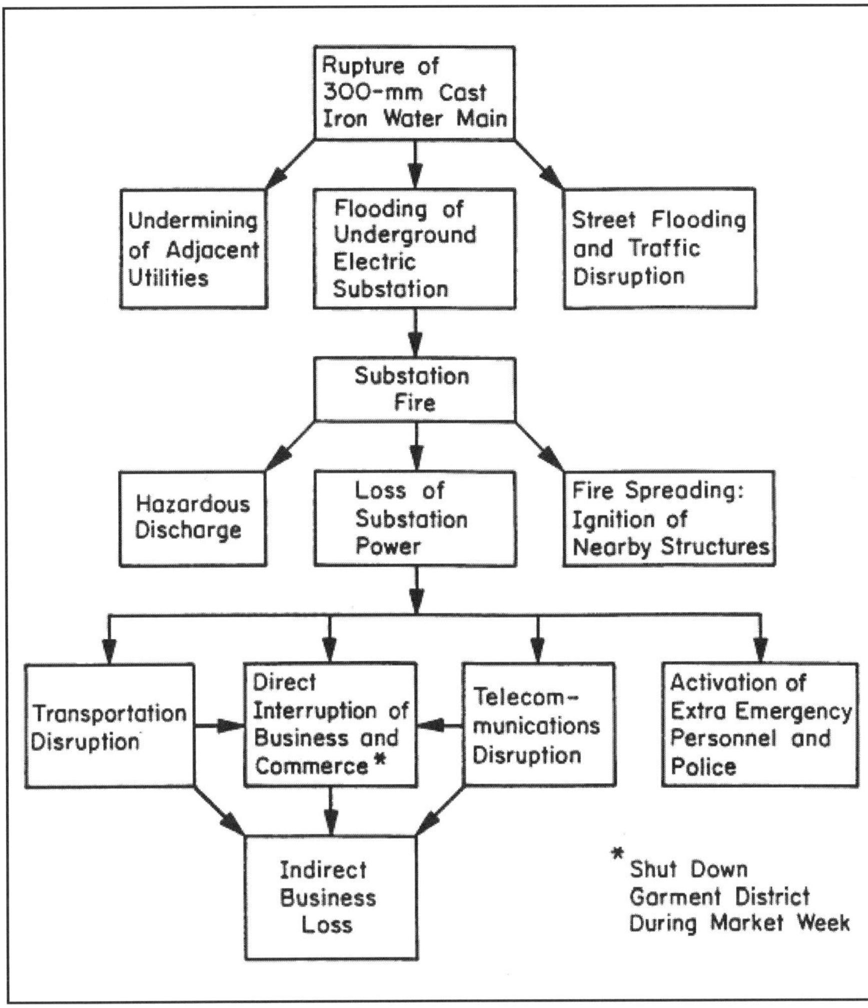

Figure 5. Event tree for Garment District incident (O'Rourke, 1993).

impact. Consequently, the areal extent of damage to critical water, telecommunications, gas, and steam conduits was confined principally to the zones of debris impact (see Figure 1). This lesson is significant for establishing planning and emergency response scenarios. It indicates that underground infrastructure damage from building collapse will likely be confined within the near vicinity of the failed structures.

Loss of Infrastructure Systems

Although physical damage to water, gas, and steam pipelines was confined principally to the area of debris impact around the World Trade Center, the temporary loss of these systems extended far beyond this perimeter. After collapse of the towers, damaged zones of the water and steam systems had to be isolated to preserve pressure and flow in undamaged sections. Gas pipelines had to be shut off to suppress leakage and potential ignition. Because the extent of damage was unknown, conservative decisions were made about the size of the isolation zones. The initially isolated portions of the water, gas, and steam systems covered approximately 0.8 square kilometer, 1.2 square kilometers, and 3.4 kilometers, respectively. One to two weeks were required to restore the isolated zones, with the exception of the damaged conduits immediately adjacent to the World Trade Center site.

In extreme events, conservative decisions under emergency conditions are likely to result in a substantial zone of temporarily lost service. For planning purposes, significant system disruptions should be anticipated.

Water Supply

Damage to water distribution pipelines surrounding the World Trade Center site was sufficiently severe that pressure losses interfered with fire fighting until isolation of damaged water lines could be achieved. Between two and four times the amount of water taken from distribution pipelines was supplied from fireboats during the critical hours after collapse of the towers.

The World Trade Center experience is similar to experience gained during the 1989 Loma Prieta earthquake. Water supplied by fire boat and relayed with a Portable Water Supply System (PWSS), consisting of hose tenders and above-ground hydrants, was responsible for suppressing the fire in the Marina and averting a much more serious conflagration (O'Rourke, 1993). The deployment of fireboats and the strategic use of bay water plays a critical role in emergency earthquake planning for San Francisco, Oakland, and Berkeley, California, as well as Vancouver, British Columbia. The September 11th experience supports a similar strategy for extreme events of non-earthquake origin. Fire boat deployment and implementation of PWSS concepts should be considered for supplying water under extreme emergency conditions. Water supply from fireboats to land-based locations is especially important where significant damage to the distribution pipeline system has occurred or where the capacity of the system to deliver water is limited.

Substantial amounts of water were supplied to World Trade Center sites at significant distances inland. Experience in the San Francisco Bay Area during and after the Loma Prieta earthquake (O'Rourke, 1993) reinforces the World Trade Center experience and demonstrates that water can be conveyed rapidly from marine locations for distances of 1 to 1.5 kilometers, provided that appropriate planning and equipment acquisition have been undertaken. In a post-September 11th world, it is important to include marine-based fire protection as part of strategic planning for disaster response.

The New York City fire boats with greatest pumping capacity were commissioned in 1938 and 1954. When decisions are made for commissioning new fire boats, it will be advantageous to provide for pumping capacity that is comparable to or exceeds the capability of the current fleet.

Telecommunications

As described above, redundant, dispersed facilities and the ability to rapidly by-pass damaged telecommunication nodes were important contributors to the performance of the telecommunications system. Even so, cell phone and landline traffic were impaired, and substantial repairs to the telecommunication service had to be made before reliable circuits were restored.

There are many central offices in lower and mid-town Manhattan. The proximity of many telecommunication facilities was a distinct asset during response and recovery. Similar configurations may not be present in other urban centers, thereby prompting consideration of additional facilities and dispersion of facilities where loss of business and commercial activity can have severe regional and national consequences. Special consideration should be given to facilities that affect urban, state, and national programs.

In general, the telecommunication equipment performed very well. The successful performance of telecommunication equipment encourages continued testing and evaluation of critical facilities, guided by observations of performance during extreme events.

The emergency procedures enacted by telecommunication companies worked well. The telecommunication industry operates with a formal, well-rehearsed protocol whereby command-and-control operations are rapidly established, with multiple communication pathways between upper management and field personnel. The U.S. telecommunications industry is served by a National Coordinating Center (NCC) that consists of government and industry representatives, who coordinate telecommunication services for national security and emergency preparedness. The NCC uses information sharing and analysis to evaluate the experience of industry participants, including lessons learned during emergency operations. The Network Reliability and Interoperability Council (NRIC) provides recommendations to the telecommunications industry and the FCC for operating telecommunication networks under various conditions, including natural disasters and both cyber (e.g., computer viruses, network intrusion, etc.) and physical attacks. The best practices, compiled by NRIC, have benefitted from lessons learned during previous natural disasters, including the Loma Prieta and Northridge earthquakes (e.g., Network Reliability Council, 2002).

Wireless e-mail proved to be a valuable asset after September 11th. The use of handheld e-mail devices greatly facilitated communications and contributed to substantial improvements in emergency response efficiency.

Electric Power Systems

Perhaps the most important observation from the World Trade Center disaster and other incidents described in this paper is that damage becomes most pervasive when it interferes with the electric power system. Both the Hellgate and Garment District incidents show that power outages were responsible for major escalation of initial damage. The electric power system is, in effect, the gateway for local damage to escalate or cascade into other systems.

The effects of local damage appear to be influenced by the proximity of damage to an electrical substation. When damage affects a New York City substation, its potential for cascading consequences increases dramatically, but nonetheless is capped by the configuration of the electrical network. Because the New York City system is compartmentalized into local distribution networks, substation loss tends to be limited to the one or two local distribution networks that are linked to that substation.

Differently configured systems will not react to damage in the same way. In other systems, for example, damage may trigger switching stations that isolate several substations, thereby resulting in power loss over a substantial area. Other systems do not have the redundancy embodied in the New York City system so that loss of one or two transmission lines or primary distribution feeders may compromise a significant part of the overall network.

The damage on September 11th was so extensive that disruption spread through five distribution networks, with a sixth impaired through the loss of six 13.8-kilovolt feeders. Extensive isolation, repairs, and new feeder installation were required for power restoration to financial centers in lower Manhattan and the resumption of New York Stock Exchange activities. Rehabilitation on such a scale within the short time frame required for the resumption of financial trading activities was only possible through extensive stockpiling of specialized parts and the availability of many trained, highly motivated, and well-equipped staff. System resilience depended on organizational resilience that was embedded in the planning and preparations that had been developed for emergency response.

Brave New World

The events of September 11th have changed not only the way we think about urban infrastructure, but the way we treat infrastructure information. In the "brave new world" after September 11th, knowledge is power both for those who wish to build and those who want destruction. It is therefore logical and responsible for the managers of infrastructure to be cautious about providing information that involves critical facilities. The notable lack of maps (except those available at the FEMA website) and graphical details in this paper is a testament to this caution. Although detailed information was collected in the preparation of this paper, it is not appropriate for the authors to provide further details until review and clearance of the information can be obtained. Such clearance may take many months or longer.

To benefit from the revolution in information technology, it is necessary to have access to information. It is ironic that the power of information technology may seriously inhibit its beneficial use because we are unable to decide the intentions of those who wish to use it. From an engineering and scientific viewpoint, there has never been a more opportune time for advancing the state-of-the-art and practice for characterizing and modeling complex infrastructure systems. Advanced geographic information systems, remote sensing, condition monitoring, model-based simulation, and systems engineering coupled with the capability of producing precise digital base maps, which can integrate the spatial characteristics of infrastructure, provide unparalleled opportunities. Legitimate concerns about security and attendant restrictions on information are equally important factors that may become barriers, unless we develop suitable procedures for information accessibility and dissemination. It is extremely important to develop a consistent policy regarding the need to know versus the need to secure information and databases about critical infrastructure systems.

In addition to improved protocols for information accessibility, there is the need to integrate and utilize the talent and resources of those working on extreme events. As described in this paper, the rapid inspection procedures for buildings adjacent to the World Trade Center complex and the resiliency of telecommunications equipment can be attributed in part to experience gained from earthquakes. Collectively, the professionals and research specialists, who focus on natural disasters, have expertise in risk assessment for the built environment, design procedures under conditions of uncertainty, strengthening of structures for multiple loadings, simulation of complex infrastructure under various damage states, emergency response, and community recovery. Such expertise is an asset when developing procedures for countering terrorism. An effective approach to homeland security requires that the parties responsible for implementing security have the vision and intelligence to engage those involved in natural hazards. Homeland security should integrate and harmonize the technologies for mitigating natural and terrorist disasters to create communities resilient to all hazards.

References

Applied Technology Council. 1989. *Procedures for Postearthquake Safety Evaluation of Buildings*. ATC 20. San Francisco, CA: ATC.

Condello, K. 2001. "Wireless Industry: Impact and Recovery Efforts Summary Report." The Network Reliability and Interoperability Council Meeting, October 2001. http://www.nric.org/meetings/meeting20011030.html.

Federal Emergency Management Agency. 2002. *World Trade Center Building Performance Study*. FEMA 403. Washington, D.C.: FEMA.

National Coordinating Center for Telecommunications. 2002. "Telecommunications Service Priority Program." http://www.ncs.gov/ncc/main.html.

Network Reliability Council. 2002. "Network Reliability—The Path Forward." Alliance for Telecommunications Industry Solutions. Washington, D.C. http://www.nric.org/pubs/index.html.

O'Rourke, T.D. 1993. "Prospectus for Lifelines and Infrastructure Research." Pp. 37–58 in B. Stenquist, ed., *The Art and Science of Structural Engineering; Proceedings of the Symposium Honoring William J. Hall*. Upper Saddle River, NJ: Prentice-Hall, Inc.

Verizon Communications. 2002. "System Integration and Testing NEBS Compliance Checklist" *SIT.NEBS.TE.NPI.2000.004*, Revision October 1, 2001. http://www.verizonnebs.com/nebs_d.doc

Acknowledgments

The authors acknowledge the support of the National Science Foundation under Award No. CMS-0207266. We thank Drs. Priscilla Nelson and Miriam Heller for their encouragement and assistance. We also express our deep gratitude for the help of talented and dedicated personnel at the Consolidated Edison Company of New York, Inc., New York City Fire Department and Department of Environmental Protection, AT&T, Verizon, Cingular Interactive, and IBM Global Services. We appreciate the support of the Institute for Civil Infrastructure Systems in conducting the research leading to this paper, and acknowledge Professor Rae Zimmerman of New York University for her leadership and continuing assistance.

Correspondence should be directed to Thomas D. O'Rourke, 273 Hollister Hall, Cornell University, Ithaca, NY 14853; (607) 255-6470; e-mail: tdo1@cornell.edu.

Part 4

Individual and Collective Behavior

A Need to Help: Emergent Volunteer Behavior after September 11th

Seana Lowe
INVST Community Studies Program
University of Colorado

Alice Fothergill
Department of Sociology
University of Vermont

Introduction

On September 11, 2001, terrorists attacked the World Trade Center in New York City and killed thousands of people. According to disaster response officials, the attacks involved unique and compounding variables that challenged all aspects of disaster planning and management. Indeed, in the 72 hours following the disaster, official response organizations such as the Office of Emergency Management (OEM), which lost its infrastructure in the attacks, were not able to respond in typical fashion. Some have called this 72-hour period a "vacuum of authority" (Halford and Nolan, 2002). For organizations like the American Red Cross, making the transition from local operations to Disaster Relief Operations (DRO) was slowed by the inability of national disaster personnel to arrive because planes and trains were not operating. In addition, having the disaster sites be crime scenes caused significant identification and screening challenges, and the complexity of the response and recovery needs had operations increasing when they normally would have been decreasing. Finally, and most relevant to this research, service organizations were overwhelmed by volunteer demand.

Immediately after the disaster, a "mass assault" (Barton, 1969) of people wanting to be of assistance converged on the disaster scene. Consistent with socially integrative responses that typically follow natural disasters (Tierney et al., 2001), New York City community members initiated and worked around the clock on a wide variety of protective response activities. Thousands of volunteers from around the country continued to want to help after the official agencies were up and running and the "vacuum of authority" no longer existed. Indeed, so many individuals wanted to be helpful that service organizations and official agencies had trouble sending them home when they were no longer needed (Halford and Nolan, 2002). According to one Red Cross professional, "We've never had this many volunteers at any disaster . . . This is the most volunteers we've ever had in the history of the Red Cross." By two and one-half weeks after the disaster, the Red Cross had received approximately 22,000 offers of assistance and had processed 15, 570 volunteers.

Our research focused on the emergent behavior of a subgroup of volunteers known as "spontaneous volunteers," a group considered understudied in the disaster literature (Tierney et. al., 2001). This paper uses data from our research in New York City to explore the phenomenon of spontaneous volunteerism at the site of the World Trade Center terrorist attacks. For the purposes of this inquiry, volunteerism is defined as the "contribution of time without coercion or remuneration" for public benefit (Smith, 1994, p. 244), and spontaneous volunteers are those individuals who contribute on impulse immediately after a disaster. Our goal is to analyze the motivations and experiences of approximately two dozen volunteers in order to increase knowledge about convergence and helping behaviors in a disaster and to offer suggestions for new directions in policy.

Background

Scholars have long been interested in the social behavior that immediately follows disasters. In a disaster, people converge on a disaster site; this includes both external convergence, moving to the disaster area, and internal convergence, moving to specific sites within the disaster area (Fritz and Mathewson, 1957). Fritz and Mathewson (1957), who coined the term "convergence" in the disaster field, theorized that there were five types of informal or unofficial convergence: the returnees, the anxious, the helpers, the curious, and the exploiters.[1] The "helpers" often include other survivors of the disaster. In fact, most response work is done by those community members who are present or nearby during a disaster. They often initiate and perform critical activities such as search and rescue and first aid immediately after events (Tierney et al., 2001). According to Scanlon (1992), although convergence behaviors by groups such as the helpers is widely accepted by disaster researchers, it has not been well studied.

Wenger (1991) points out that disaster researchers have observed and documented the important role of volunteer citizens and organizations in disaster aftermaths since the early 1950s. For example, after the 1985 Mexico City earthquake, it was estimated that 2 million people volunteered in the disaster aftermath, performing tasks such as search and rescue, clearing debris, offering medical aid, providing shelter and transportation, and serving as translators (Dynes et al., 1990). As O'Brien and Mileti state, past disaster research has long shown that "community wide disasters elicit a therapeutic community response" (1992, p. 87) in which disaster victims assist other victims (see also Barton, 1969; Fritz, 1961a, 1961b; Wilmer, 1958). Mileti (1999) points out that volunteer behavior at the time of a disaster impact, as well as during the emergency period, may emerge spontaneously or be institutionalized as part of an organization such as the American Red Cross. Mileti (1999) argues that although the culture in the United States is based on individualism, the country also has an "altruistic orientation" that fosters volunteerism and involvement in community activities (p. 145). This orientation is amplified in disasters; research has shown that individuals put their self-interest aside to volunteer to help others in need (Tierney et al., 2001).

We are interested in the idea, put forth by Mileti and other researchers, that an altruistic orientation exists after a disaster and that as a result people volunteer. Thus, in our study we examine spontaneous volunteers' initial reactions and motivations to serve in an effort to understand their primary reasons for giving their time, energy, and blood—literally—to people in need after the disasters. Then, we investigate both their emergent volunteer behaviors and the consequences or impacts of those behaviors. Finally, we suggest policy implications.

Methods

We conducted a qualitative study of volunteer behavior immediately after the September 11th terrorist attacks on the World Trade Center. Our sampling technique was purposive, meaning that we intentionally sought people who volunteered or tried to volunteer within days of the disaster. Using several initial New York City contacts, we employed snowball sampling that involved asking individuals from the disaster area to direct us to volunteers who fit the criteria to be solicited for interviewing. Interviews took place between 15 and 30 days after the attacks, and the majority took place in New York City. We conducted 23 in-depth interviews that ranged in length from one to two hours. All interview data were taped and then transcribed into the Non-Numerical Unstructured Data Indexing and Searching and Theorizing (NUD*IST) program designed specifically for qualitative data analysis. Once the data were input, we ran reports and reviewed them for analysis and coding.

Our final sample of 23 individuals was fairly diverse, including 11 women and 12 men, ranging in age from 23 to 66 years old. Their income levels ranged from low to very high, and all had graduated from college. While the majority of participants were European-Americans, there were also several Asians, several Latinos, and a Native American. In terms of sexual orientation, the sample included 20 heterosexuals and three homosexuals. Interestingly, 19 participants were single while only four were married or partnered. The participants' occupations included artists, financial analysts, teachers, counselors, producers, a waiter, a publisher, a salesperson, and retired and unemployed persons. Respondents declared religious affiliations as Buddhism, Judaism, various forms of Christianity, and no organized religion. Two had served in the armed forces. Of the respondents, 20 were spontaneous volunteers and three were Red Cross professionals working with spontaneous volunteers.

Significantly, most of the participants in our study stated that they had never engaged in volunteer behavior of any kind before taking action after September 11th. Those participants in our study who successfully volunteered after the Word Trade Center attacks participated in a variety of helping behaviors within 96 hours of the disaster. Volunteer activities included translating for families, delivering and moving supplies, removing debris, cheering for rescue workers, helping with crowd control, donating blood, counseling, preparing and serving food, fundraising, and giving massages to rescue workers. Understandably, the interviews and the fieldwork were emotional undertakings; both the researchers and the participants were deeply affected by what they saw, heard, and felt. We created pseudonyms for all individuals mentioned in the study.

Findings

This section presents the findings from our qualitative research project on volunteers in the aftermath of the World Trade Center attacks. First, we discuss the initial reactions of the volunteers, including where they were and how they responded to the news of the disaster. Second, we explore the participants' motivations for volunteering after the attacks. Next, we examine the actual volunteer behaviors that participants engaged in or attempted to engage in during the emergency, as well as some of the barriers to volunteerism that they encountered. Finally, we describe how the participants were personally affected by their volunteer activities.

Initial Reactions of Spontaneous Volunteers

In our study, we asked participants to tell us about their experiences when they first saw or heard about the attacks. We thought that to understand fully their experiences in the disaster, it was important to learn about their initial reactions. We wanted to know where they were, what they saw, and how they felt when they first learned about the attacks or saw the destruction. As members of the "community of people affected" by the terrorists (Omoto and Snyder, 2002), the study participants expressed feelings of victimization from the loss and suffering they experienced. They described themselves as shocked, sad, powerless, hopeless, helpless, fearful, and confused.

Many participants witnessed the events unfold. Catherine, a European-American businesswoman in her 30s, explained what she saw on September 11th:

> I was in a coffee shop and saw a whole bunch of sirens going by, undercover police cars. I said to the girl in the coffee shop, "Wow, there must be an undercover drug bust." . . . I started driving and there were all these people staring up at the sky so I turned and looked and I saw the first tower billowing with smoke. I rolled down my window and yelled to a guy on the corner, "What's going on?" He said a plane had just hit, they think maybe it's a terrorist attack . . . I started shaking so I turned my car back towards home and found myself right in front of one of my best friend's building. . . . They were on the roof which has a clear view of the twin towers. We watched everything. We saw a lot of people jumping out, saw the first tower come down, watched the fire spread through the second one, and saw that come down. Right away I was in shock. . . . I watched the whole thing through binoculars. Looking back I think I just wasn't really reacting. . . . The first thing I felt was, I felt really helpless.

Lucy, a single unemployed European-American in her 30s, also witnessed people jumping to their deaths:

> [After watching the first tower fall], I came back inside in tears and thinking "Oh my God, I'm so sorry," and not really being clear who exactly I was sorry to, or what I was sorry for. That seemed to be the refrain in my mind. . . . Then I went back outside and had the unfortunate experience of watching people jump out of the building at which point I think I said, "I think I've seen enough." [My feelings were] mostly it was horror and disbelief. How can this be happening here, I can't believe this actually happened. I can't believe I witnessed people making the decision between what way they were going to die. And shock and sadness. Everything all at once. It was completely overwhelming.

Witnessing the event greatly affected Catherine, Lucy, and the other volunteers who actually saw the harm inflicted on those in the World Trade Center buildings. Some participants feared for their own family members. Roger, a Jewish therapist in his 50s, explained his reaction to the news of the attacks:

> We were sort of in shock. . . . [I felt] mostly disbelief and incomprehension . . . fear, terror, panic a little bit. My younger boy had just started school

> in Manhattan and we hadn't heard from him. As it turns out, we didn't hear from him until 4:00 or 5:00 p.m. that day. . . . Personally it was a worrisome day because we didn't hear from him . . . so I have to say that put a personal spin on everything. Then I was wondering who I knew in those buildings, which of my patients might be down there, or patients' families, or friends, or neighbors.

Other volunteers described what they witnessed at Ground Zero in the aftermath of the disaster. Mermaid, a 30-year-old Korean writer, explained the confusion she experienced:

> There's 168 hospitals or something and all of the families were going to every one of them, so there were these crazed people crying on the streets. The family center is, really to me, the Ground Zero. Ground Zero is happening over there, but this is in a way bigger. You're seeing women and kids. It's insane. Because each missing person comes with two to seven people roaming for them. So that number becomes . . . you could easily have 50,000 people, 40,000 people, families in devastation.

All of these accounts illustrate the enormity of the crisis for the volunteers. These findings indicate that the volunteers were emotionally impacted by the disaster and personalized the attacks as members of the community affected, which appeared to have contributed to their heightened feelings of victimization. Many of them explained that when they saw the destruction they knew that they had to do something—they *needed* to help. Indeed, Sandy, a Red Cross Local Disaster Volunteer Officer, observed:

> In my history with Red Cross (11 years), this is the first time that every single person that I have dealt with has been personally affected. The fact that they are Americans and concerned and confused. In this particular situation, the entire country wants to help. . . . They were coming from all over the world spontaneously.

Motivations of Spontaneous Volunteers

In this study we are concerned with many aspects of convergent volunteer behavior, including what motivates an individual to help in a disaster. The participants were asked to explain why they volunteered in the aftermath of September 11th. We found from the interviews that the primary motivation for volunteering was a compelling need to help in some way, particularly a need to assist victims, and a desire—even obsession—to "do *something*" in order to contribute something positive and find something meaningful in the midst of a disaster characterized by cruelty and terror.

Most expressed that the need to volunteer was overpowering. For Catherine and the other spontaneous volunteers, it was not a question of whether or not to volunteer, it was only a matter of finding ways to volunteer. Catherine, who watched the World Trade Center collapse from a rooftop, explained her motivations for deciding to volunteer with her friend:

> I just felt really helpless, and before the television started saying it, we said, "Let's go give blood. They're going to need blood. That's what we can do." Jane's husband and brother really just wanted to watch TV. It was interesting that we felt that we needed to, we both felt that we needed to get outside and do something and contribute. And they both felt that they needed to know what was going on, they needed the information. My mother is really altruistic, volunteers a lot, or maybe it's a nurturing instinct that I felt.

Lucy, who had "seen enough" when she witnessed people jumping to their deaths, explained her feelings about wanting to pursue some type of action:

> I called a friend who was up the street in the West Village. So I walked up to his house and grabbed him to go give blood. At that point we really thought that a lot of people were going to need blood. We were right near a hospital and it was just anything to keep you moving and you were feeling like you were potentially helping. There was definitely a feeling of desperation to do something no matter what.

As Lucy explained, there was a "feeling of desperation" to help others, a sentiment expressed by many of the volunteers in our study. Joy, a 23-year-old counselor and writer who saw the second plane hit the World Trade Center, expressed her motivation to do something to alter the negativity:

> I think wanting to be connected to the hope that's come out of this. Wanting to walk away with some positive life teaching realization. I don't want to walk away from this thinking there was nothing I could do and it's just horrible, horrible, horrible. There needs to be some positive that comes out of it and the only way it's positive is if I and other people make it positive. Knowing that there are so many people out there that just have no one they can call when they are at the end of their rope or they're just freaked out completely and they're just paralyzed. It's a need. I want to do it and I need to do it.

Joy's quote illustrates the dynamic of both self-oriented and other-oriented motives to serve. Charlie, a single paramedic in his 20s, had been a volunteer with Red Cross for eight years. In his experience, people volunteered because it was "their form of therapy—we're allowing them that therapy to help." It appears that if the volunteers could do something altruistic, then as Joy described, they could transform the negative into something positive. They were compelled by a need to take some action for change that could create hopeful outcomes from disastrous circumstances.

Behaviors of Spontaneous Volunteers

As described in the previous section, volunteers primarily were motivated by a compelling *need* to do something. In this section we briefly discuss some of the activities and behaviors of the volunteers. We found that those who were successful at volunteering demonstrated a tireless resolve to "either act or be paralyze(d)." As mentioned before, they did a wide variety of tasks: they cheered on the Ground Zero workers[2], gave massages, prepared food, made beds, gave blood, worked as translators, vacuumed rooms, fed workers, counseled, did research, and helped with supplies for the Red Cross. For example, Joy was a 23-year-old European-American woman who decided to use her life and professional experience to serve as a crisis counselor:

> They were a little worried about me coming (to the counseling center) because I've been diagnosed with a terminal illness and I just said, "You know, I think I'm the best person to come down there because I have been thinking about death for the last couple of weeks. No one went to that building thinking about death that morning. None of their families were thinking about it. I was thinking about it, so I've had time to process it. . . . I'm going to use my experience." There were so many people coming over because it was so close to the hospital. People just sitting there needing to talk to somebody, needing to sit silently with someone. At that point I just had to give hope even though I knew that logically not everyone would survive . . . and it was long, beginning about 7:00 pm on September 11 . . . for about 40 hours.

Jack was a 32-year-old financial analyst who was not permitted give blood because he was gay. Out of his resolve to volunteer, he explained:

> I said, "I am not leaving, there has to be something that I can do." So they got me out on a truck, an emergency rescue vehicle, and we took some supplies over to the pier where they had the manpower set up. We came back and some company had donated like 50 roll-away beds . . . and we loaded the truck and we were distributing them to some of the first stations up town . . . and they were so thankful to us for doing this little thing of bringing them a bed, whereas they are out there putting their life on the line.

Like Jack, many of the spontaneous volunteers faced barriers to serving, and some of them demonstrated opportunistic and sometimes deviant behaviors in order to overcome those barriers. Participants described difficulties in having useful skills to offer, in finding needs to fulfill, and in dealing with overwhelmed response systems. The spontaneous volunteers in this study described frustrations of long lines, uncoordinated leadership, disorganized lists, and unclear information about what to do immediately after the attacks. In addition, there was such significant response by people wanting to help that needs like blood donations were saturated very quickly. Most volunteers spoke of the difficulties that they had finding volunteer work. In one instance, Taylor, a graphic designer in her 20s from Hong Kong, volunteered at a triage center, but because there were so few injured survivors there was nothing for her and the other volunteers to do. Taylor explained what happened:

> We were like one of the first volunteers at Chelsea Piers. . . . that is where the triage was set up. The bodies would be marked with black, which means they expired, or red, almost dead, or green or yellow . . . I think yellow was you could walk. . . . The one guy who was training us kept repeating the information over and over what we had to do when a black came in. . . . We would sit down and relax for a while and then he would say 50 police officers are going to come in and then no one would come. I probably saw five people come in, tops. . . . There were no survivors. Everyone was disintegrated. There was no survivors. The triage was pointless but no one knew that because everyone kept thinking that there were going to be survivors . . . we were just standing there with nothing to do.

Taylor, like many of the participants of this study, wanted to be able to do more for the victims of this disaster.

Participants had other troubles accessing volunteer work. Some had to find ways to gain entry into areas where people were needing help. For example, Mermaid, a 30-year-old Korean writer, shared her experience on September 12:

> There was no organization, we were all waiting. I kept going up to the policeman and I was like, "Listen, you need my skill, you need somebody that speaks Korean." We just realized we just have to go ourselves to the family center. We just walked to the armory on the other side of town. I just lied at the gate, I said, "Listen, Jacob Javitz sent me. You guys need Korean translator and somebody is waiting for me inside." So I just went in. It was so unorganized.

Mermaid thought that her lying was necessary because she felt certain that her skills could be of use. Josh, a 31-year-old European-American actor who had worked on a bucket brigade September 12, explained how he pursued volunteering on September 13:

> So I went up to the Javitz Center and filled out paperwork to register there and when I got there, you are talking thousands and thousands of people. I actually cut the line, which is probably an immoral thing to do, but I had already seen the location and knew that I was capable of doing work there. I lived with the guilt of cutting the line.

Josh, like Mermaid, thought that his skills were needed, and thus felt it was necessary to cut into line. However, unlike Mermaid, Josh expressed feeling badly about his "immoral" behavior when reflecting on it. Sung, a Korean-American non-profit associate in her 20s, also broke some rules in order to help victims. Sung was a volunteer translator at a crime victims' advocacy organization, and she had been instructed simply to translate various forms for aid for grieving families. She and another volunteer translator quickly realized that the families were not aware of all the benefits that they deserved. On their own, they drafted a document listing all the different services that were available to families of victims and made sure it got printed in Korean newspapers, on the radio, and on websites. They felt, however, that they might get in trouble for overstepping their role as volunteer translators.

Some of the volunteers also found it difficult to have anything to offer in the unusual circumstances. Catherine felt she was on a "quest" to find volunteer work. Although she had gone to Harvard Business School, she discovered that her assets were not useful after a disaster:

> For the first time I really feel like my accomplishments don't mean anything. Sure, I went to the best business school in the world, but what does that do in a crisis? It was really sad, I spent all this money and all this time and all I really want to do is be a firefighter, or be a doctor. Not having something practical to do was really difficult. . . . Literally I was running around: we went to the Javitz Center, went to the Salvation Army, we put our name on lists everywhere but it wasn't enough. We did the candlelight vigils but it wasn't enough. . . . It was very strange to feel like, you're in demand for every job in management, and yet you can't even volunteer. It was difficult. That's still something I think about, I really question what I've chosen to do.

Catherine eventually found volunteer work helping with a cruise ship that served food. Like Catherine, other volunteers also felt inadequate about their skills and backgrounds during the response and recovery. Lucy, the single unemployed European-American in her 30s, explained her frustration with not having the right skills:

> I rode around trying to find places to volunteer and I realized I had no skills, which was totally disappointing because they had forms to fill out for things that they needed and everybody in all of New York was trying to volunteer and they were just completely overwhelmed with everything. So I filled out this form but they ask, "Can you weld?" No. "Can you drive a semi?" No. Basically I had no talents that were relevant.

Despite having some trouble finding volunteer work, all but two of our study participants ultimately were able to volunteer in New York City after the attacks. As the next section will outline, these volunteer activities were significant and meaningful experiences.

Impacts of Spontaneous Volunteerism

The efforts of the spontaneous volunteers had positive impacts both for the local community and for the volunteers themselves. Sandy, a Local Disaster Volunteer Officer, discussed how the activities of over 15,000 volunteers enhanced the effectiveness of meeting emergency response and recovery needs:

> They are feeding people, they are cooking food, they are moving food, they are moving the products in and out of warehouses. . . . They work with the emergency workers, they provide massage therapy, mental health counseling. . . . They come in, they work their time, and they go home. Health services are out providing emergency first aid, anything that the Red Cross does takes volunteers because that is *how* we help with catastrophic events. The successes are for me that . . . we are honoring our commitment to the American public in that our commitment is that we will be there and we are.

Many of the volunteers concurred that their volunteer work significantly helped the response efforts in New York City. We believe, however, that the most noteworthy impact of their volunteer work was how the experience positively affected the volunteers themselves. The interview data suggest that spontaneous volunteers experienced increased feelings of interconnection, healing, and empowerment.

One positive outcome was that the volunteers found that by working with new groups of people during their volunteering, they experienced a sense of solidarity with different community members. Steve, a waiter in his 50s and of Irish and Native American descent, shared his thoughts on this:

> In the beginning we were doing 12-hour shifts and you just go and go and go and when you go home you are just totally exhausted. So there's a lot of fun, there's a lot of work, there's a lot of healthiness to it, there's a lot of interaction with a lot of different kinds of people from every walk of life and who knows how many nationalities.

Olivia, a 46-year-old Jewish single mother, explained what change she experienced as a result of giving massages to rescue workers:

> There was a breakdown of a certain isolation and connecting to the larger community that happened as a result of living through this and volunteering. My relationship to the world was expanded by serving and relating to new groups of people like the police.

Like Olivia, other volunteers also expressed new relationships with firefighters, police, and hospital personnel. Because of their volunteer work, many of the volunteers were able to see these personnel in action. They often praised these workers for their preparedness and well-organized response strategies. Many of the volunteers expressed a deepened understanding and appreciation of emergency responders with comments like, "I now see what disaster groups do and how well society is organized to address this kind of thing," and, "It was amazing to think that these are the men that dedicate their lives to running into the mouth of the fire while other people are running for their lives."

The most significant outcomes for volunteers that we found in our study were the positive emotional experiences that the participants had when they were able to help other people. According to individuals who successfully helped, the experiences of volunteering were "empowering" and "healing" for them. They described feeling like authentic contributors to the response efforts. The acts of helping to transform the physical and emotional spaces of the disaster site changed their roles from passive victims of loss to active participants in the recovery of their community. Jack, a 32-year-old financial analyst, shared the following:

> [Volunteering] was one of the best days of my life. I wish they could have found a way to let more people help who wanted to help . . . even if it was busy work . . . even if it was filing that needs to be done and it might not have anything to do with the tragedy but someone could have said to me, "You know, we have had this filing stacked up for about 2 months now and we are really backed up and if you help us file that frees somebody else who is more experienced in this situation to go and do something else." And I would be helping, even though I wasn't helping with the specific disaster. . . . I think a lot of people who wanted to volunteer would have had the healing experience that I had that way.

Lucy, who had been discouraged by her lack of skills, found a volunteer job serving food. She described the experience to be emotionally positive for several reasons:

> It was a huge kind of therapeutic process for me. Definitely a relief to be occupied with something to do. It was really frustrating to not work and not be able to do anything to distract myself. It was awesome to get down to Ground Zero and get a sense for the reality rather than the movie version that we've all kind of got inside our minds. It was really amazing to help all the relief workers from the human connection. Helping really strong, but totally fatigued guys who can barely hold their plate of food. It was really an emotional thing but it was much better than sitting at home.

It should be noted how Lucy provided both self-oriented and other-oriented explanations for why the experience was emotionally positive for her. Steve also explained how the most basic tasks took on great importance:

> I think psychologically [volunteering is] a very good thing to do because you're able to work through a lot of the problems in a very constructive way and in a way you help affect the recovery, even if it's just making beds. . . . I've unloaded trucks, we've scrubbed floors, we've vacuumed, I vacuumed for 5 hours one day. . . . It's not that glamorous or glorious or anything but it's healthy, it's very healthy.

His statement highlights again the benefits of volunteering for the volunteers themselves; not only were they doing something helpful for others, but they were doing something "healthy" for themselves.

As a result of participating actively in the recovery of the community, volunteers also experienced heightened personal power. Joy, who was the 23-year-old crisis counselor battling a terminal illness, expressed the impact of her volunteer activity:

> I experienced so much good out of this . . . because I was working on the positive end of the process, I was able to walk away with a stronger sense of who I am and what I can do, . . . I really do feel lucky that I was able to focus my shock and anger and my pain, sadness, and grief, all those other negative feelings into something positive. If you can't do that, it just eats you up. I think that's why I'm able to move on. I think that I have a lot more faith in other people. . . . I'm happier in some ways about the state of the society we live in, the community that we live in. But realizing that [I], and everyone else, we are so much stronger than I thought we were.

Like Joy, Josh was empowered by his experiences working on the bucket brigade and delivering supplies to official rescue workers:

> [There was a] confirmation of what I believed of myself, that I'm a good man. I just feel now, more than ever, that I don't need to waste energy apologizing for being involved with things. There are some things in life that just have to get done. It's that simple, and somebody has got to do them. There are (issues) in our social lives that need to be addressed: forms of prejudice, famine throughout the world, planes flying into buildings. Of all the good that is in my country, I am deeply in love with that. Of that which is darker, of the mistakes we've made, or the ones we continue to make, I feel a greater resolve now to speak the truth that I know in the hopes that that can be changed, and speak it from a place of love. To know that I've had some piece of that legacy in my life, then it will be a life well lived.

The majority of those who had not volunteered before responding to this disaster expressed a likelihood of taking action to serve their community in the future. Other research (Fothergill, forthcoming) also has found that many first-time disaster volunteers plan to volunteer in future disasters, and some even plan to devote their retirement to disaster volunteer activities.

Discussion and Conclusion

Research on the September 11th attacks can help us to understand response and recovery in various disaster settings. While many aspects of this disaster were unique, it was a large-scale crisis event that had similarities to other disasters, including many natural disasters. This study was intended to increase our knowledge about volunteer convergence and helping behaviors in a disaster setting and to offer suggestions for new directions in policy. The data reveal interesting and useful information about the reactions, motivations, and behaviors of spontaneous volunteers. In addition, and most significant from our perspective, the findings suggest positive outcomes for the volunteers themselves. As we explore the motivations, behaviors, and changes experienced by spontaneous volunteers, it is important to consider the extant literature on volunteerism and disaster-related volunteerism.

Community Volunteerism

Sociologists have explored variables influencing volunteer activities, motives for volunteering, and consequences from volunteer participation. In his summary, Smith (1994) noted that most research finds that participation is influenced by context, social background, personality, and attitudinal and situational variables. Social backgrounds of higher education and higher income are the most consistent predictors of volunteerism, while female gender and higher occupational status also have positive relationships to volunteering (see also Okun, 1993). In the area of personality, those with such social orientations as empathy and morality are more likely to volunteer, and in the area of attitudes, positive feelings toward the group to be served and feelings of civic duty have positive impacts. In terms of context variables, environments that are large and urban tend to be less conducive to volunteering (Smith, 1994).

Most research on volunteer motives fits within a functionalist framework theorizing that volunteer activity is sought to fulfill social-psychological goals. Goals fulfilled by volunteerism may be extrinsic, with a desired outcome, or intrinsic, with benefit derived from the activity itself (Isley, 1990). For example, AIDS volunteers were motivated by both the extrinsic reward "to be of service to others" and the intrinsic reward of doing something "that could be 'transformative' for the volunteers themselves" (Chambré, 1991, p. 284). Research shows that most volunteers are motivated by altruistic and humanitarian values, desires to increase understanding, needs to enhance esteem, and interests in fitting in socially (Clary et al., 1996; Omoto and Snyder, 2002). Altruistic motives, or those motives intended for the benefit of others with no benefits to oneself (Kohn, 1990; Oliner, 2000; Piliavin et al., 1981), appear to be consistently strong predictors of volunteer behaviors (Penner and Finkelstein, 1998). Recent research, however, also identifies a psychological sense of community as a significant impetus for volunteering (Omoto and Snyder, 2002). For those who pursue volunteering, there is evidence of positive outcomes such as enhanced well-being (Wheeler et al., 1998), increased commitment to community (Astin et al., 1999), and heightened social responsibility (Hamilton and Fenzel, 1988) for participants, among other benefits.

There is also some research, although limited, that has addressed community volunteerism in natural disaster situations. Dynes and Quarantelli (1980) suggest that there are many different types of disaster volunteers, including organizational volunteers who volunteer because their member organization mobilized them (Quarantelli, 1994). In his review of research findings on volunteer behavior, Wenger (1991) reports that there is "conflicting evidence" regarding who is most likely to participate in emergent volunteerism after a disaster. Based on some research, volunteers are most likely to be males, people of ages 18–45, members of the upper class, and those with high levels of formal education (Wenger, 1991). However, Wenger (1991) states that other research has found that socioeconomic status and gender do not influence volunteering. From his analysis, he determined that more research is needed on the factors that facilitate volunteerism in a natural disaster. Tierney et al. (2001) also argue that "not much is understood about which social groups volunteer and why" (p. 113), and that the disaster field knows relatively little about spontaneous volunteers.

Emergent Volunteer Behavior after September 11th

This exploratory study of volunteerism after September 11th has increased understanding about disaster-related volunteering by expanding the knowledge base on convergence behaviors of the group Fritz and Mathewson (1957) call the "helpers." In particular, these findings contribute to sociological knowledge about the factors that influence emergent volunteer behavior. Our study focused on the influences, motives, and impacts of spontaneous volunteerism in an effort to understand why the "helpers" decided to help, what actions they took, and how those behaviors affected community members or the community itself.

Our analysis of the motivations of spontaneous volunteers suggests that victimization affects emergent volunteerism. In New York City, it is clear that the attacks were personalized. Community members, as New Yorkers and

Americans, felt targeted by the terrorists. Indeed, it appears from our fieldwork that people internalized the loss and suffering of their fellow residents in the disaster even if they did not have any family or friends hurt or killed by the attacks. Research on other large-scale disasters, such as the Loma Prieta earthquake, also found that residents experienced both personalization and identification with the disaster even if they were not victims (O'Brien and Mileti, 1992). With those findings in mind, we believe that the nature of this disaster—the murder of thousands of people at the hands of terrorists—did have an effect on the degree to which people felt victimized by the attacks and the reasons that they volunteered during the response and recovery.

Our data reveal that heightened feelings of victimization played a strong role in the decisions to converge on the disaster scene and take part in emergent volunteer behaviors. The spontaneous volunteers in this study engaged in helping behaviors both because of compelling altruistic needs to serve members of their community and compelling personal needs to serve themselves. Consistent with other research on volunteerism (Clary et al., 1996; Omoto and Snyder, 2002), their altruistic motives were oriented toward outcomes that would better the situation for those who were directly affected by the brutality of the attacks. Overwhelmingly, the data demonstrate volunteers being extrinsically motivated to help others in order to relieve their pain and suffering. The intrinsic motivations, however, also appeared to be directed toward helping victims—the volunteers themselves. As mentioned earlier, the spontaneous volunteers were compelled by needs to take some action for change that could create hopeful outcomes from disastrous circumstances. It appears that the spontaneous volunteers were intuitively aware of their power to redefine the situation (Thomas, 1928/1994) of their lived experience as actors interacting with circumstances and constructing the meaning of the consequences (Collins, 1985). In other words, they recognized that they could make choices that would affect how they felt after having been attacked. Many of the spontaneous volunteers, by recognizing their choice "to act or be paralyzed" in the face of the attacks, illustrate the symbolic interactionist paradigm (Blumer, 1969). As agents capable of redefining the situation through their actions and interpretations (Blumer, 1969), they were motivated to take steps to alter their negative experiences of victimization through volunteering to "make it [the situation] positive."

Last, our study found that, in addition to the benefits typically yielded for community recovery (Tierney et al., 2001), the helping behavior of volunteerism during disaster convergence was an effective strategy for individual recovery. Specifically, the volunteers reported emotional impacts of healing and empowerment. It appears that the positive experiences of agency through volunteerism helped heal the victimization by transforming feelings of helplessness to feelings of efficacy. The presence of efficacy, or the belief that one has the ability to influence one's environment (Berman, 1997; Gecas, 1989), reflected a positive social-psychological state change for volunteers who had felt victimized by having been members of the community affected by the attacks. As mentioned earlier, the acts of helping to transform the physical and emotional spaces of the disaster site changed their role from passive victims of loss and suffering to active participants in the recovery of their community. By doing something altruistic that benefitted others, they transformed the negative into something positive, thereby experiencing their power to influence their environment.

Policy Implications

The most significant implications from this study are drawn from our findings about the motives for and impacts of spontaneous volunteerism. It is important for both disaster researchers and practitioners to understand the need to volunteer and the positive impacts provided by volunteering. Disaster researchers have begun to document the enormous protective response benefits of volunteer activities. In many disasters, the first people on the scene are members of the affected community who spontaneously volunteer before official responders arrive (Mileti, 1999; Tierney et al., 2001). Indeed, "key response tasks typically are performed by community residents themselves" (Tierney et al., 2001, p. 110). Some of these volunteer activities include search and rescue efforts that save many lives. As discussed earlier, large disasters often elicit a therapeutic community response where victims help other victims (Barton, 1969; Fritz, 1961a, 1961b; O'Brien and Mileti, 1992; Wilmer, 1958). It is clear that, in addition to recognizing those benefits, we need to consider the intrinsic benefits for the "helpers." Given the generalized experience of victimization described by respondents in this study who were members of the community affected by the disaster, the significant benefits to spontaneous volunteers need to be taken into account.

Trends about convergence and the tendency to do so en masse (Barton, 1969; Dynes et al., 1990) are known and accepted (Scanlon, 1992). Disaster researchers have documented the large-scale convergence of volunteers in natural disasters in general (O'Brien and Mileti, 1992), and specifically in New York City after the September 11th terrorist attacks (Kendra and Wachtendorf, 2002; Lien, 2002; National Academy of Sciences, 2002). Based on the findings in our study, spontaneous volunteers are converging because they are compelled by needs to help others by being of service and to help themselves by constructing new meaning. While some research has found that the convergence of volunteers can be problematic to relief efforts (Wenger, 1991) and even disruptive (Halford and Nolan, 2002), we urge a reframing of spontaneous volunteers as both victims and community resources. It is clear that a balance needs to be found between the emotional needs of community residents who want to volunteer and the needs of official response agencies that may be hindered and overwhelmed by too many volunteers.

In the spirit of finding that balance, we offer several suggestions. First, we suggest that disaster response strategies make a commitment during the disaster planning stages to serve all members of a community who feel victimized by a disaster. Second, we suggest that an effective way to serve those indirectly affected is to design emergency response plans in anticipation of the "need to do something." A plan could include established on- and off-the-scene work that allows those with and without skills to be of service to the community and thereby heal themselves. A strategy also could train community members in areas at high risk from terrorism to be "a 'resource' rather than a 'victim'" (Lichterman, 2000, p. 262). The "community as a resource" model can contribute significantly to the overall health of a community by preparing citizens to be soft mitigation resources (Lichterman, 2000) and proactively engaging them as agents in their own and their community's response and recovery.

Notes

1. Kendra and Wachtendorf (2002) suggest in New York after September 11th another category of "supporters/fans" was observed.

2. Although Kendra and Wachtendorf (2002) categorize this convergence behavior not as "helper" but as "supporter" or "fan," we believe that those who cheered on the Ground Zero workers felt as though they were indeed "helping" the recovery efforts by showing their support.

References

Astin, A.W., L.J. Sax, and J. Avalos. 1999. "Long-term Effects of Volunteerism During the Undergraduate Years." *The Review of Higher Education* 22(2): 187–202.

Barton, A.H. 1969. *Communities in Disaster: A Sociological Analysis of Collective Stress Situations.* Garden City, NY: Doubleday and Co.

Berman, S. 1997. *Children's Social Consciousness and the Development of Social Responsibility.* New York: State University Press.

Blumer, H. 1969. *Symbolic Interactionism: Perspective and Method.* Englewood Cliffs, NJ: Prentice-Hall.

Chambré, Susan. 1991. "The Volunteer Response to the AIDS Epidemic in New York City: Implications for Research on Voluntarism." *Nonprofit and Voluntary Sector Quarterly* 20(3): 267–287.

Clary, E. Gil, Mark Snyder, and Arthur Stukas. 1996. "Volunteers' Motivations: Findings From a National Survey." *Nonprofit and Voluntary Sector Quarterly* 25(4): 485–505.

Collins, R. 1985/1994. *Four Sociological Traditions.* New York: Oxford University Press.

Dynes, R. and E.L. Quarantelli. 1980. "Helping Behavior in Large Scale Disasters." Pp. 339–354 in D. Smith and J. Macauley, eds., *Participation in Social and Political Activities.* San Francisco, CA: Jossey-Bass Publishers.

Dynes, Russell, E.L. Quarantelli, and Dennis Wenger. 1990. *Individual and Organizational Response to the 1985 Earthquake in Mexico City*. Newark, DE: Disaster Research Center, University of Delaware.

Fothergill, Alice. Forthcoming in 2004. *Heads Above Water: Gender, Class, and Family in the Grand Forks Flood.* Albany, NY: The State University of New York Press.

Fritz, C.E. 1961a. "Disasters." Pp. 651–694 in R.K. Merton and R.A. Nisbet, eds., *Contemporary Social Problems.* New York: Harcourt.

Fritz, C.E. 1961b. Published as Fritz, C.E. 1996. *Disasters and Mental Health: Therapeutic Principles Drawn from Disaster Studies.* Newark, DE: University of Delaware.

Fritz, Charles E. and J.H. Mathewson. 1957. *Convergence Behavior in Disasters: A Problem in Social Control.* Committee on Disaster Studies, Disaster Research Group. Washington, D.C.: National Research Council, National Academy of Sciences.

Gecas, V. 1989. "The Social Psychology of Self-Efficacy." *Annual Review of Sociology* 15: 291–316.

Halford, Elizabeth and Shawn Nolan. 2002. "Rogue Volunteers: Response to the WTC Attacks by Volunteers who Refused to Leave." Paper presented at the American Sociological Association, International Committee on Disasters Session, August, Chicago, IL.

Hamilton, S. and L.M. Fenzel. 1988. "The Impact of Volunteer Experience on Adolescent Social Development: Evidence of Program Effects." *Journal of Adolescent Research* 3(1): 65–80.

Isley, Paul. 1990. *Enhancing the Volunteer Experience*. San Francisco: Jossey-Bass Publishers.

Kendra, James M. and Tricia Wachtendorf. 2002. "Rebel Food . . . Renegade Supplies: Convergence after the World Trade Center Attack." Paper presented at the American Sociological Association, August, Chicago, IL.

Kohn, A. 1990. *The Brighter Side of Human Nature: Altruism and Empathy in Everyday Life.* New York: Basic Books.

Lien, Onora T. 2002. "Emerging Volunteers: The Public's Response to the September 11th Attacks on the World Trade Center." Paper presented at the annual meeting of the Southern Sociological Society, Baltimore, MD.

Lichterman, J.D. 2000. "A 'Community As Resource' Strategy for Disaster Response." *Public Health Reports* 11: 262–265.

Mileti, Dennis S. 1999. *Disasters by Design.* Washington, D.C.: Joseph Henry Press.

National Academy of Sciences. 2002. "Countering Terrorism: Lessons Learned from Natural and Technological Disasters." Summary to the Natural Disasters Roundtable. Washington, D.C.: National Research Council.

O'Brien, Paul and Dennis S. Mileti. 1992. "Citizen Participation in Emergency Response Following the Loma Prieta Earthquake." *International Journal of Mass Emergencies and Disasters* 10(1): 71–89.

Okun, Morris. 1993. "Predictors of Volunteer Status in a Retirement Community." *International Journal of Aging and Human Development* 36(2): 57–74.

Oliner, S.P. 2000. *The Altruistic Personality: Rescuers of Jews in Nazi Europe.* New York: The Free Press.

Omoto, Allen and Mark Snyder. 2002. "Considerations of Community." *American Behavioral Scientist* 45(5): 846–867.

Penner, Louis and Marcia Finkelstein. 1998. "Dispositional and Structural Determinants of volunteerism." *Journal of Personality and Social Psychology* 74(2): 525–537.

Piliavin, J.A., J.F. Dovidio, S.L. Gaertner, and R.D. Clark III. 1981. *Emergency Intervention.* New York: Academic Press.

Quarantelli, E.L. 1994. "Emergent Behaviors and Groups in the Crisis Time Periods of Disasters." Preliminary Paper #206. Newark, DE: Disaster Research Center, University of Delaware.

Scanlon, Joseph. 1992. *Convergence Revisited: A New Perspective on a Little Studied Topic.* Working Paper #75. Boulder, CO: Natural Hazards Research and Applications Information Center, University of Colorado.

Smith, Horton. 1994. "Determinants of Voluntary Association Participation and Volunteering: A Literature Review." *Nonprofit and Voluntary Sector Quarterly* 23(3): 243–263.

Thomas, W.I. 1928/1994. "The Regulation of the Wishes." Pp. 156–161 in N. Herman and L. Reynolds, eds., *Symbolic Interaction.* New York: General Hall, Inc.

Tierney, Kathleen J., Michael K. Lindell, and Robert W. Perry. 2001. *Facing the Unexpected: Disaster Preparedness and Response in the United States.* Washington D.C.: Joseph Henry Press.

Wenger, Dennis E. 1991. "Emergent and Volunteer Behavior During Disaster: Research Findings and Planning Implications." College Station, TX: Hazard Reduction Recovery Center, Texas A&M University.

Wheeler, J., K.M. Gorey, and B. Grenblatt. 1998. "The Beneficial Effects of Volunteering for Older Volunteers and the People They Serve." *International Journal of Aging and Human Development* 47(1): 69–79.

Wilmer, H. 1958. "Toward a Definition of the Therapeutic Community." *American Journal of Psychiatry* 114: 824–834.

Acknowledgments

We would like to thank Jenna Peck for her extraordinary service to this project as our Research Assistant. Without Jenna's dedication in the field, this study would not have been possible. We also thank Greg Murr for his invaluable support, Lori Peek for her assistance, and the anonymous reviewers for their helpful comments. This project was funded by a Quick Response grant from the Natural Hazards Research and Applications Information Center in Boulder, Colorado, which is gratefully acknowledged.

Please direct all correspondence regarding this paper to Seana Lowe, INVST Community Studies Program, University of Colorado, 471 UCB, Boulder, CO, 80309-0471, e-mail: lowes@Colorado.edu or to Alice Fothergill, Department of Sociology, University of Vermont, Burlington, VT 05405-0176; (802) 656-2127; e-mail: alice.fothergill@uvm.edu.

Resiliency, Posttraumatic Growth, and Psychological Distress after the Attacks on America

David N. Sattler
Department of Psychology
Western Washington University

Introduction

During the first few weeks after the terrorist attacks on the United States on September 11, 2001, the country was in a period of great uncertainty. When would the responsible persons and organization(s) be identified, and could they be brought to justice? What was the motive? Would additional attacks occur, and if so, how and when? What actions should the U.S. government take to prevent future attacks?

Mental health professionals and those in the human services professions, such as psychology and social work, were especially concerned not only about the well being of those most directly affected by the attacks but also of other Americans across the country. What types of distress reactions were they experiencing, what services were they in most need of receiving, and which interventions would be most helpful? In order to address these and other essential questions, three weeks after the attacks I conducted a study in New York, South Carolina, Colorado, and Washington. The study grew out of more than a decade of research and experience gained by traveling to disaster sites and studying human responses to large-scale traumatic events, including hurricanes and earthquakes.

Because of the important parallels between the attacks and other types of catastrophic stressors, and the lack of research on responses to terrorist attacks, it will be helpful to briefly review existing theory and research on responses to other types of large-scale traumatic events. This review will be followed by a discussion of the study, an overview of how it was conducted, and its primary findings. The conclusion considers the implications of the findings and explores directions for future work.

Responses to Traumatic Events

Theory and research examining responses to traumatic events and catastrophic stressors, such as natural and technological disasters, military combat experience, and life-threatening assaults, provide a useful guide to understanding responses to the September 11th attacks. The attacks share a number of important characteristics with many of these events, but also have several unique features. Like other large-scale catastrophic events, the attacks affected a large number of people, occurred suddenly, were unpredictable, exposed many people to horror, appeared to be beyond the control of any one person, threatened life and the lives of family members and friends, and placed excessive demands on coping (Baum, 1991; Lazarus and Cohen, 1977; Sattler et al., 1997). Research suggests that many of these characteristics are associated with survivors' reporting that they feel some loss of their sense of control, predictability, safety, and trust (Baum, 1991; Updegraff and Taylor, 2000). The unique characteristics of the attacks include being deliberate acts intended to inflict harm; creating uncertainty concerning future attacks and when the responsible individuals and organizations would be brought to justice; prompting an extensive response from the U.S. government both domestically and internationally to locate members of terrorist organizations and to prevent future attacks; and causing citizens throughout the United States to feel directly affected and at risk.

Responses to large-scale traumatic events tend to occur in stages, or time periods. The acute stage occurs within the first few months, and the chronic stage occurs several months to years after the event. Research has documented a variety of psychological responses, including anxiety (Canino et al., 1990), acute stress disorder symptoms (Sattler et al., 2002; Waelde et al., 2001), posttraumatic stress disorder (Norris, 1992; Waelde et al., 2001), depression (Kaiser et al., 1996; Shore et al., 1986), problems in cognitive functioning (Freedy et al., 1994; Sattler et al., 1995), sleep disturbances (Wood et al., 1992), relationship difficulties (Adams and Adams, 1984), and substance abuse (Gibbs, 1989). Distress reactions that followed the bombing of the Federal Building in Oklahoma City—a domestic act of terrorism—were similar to those experienced after other types of traumatic events, and included fear, depression, and posttraumatic stress disorder symptoms (Benight et al., 2000; North et al., 1999; Shalev,1992; Tucker et al., 1997, 2000). Fortunately, most distress symptoms that follow natural disasters last for only a few weeks or months and may not constitute severe mental illness. However, a relatively small percentage of persons may develop more serious problems (Rubonis and Bickman, 1991). Additional research is needed to explore how distress responses to terrorism may vary over time.

Recent research also has shown that people may have positive experiences during the recovery that can help them cope and to move forward. For example, during the acute stage, people tend to pull together and help one another (Baum, 1991; Sattler et al., 1995). Social support provided by friends and family can play a vital role in helping people cope with tragedy (Kaniasty and Norris, 1995).

The conservation of resources stress theory provides a useful framework for understanding responses to the attacks (Hobfoll, 1989, 1998). The theory suggests that people build and retain resources to enhance the self and maximize positive reinforcement. The theory predicts that psychological stress occurs when there is a **threat** of resource loss, **loss** of resources, or **lack** of resource gain after investment of resources (Hobfoll, 1989). Four resource types are identified: condition (e.g., marriage, employment, or other social roles), personal characteristic (e.g., age, knowledge, locus of control, self-esteem, skills), energy (e.g., money, insurance), and object (e.g., house, car, or other physical possessions). The theory also predicts that resource gains may have positive effects. For example, survivors of a natural disaster or other traumatic event may learn about the value of preparation, learn new coping skills, and develop an enhanced sense of self-efficacy (Calhoun and Tedeschi, 1998; Monnier and Hobfoll, 2000; Sattler et al., 2000).

The conservation of resources stress theory has received direct and indirect support. Several studies show that resource loss accounts for a greater portion of psychological distress variance than variables such as coping style, sense of coherence, and general anxiety after a natural disaster (Freedy et al., 1994; Kaiser et al., 1996; Sattler et al., 2002). For example, four to five weeks after Hurricane Georges struck the U.S. Virgin Islands, Puerto Rico, the Dominican Republic, and the U.S. Gulf Coast, Sattler et al. (2002) found that in each of these locations, resource loss (especially loss of personal characteristic resources) was a more important predictor of psychological distress than prior exposure to traumatic events, stressful life events, or social support. The contribution of personal characteristic resources to distress may be due, in part, to changes associated with the significant disruption of daily routines and activities during recovery.

Taylor's (1983) theory of cognitive adaptation also suggests that traumatic events can challenge people's sense of meaning, mastery, and self-esteem. According to the theory, people may try to counter feelings of loss of meaning, mastery, and self-esteem by generating thoughts and ideas that enhance the self. For example, to regain a sense of meaning, an individual might reevaluate his or her attitudes and life priorities in relation to the event. To reestablish or maintain a sense of control and mastery, an individual might focus on areas in which he or she has control (Taylor et al., 1991). To maintain self-esteem, a person may focus on aspects of the self that are "relatively unaffected or improved, or by comparing oneself to less fortunate others in an effort to cast oneself in a more positive light" (Updegreff and Taylor, 2000, p. 7). Through these processes, individuals may learn about resiliency and useful coping strategies, clarify values and life priorities, and experience some positive outcomes that offset—to some degree—the negative outcomes (see Janoff-Bulman and Berger, 2000).

The theory of posttraumatic growth also explores how persons adapt to traumatic events and ways in which they might "perceive at least some good emerging from their struggle" (Tedeschi, 1996, p. 455). Growth may occur in any or all of three areas: the self, relationships with others, and philosophy of life. Growth in the self may include learning about one's vulnerabilities, the value of preparation, and new problem-solving skills; and developing an enhanced sense of self-efficacy or self-reliance. Growth in relationships may involve a deepening appreciation of relationships, increased self-disclosure and emotional expressiveness, and increased willingness to accept help from others. Growth in philosophy of life may include an increased appreciation for life and stronger spiritual beliefs (Calhoun and Tedeschi, 2001; Monnier and Hobfoll, 2000; Sattler et al., 2000; Tedeschi and Calhoun, 1995).

It is likely that the resource gains, adaptive thoughts, and posttraumatic growth reactions, discussed in the three theories just reviewed, can help survivors balance out and cope with some of the negative outcomes after a traumatic event. Calhoun and Tedeschi (2001) speculate that the changes are fundamental, represent a new way of viewing the world, and can endure for years (see Updegreff and Taylor, 2000, for a review of this literature).

Given the unique nature of the attacks, consideration of how Americans responded when the country was attacked in 1941, and informal conversations with colleagues, I was especially interested in studying posttraumatic growth and resiliency in addition to psychological distress.

Overview of the Study

Three weeks after the attacks, I conducted a study to examine psychological distress, posttraumatic growth, resiliency, and coping. Based on research, theory, anecdotal evidence, and the nature of the threat posed by the attacks, I speculated that in addition to experiencing fear and some degree of distress, citizens were reflecting on and/or reassessing their own lives and showing resiliency. The study also was designed to examine if and how responses varied as a function of distance from the areas struck. As a professor and educator, I was especially concerned about the welfare of college students and how they were responding to the attacks. Information about the needs of students and how they respond to such traumatic events can help mental health professionals in college counseling centers and other facilities provide appropriate services and design effective interventions to minimize or prevent subsequent mental health problems. For these reasons, this study includes samples of college students from four regions of the country (the Northeast, Southeast, Midwest, and Northwest). In order to make comparisons across the four regions, the study was designed so that the samples had similar demographic characteristics. Participants completed the same questionnaire in a similar environment (their classrooms) three weeks after September 11th. During class, participants were asked if they would be willing to complete the questionnaire.

The questionnaire, which was confidential and anonymous, had five sections. The first section asked about demographic characteristics (e.g., gender, age). Questions in the second section asked about losses and gains in resources, and were based on the conservation of resources stress theory. These items asked if, since the attacks, participants had experienced decreases, increases, or no changes, in their personal characteristic, energy, and condition resources as a result of the attacks. Participants used a 7-point scale, where –3 = quite a decrease, 0 = no change, and +3 = quite an increase, to indicate their answers. Items asking about loss of object resources were not included, because it was assumed that the vast majority of participants did not lose any personal property in the attacks. The third section assessed psychological distress with a measure designed to assess symptoms associated with acute stress disorder (Sattler et al., 2002). Acute stress disorder is related to posttraumatic stress disorder, but it lasts for a minimum of two days and a maximum of four weeks after the traumatic event. Several studies have documented symptoms associated with acute stress disorder in the first months following exposure to a traumatic event (Classen et al., 1998; Sattler et al., 2002; Waelde et al., 2001) (examples of items can be seen in Table 3). Participants used a 4-point scale, where 1 = not at all to 4 = very much, to indicate their answers. The fourth section assessed depression (adapted from Berdnt, 1986), concern for future attacks (adapted from Sattler et al., 2002), posttraumatic growth (adapted from Tedeschi and Calhoun, 1996), and coping (adapted from Sattler et al., 1995) (examples of the items can be seen in Table 2). Participants used a 4-point scale, where 1 = not at all to 4 = very much, to indicate their answers. The final section asked about social support and prior experience with traumatic events (adapted from Sattler et al., 2002). Participants used a 2-point scale answer the traumatic events items.

With the assistance of colleagues and graduate students, the questionnaires were administered to students in New York, South Carolina, Colorado, and Washington. In each location, the questionnaires were administered according to the same guidelines. There were a total of 1,283 college student participants (426 men, 857 women) who were attending universities in New York, New York; Charleston, South Carolina; Boulder, Colorado; and Bellingham, Washington. They did not receive any inducements to participate. Table 1 presents the demographic characteristics

Table 1. Demographic characteristics.

Characteristic	New York (*n* = 414)	South Carolina (*n* = 259)	Colorado (*n* = 280)	Washington (*n* = 329)
Gender				
Women	76	76	48	74
Men	24	24	52	26
Ethnicity				
White	64	91	86	87
Latino American	13	1	4	2
African American	8	6	1	1
Asian American	3	1	3	7
Other	1	1	7	3
Marital Status				
Single	91	93	97	98
Married	4	4	0.3	1
Separated/ Divorced	2	1	0.3	0.3
Widowed	1	1	0.3	0.3
Other	2	1	2	0.2
Age (*M* years)	21	22	20	19

Note: Gender, ethnicity, and marital status are percentages.

of each sample, including the number of participants at each location. Most of the participants were white and single, and the average age ranged from 19 to 22 years. Most persons asked to complete the questionnaire did so, resulting in a response rate of 99%.

Results

The results are presented in two sections. The first section consists of descriptive analyses that examine concern about future attacks, posttraumatic growth and resiliency, actions taken since the attacks, and prevalence of psychological distress symptoms. The second section contains analyses that examine factors that are associated with psychological distress.

Concern about Future Attacks and Safety

Items dealing with concern about future attacks and safety are listed in Table 2. The table shows that when considering the possibility of a future attack, about three-quarters of participants in each location were concerned the safety of a family member, and about half to three-quarters were concerned about their own safety. Between one-third and about two-thirds of the participants believed that the safety of a family member or friend might be at risk as a result of the response of the United States.

Posttraumatic Growth and Resiliency

Table 2 also presents items asking about growth and resiliency since the attacks. Most participants at each location reported experiencing posttraumatic growth and resiliency. About three-quarters of the participants reported that since the attacks they had new priorities about what is important in their lives, had new respect for people in their community, appreciated each day more, discovered that they are stronger than they thought they were, and learned that they can count on others in times of trouble. Participants also reported substantial increases in patriotism, and spending time with loved ones. In considering these findings, two points are especially noteworthy. First, a large percentage of participants—about three-quarters—reported increases in growth and resiliency. Second, the percentages of persons reporting growth and resiliency were vastly similar, irrespective of distance from the attacks.

Actions since the Attacks

Between half and almost three-quarters of the participants reported that since the attacks they had displayed the American flag (Table 2). Between one-third and half donated money to a charity.

Psychological Distress: Symptoms Associated with Acute Stress Disorder

Symptoms that are associated with acute stress disorder are presented in Table 3. The table shows that the prevalence of symptoms was relatively low (in the single digits to mid-teens), but higher in New York and South Carolina compared to Colorado and Washington. The most common symptoms in New York and South Carolina were avoiding things that reminded the person of the attacks, feeling anxious, having difficulty sleeping, having nightmares, feeling emotionally numb, and feeling irritable or on edge.

Table 2. Concern about future attacks, posttraumatic growth and resiliency, and actions.

Concern About Future Attacks	NY (*n*=414)	SC (*n*=259)	CO (*n*=280)	WA (*n*=329)
Afraid that a family member might lose his/her life or be seriously injured due to future terrorist attack	82	69	61	76
Afraid that I might lose my life or be seriously injured due to future terrorist attack	71	54	48	60
Have a family member or friend who will be at increased risk due to the U.S. response	65	42	47	52
Posttraumatic Growth and Resiliency and Feeling Patriotic *				
Feeling patriotic	82	85	80	84
Have new priorities about what is important in my life	77	81	73	76
Have new respect for people living in my community	77	83	78	87
Appreciating each day	77	78	75	82
Feeling closer to one or more family members	76	80	78	77
Discovered that I am stronger than I thought I was	75	74	66	65
Knowing that I can count on people in times of trouble	73	74	73	75
Feeling valuable to others	68	66	65	67
Showed concern for someone I did not know	60	61	60	57
Feeling that my life has purpose	60	46	35	50
Spending time with loved ones	53	43	42	50
Actions since the attacks				
Following news reports about the situation	98	99	98	99
Displaying American flag	81	54	51	71
Praying more	53	49	34	53
Donated money to charity	48	52	36	40

* Posttraumatic growth and resiliency numbers reflect the percentage of participants indicating an increase since the attacks.

Predicting Psychological Distress

To examine which variables were associated with psychological distress, a hierarchical multiple regression analysis was conducted for each location. For each analysis, the variables were entered in nine predictor blocks: demographic characteristics, fear of future attacks, changes in daily routines and activities as a result of the attacks, feeling angry, depression, posttraumatic growth, social support, actions taken to support the community since the attack, and somatic problems.

Table 3. Examples of symptoms associated with acute stress disorder.

Item	New York (n = 414)	South Carolina (n = 259)	Colorado (n = 280)	Washington (n = 329)
Avoiding things that remind me of the attacks	18	14	12	8
Feeling anxious	17	19	11	7
Having difficulty sleeping	16	21	11	11
Having nightmares	12	17	10	6
Feeling emotionally numb	11	13	6	8
Feeling irritable or on edge	10	13	4	2
Having difficulty remembering important things about the situation	5	5	2	3

Note: Data presented are percentages.

New York—The predictor blocks accounted for 69% of acute stress disorder symptom variance, $F(14, 360) = 58.33$, $p < .001$. Every block accounted for a significant portion of the acute stress disorder symptom variance (Table 4). For blocks that had more than one variable, the beta coefficients indicate that acute stress disorder symptoms were associated with being female, safety fears, and posttraumatic growth (life priorities and strength, motivation and purpose).

South Carolina—The predictor blocks accounted for 70% of acute stress disorder symptom variance, $F(14, 215) = 35.00$, $p < .001$. All blocks except social support accounted for a significant portion of the variance (Table 4). For blocks that had more than one variable, the beta coefficients indicate that acute stress disorder symptoms were associated with being female, prior exposure to traumatic events, safety fears, and posttraumatic growth (life priorities and strength, motivation and purpose).

Colorado—The predictor blocks accounted for 59% of acute stress disorder symptom variance, $F(14, 247) = 25.42$, $p < .001$. All blocks except

Table 4. Prediction of acute stress disorder symptoms by demographics, fear of future attacks, change in daily routine, anger, depression, posttraumatic growth, social support, community involvement, and somatic problems.

Variable	New York[2] (*n* = 414) *B*	*SE B*	β	South Carolina[3] (*n* = 259) *B*	*SE B*	β	Colorado[4] (*n* = 280) *B*	*SE B*	β	Washington[5] (*n* = 329) *B*	*SE B*	β
Step 1: Demographics												
Gender	.23	.06	.21***	.29	.07	.26***	.19	.05	.23***	.17	.04	.24***
Age	.02	.00	.03	-.03	.00	-.02	-.01	.01	-.01	-.02	.00	-.02
Prior Traumatic Exposure[1]	.01	.03	.03	.12	.03	.25***	.020	.03	.04	—	—	—
Step 2: Fear												
Fear for Safety	.35	.03	.56***	.35	.04	.50***	.25	.03	.43***	.20	.02	.43***
Fear of Future Attacks	.04	.04	.05	-.02	.05	-.03	-.02	.04	-.03	-.20	.02	-.04
Step 3: Change in Routine	.16	.03	.29***	.22	.03	.34***	.17	.03	.34***	.09	.02	.24***
Step 4: Anger	.15	.02	.31***	.16	.03	.29***	.09	.02	.22***	.10	.02	.22***
Step 5: Depression	.20	.05	.16***	.33	.05	.29***	.30	.04	.34***	.16	.04	.21***
Step 6: Posttraumatic Growth												
Life Priorities and Strength	.23	.03	.34***	.14	.05	.19**	.09	.03	.17**	.10	.02	.25***
Importance of Relationships	.06	.01	.06	.01	.01	.01	.10	.01	.13*	.10	.00	.13*
Motivation and Purpose	-.02	.01	-.17***	-.02	.01	-.17**	-.02	.01	-.13*	-.10	.01	-.09
Step 7: Social Support	.06	.03	.09*	.02	.04	.04	-.02	.03	-.05	-.35	.02	.09
Step 8: Community Involvement	.05	.02	.12**	.05	.02	.13**	-.08	.02	-.03	.15	.01	.05
Step 9: Somatic Problems	.44	.04	.41***	.43	.05	.43***	.37	.05	.35***	.32	.04	.34***

* = $p < .05$ ** = $p < .01$ *** = $p < .001$

[1] Prior exposure to traumatic events was included in the analyses for New York, South Carolina, and Colorado. Due to an oversight, participants in Washington did not complete these items.

[2] Step 1: $R^2 = .04, p < .001$; Step 2: $R^2 = .29, p < .001$; Step 3: $R^2 = .07, p < .001$; Step 4: $R^2 = .08, p < .001$; Step 5: $R^2 = .02, p < .001$; Step 6: $R^2 = .07, p < .001$; Step 7: $R^2 = .01, p < .05$; Step 8: $R^2 = .01, p < .001$; Step 9: $R^2 = .10, p < .001$.

[3] Step 1: $R^2 = .10, p < .001$; Step 2: $R^2 = .22, p < .001$; Step 3: $R^2 = .09, p < .001$; Step 4: $R^2 = .07, p < .001$; Step 5: $R^2 = .07, p < .001$; Step 6: $R^2 = .03, p < .001$; Step 7: $R^2 = .00, p > .05$; Step 8: $R^2 = .02, p < .01$; Step 9: $R^2 = .10, p < .001$.

[4] Step 1: $R^2 = .05, p < .01$; Step 2: $R^2 = .17, p < .001$; Step 3: $R^2 = .10, p < .001$; Step 4: $R^2 = .04, p < .001$; Step 5: $R^2 = .10, p < .001$; Step 6: $R^2 = .05, p < .001$; Step 7: $R^2 = .00, p > .05$; Step 8: $R^2 = .00, p > .05$; Step 9: $R^2 = .08, p < .001$.

[5] Step 1: $R^2 = .06, p < .001$; Step 2: $R^2 = .18, p < .001$; Step 3: $R^2 = .05, p < .001$; Step 4: $R^2 = .05, p < .001$; Step 5: $R^2 = .04, p < .001$; Step 6: $R^2 = .07, p < .001$; Step 7: $R^2 = .01, p > .05$; Step 8: $R^2 = .00, p > .05$; Step 9: $R^2 = .08, p < .001$.

social support and community actions accounted for a significant portion of the variance (Table 4). For blocks that had more than one variable, the beta coefficients indicate that acute stress disorder symptoms were associated with being female, prior exposure to traumatic events, safety fears, and posttraumatic growth (life priorities and strength, importance of relationships, and motivation and purpose).

Washington—The predictor blocks accounted for 54% of acute stress disorder symptom variance, $F(13, 315) = 27.94, p < .001$. All blocks except social support and community actions accounted for a significant portion of the variance (Table 4). For blocks that had more than one variable, the beta coefficients indicate that acute stress disorder symptoms were associated with being female, prior exposure to traumatic events, safety fears, and posttraumatic growth (life priorities and strength, importance of relationships).

Conclusions, Implications, and Future Directions

The results show four key findings. First, most participants were concerned about the possibility of future attacks, and concerned for their own safety and the safety of family and friends due to future attacks. Second, the prevalence of acute stress disorder symptoms (i.e., psychological distress) was relatively low overall (in the single digits and teens), but somewhat higher in New York and South Carolina than in Colorado and Washington. Third, acute stress disorder symptoms were associated with concern about safety, changes in daily routines and activities as a result of the attacks, feeling angry, depression, posttraumatic growth, and somatic problems. Further, in New York, social support and actions taken to support the community since the attack were associated with acute stress disorder symptoms. Fourth, nearly three-quarters of the participants at each location reported posttraumatic growth, cognitive adaptation, resource gain, and resiliency. The areas of growth or resiliency included the self, relationships with others, and philosophy of life. Participants reported reflecting on and reassessing their lives (e.g., having new priorities about what is important in their lives; trying to grow as a person as a result of the experience), and being resilient (e.g., discovering that they were stronger than they thought they were). Many participants were taking positive action by showing concern for someone they did not know and showing unity by displaying the American flag, and donating money to charity.

These findings are especially important, given that the study was conducted only three weeks after the attacks, the participants were college students, and the samples comprised persons living in four distinct areas of the country. Together, these findings support the conservation of resources stress theory, posttraumatic growth theory, and the theory of cognitive adaptation, and extend previous research (e.g., Freedy et al., 1994; Sattler et al., 2002; Smith and Freedy, 2000).

One influential factor associated with growth and vulnerability after the experience of a traumatic event is coping style. Three general styles of coping include active or problem-solving coping, acceptance and positive reinterpretation, and avoidance (Carver et al., 1989). Active coping involves responses that attempt to solve problems by taking direct action in order to reduce negative consequences of the experience. Acceptance and positive reinterpretation refers to accepting the stressor as unavoidable and focusing on the positive aspects of the situation. Avoidance coping generally refers to emotion-focused strategies, and may involve denial or withdrawal from the situation in an attempt to reduce distress.

Active coping has been shown to be effective in handing severe stressors (Taylor and Clark, 1986), and associated with lower levels of depression (Aldwin, 1991) and symptoms associated with posttraumatic stress disorder (Solomon et al., 1988). In situations where direct action is not possible, positive reinterpretation also may be an effective coping strategy. In these situations, it may only be possible to accommodate, rather than change, the stressor. "Positive reinterpretation can be used to manage one's emotions in an uncontrollable situation and to motivate the use of active coping strategies in a controllable situation. Most important, positive reinterpretation and acceptance coping strategies appear to be significant determinants of stress-related growth. By allowing individuals to accept a situation and focus on its positive aspects and implications, these coping strategies may be the most responsible for contributing to people's beliefs that they have benefitted from a stressful life experience" (Updegraff and Taylor, 2000, p. 13). It is likely that the types of growth and resiliency shown in the present study involve both active coping and positive reinterpretation. It is also possible that the growth shown in this study is a form of coping occurring during the acute stage. Future research needs to examine how such responses are associated with long-term psychological functioning.

Secondary stressors have continued since the attacks, and at the time this paper was written—nine months after the attacks—the stressors continue for many Americans. Secondary stressors include stressful life events, strains, and hassles that develop in the wake of a disaster, and can include delays in obtaining resources, employment difficulties, financial difficulties, and threats to one's safety. Secondary stressors can tax personal characteristic, energy, and condition resources; exacerbate the influence of preexisting stressors; and contribute to psychological distress and relationship difficulties (Baum, 1991; Norris and Uhl, 1993).

Americans have endured a number of secondary stressors: lengthy clean-ups at the World Trade Center site and Pentagon; uncertainty concerning whether persons and organization(s) responsible for the attacks would be brought to justice; the U.S. government's declaring that the country is at war and on heightened alert; the U.S. government's warning of possible future attacks; anthrax sent in letters to government officials and news personnel; and possible military action in countries other than Afghanistan (e.g., Iraq).

As a direct result of the attacks, many companies have faced financial hardships, especially the travel and airline industries. Short-term, long-term, and permanent job layoffs have occurred, and it is not clear whether certain companies and industries will be able to recover. Because these secondary stressors have been prolonged, it is possible that delayed mental health problems, as well as additional growth, may develop many months after the event.

Understanding the nature and role of secondary stressors is critically important to intervention and recovery programs. Since these stressors continue to exist, research should continue to examine the mental health and psychological implications of the terrorist attacks and threats. Especially important issues are identifying which persons are most at risk of adjustment difficulties, which interventions may minimize or prevent adjustment problems, and which interventions are most helpful. It would be useful for studies to examine factors associated with growth and distress, including optimism, perceptions of control over life events, sense of self, preexisting vulnerabilities, and the characteristics of the situation.

It is likely that some participants have had more exposure to regional large-scale disasters than others, and that such exposure might modify responses to subsequent stressors. For example, Charleston, South Carolina, had experienced more large-scale disasters and disaster threats in recent years than the other three locations included in this study. In 1989, Hurricane Hugo, a category four storm, struck Charleston and caused $7 billion in damage. During the 1990s, about one-half dozen storms seriously threatened the area. In 1999, a large and powerful storm, Hurricane Floyd, was predicted to strike Charleston, and almost three-quarters of the residents fled their homes (Sattler, 2001). Research and theory suggests that prior and repeated exposure to large-scale traumatic events and threats can sensitize persons to future traumatic events (Sattler et al., 2000). This might explain, in part, the comparable levels of distress in New York and South Carolina. The higher levels in those two locations also may reflect the fact that these communities are closer to the areas attacked than Colorado and Washington. This idea is supported by the finding that more persons in New York and South Carolina were concerned about future attacks than in Colorado and Washington.

This correlational study has several limitations. First, because participants were college students at select campuses in four states, the findings may not generalize to all college students or to all Americans. The findings offer detailed information about how persons within a specific demographic group responded to the tragic event. Future research should extend these findings by examining individuals who represent other groups. Second, we do not know about preexisting psychopathology among the participants. It is possible that a small proportion of participants were experiencing distress or living with a mental health issue prior to the attacks. If so, these issues may have been reflected, to some degree, in the participants' answers, and we cannot conclude with complete certainty that the levels of distress shown in the results are solely due to the attacks. Obtaining information about psychological functioning before the traumatic event is the best way to address this problem. However, since the attacks were not foreseen, it was not possible to obtain this information. Third, this study relied on self-report and it is possible that some participants may have tried to present themselves in a favorable or unfavorable light. However, special care was taken to minimize such response biases, including making participation completely anonymous and confidential. In reviewing participation in other disaster studies, Norris (1992) suggests that self-report data following disasters appear to be reliable. A longitudinal study is underway to examine any delayed mental health problems as well as growth and resiliency since the initial assessment reported here.

References

Adams, P.R. and G.R. Adams. 1984. "Mount Saint Helens's Ashfall: Evidence for a Disaster Stress Reaction." *American Psychologist* 39(3): 252–260.

Aldwin, C.M. 1991. "Does Age Affect the Stress and Coping Process? Implications of Age Differences in Perceived Control." *Journal of Gerontology* 46: 172–180.

Baum, A. 1991. "Toxins, Technology, and Natural Disasters." Pp. 97–139 in A. Monat and R.S. Lazarus, eds., *Stress and Coping: An Anthology*, 3rd edition. New York: Columbia University Press.

Benight, C.C., R.W. Freyaldenhoven, J. Hughes, J.M. Ruiz, T.A. Zoschke, and W.R. Lovallo. 2000. "Coping, Self-efficacy, and Psychological Distress Following the Oklahoma City Bombing." *Journal of Applied Social Psychology* 30(7): 1331–1344.

Berndt, D.J. 1986. *Multiscore Depression Inventory Manual*. Los Angeles: Western Psychological Services.

Calhoun, L.G., and R.G. Tedeschi. 2001. *Posttraumatic Growth: the Positive Lessons of Loss*. Washington, D.C.: American Psychological Association.

Calhoun, L.G., and R.G. Tedeschi. 1998. "Beyond Recovery from Trauma: Implications for Clinical Practice and Research." *Journal of Social Issues*, Special Issue—Thriving: Broadening the Paradigm Beyond Illness to Health 54(2): 357–371.

Canino, G., M. Bravo, M. Rubio-Stipec, and M. Woodbury. 1990. "The Impact of Disaster on Mental Health: Prospective and Retrospective Analyses." *International Journal of Mental Health* 19: 51–69.

Carver, C.S., M.F. Scheier, and J.K. Weintraub. 1989. "Assessing Coping Strategies: A Theoretically Based Approach." *Journal of Personality and Social Psychology* 56: 267–283.

Classen, C., C. Koopman, R. Hales, and D. Spiegel. 1998. "Acute Stress Disorder as a Predictor of Posttraumatic Stress Symptoms." *American Journal of Psychiatry* 155: 620–624.

Freedy, J.R., M.E. Saladin, D.G. Kilpatrick, H.S. Resnick, and B.E. Saunders. 1994. "Understanding Acute Psychological Distress Following Natural Disaster." *Journal of Traumatic Stress* 7: 257–273.

Gibbs, M.S. 1989. "Factors in the Victim that Mediate between Disaster and Psychopathology: A Review." *Journal of Traumatic Stress* 2: 489–514.

Hobfoll, S.E. 1989. "Conservation of Resources: a New Attempt at Conceptualizing Stress." *American Psychologist* 44: 513–524.

Hobfoll, S.E. 1998. *Stress, Culture, and Community: the Psychology and Philosophy of Stress*. New York: Plenum Press.

Janoff-Bulman, R., and A.R. Berger. 2000. "The Other Side of Trauma: Towards a Psychology of Appreciation." Pp. 29–44 in J.H. Harvey and E.G. Miller, eds., *Loss and Trauma: General and Close Relationship Perspectives*. New York: Brunner-Routledge.

Kaiser, C.F, D.N. Sattler, D.R. Bellack, and J. Dersin. 1996. "A Conservation of Resources Approach to a Natural Disaster: Sense of Coherence and Psychological Distress." *Journal of Social Behavior and Personality* 11: 459–476.

Kaniasty, K., and F.H. Norris. 1995. "In Search of Altruistic Community: Patterns of Social Support Mobilization Following Hurricane Hugo." *American Journal of Community Psychology* 23: 447–477.

Lazarus, R.S., and J.B. Cohen. 1977. "Environmental Issues." Pp. 89–127 in L. Altman and J. F. Wohlwill, eds., *Human Behavior and the Environment: Current Theory and Research*. New York: Plenum.

Monnier, J., and S. Hobfoll. 2000. "Conservation of Resources in Individual and Community Reactions to Traumatic Stress." Pp. 325–336 in A. Shalev, R. Yehuda, and A. C. McFarlane, eds., *International Handbook of Human Response to Trauma*. New York: Kluwer Academic/Plenum Publishers.

Norris, F.H. 1992. "Epidemiology of Trauma: Frequency and Impact of Different Potentially Traumatic Events on Different Demographic Groups." *Journal of Consulting and Clinical Psychology* 60: 409–418.

Norris, F.H., and G.A. Uhl. 1993. "Chronic Stress as a Mediator of Acute Stress: the Case of Hurricane Hugo." *Journal of Applied Social Psychology* 23: 1263–1284.

North, C.S, S. Nixon, S. Shariat, S. Mallonee, J.D. McMillen, E.L Spitznagel, and E.M. Smith. 1999. "Psychiatric Disorders among Survivors of the Oklahoma City Bombing." *Journal of the American Medical Association* 282: 755–762.

Rubonis, A.V., and L. Bickman. 1991. "Psychological Impairment in the Wake of Disaster: the Disaster Psychopathology Relationship." *Psychological Bulletin* 109: 384–399.

Sattler, D.N. 2001. "Repeated Exposure to Hurricanes and Willingness to Evacuate: Implications from the Hurricane Floyd Experience." Pp. 69–76 in J. R. Maiolo, J. C. Whitehead, M. McGee, L. King, and J. C. Johnson, eds., *In the Aftermath of Hurricane Floyd: Recovery in the Coastal Plain*. Wilmington, NC: Coastal Carolina Press.

Sattler, D.N., J.F. Freedy, K. Anderson, and C.F. Kaiser. 1997. "Natural Disasters and Psychological Adjustment: Implications of Research for Intervention Efforts." *Journal of Psychological Practice* 3: 113–127.

Sattler, D.N., C.F. Kaiser, and J.B. Hittner. 2000. "Disaster Preparedness: Relationships among Prior Experience, Personal Characteristics, and Distress." *Journal of Applied Social Psychology* 30: 1398–1420.

Sattler, D.N., A. Preston, C.F. Kaiser, V.E. Olivera, J. Valdez, and S. Schlueter. 2002. "Hurricane Georges: a Cross-national Study Examining Preparedness, Resource Loss, and Psychological Distress in the U. S. Virgin Islands, Puerto Rico, Dominican Republic, and the United States." *Journal of Traumatic Stress* 15: 339–350.

Sattler, D.N., J.M. Sattler, C. Kaiser, B.A. Hamby, M. Adams, L. Love, J.M. Winkler, C. Abu-Ukkaz, B. Watts, and A. Beatty. 1995. "Hurricane Andrew: Psychological Distress among Shelter Victims." *International Journal of Stress Management* 2: 133–143.

Shalev, A.Y. 1992. "Posttraumatic Stress Disorder among Injured Survivors of a Terrorist Attack: Predictive Value of Early Intrusion and Avoidance Symptoms." *Journal of Nervous and Mental Disease* 180 (8): 505–509.

Shore, J.H., E.L. Tatum, and W.M. Vollmer. 1986. "Psychiatric Reactions to Disaster: the Mount St. Helens Experience." *American Journal of Psychiatry* 143: 590–595.

Smith, B.W., and J.R. Freedy. 2000. "Psychosocial Resource Loss as a Mediator of the Effects of Flood Exposure on Psychological Distress and Physical Symptoms." *Journal of Traumatic Stress* 13: 349–357.

Solomon, Z., M. Mikulincer, and H. Flum. 1988. "Negative Life Events, Coping Responses, and Combat-related Psychopathology: a Prospective Study." *Journal of Abnormal Psychology* 97: 302–307.

Taylor, S.E. 1983. "Adjustment to Threatening Events: a Theory of Cognitive Adaptation." *American Psychologist* 38: 1161–1173.

Taylor, S.E., and L.F. Clark. 1986. "Does Information Improve Adjustment to Noxious Events?" In M. J. Saks and L. Saxe, eds., *Advances in Applied Social Psychology* 3: 1–28. Hillsdale, NJ: Erlbaum.

Taylor, S.E., V.S. Helgeson, G.M. Reed, and L.A. Skokan. 1991. "Self-generated Feelings of Control and Adjustment to Physical Illness." *Journal of Social Issues* 47: 91–109.

Tedeschi, R.G., and L.G. Calhoun. 1996. "The Posttraumatic Growth Inventory: Measuring the Positive Legacy of Trauma." *Journal of Traumatic Stress* 9 (3): 455–472.

Tedeschi, R.G., and L.G. Calhoun. 1995. *Trauma and Transformation: Growing in the Aftermath of Suffering*. Thousand Oaks, CA: Sage Publications, Inc.

Tucker, P., W. Dickson, B. Pfefferbaum, N.B. McDonald, and G. Allen. 1997. "Traumatic Reactions as Predictors of Posttraumatic Stress Six Months after the Oklahoma City Bombing." *Psychiatric Services* 48 (9): 1191–1194.

Tucker, P., B. Pfefferbaum, S.J. Nixon, and W. Dickson. 2000. "Predictors of Post-traumatic Stress Symptoms in Oklahoma City: Exposure, Social Support, Peritraumatic Responses." *Journal of Behavioral Health Services and Research* 27 (4): 406–416.

Updegraff, J.A., and S.E. Taylor. 2000. "From Vulnerability to Growth: Positive and Negative Effects of Stressful Life Events." Pp. 3–28 in J. H. Harvey and E.D. Miller, eds., *Loss and Trauma: General and Relationship Perspectives*. Philadelphia, PA: Brunner-Routledge.

Waelde, L.C., C. Koopman, J. Rierdan, and D. Spiegel. 2001. "Symptoms of Acute Stress Disorder and Posttraumatic Stress Disorder Following Exposure to Disastrous Flooding." *Journal of Trauma and Dissociation* 2: 37–52.

Wood, J.M., R.R. Bootzin, D. Rosenhan, S. Nolen-Hoeksema, and F. Jourden. 1992. "Effects of the 1989 San Francisco Earthquake on Frequency and Content of Nightmares." *Journal of Abnormal Psychology* 101: 219–224.

Acknowledgments

This project was supported by a grant to David N. Sattler from the Natural Hazards Center, University of Colorado, Boulder, and by Western Washington University. I would like to express my sincere appreciation to the Natural Hazards Center and Western Washington University for their support, and to the many persons whose generous assistance made this project possible: the participants, Mary Fran Myers, Virginia Shabatay, David Koch, Kim Edel, Sidney Hochman, Elke Weber, Charles Kaiser, Lori Peek-Gottschlich, Lori Hunter, Robert M. Thorndike, Dale Dinnel, Ron Kleinknect, Larry Symons, Kristi Lemm, Geri Walker, Leslie Croot, and Peter Ballantyne.

Correspondence should be addressed to David N. Sattler, Department of Psychology, Western Washington University, 516 High Street, Bellingham, WA 98225-9089; e-mail: David.Sattler@wwu.edu.

Community Isolation and Group Solidarity: Examining the Muslim Student Experience after September 11th

Lori A. Peek
Department of Sociology
University of Colorado–Boulder

As news spread of the events of September 11, 2001, people around the world reacted with shock, horror, and grief. However, in addition to reports of the immense destruction and suffering caused by the terrorist attacks, there was also a significant amount of media coverage regarding the pro-social behavior that subsequently occurred. For example, thousands of individuals lined up at blood donation stations across the United States; volunteers converged at the disaster sites (Kendra and Wachtendorf, 2001; Webb, 2002); individuals and organizations donated unprecedented amounts of money and physical goods to charities (Turkel, 2002); and a multitude of candlelight vigils, communal interfaith services, teach-ins on university campuses, and other such events were organized to help people cope in the aftermath of the disaster. Radio and television announcers discussed this unity and positive behavior, and they frequently uttered the phrases, "Today we are all Americans. Today we are all New Yorkers." Upholding this national identity offered many Americans a way to begin to deal with the devastation caused by the attacks.

Some scholars asserted that the widespread public consensus and outpouring of pro-social behavior that followed the attacks closely resembled reactions which have been documented following most natural disasters (Alexander, 2002; Mileti, 2001; Tierney, 2002; Webb, 2002). That is, the aftermath of September 11th was compared to what social scientists have labeled as the creation of an "altruistic" (Fritz, 1961) or "therapeutic community" (Barton, 1970). Essentially, these terms characterize the coming together of a community in the immediate aftermath of a disaster, which ultimately provides disaster survivors the physical and emotional support that is necessary to create a sense of solidarity and unity of purpose (Cuthbertson and Nigg, 1987).

While undoubtedly there was a tremendous increase of pro-social behavior after September 11th, there was also a general need to focus a sense of blame (Alexander, 2002). The call to find those who were responsible was understandable, given the intentional, violent, and criminal nature of the attacks. However, the attribution of blame, and subsequent scapegoating (see further Barton, 1970; Bucher, 1957; Drabek and Quarantelli, 1967; Quarantelli and Dynes, 1976) that followed the attacks resulted in some ethnic and religious minority groups in the United States becoming disconnected from the national therapeutic community that emerged. Indeed, while the events of September 11th brought many together and led to increased feelings of patriotism and national unity (Turkel, 2002), the attacks isolated certain marginalized members of U.S. society as well. Truly, we were not "all Americans" on that day.

This paper examines the Muslim experience after the events of September 11th. I begin by reviewing the literature regarding post-disaster communities, with a specific focus on community cohesion and isolation after natural and technological disasters and intentional acts of violence. Next, I discuss the setting in which this study was conducted, the research participants, and the methods that were used. I then explicate several reasons why Muslim university students in New York City often felt excluded from the larger therapeutic community that emerged after the September 11th attacks. The group solidarity that developed among Muslims in response to this exclusion is detailed. The report concludes with a discussion of the sociological implications of post-disaster community isolation as well as suggestions for future disaster research.

The Post-disaster Community

Scholars recognize that disasters and large-scale catastrophes provide unique occasions to examine how individuals and groups behave in crisis situations (Quarantelli, 1998). Fritz (1961) posits that disasters offer a realistic laboratory for testing the integration, stamina, and recuperative power of large-scale social systems and communities. Subsequently, there is general consensus in the social science literature that along with destruction and devastation, opportunities also arise out of disastrous events (Quarantelli and Dynes, 1977).

Natural disasters represent a definitive risk, are often sudden in their onset, and are largely viewed by the public as indiscriminate and external to social systems. Hence, heightened levels of community solidarity and social cooperation are repeatedly reported in the aftermath of natural disasters

(Barton, 1970; Dacy and Kunreuther, 1969; Drabek, 1986a; Dynes, 1970; Fritz, 1961) as people come together to react to unforeseen and threatening situations. After the disaster strikes, there is a temporary focus on the urgent needs of the victims and their families, which precipitates a pulling together of the community, albeit typically short-lived (Demerath and Wallace, 1957; Drabek, 1986a; Oliver-Smith, 1979).

Although there is much evidence that natural disasters and catastrophes bring people together, there has been less empirical or theoretical examination of why communities break down in certain disaster situations (Barton, 1970; Erikson, 1976), or why some groups become isolated from emergent therapeutic communities. Research has shown that the type of disaster agent may directly correlate with whether community consensus or conflict follows an event (Couch and Kroll-Smith, 1985; Drabek, 1970; Drabek, 1986b; Quarantelli and Dynes, 1976). Specifically, crises that involve some sort of perceived human culpability are more likely to result in the demise of communal solidarity. For example, we know that civil disturbances (Hewitt, 1997; Warheit, 1976), riots (Quarantelli, 1993; Scanlon, 1988), and technological disasters (Couch and Kroll-Smith, 1985, 1991; Erikson, 1994; Kroll-Smith and Couch, 1990; Neal, 1984) are more likely to result in conflict, blame attribution, long-term negative impacts, and anti-social behavior.

Erikson (1994) posited that "a new species of trouble" emerged with the development of disasters that represent the work of humankind, versus those previously studied as acts of nature. In their examination of an underground mine fire, Kroll-Smith and Couch (1990) discovered that the type of extreme environment created by this chronic disaster resulted in the destruction of the social bond within the community. Edelstein (1988) reported that victims of residential toxic exposure came to feel stigmatized and isolated from friends, relatives, and co-workers living outside their contaminated community. Using case study data from two separate technological catastrophes, Cuthbertson and Nigg (1987) found that the factors conducive to the emergence of a therapeutic community were not present in these instances, and therefore conflict, rather than solidarity, followed the events. Baum et al. (1983) concluded that technological catastrophes are more likely to cause chronic stress and therefore have more widespread and long-term social impacts than natural disasters.

The human tendency to blame the devastation caused by natural disasters on nature, fate, or some divine force often leads to higher levels of post-disaster, pro-social behavior.[1] Conversely, because human beings generate technological disasters, these events are more likely to result in conflict and other negative communal responses. Yet, even during technological disasters, there is a tendency to attribute the devastation to human error and to label the event an "accident," allowing consensus-oriented behavior to occur.

Intentional Acts, Blame, and Community Isolation

A final type of disaster to be considered differs from those that are of natural or technological origin in that under no circumstance is the event attributed to natural forces or human error. These acts—most often labeled as terrorism—are human-conceived, violent, and purposely designed to cause widespread fear, psychological and social trauma, and physical destruction.

Hence, one of the most obvious characteristics that distinguishes September 11th from other large-scale catastrophes that have impacted the United States is that a group of individuals was directly responsible for the destruction and loss of life. The disaster agent—an intentional, human-conceived attack—certainly influenced the response, both positively and negatively, that followed the events (Gunn, 2001; Hewitt, 1997).

The demonstrative nature of terrorist acts elicits feelings of revulsion and anger in those who witness them (Juergensmeyer, 2000), and the ensuing anger may become manifest in various forms. In the days, weeks, and months after September 11th, members of various ethnic and religious groups were targeted for blame. Thousands[2] of Arabs, Muslims, Sikhs, and individuals who appeared to be of Middle Eastern descent were victims of discrimination, harassment, racial and religious profiling, and verbal and physical assault (American-Arab Anti-Discrimination Committee, 2002; Council on American-Islamic Relations, 2002; Federal Bureau of Investigation, 2002; United States Department of Justice, 2002). Over the past two decades, Arabs and Muslims living in the United States have been victims of backlash violence, largely triggered by conflict in the Middle East and acts of terrorism associated with Arabs and Muslims (Human Rights Watch, 2002). However, the victimization of these and other minority groups in the aftermath of September 11th was unique in the magnitude and severity. The devastation of September 11th, combined with the ensuing attribution of blame to groups perceived as exhibiting ethnic or religious characteristics similar to those of the hijackers, led to a significant perception of communal isolation for the Muslims studied herein.

Setting and Methods

Less than three weeks after September 11th, I traveled to New York City to interview Muslim students enrolled in area colleges and universities. I contacted the participants for this study through the Muslim Student Associations (MSAs) on their respective campuses.[3] After the first trip to New York City, I conducted follow-up interviews with the same group to gain a more nuanced understanding of the long-term impacts of September 11th. I returned for a second site visit in December of 2001 and a third visit in April of 2002. Over five weeks of direct study and observation were completed during those three visits.

I employed three descriptive fieldwork methods for this research project: focus group interviews, individual interviews, and participant observation. I used focus group interviews as the primary means of data collection during the initial phases of this study. In total, I conducted 19 focus groups, which ranged in size from three to 15 participants. The focus groups lasted from one to four hours. Conducting interviews in a focus group setting allowed several students to be heard, effectively utilizing limited time and resources to gather a breadth of perishable data (Krueger, 1988; Morgan, 1997).

During subsequent trips, I used one-on-one, semi-structured interviews as the primary form of data collection. This shift in method elicited richer and more in-depth data (Lofland and Lofland, 1995; Reinharz, 1992). In total, I conducted 55 individual interviews, lasting between one and three hours. Combining the focus groups (n = 19) and individual interviews (n = 55), yielded a total of 76 qualitative interviews, which were audio-taped, transcribed, and coded for analysis.

I utilized participant observation throughout the project to collect additional data (Adler and Adler, 1987; Lofland and Lofland, 1995), and recorded extensive sets of field notes. I engaged in a variety of activities with the Muslim students, which helped me to connect with them and better understand their perspectives. For example, I observed Friday prayers at mosques, ate Ramadan[4] dinners at various religious centers, attended religious and political speeches and informational events, and attended a Muslim wedding. I also visited dorms and apartments that were evacuated after the collapse of the World Trade Center towers, walked around the campuses and traveled on the subways and buses with the participants, and often shared a cup of coffee, lunch, or dinner with the students. Participating in these various activities, as well as simply spending casual time with the Muslim students, served to further verify the experiences discussed in their interviews (Bogdan and Taylor, 1998; Lofland and Lofland, 1995). For example, one of the greatest fears expressed by the participants immediately after September 11th, particularly by the women, was traveling on the subway alone (Peek, 2002). As I walked through subway stations and sat on trains with these young women, it quickly became clear that "the looks" they reported were unquestionably occurring.

In between site visits to New York City, I continued to talk on the phone, send cards and letters, and correspond via e-mail with most of the participants. Staying in contact between trips helped to secure follow-up interviews when I returned to the field, provided new themes for exploration, and resulted in a more personal researcher role.

Muslim Student Sample Information

In total, 99 students from seven different colleges and universities in the New York City area were interviewed between September 30, 2001, and April 11, 2002. Most of the students were interviewed on two or three separate occasions over the seven months following September 11th.

Sixty-eight of the students interviewed were female and 31 were male.[5] Eighty-five were undergraduates and 14 were graduate students. The participants ranged in age from 18 to 31. Ninety of the students were single and nine were married. Sixty-eight of the interviewees were U.S. citizens, and the other 31 had student visas or were permanent residents. All of the participants were fluent in English, and over 75% spoke at least one other language (such as Arabic, Bengali, Cambodian, Farsi, French, Hebrew, Indonesian, Japanese, Persian, Punjabi, and Urdu). Most of the participants in this study were first- or second-generation immigrants who were Muslim by birth. Converts to Islam were interviewed as well.

The interviewees reported a wide range of ethnic backgrounds. The majority of participants (53 students) was of South Asian or Asian descent. Thirty students identified themselves as Arab or Arab-American. Eight students were Caucasian, four were African-American, three were Latino, and one was a Pacific Islander. However, these standard ethnic categories do not depict the true diversity of the sample population. The following are just some of the national backgrounds reported by the interviewees: Afghanistan, Albania, Bangladesh, Cambodia, Canada, Egypt, Germany, Great Britain, Greece, Guinea, Guyana, India, Indonesia, Iran, Japan, Kuwait, Lebanon, Morocco, Pakistan, Palestine, Puerto Rico, Saudi Arabia, Syria, Trinidad, Turkey, United States, Yemen, and Zimbabwe.

Finally, it is important to note that this sample population was self-selected and highly religious. This religious identification is based on my own observations, as well as the self-characterization of the majority of students. Almost all of the participants reported praying five times a day (one of the five fundamental pillars of Islam), fasting during the Muslim Holy Month of Ramadan (another fundamental pillar of Islam), being active members of religious organizations, and abstaining from religiously prohibited activities (such as drinking alcohol or eating pork). Most of the interviewees also chose to dress in an Islamic (modest) manner. The vast majority of women interviewed (over 90%) wore the *hijab,*[6] and two of the women wore the *niqaab.*[7] The men also dressed modestly. The level of religiosity and manner of dress of most of the interviewees—the fact that they were visibly identifiable as Muslim—certainly affected their individual experiences, and their interpretations of those experiences, after September 11th.

Isolation from the Therapeutic Community

This section uses qualitative data to examine the feelings of community isolation reported by the Muslim students, and the consequences of those perceptions. In analyzing the data, four primary factors emerged (Glaser and Strauss, 1967), which represent the reasons why many of the students felt excluded from the post-September 11th national therapeutic community.

Muslims as Blamed Category

Although disasters often motivate individuals and groups to behave in altruistic ways, it is also widely recognized that catastrophic events can result in a search for scapegoats to blame for the destruction and loss of life (Drabek and Quarantelli, 1967). As stated previously, disasters that are caused by intentional human acts seem most likely to produce the negative behaviors associated with blame attribution, such as social isolation, discrimination, harassment, and violence.

The Muslims interviewed after September 11th did not feel a part of the larger community because they were quickly portrayed as "the enemy" or "the other." Therefore, the majority felt as if they were being blamed for the horrific events. On the morning of the attacks, some of the students had yet to learn what had transpired, but they were already implicated. Habeel,[8] whose family immigrated to the United States from Pakistan when he was a child, discussed what happened to him on September 11th:

> I heard it from the bus operator. The first thing he told me was, "Your people have done this thing." I was like, "What?" I didn't even know anything at that point. I could see the clouds of dust. He said, "They took the planes and hit the World Trade Center and the Pentagon. Your people did that. The President is going to attack." I thought he was just making fun of me. Then I went home and saw the TV. From the first minute I felt like they have cut off the Muslims.

As Allport (1954, p. 258), in his discussion of scapegoating so succinctly claimed, "Anger wants a personal victim, and it wants it now." Some people in the United States directed their anger toward innocent civilians. It was this misdirected anger, and blaming, that led Muslims to feel disconnected from others. As time passed, some of the students indicated that they still felt as if others were holding them responsible for the events of September 11th. Fatima, a native of New York City, said the following:

> Even now, weeks after it happened, we still get looks. People still look at you like, you know, you did something wrong. You just have to smile back and then they're surprised. . . . I'm a person just like you. I didn't do anything.

Exclusion from Mourning, Bonding, and Helping Behaviors

Disasters may strengthen community identification by allowing extensive opportunities for participation in community-relevant activities (Quarantelli and Dynes, 1976). The acts of providing and receiving emotional support, in conjunction with assisting fellow victims, are a crucial aspect of the healing process and in large part characterize the therapeutic community (Barton, 1970). After the September 11th attacks, Muslims in this study reported feeling excluded from the process of mourning and social bonding. They also felt unable to engage in helping behaviors. The perceived or actual inability to participate in such emotional processes and behaviors naturally resulted in feelings of individual and group isolation.

Interviewees commented that they could not talk about the events of September 11th as "normal" people, but rather had to respond as a "Muslim" person. Therefore, the students often did not have the opportunity to simply grieve or tell their stories of what happened on that day. They were busy answering questions about their faith, the meaning of *jihad,*[9] and whether or not they personally condone violence. In the United States, Muslims had to immediately mobilize to defend themselves and their religion, and many Islamic advocacy groups and scholars issued statements of Islam's condemnation of attacks such as those perpetrated on September 11th (Council on American-Islamic Relations, 2002). Iffat, a native of New York City, discussed the inability to adequately express her grief:

> I'm so sad about it, but you can't really show that you're sad because you have to be more defensive. We can't express how we were just as sad as anyone else. I don't think I felt included as one of them who were affected by it. I had to defend myself and defend my religion, more than being a New Yorker who was affected by it.

Many of the interviewees said that they wanted to participate in post-September 11th memorial services and volunteer activities, so that they could experience the opportunity to grieve for the victims, show support for the rescue workers, and connect with other survivors. However, Muslims reported a certain level of discomfort being a part of such events. The interviewees either were worried for their own safety, or were concerned that their visible presence might upset others. For example, Raja, whose family moved to the United States from Pakistan when she was 15, was acutely aware of her Muslim appearance, and the feelings that she might provoke:

> In my neighborhood, there's a fire station, and every time I would pass by it, there was a picture of one of the firefighters who passed away. It was so sad. All I wanted to do was go inside and hug every firefighter and tell them, "You guys are wow, amazing. I'm supporting and praying for you." But then

> I was like, if I do that . . . Every time they see me, probably I remind them of September 11th. Every time. Even every time I pass by, I'm like, oh God, I hope I don't remind them. And I know I do with my headscarf. Sometimes I feel ashamed. I really do.

Because of their apprehension, many of the Muslim students stayed at home or away from the disaster site, at least initially. Henna, a female graduate student whose educational background qualified her to volunteer at the World Trade Center, declined out of fear for her personal safety:

> For me, I wanted to join those people who were volunteering downtown and do stuff. My undergraduate degree was in engineering and they needed engineers there to help with excavation at the site. They also needed people who could translate. To me, that was the American community coming together and trying to do what they can. But I didn't feel like I could for my own safety. I wear a headscarf. I wanted to be a part of that community, but I'm not really.

Muslims had a need, just as their fellow Americans did, to heal the psychological wounds of the attacks (Council on American-Islamic Relations, 2002). However, through actual or perceived exclusion, many of the Muslims interviewed did not have the same opportunities to recover from the devastation through normal post-disaster outlets.

Concern for Safety

After September 11th, Muslims were worried about their personal safety and the security of the United States, as were other Americans. No one knew whether subsequent attacks were planned, and feelings of uncertainty and fear pervaded. However, Muslims also experienced another level of fear. Reports were soon released that the September 11th attacks had been perpetrated by "Muslim" men. In light of the history of backlash violence against Muslims in the United States, the interviewees recognized the imminent possibility of hostile repercussions against groups perceived as similar to the hijackers.

During interviews, participants described their intense concern for the safety of their friends and family members who were Muslim, as well as for their own safety. Sadiyah, a native of Syria, talked about her response to learning of the attacks:

> I was on campus. I heard about it through a professor who happened to pass by our classroom. There was panic. Everybody got up and went to make phone calls. My immediate reaction was first, I couldn't believe it. I was shocked. Then I was like, safety. What's going to happen to me? I'm walking around the halls and people are giving me really dirty looks. So I came down here [to the MSA office]. There were Muslims gathered and a

> couple of other groups like the Puerto Rican Alliance. They came over to us and said, "We advise that you guys go home. We don't know what's going to happen around campus."

Janan, the daughter of a West African immigrant and an African American convert, was born and raised in the United States. Because Janan wore a headscarf, and therefore was highly visible, she talked about how frightened she was to leave her home after learning of the attacks:

> I was getting a lot of e-mails that were saying "Stay in the house, don't go out if you don't have to." It was freaking me out. I remember somebody was on the news talking about how they should put all the Muslims in concentration camps. When people are saying stuff like that, they were talking about it like it was a valid viewpoint, when people are talking about things like that, you're like . . . How can I leave the house? How can I go anywhere? I'm scared out of my wits.

This heightened concern for personal safety obviously differs from the fears expressed by survivors of natural disasters. For example, after an earthquake, victims may need to worry about aftershocks or lack of electricity. They do not need to fear being verbally or physically assaulted. These additional and unprecedented safety concerns contributed to increased feelings among Muslims of exclusion from the larger American community.

Discrimination

The apprehension that Muslims expressed regarding their safety was not unfounded. Since the attacks, increased incidents of harassment and physical violence have been reported across the nation. The September 11th disaster was followed by the most dramatic rise in anti-Muslim hate crimes to date (Council on American-Islamic Relations, 2002; Federal Bureau of Investigation, 2002; Human Rights Watch, 2002).

Whether the students reported being personally targeted or not, they were emotionally disturbed and frightened simply knowing that family members, friends, or members of their community had been the victims of some form of backlash. Indeed, one aspect that characterizes bias-motivated crimes is that not only does the direct victim of the crime suffer, but non-victims of the targeted group do as well (Cogan, 2002; Cogan and Marcus-Newhall, 2002). Recognition of the magnitude of anti-Muslim acts further served to isolate and alienate this already marginalized minority group.

The Muslim students reported a variety of discriminatory impacts in the aftermath of September 11th. A small number of participants had a difficult time finding or keeping a place to live. This was particularly true for those students who arrived in the United States just before the attacks. Khalid, a graduate student from Turkey who had been in New York City for only a few days before September 11th, was unable to secure a place to live for several weeks after the disaster. Thus, he was forced to sleep in the library at his university (which fortunately was open 24 hours a day), and to shower at the on-campus gym. Although Khalid initially blamed his problem on the general lack of affordable housing in the city, he later revealed that several of the people he had contacted in his attempt to find an apartment had hung up on him when they heard his Middle Eastern accent.

The majority of the interviewees reported receiving some form of verbal harassment in the days and weeks following the attacks. This verbal discrimination ranged from someone muttering "terrorist" as they passed by on the street, to shouts and threatening comments. Additionally, some of the respondents described physical incidents, such as being spit on, shoved in the subway, or having things thrown at them while walking down the street. Some of the women reported having their headscarves yanked or pulled off. Being stared at or given "dirty looks" was the most common form of discrimination reported by the Muslim students. Janan discussed her experiences following September 11th:

> I was so upset. I said, Please, God, don't let it be a Muslim. I think it was really hard for me in my freshman year, because this was my second week of being a freshman in college. I was adjusting. I was really loving being here. And all of the sudden, having to defend myself in classrooms, getting dirty looks from people, even my neighbors. I remember people staring at me. There were these guys who kept driving past me and giving me these looks, just for no reason, just to freak me out when I was walking in town. So I was even scared to go out of my house. I was thinking about taking off my headscarf. That was the initial thing.

Perhaps most difficult to measure, fear, tension, depression, and psychological strain were reported by the students. Some of the participants missed several days or even weeks of their college classes. Physical manifestations of stress included problems with sleeping, weight loss, and recurring headaches. When asked about the effects of feeling discriminated against, Najah, a first generation immigrant from Trinidad, said:

> It ruins my day. Sometimes it ruins my week. Last semester I got extremely, extremely depressed. When I hear the stuff on the news that's happening, the portrayal that the media is giving of Muslims, I wish I could have a microphone and tell people, "We're not like that." It would really, really aggravate me. I was so tired, I was having headaches every day. It definitely hurts. It made me feel like, I wanted to do something, tell people more.

It is impossible to understand the feelings and perceptions of the Muslims interviewed without recognizing that these discriminatory acts occurred after the September 11th disaster. Even so, the Muslim students often remarked that although the backlash was bad, it was not nearly as bad as it might have been.[10] However, the definition of "bad" must be placed in context. For example, I considered the harassment, verbal threats, and physical violence reported by the interviewees as significant enough to warrant concern. On the other hand, given the larger fears that many interviewees expressed—such as the possibility of being detained, deported (even though most were U.S. citizens), or "placed in internment camps like the Japanese during World War II"—less substantial acts of discrimination perhaps seemed insignificant in comparison.

Increased Group Solidarity

After September 11th, one of the major themes conveyed by the interviewees was that their isolation from the larger U.S. community led to an increased sense of group solidarity among Muslims. This response is not necessarily unique. Several scholars have noted that one reaction to real or perceived group threats is increased group solidarity (Bozorgmehr, 2000; Coser, 1964; Doosje and Ellemers, 1997; Durkheim, [1893] 1984; Portes and Sensenbrenner, 1993; Simmel, 1955; Turkel, 2002; Yancey et al., 1976). Indeed, social scientists have documented ethnic and religious minority groups coming together in the face of adversity throughout U.S. history, such as Japanese citizens during and following World War II (Bonacich and Modell, 1980; Fujita and O'Brien, 1985; Montero, 1981), and among the Irish who immigrated to America during the potato famine (Ignatiev, 1995).

Similarly, Goffman (1963) argues that members of a stigmatized group have a tendency to come together into small social groups whose membership derives from the stigmatization. Malik, who was raised in Saudi Arabia and moved to the United States for college, anticipated that alienation from the larger community would bring Muslims together:

> You can see that it's only natural that something like this would foster group cohesion. You're being identified in a particular way. When that is the case, in order to feel stronger, you'll identify with that group. You want to get closer to that group. It's a natural urge.

The internal cohesion and structure of a minority group may sometimes arise as a result of the development of an awareness of being considered different (Tajfel, 1978). Likewise, a distinct form of "therapeutic community" emerged among Muslims following the attacks. Because they felt, and often were, excluded from much of the post-September 11th nationalistic sentiment, they began to rely more heavily on fellow Muslims for social support.

Muslim Community Cohesion

During the interviews, the students reported that although they were close friends with other Muslims before September 11th, there was an increased level of bonding within the community after the disaster. The participants said that this was primarily due to a common level of understanding among fellow Muslims, as many Muslims came to share similar fears and uncertainties, on personal, local, national, and international levels, after the disaster. Hanan explained how her relationships changed after September 11th:

> Everyone was so concerned and so understanding, especially my Muslim friends. They were scared, too. That made me realize that they're really there for me. Because of that I feel a lot closer to them than I did before. I think whenever people go through something like that, it's a bonding process. Now I see them a lot more than I used to.

Because the students felt isolated from the non-Muslim community, other Muslims served as an important source of emotional support. Muslim student groups offered a safe space to express emotional responses, fear, and anger, and to receive empathy.[11]

An enhanced sense of community within the Muslim group also resulted from increased amounts of time spent together. Much of this time was related to additional efforts to educate the U.S. community about Islam and Muslims in America. The interviewees described a multitude of activities that they had developed and implemented to improve the public's cultural awareness and general understanding of Islam. Most of the students helped organize or chose to participate in one or all of the following activities: presenting guest lectures to their university classes; visiting local churches, synagogues, and schools through community outreach projects; or speaking at campus-wide educational events.

When I returned to New York City to conduct a third set of interviews, almost seven months after the attacks, the students were more hopeful about their futures as Muslim Americans. This increased optimism seemed directly related to the extensive efforts that they had undertaken as a community following September 11th, as well as the positive reception that they received from non-Muslims.

Along with incidents of harassment, the attacks were also followed by a surge of public interest in Islam and Muslims. One of the most encouraging aspects reported by the students was that they believed that people were genuinely interested in learning about Islam and understanding them and their faith. The participants talked about how they believed much of the anger and blame they initially received from others was beginning to be replaced by interest and engagement. Sanae, who moved from Pakistan to New York City to attend college, talked about this shift:

> Initially things were very rough, like when you came last time. People were cursing you out. Every time they saw us they'd make this face. Then it slowly got better, but it's still not back to normal. Even now, sometimes when you're sitting in the subway, you do get looks. But now it's more interest than anything else. Before it was fear, like, "Oh, my God, I hope she doesn't have a bomb." Now it's more like interest, "Okay, so she's a Muslim." They observe what I'm wearing, how I'm talking. If I'm with my friend and I'm talking English, or if I talk about something, we're business majors, we talk about finance and this and that all the time. They're kind of, I can see surprise on their face, like, "Oh, these really are normal people." It's getting better now.

The interviewees agreed that many of their concerns and fears had gradually decreased over time, particularly after the initial ethnic and religious backlash subsided. However, some wondered whether things would ever "return to normal" for the Muslim community, or if the aftermath of September 11th would continue to impact their lives, relationships, and futures in America. Ahmad, a second-generation immigrant of Palestinian descent, worried not only about his future, but also about the prospects for his children:

> Our main concern here is, what is our future? That's what it comes down to. You get these new history books: "Muslims Attack America." What's going to happen when kids in school read them? Your people attacked my culture. How do you raise children in such an environment? It all comes into effect in the long run. You've really got to look into the future perspective rather than what's just now, about what's going to happen a month, a year, ten years, a few generations from now. What is this country going to fall back on?

Conclusion

The Muslims interviewed for this study felt isolated from the national therapeutic community that emerged after the events of September 11th for four primary reasons: feeling blamed for the attacks; being excluded from mourning, bonding, and helping behaviors; having to manage increased concerns for individual safety; and witnessing acts of discrimination and violence perpetrated against Muslims in the United States. This social exclusion led to increased group solidarity and identification with the Muslim community. Like all citizens, Muslim Americans were significantly impacted by the events and aftermath of September 11th. As a group, they suffered the same shock, anger, and grief as the rest of the nation. However, because they were immediately placed on the defensive after the attacks, they were victimized not only by the disaster, but also by the public reaction that followed.

A recurring theme of sociological investigations of natural disasters has been that mass emergencies create a sense of solidarity out of which emerges a therapeutic community response. In a therapeutic community, it is rare for segments of the impacted population to be excluded from the post-disaster unity, such as occurred with the Muslim community after September 11th. Therefore, when an intentional, human-conceived disaster is considered, traditional ways of understanding and responding to emergent social processes must be re-evaluated.

It is obvious that a richer and more sophisticated exploration of post-disaster therapeutic communities is necessary. Immediately after September 11th, we witnessed communities develop through both inclusion and exclusion. The existence of these types of emergent communities warrants inquiry into how violent and intentional acts, including terrorist attacks, are likely to encourage anger and blame, creating an "us" sentiment that can only take shape in relation to a "them" or "other" sentiment. Consideration of this issue is important because the resultant division may lead to illegal acts of discrimination and bias-motivated crimes, which are socially disruptive to the larger community and may be emotionally devastating to minority communities (Cogan, 2002).

Furthermore, the focus on the physical and economic devastation caused by the September 11th attacks, and their potential psychological impacts on U.S. citizens, should be extended to explore their specific impacts on ethnic and religious minority groups who have been victimized twice by disaster: first by the actual attacks themselves, and second by the backlash that followed. It has often gone unrecognized that the emotional devastation from a disaster can be at least as great as the physical devastation (Clizbe, 2003). Therefore, this longitudinal research is necessary to understand the unique social and psychological impacts for dual-victimized minority communities.

Outreach efforts should be offered to vulnerable communities to provide opportunities to enhance public dialogue—before disaster strikes. Moreover, local and federal policy should be used to help promote tolerance, build community, and decrease social isolation among individuals and groups, particularly in light of the unprecedented severity of the backlash that followed the September 11th attacks. It is imperative that future social research and public policy address issues of religious and ethnic discrimination and social exclusion after terrorist attacks and other intentional, human-conceived disasters.

NOTES

1. Conflict certainly may occur after natural disasters, although it is much less widespread—at least during the initial aftermath—than that which has been

documented after technological disasters or intentional acts of violence (see Quarantelli and Dynes, 1976).

2. Advocacy groups for Arab and Muslim Americans documented a combined total of over 2,000 reported incidents of harassment, violence, and other discriminatory acts in the year following the September 11th attacks (American-Arab Anti-Discrimination Committee, 2002; Council on American-Islamic Relations, 2002; Human Rights Watch, 2002; Ibish, 2002).

   The Federal Bureau of Investigation (FBI) reported a dramatic rise in hate crimes against Muslims following the September 11th attacks. In 2000, the FBI received reports of only 28 hate crimes against Muslims in the United States. In 2001, that number increased to 481, which represents a 1700% increase. As of January 2003, the U.S. Department of Justice was investigating approximately 380 hate crime cases that were directly linked to the backlash that followed September 11th. Prosecutors contend most of the crimes—including threats, assaults, bombing plots, acts of vandalism, arson, shootings, and murders—were in retaliation for the attacks on the World Trade Center and the Pentagon.

3. Because there is no official record of Muslims living in the United States, one of the most viable ways to find a large population of Muslims in New York City was through campus Muslim Student Associations (MSAs), a nationally recognized student organization. Within the MSAs, I relied on key contacts who helped to organize interviews and recruit other students. Initially, to gain acceptance and permission to conduct the interviews, I faxed or e-mailed a detailed description of the project to each of the contacts at the colleges and universities. This description defined the purpose and goals of the research project, the sample population I was seeking, and the broad themes that I would be exploring. I included a cover letter verifying my credentials as a doctoral student and research assistant. Once I had received approval to interview the students, I sent a shortened version of the research description that could be posted to e-mail listservs and flyers to be placed in the MSA office. The e-mails, flyers, and support from the student contacts helped with recruitment for the study. All those who responded to the recruitment efforts were interviewed.

4. One of the five fundamental pillars of Islam requires that adherents fast during the Holy Month of Ramadan. From before dawn until sunset, those who are observing the fast are forbidden to eat, drink, smoke, or have sexual relations. The Muslim lunar year is shorter than the solar year by approximately 11 days. Thus, Ramadan occurs at slightly different times each year, according to the solar calendar followed in the United States (Denny, 1994, p.126-130). My second site visit, in December 2001, coincided with Ramadan. Therefore, I was able to break the fast, listen to the calls to prayer, and eat Ramadan dinners with the interviewees.

5. More females than males were interviewed for two reasons. First, as with many on-campus student organizations, most of the MSAs that I visited had a majority

female membership. Secondly, my position as a female researcher allowed me more access to the females than the males.

6. The word *hijab* comes from the Arabic word for hiding or concealing, and for women, also denotes covering the body completely with loose clothing to ensure modesty. The head covering typically drapes around the neck and covers the bosom as well.

7. The *niqaab* is a face veil that covers the hair, neck, and face, leaving only the eyes visible.

8. For the purposes of anonymity, all interviewee names have been changed. Pseudonyms are used throughout the paper.

9. The word *jihad* has been frequently referenced in the U.S. media—particularly since September 11th—and is often translated as "holy war." The word *jihad* actually means "striving" or "to strive for a better way of life." In its primary sense, it is an inner striving to rid the self from debased actions or inclinations, and to exercise constancy and perseverance in achieving a higher moral standard.

10. It is also important to note that after recounting negative experiences, the interviewees would often focus on positive, kind responses from friends, classmates, professors, and strangers after the disaster. The students regularly commented that they recognized that hateful acts were only being committed by a small minority of ignorant individuals who did not represent the United States as a whole. This topic warrants further consideration, but is beyond the scope of this paper.

11. This is similar to the findings reported by Edelstein (1988, pp. 110-117), in his study of a community contaminated by residential toxic exposure. He found that as the victims came to feel increasingly isolated from friends, relatives, and co-workers living outside the community, they became more reliant on fellow victims as sources of social support. Edelstein refers to this phenomenon as "outsiders don't know what we went through," and reports that this sentiment is echoed in virtually every toxic exposure incident he has investigated.

References

Adler, Patricia A. and Peter Adler. 1987. *Membership Roles in Field Research.* Newbury Park, CA: Sage Publications.

Alexander, David. 2002. “Nature’s Impartiality, Man’s Inhumanity: Reflections on Terrorism and World Crisis in a Context of Historical Disaster.” *Disasters* 26(1): 1–9.

Allport, Gordon W. 1954. *The Nature of Prejudice.* Cambridge, MA: Perseus Books.

American-Arab Anti-Discrimination Committee. 2002. http://www.adc.org/.

Barton, Allen H. 1970. *Communities in Disaster: A Sociological Analysis of Collective Stress Situations.* Garden City, NY: Doubleday and Company, Inc.

Baum, Andrew, Raymond Fleming, and Laura M. Davidson. 1983. “Natural Disasters and Technological Catastrophe.” *Environment and Behavior* 15(3): 333–354.

Bonacich, Edna and John Modell. 1980. *The Economic Basis of Ethnic Solidarity: Small Business in the Japanese American Community.* Berkeley, CA: University of California Press.

Bozorgmehr, Mehdi. 2000. “Does Host Hostility Create Ethnic Solidarity? The Experience of Iranians in the United States.” *Bulletin of the Royal Institute for Inter-Faith Studies* 2(1): 159–178.

Bucher, Rue. 1957. “Blame and Hostility in Disaster.” *American Journal of Sociology* 62(5): 467–475.

Clizbe, John. 2003. “Emerging Research and Practice in Disaster Mental Health.” *Natural Hazards Observer* 27(3): 12–13.

Cogan, Jeanine C. 2002. “Hate Crime as a Crime Category Worthy of Policy Attention.” *American Behavioral Scientist* 46(1): 173–185.

Cogan, Jeanine C. and Amy Marcus-Newhall. 2002. “Introduction.” *American Behavioral Scientist* 46(1): 8–13.

Coser, Lewis A. 1964. *The Functions of Social Conflict.* 2nd edition. Glencoe, IL: Free Press.

Couch, Stephen R. and J. Stephen Kroll-Smith. 1985. “The Chronic Technical Disaster: Toward a Social Scientific Perspective.” *Social Science Quarterly* 66: 564–575.

Couch, Stephen R. and J. Stephen Kroll-Smith, eds. 1991. *Communities at Risk: Collective Responses to Technological Hazards.* New York: Peter Lang Publishing.

Council on American Islamic Relations. 2002. *American Muslims: One Year After 9-11.* Washington, D.C.: Council on American-Islamic Relations Research Center.

Cuthbertson, Beverley H. and Joanne M. Nigg. 1987. "Technological Disaster and the Nontherapeutic Community: A Question of True Victimization." *Environment and Behavior* 19(4): 462–483.

Dacy, Douglas C. and Howard Kunreuther. 1969. *The Economics of Natural Disasters.* New York: The Free Press.

Demerath, Nicholas J. and Anthony F.C. Wallace. 1957. "Human Adaptation to Disaster." *Human Organization* 16: 1–2.

Denny, Frederic Mathewson. 1994. *An Introduction to Islam.* 2nd edition. New York: Macmillan Publishing Company.

Doosje, Bertjan and Naomi Ellemers. 1997. "Stereotyping Under Threat: The Role of Group Identification." Pp. 257–273 in R. Spears et al., eds., *The Social Psychology of Stereotyping and Group Life.* Oxford: Blackwell.

Drabek, Thomas E. 1970. "Methodology of Studying Disasters: Past Patterns and Future Possibilities." *American Behavioral Scientist* 13(3): 331–343.

Drabek, Thomas E. 1986a. *Human System Responses to Disaster: An Inventory of Sociological Findings.* New York: Springer-Verlag.

Drabek, Thomas E. 1986b. "The Problem of Taxonomy in Disaster Research." Presentation at the annual meeting of the American Sociological Association, New York City, August.

Drabek, Thomas E. and Enrico L. Quarantelli. 1967. "Scapegoats, Villains, and Disasters." *Trans-action (*March): 12–17.

Durkheim, Emile. [1893] 1984. *The Division of Labor in Society.* Translated by W. D. Halls. New York: The Free Press.

Dynes, Russell R. 1970. *Organized Behavior in Disaster.* Lexington, MA: Heath Lexington Books.

Edelstein, Michael R. 1988. *Contaminated Communities: The Social and Psychological Impacts of Residential Toxic Exposure.* Boulder, CO: Westview Press.

Erikson, Kai. 1994. *A New Species of Trouble: The Human Experience of Modern Disasters.* New York: W.W. Norton & Company.

Erikson, Kai T. 1976. *Everything in Its Path: Destruction of Community in the Buffalo Creek Flood.* New York: Simon & Schuster.

Federal Bureau of Investigation. 2002. "Crime in the United States–2001." http://www.fbi.gov/ucr/01cius.htm.

Fritz, Charles E. 1961. "Disaster." Pp. 651–694 in R.K. Merton and R.A. Nisbet, eds., *Contemporary Social Problems.* New York: Harcourt, Brace & World, Inc.

Fujita, Stephen S. and David J. O'Brien. 1985. "Structural Assimilation, Ethnic Group Membership, and Political Participation among Japanese Americans: A Research Note." *Social Forces* 63(4): 986–995.

Glaser, Barney, and Anselm Strauss. 1967. *The Discovery of Grounded Theory.* New York: Aldine de Gruyter.

Goffman, Erving. 1963. *Stigma: Notes on the Management of Spoiled Identity.* New York: Simon & Schuster Inc.

Gunn, S.W.A. 2001. "On Man-Conceived Disasters." *Journal of Humanitarian Medicine* 1(2): 7–8.

Hewitt, Kenneth. 1997. *Regions of Risk: A Geographical Introduction to Disasters.* Boston: Addison Wesley Longman, Inc.

Human Rights Watch. 2002. "We Are Not the Enemy: Hate Crimes against Arabs, Muslims, and Those Perceived to be Arab or Muslim after September 11." *Human Rights Watch Report* 14(6): 1–42.

Ibish, Hussein. 2002. Personal communication with Lori Peek on September 25.

Ignatiev, Noel. 1995. *How the Irish Became White.* New York: Routledge.

Juergensmeyer, Mark. 2000. *Terror in the Mind of God: The Global Rise of Religious Violence.* Berkeley, CA: University of California Press.

Kendra, James M. and Tricia Wachtendorf. 2001. "Rebel Food . . . Renegade Supplies: Convergence after the World Trade Center Attack." Preliminary Paper #316. Newark, DE: Disaster Research Center, University of Delaware.

Kroll-Smith, J. Stephen and Stephen Robert Couch. 1990. *The Real Disaster is Above Ground: A Mine Fire & Social Conflict.* Lexington, KY: The University Press of Kentucky.

Krueger, Richard A. 1988. *Focus Groups: A Practical Guide for Applied Research.* Newbury Park, CA: Sage Publications.

Lofland, John and Lyn H. Lofland. 1995. *Analyzing Social Settings: A Guide to Qualitative Observation and Analysis.* 3rd edition. New York: Wadsworth Publishing Company.

Mileti, Dennis. 2001. "Rising from the Ashes: In Disasters, We Are All the Same." *The Denver Post* (September 16): E-01.

Montero, Darrel. 1981. "The Japanese Americans: Changing Patterns of Assimilation Over Three Generations." *American Sociological Review* 46(6): 829–839.

Morgan, David L. 1997. *Focus Groups as Qualitative Research.* 2nd edition. Thousand Oaks, CA: Sage Publications.

Neal, David M. 1984. "Blame Assignment in a Diffuse Disaster Situation: A Case Example of the Role of an Emergent Citizen Group." *International Journal of Mass Emergencies and Disasters* 2(2): 251–266.

Oliver-Smith, Anthony. 1979. "Post Disaster Consensus and Conflict in a Traditional Society. The 1970 Avalanche of Yungay, Peru." *Mass Emergencies* 4: 39–52.

Peek, Lori A. 2002. *Religious and Ethnic Issues After September 11, 2001: Examining Muslim University Student Experiences*. Quick Response Report #156. Boulder, CO: Natural Hazards Research and Applications Information Center, University of Colorado. http://www.colorado.edu/hazards/qr/qr156/qr156.html.

Portes, Alejandro and Julia Sensenbrenner. 1993. "Embeddedness and Immigration: Notes on the Social Determinants of Economic Action." *American Journal of Sociology* 98: 1320–1350.

Quarantelli, E.L. 1993. "Community Crises: An Exploratory Comparison of the Characteristics and Consequences of Disasters and Riots." *Journal of Contingencies and Crisis Management* 1(2): 67–78.

Quarantelli, E.L., ed. 1998. *What is a Disaster? Perspectives on the Question.* New York: Routledge.

Quarantelli, E.L. and Russell R. Dynes. 1976. "Community Conflict: Its Absence and Its Presence in Natural Disasters." *Mass Emergencies* 1: 139–152.

Quarantelli, E.L. and Russell R. Dynes. 1977. "Response to Social Crisis and Disaster." *Annual Review of Sociology* 3: 23–49.

Reinharz, Shulamit. 1992. *Feminist Methods in Social Research.* New York: Oxford University Press.

Scanlon, J. 1988. "Planning for Peace and War Emergencies – Learning from 70 Years of Disaster Research." *Disaster Management* 1(2): 3–8.

Simmel, Georg. 1955. *Conflict and the Web of Group Affiliations.* Glencoe, IL: Free Press.

Tajfel, Henri. 1978. "The Social Psychology of Minorities." Report #38. New York: Minority Rights Group.

Tierney, Kathleen. 2002. "September 11 and Beyond: Shifting Concerns and Cross-Cutting Lessons." Plenary Session Presentation at the 27th Annual Workshop on Hazards Research and Applications. Boulder, Colorado, July.

Turkel, Gerald. 2002. "Sudden Solidarity and the Rush to Normalization: Toward an Alternative Approach." *Sociological Focus* 35: 73–79.

United States Department of Justice. 2002. http://www.usdoj.gov/.

Warheit, George J. 1976. "A Note on Natural Disasters and Civil Disturbances: Similarities and Differences." *Mass Emergencies* 1(2): 131–137.

Webb, Gary R. 2002. "Sociology, Disasters, and Terrorism: Understanding Threats of the New Millennium." *Sociological Focus* 35(1): 87–95.

Yancey, William L., Eugene P. Ericksen, and Richard N. Juliani. 1976. "Emergent Ethnicity: A Review and Reformulation." *American Sociological Review* 41(3): 391–403.

Acknowledgments

This work was supported by the Graduate School at the University of Colorado, the Natural Hazards Research and Applications Information Center, and the National Science Foundation, Grant Number 0200.05.0203B, which is gratefully acknowledged. However, only the author is responsible for the interpretations and conclusions made in this paper.

I would like to thank Syed Ali, Gary David, Julie Gailus, Sayaros Mohamed, Mary Fran Myers, and Joyce Nielsen for commenting on earlier versions of this paper. I am also grateful for the suggestions provided by two anonymous reviewers. Wendy Steinhacker offered valuable editorial advice, and Sandy Adler assisted with the interview transcriptions for this project. Norbert Baer and Dean Savage kindly assisted with my research efforts in New York City. Jacki Monday provided sound editorial and intellectual guidance, which I sincerely appreciate. Finally, I wish to thank the Muslim students who made this research possible through their willingness to share their time, thoughts, and experiences.

Correspondence should be addressed to Lori A. Peek, Doctoral Student, Department of Sociology, Research Assistant, Natural Hazards Research and Applications Information Center, University of Colorado–Boulder, 482 UCB, Boulder, CO 80309-0482; (303) 492-1028; e-mail: lori.peek@colorado.edu.

Risk Communication and Public Warning Response to the September 11th Attack on the World Trade Center

Paul W. O'Brien
Department of Sociology & Criminal Justice
California State University, Stanislaus

Introduction

This research was conducted over a six-day period after the terrorist attacks on the United States on September 11, 2001. Research included conducting interviews with a wide variety of populations in assessing the impact of the event and risk communication. A model of risk communication was used in an attempt to understand some of the actions and behaviors of those affected by the attack. Risk communication proved useful in providing a basic behavioral model to this study, but the September 11th event was different from the usual context to which the model is applied (natural hazards) and needs further development and refinement for use in future events. These interviews were administered on an "as available" basis with interviewees with the City of New York still on emergency alert. Two additional days of interviewing were conducted two and one-half months later. Data collection is ongoing by the author and student assistants. This report represents the first phase of a larger project. The purpose of this phase was to better understand the immediate human dynamics of the event. Follow-up research in phase two will use quantitative methods to investigate the event.

The qualitative data presented here represent feelings, opinions, and activities in the immediate aftermath of the event. Subsequent research in phase two will apply rigorous social science research methods for greater generalizability. Structured interviews, structured focus groups, survey research using representative samples of the population, and an elaborate detailing of events, actions, and warnings need to be undertaken to fully address critical public policy issues and future emergency planning.

A wide cross-section of the public and emergency personnel was interviewed for this project. One of the focus populations was the kindergarten through 12th grade (K-12) public school system. Many schools were affected by the event. The New York City school system is responsible for over one million children on a daily basis. The disaster had a profound impact on its administrators, staff, teachers, and pupils.

This paper includes the results of those initial interviews. Based on the interviews conducted, it is believed that considerable effort needs to be accomplished in the theoretical area of terrorist attacks to better explain the human actions. The long-term objective of this research is to better understand the risk communication model dynamics that occur in this new context. This understanding should enhance the ability of both the private and governmental sectors to deal with future events.

Description of Event

On the morning of September 11, 2001, two hijacked jetliners were flown into the World Trade Center Towers in lower Manhattan in New York City. The first impact occurred at 8:48 a.m., bringing a wide array of emergency personnel and vehicles to the aid of victims. At 9:07 a.m. the second tower was struck and emergency personnel were coming to the realization that this was not an accident.

At 9:50 a.m. the south tower of the World Trade Center collapsed, followed at 10:29 a.m. by the collapse of the north tower. At the time it was believed that upwards of 20,000 lives were lost. Presumed victims included those working, visiting, and running the towers and hundreds of emergency workers including police, fire, emergency medical, and Port Authority personnel. Since the initial projections the number of fatalities has been drastically reduced and now hovers at approximately 3000.

Sampling Frame

This investigation used a purposeful sample chosen on the basis of availability, willingness to be interviewed, and being in a position to have knowledge on the topic. There were three types of interviewees. The first was emergency personnel and included

- Police
- Fire department personnel
- Military personnel posted around the World Trade Center site
- Officials of the public school system.

The second set of interviewees were citizens in the area with more than a passing knowledge of the situation, but not having an official emergency function. They included

- K-12 teachers
- Business persons, e.g., restaurateurs, former World Trade Center employees
- Wall Street traders
- Support personnel of the World Trade Towers
- Religious leaders including priests and ministers.

The third sample category was conducted with people on the street, including

- Citizens in stores, restaurants, subway stations, and anywhere people congregated to discuss events.

The interviewees were 60% female and 40% male; 40% Anglo, 30% African-American, 15% Asian, and 15% Latino. Although the information they provided is anecdotal, the citizens of New York City gave their time in the hope that this research would aid future victims.

Risk Communication Theoretical Model

Risk communication informed the collection of data. Risk communication in hazards is a well-established model (Mileti, 1975; Mileti and Sorensen, 1990). The model has been used by both disaster researchers and practitioners in the field as a method of understanding the complex human behavior of hearing and responding to warning messages. Until this event, the risk communication model has been used extensively in the area of disaster research, primarily for natural disasters such as earthquakes, hurricanes, and tornados.

Risk communication has been used for both short- and long-term warnings. It can be broken down into three broad categories of variables: pre-event factors, warning information, and demographic variables.

Pre-event Factors

There are three sub-categories of pre-event factors. A person's pre-warning perceptions or biases can determine, in part, how a particular person responds or fails to respond to a given targeted warning message. These pre-warning perceptions can thus influence the subsequent decision-making behavior in which a person engages (Mileti et al., 1981). The pre-event factors are described below.

Pre-event Salience

This category measures the degree to which a hazard resides in the conscious awareness of the public. Those living in coastal areas may have a high level of awareness of hurricanes but perhaps little for tornadoes. People in California have a high degree of awareness of earthquakes but little for hurricanes. This variable has been shown to be important because the more salient a disaster type is for a public, the more likely that public is to respond to warnings regarding that particular hazard (Turner et al., 1981).

Pre-event Knowledge

This category measures what people know about disasters. This knowledge can be general and/or specific. It is important to know this in fashioning a clear warning message. If the public is told that a carload of radon gas has just escaped, what does this message truly convey? If few know what radon gas is, the warned public will not understand the message and no response will be forthcoming. In such a circumstance it is important that knowledge of what the threat means also be explained for any significant public response to occur. Only then can the public be expected to respond to a warning in a meaningful way (Perry and Lindell, 1986).

Pre-event Experience

Prior experience with a particular disaster type ties in directly with how the public will respond to a given warning. Those people with prior experience will respond more readily to warnings than will those without the experience. This concept is important in deciding how to inform such a public in the event of a disaster (Perry and Lindell, 1986).

Warning Information

The second set of variables includes information factors. This is the warning message itself that must be evaluated to see if the message was a coherent one. Research has shown that the following variables are important.

Consistency

Consistency is an important characteristic of the warning message. Although it is good to have multiple sources, it is critical that they be consistent with one another, i.e., that the messages are conveying the same warning (Quarantelli, 1984).

Specificity

Specificity is how detailed the message is. The more specific the message is about the type of danger and what to do about it, the greater the personal response on the part of the public (Nigg, 1987).

Source Credibility

Source credibility is an essential characteristic of any good warning message. The warning must come from a source that is deemed credible by the public. Different members of the public have different perceptions of who is a credible source. Given this diversity, it is best to have a mix of sources. This could include spokespersons from the government, scientists, or local officials all giving a similar warning to deny the public the ability of claiming that the message itself is not believable (Mileti et al., 1981).

Frequency

Frequency refers to the number of times that a warning message is given. Research has shown a correlation between the number of warnings and the public's response to act on those warnings. Frequency of the warning must also be addressed in understanding the public's response to it. How many times a message is disseminated to the public will affect their hearing, understanding, and belief of the message. The public will respond more readily to a message that is heard numerous times (Mileti and O'Brien, 1992). Frequency of message has several effects on the warned public. Some of these include the reduced potential for such public misperceptions as the spread of rumor and not focusing on the official warnings. Hearing a message over and over helps to reinforce to people that protective actions need to be taken.

Channel

The channel is the vehicle by which the message is conveyed, such as television, radio, newspapers, and the internet. Research has shown that warnings transmitted over multiple channels yield the best results (Turner et al., 1981).

Demographic Variables

The final set of variables in the model deals with the demographic make-up of the population hearing the warnings. Relevant demographic variables include sex, age, socioeconomic status, and ethnicity. Research has demonstrated that the background of the people hearing the warning affects their hearing and responding to the warning. For example, older people, women, and persons with a higher socioeconomic status are more likely to respond to warnings.

Field Questions

Along with the risk communication model, a series of questions guided this research. They focused on

- What warning messages were heard
- Sources of warning information

- Content of messages
- Procedures for dissemination of information to children and parents
- Protective actions taken for school populations
- Warning messages heard by both parents and children
- Languages used in warning messages
- Public responses to warnings.

Findings

Findings from this investigation must be seen as preliminary. Whereas every effort was made to interview a representative sample of all populations, i.e., victims, emergency personnel, and the general public, it nevertheless was not possible to use the widely accepted social science research methods that would allow for greater generalizibility. Nevertheless, much was learned from this initial data collection phase, and will be detailed in this section.

Warning Background

Cellular phones, beepers, the internet, and 24-hour news coverage are changing how humans interact and receive information. In addition, much of the country's news industry is based in New York City. There was no shortage of networks, reporters, or media infrastructure to report and analyze events as they unfolded on and after September 11th. Many respondents spoke repeatedly about information overload and the difficulty of making sense of all the information.

Typical was a 28-year-old schoolteacher whose class saw the planes hit the towers and the subsequent collapse from the classroom across the Hudson River. As her students watched out the window, she received instructions via the classroom phone, while some kids were already on the internet getting additional information. She stated, "We were inundated with advice and directives on what to do." Another teacher from Leadership High School in lower downtown explained, "Our concern was to get the students out of the building and area immediately, and we didn't have time to watch television—we just knew something was terribly wrong." Where people were located in relation to the unfolding events had an impact on whether they heard warnings. One could categorize people as "spectators" or "participants" with regard to hearing messages, with the former hearing and remembering warnings issued and the latter reacting like others in their immediate surroundings. One respondent expressed it perfectly: "I ran like everyone else."

Respondents reported an eerie silence immediately after the attack, as people tried to understand what had happened. After the second plane impact, however, warnings began to be issued by a host of government agencies. Initial warnings related to possible building collapse and concern about structural integrity of remaining buildings in the immediate area of the World Trade Center. National news reporting agencies were focused on the collapse of the buildings and the grounding of all commercial aircraft nation-wide. Local warnings were targeted on the safety of children in the school system (see discussion, below), crowd control, what areas were cordoned off (a typical message was not to travel south of Canal Street in lower Manhattan). The third wave of messages evolved over the afternoon of September 11th to become more informational and less specific warnings, i.e., where to go to locate family members, where to go for help, hospital hotlines, messages from Mayor Rudolf Giuliani, and public transit and bridge/tunnel warnings.

Pre-event Variables

The pre-event variables did not have applicability in this situation. The airplanes struck the World Trade Center without any warning. Given the nature of the event no prior experience could be called upon by the public to help form their perceptions in preparing them to undertake protective actions. Although the World Trade Center was the target of a terrorist attack in 1993 that resulted in the evacuation of the towers, none of the people interviewed made reference to the 1993 attack as being similar in any way.

The first question asked of all respondents was where they were when the planes struck the towers. The question served to allow the respondents to collect their thoughts for further questions. All the respondents could state exactly where they were and what they were doing when they received the news. Many respondents spoke of the surreal nature of the event. A typical comment was, "We were out in the street calmly drinking our coffee, watching smoke come out of the building, wondering what had happened." Those in lower downtown thought it was all a terrible accident until the second plane hit. Thus pre-event variables were irrelevant in explaining whether people heard warnings or not.

Warning Information

The second set of informational variables included source, consistency, and frequency of warning messages. As noted above, in the immediate aftermath of the event emergency personnel and the public were inundated with warnings. Many warnings were to take protective actions (such as leaving the area) or to watch for falling debris. These warnings came from emergency personnel in the affected area. The magnitude of this event made it impossible to escape both official and unofficial warnings that were being disseminated by federal, state, and local officials.

In attempting to investigate the consistency of warnings disseminated, the model assumes that there are protective steps that can be taken. In the natural hazards example, one is warned to seek higher ground with heavy rain or not to run outside the building in an earthquake. What, however, is one to do in a terrorist attack? For days and weeks after the event the media showed pictures of the towers being hit by airplanes, people running down streets, and the collapse of the towers. Many of those interviewed spoke of an intense need for a structured (thought-out) response to the events. Said one respondent, "I was scared but what could I do—my child was missing." Specific guidance and help was needed with less dramatic media frenzy. Questions such as what to do immediately after such an event and what protective actions families, businesses, and governments should take need to be addressed so that future warnings are not only targeted to those needing to receive them, but also focused on actions that are feasible and practical.

It is premature at this stage to attempt to have a specific emergency plan in place for future possible terrorist attacks. One sees, for example, that in natural disasters, emergency managers explicitly state what the public should do. Emergency practitioners sorely need this for future terrorist threats and deeds. The emergency community now has the Oklahoma City bombing, and the World Trade Center attacks of 1993 and 2001 as baseline events. A critical question at this juncture is what specific protective actions can be gleaned from these events. Research needs time to discover the best protective actions to take, and then have that information work its way through the emergency community. This will require multiple methods by a variety of disciplines to come to a consensus on the most productive course of actions.

Based on this preliminary research, it appears that a new variable of magnitude must be added to the model for similar events. This event was perceived as so extraordinary that trying to get the public to take protective actions (as, for example, with earthquake aftershocks) was not necessary. At times the public's perceptions and response was in front of official warning messages. This event was of such a magnitude that it created a new reality. Events were unfolding that have not been experienced by this society.

The fact that this event was not only a human disaster but also a crime scene made for some special dynamics. Local, state, and federal law enforcement agencies were involved. Instead of the usual crowd control concerns of keeping on-lookers at a distance (Quarantelli, 1984), these agencies were charged with the arrest of anyone who did not follow specific instructions. This raised the level of urgency and seriousness to a much higher level than in many natural disasters. This higher level of urgency made the warnings to evacuate and take protective actions impossible to discount.

Situational Variables

Prior research has shown that situational variables such as environmental cues, social setting, and social ties have an impact on whether people take protective action (Drabek, 1969; Mileti and O'Brien, 1992). Given the nature of this event, situational variables had an acute salience. The entire region and country were affected by the events. It was impossible to deny their gravity, magnitude, seriousness, and reality. All of these variables coalesced to create a public that was simultaneously scared, angry, sad, and giving. In addition, the entire country and world was witnessing the event live on television. Environmental cues and social setting were conducive to taking protective actions. One stark example of an environmental cue in the days after the event was the posting of World Trade Center victims' pictures in the subway. Many stations were filled with thousands of pictures of lost and/or missing people. The scope of loss and its real life consequences were unavoidable in the entire region. Social ties also played a pivotal role in getting the public to take protective actions. Not only were many people tied directly to the event—as emergency responder or victim—but their family and friends across the country felt an indirect connection to the disaster as well.

Demographic Variables

The severity of the event brought all types of people together regardless of their demographic characteristics (sex, age, socioeconomic status, or ethnicity). New York's history is that of a "melting pot." That tradition continues today with a highly diverse population along all the variables in this category. The event cut across all of the traditional factions that divide a society. All populations had friends, acquaintances, and family members involved in the subsequent response. The public behaved very similarly to what would be expected after a natural disaster where one sees the emergence of a "therapeutic community" (Barton, 1969).

Warning Information Discussion

The World Trade Center attack was extraordinary in this country, based on the magnitude of loss of life. This investigation aimed to identify lessons that could aid in minimizing and/or responding to future possible events, regardless of the low probability of their occurrence. The following discussion is based on interviews and impressions of being in New York City after the September 11th disaster.

As noted throughout this paper, the model of risk communication was adapted for use to understand the on-going emergency. The theoretical model needs to be modified to include extreme events such as the World Trade Center attack. Among the core components of the model are the pre-event variables, which were not applicable in this event. Little prior experience could be called upon to guide behavior in this event.

One question raised by the September 11th disaster in New York City is that of the historical role of fire fighters and others. The first responder community may need to re-evaluate the risks of entering buildings to rescue people and make a judgment based not only on the emergency response needed (as in the past) but also on a new set of criteria of additional dangers unforeseen by earlier generations of emergency personnel. Over 300 fire fighters and emergency personnel lost their lives in the collapse of the World Trade Towers. A public outpouring of emotion illustrated the nationwide mourning of the loss of so many (Figure 1).

Warnings targeted at the emergency responders themselves also may need to be evaluated. Instead of the historical perspective of protecting the public, emergency warnings might need to be disseminated to all people in the area regardless of their role in the activities. This will make the function of warnings much more complex and difficult.

The traditional perspective of measuring source, consistency, and frequency of warning messages in a simple linear model needs to be rethought. Emergency personnel and the public were bombarded with information 24 hours a day, seven days a week. Information technology has changed dramatically since the days of the 1950s when sirens warned of impending dangers. In those earlier times, the concern was how to get warning messages out to all those who could be affected; the concern was for greater dissemination of the warnings. From interviews conducted for this investigation, the opposite appeared to be the case. Many interviewees complained of information overload as opposed to an information void. The numerous warnings and the subsequent media coverage were so intense that what was a warning and what was media coverage was intermixed. The City of New York, the state, and the Federal Emergency Management Agency all used, for example, internet web sites to provide the public with a lot of information, ranging from what schools were closed to how to request a death certificate. The level of technology in a given society must be included in the risk communication model because it will have an impact on warning messages heard and the public's response. It will be necessary to measure or account for the level of internet use, 24-hour news coverage, personal beepers, and cellular phones. Warning messages can no longer be seen as a simple linear model of "issue warning, public hears warning, and response to warning."

The model's situational variables of environmental cues, social setting, and social ties were the most relevant to understanding public perceptions and warnings heard. First, the physical environment had changed dramatically with the collapse of the towers and portions of the city being off-limits to the

Figure 1. A typical scene at a fire station in New York City after the September 11th disaster.

public (Figure 2). Priority was given to the restoration of public infrastructure(Figure 3). Much of the financial district's infrastructure was either damaged, destroyed, or needed to be redirected. Major portions of the subway system coming into the financial district from New Jersey (the PATH system) were no longer functioning. All traffic that passed through the tunnels into the city was searched. Hundreds of photographs of victims were displayed in the subway stations (Figure 4). These all were stark environmental cues that something catastrophic had occurred and people could not simply dismiss them and disregard warnings. Denial of what happened was impossible.

Social ties were pivotal in getting the public to take protective actions. Respondents spoke of wide use of their cellular phones to connect to significant others. One respondent reported, "I called my mom in California and she told me what was going on." Although communication technology has developed dramatically in the past two decades, calling spouses, parents, friends, and family members appears to be as important as it was 40 years ago in earlier disaster research (Drabek, 1969).

Figure 2. Portions of lower downtown New York City were blocked off both to protect the public and to protect the crime scene.

Warning Information and the Public School Impact

One focus of this investigation was warning information dissemination in the New York City school system. The system is responsible for over one million children daily, with 79,924 teachers in 1,198 schools (New York Board of Education, 2001). Numerous interviews were completed with administrators and teachers. One of the "fortunate" aspects of this event was that it occurred in a city with tremendous resilience and infrastructure. The New York Board of Education accomplished its responsibility of warning and caring for the region's school-age children. In addition, it used its resources and its intellectual capital in the event. Curricular changes helping children cope with the situation were instituted immediately. An already strong diversity curriculum was reworked and offered to teachers within days of the attack for classroom use in helping students work through the aftermath.

The Board of Education had an elaborate emergency response plan in place. This was located on a compact disk in each and every school. The CD

Figure 3. Work to restore public infrastructure was much in evidence.

allowed administrators to use a cascading scenario approach to select the best course of action, be it a bombing, shooting, or hostage situation. One of the major problems that occurred was the wide area impacted. All of lower downtown's schools were affected. Schools evacuated included PS 234, PS 150 and P/I 89. Evacuated high schools included High School for Leadership and Public Service, Economics and Finance, Stuyvesant High School, and Murray Bergtraum High School. The emergency plan, for example, foresaw many school populations evacuating to Stuyvesant High School which, in this event, was also being evacuated. Specific warnings on school closings were done via telephone and in some cases in person by school safety officers physically going to schools. Thus, pictures on national television showed children and their teachers running through the financial district, fleeing the area. The teachers' primary concern was for their pupils and getting them to safety. Many of the children were transported by ferries to New Jersey taking them out of harm's way. This led to desperate parents' seeking their children at several locations. This suggests the importance of having an all-hazards plan in place that plans for a larger impact area.

Figure 4. Pictures of those lost and/or missing were displayed in public places.

The few interviews with teachers made note of the potential long-term psychological damage of the event and the information overload that occurred. Emergency procedures and warning messages were followed as closely as possible for the given situation. Teachers spoke more definitively of the aftermath of the event on their students rather than specific warning messages heard.

The information overload problem was addressed in part by a New York City school publication. It offered many suggestions on how to help children with the aftermath of the event. Of note is item J, which stated:

> One thing to avoid: Don't show videos of the attack or keep a television on for live coverage of the latest developments. The kids have seen enough. The most important thing for schools to do is to help them come to terms with what they have experienced already

(Roderick, 2001).

Discussion of Study

As noted throughout this document, additional research needs to be conducted. A formalized investigation needs to be undertaken from which generalizations can be drawn for the region and nation. Whereas the model of risk communication proved useful in attempting to understand this event, the model needs to be refined for a better fit with the reality of similar disasters.

The pre-event variables were not applicable to this situation, because this was the first major event of its kind in the United States. The American public now has experienced the attack on the World Trade Center and has the requisite "experience" to be better prepared for future events. Just as in natural hazard events, however, this prior experience will slowly dissipate as time passes. Parallels likely exist with natural hazards in keeping the public's attention on the lessons learned.

The second set of information variables helped with the understanding of the public's actions. The public was in the midst of on-going emergency and formed a "therapeutic community" of helping each other very similar that noted in natural disasters. This sense of "oneness" with other members of the community lasted for several weeks. *Newsweek*'s cover page of November 12, 2001, was entitled "Generation 911—Terror, War, and Recession Hit Home on Campus." One of the problems with the model's information variables was attempting to observe and note the emergency messages disseminated by public officials. Without prior incidents as a guide, there were no clearly formulated steps of protective actions except for simply leaving the area. Future research needs to address not only information dissemination, but also potential filtering mechanisms for a meaningful digestion of information. Any discussion of filtering in the American culture may be criticized as being undemocratic, but, if future terrorists acts do occur, the public might be willing to give up some of those traditional freedoms.

Another unusual aspect of the September 11th disaster was the dynamic of patriotism. To this author's knowledge, this has not been recorded or written about in earlier hazards research. American flags began to appear everywhere. A strong sense of patriotism was prevalent from lower Manhattan to ethnic restaurants in mid-town. Citizens had a sense of pride and purpose in the enterprise of responding and reacting. A possible parallel could be the immediate aftermath of the attack on Pearl Harbor in 1941, when the country was of one mind. This phenomenon had a palpable impact on warnings that were issued. They were not disregarded or dismissed. Everyone was paying attention to government officials and the media. This can be seen in the almost cult-like figure Mayor Guiliani became in the aftermath of the disaster. For a short period the city and state of New York, along with the rest of the country, became Barton's "therapeutic community."

One of the issues arising out of this investigation is dealing with low probability events. How is a society to deal with events that might never happen again? As in the case of the New York City school system, how many personnel and resources should be devoted to updating an excellent emergency plan to include another event on the scale of the World Trade Center attack? If no additional planning is completed with the modification of current plans, and another low probability does occur, who receives the blame? On the other hand, if personnel and resources are diverted in creating an enhanced plan that is never used, who bears the political costs of such a decision? This conundrum speaks loudly for the need for all-hazards planning to cover an entire range of possible events.

Theoretical development in the area of terrorist attacks should be given high priority in future research. This project attempted to "test" the risk communication model. Research targeting low probability events is also called for and has the potential to help in other areas of warnings, such as earthquakes, where the state-of-the-art prediction is still limited to 30-year probability time lines. Another area that might be focused on is bio-hazards. Can risk communication be modified or evolve to help explain behavior in that context?

The use of technology in warning research needs further investigation as well. As noted in this piece, local, state, and federal agencies all used the internet for information dissemination. Many respondents reporting getting much if not most of their information from the internet, as opposed to radio and television as in the past. One of the exciting aspects to this technology for emergency managers is its interactive nature. Information can be both disseminated and collected through one medium. It is widely known in earthquake-prone areas, for examplc, that it makes sense to use local populations to help dig out victims because they are ready, willing, and able. Is government ready to use the public as information gatherers? Where are water mains broken, where are trees down, are there looters in the area, who is buried and where? To not engage the public in information gathering may well be a lost opportunity.

Finally a deeper understanding of "social ties" needs to be reached. America continues to be a melting pot of different races, ethnic backgrounds, and religions. Some excellent research has focused on this aspect (e.g., Enarson and Morrow, 1998 or Bolin and Bolton, 1986), but issues of diversity need to remain at the forefront of the research agenda, if viable and believable warnings are to be acted upon by a diverse public.

Careful design of future research is needed to address the issues raised by this project and by other investigations undertaken immediately after the disaster. It will take considerable time to fully understand all the events and dynamics of September 11, 2001. That work needs to begin now.

References

Barton, Allen H. 1969. *Communities in Disaster: A Sociological Analysis of Collective Stress Situations.* Garden City, NY: Anchor Books.

Bolin, Robert C. and Patricia Bolton.1986. *Race, Religion, and Ethnicity in Disaster Recovery.* Environment and Behavior Monograph #42. Boulder, CO: Institute of Behavioral Science, University of Colorado.

Drabek, Thomas E. 1969. Social Processes in Disaster: Family Evacuation. *Social Problems* 16 (Winter): 336–349.

Enarson, Elaine and Betty Morrow. 1998. "Women Will Rebuild Miami: A Case Study of Feminist Response to Disaster." Pp. 185–199 in Elaine Enarson and Betty Hearn Morrow, eds., *The Gendered Terrain of Disaster: Through Women's Eyes.* Westport, CT: Praeger.

New York Board of Education. 2001. *Fact Sheet New York City School System Vital Statistics.* New York: New York Board of Education.

Mileti, Dennis. S. 1975. *Natural Hazard Warning Systems in the United States: A Research Assessment.* Boulder, CO: Institute of Behavioral Science, University of Colorado.

Mileti, Dennis S., Janice R. Hutton, and John H. Sorensen. 1981. *Earthquake Prediction Response and Options for Public Policy.* Environment and Behavior Monograph #31. Boulder, CO: Institute of Behavioral Science, University of Colorado.

Mileti, Dennis S. and John H. Sorensen. 1990. *Communication of Emergency Public Warnings: A Social Science Perspective and State-of-the-Art Assessment.* Oak Ridge, TN: Oak Ridge National Laboratory.

Mileti, Dennis S. and Paul W. O'Brien. 1992. "Warnings During Disaster: Normalizing Communicated Risk." *Social Problems* 39: 40–57.

Nigg, Joanne M. 1987. "Communication and Behavior: Organizational and Individual Response to Warnings." Pp. 103–117 in R.R. Dynes, B. DeMarchi, and C. Pelanda, eds., *Sociology of Disasters.* Milan, Italy: Franco Angeli Libri.

O'Brien, Paul and Dennis S. Mileti. 1992. "Citizen Response in Emergency Response During the Loma Prieta Earthquake." *International Journal of Mass Emergencies* 10(1): 71–91.

Perry, Ronald. W. and Michael K. Lindell. 1986. *Twentieth-Century Volcanicity at Mt. St. Helens: The Routinization of Life Near an Active Volcano.* Tempe, AZ: Arizona State University School of Public Affairs.

Quarantelli, E.L. 1984. "Perceptions and Reactions to Emergency Warnings of Sudden Hazards." *Ekistics* 309: 87–106.

Roderick, Tom. 2001. *Responding to Children's Concerns about the Tragic Attacks on the World Trade Center and the Pentagon: Suggestions for Teachers*. New York: Educators for Social Responsibility, New York School of Education. http://www.esrmetro.org/responding.html.

Turner, Ralph. H., Joanne M. Nigg, Denise Heller Paz, and Barbara Shaw Young. 1981. *Community Response to Earthquake Threat in Southern California*. Los Angeles: University of California Press.

Acknowledgments

This paper is a result of research conducted in the aftermath of terrorist attack on the World Trade Center on September 11, 2001. It is a product of the Quick Response Research Program funded by the National Science Foundation, and administered by the Natural Hazards Research and Applications Information Center housed at the University of Colorado–Boulder. Their support is gratefully acknowledged.

Correspondence should be addressed to Paul O'Brien, Department of Sociology and Criminal Justice, 801 West Monte Vista Ave., Turlock, CA 05382; e-mail: POBrien@csustan.edu.

Impacts of Extreme Events on Intercity Passenger Travel Behavior: The September 11th Experience

José Holguín-Veras
Department of Civil and Environmental Engineering
Rensselaer Polytechnic Institute

Robert E. Paaswell
Department of Civil Engineering
City College of New York

Ann-Marie Yali
Department of Psychology
City College of New York

Introduction

The tragic events of September 11, 2001, have had a profound impact on activities in all of lower Manhattan, and the New York City metropolitan region—profound, of course, because of the huge loss of life and the continuing sense of trauma of the survivors. These events have also provided a major challenge to transportation and city planners because there are few guidelines in the technical literature on how to manage after such an event occurs.

The scale of the event was enormous: 13.4 million square feet of office space were lost in the World Trade Center, while 12.1 million square feet were rendered temporarily unusable in the adjacent properties (Holusha, 2002, p. 1). Over 100,000 jobs were displaced. Tens of thousands of additional jobs have been lost or interrupted because they serve the World Trade Center and its neighborhoods. Additional acts and threats of terror, such as the anthrax attacks, have made New Yorkers cautious about where and how they travel to participate in activities.

The changing perception of the safety of transportation modes is, in particular, affecting the way in which the traveling public makes choices concerning mode of transportation, place of work, and location of residence. On September 10, most travel analysts would have said that reliability, travel time, and cost were the primary determinants of mode choice. On September 12, personal security became, and still remains for many New Yorkers, a key concern. As a result of the September 11th disaster, businesses and individuals are making choices that will impact whether or not (1) they remain in their jobs in a new location, outside the impacted site; (2) they change jobs; (3) they change travel mode or its route; (4) they move from the New York region; among many other possibilities. While all of these choices are extremely complex, closely inter-related, and changing over time, two dimensions of choice stand out. The first is the individual's overall response to the tragedy, and his or her personal relationship to it. The second is the individual's sense of security as it applies to each mode available for a given trip. Travel choices will vary according to the individual and his or her personal response to the tragedy. Much can be learned by studying and evaluating these impacts over time.

The September 11th disaster has also affected the real estate and commerce systems in rather subtle and indirect ways. Before September 11th, the business location process was determined by variables such as transportation accessibility to markets, economies of agglomeration, and accessibility. Mathematical formulations based on such assumptions have successfully been employed to model the business location process. Figure 1 and Equation 1 show a model estimated by the research team for northern New Jersey as a function of the economic attractiveness of the markets of New York City and Philadelphia (Holguín-Veras et al., 2002). The mathematical formulation used assumes that the probability that a business chooses a given location i, is a function of the proximity to the economic poles (i.e., New York City and Philadelphia), and the corresponding market sizes. As shown in Equation 1, simple assumptions of economic rationality lead to a powerful explanatory model (t-statistics in parentheses).

$$P(i) = \sum_k \alpha_k \Pi_k e^{-\beta C_{ki}} = 0.1132\Pi_{NY} e^{-0.070 C_{NY,i}} + 0.05569\Pi_{PA} e^{-0.1303 C_{PA,i}} \quad (1)$$

(19.373) (-22.100) (6.984) (-4.255) F=361.46 R^2=0.98

Where:

Π_k = Market size at economic pole k, and $P(i)$ = Probability of choosing location i, $C_{NY,i}$ and $C_{PA,i}$ are the distances to location i, from New York and Philadelphia, respectively.

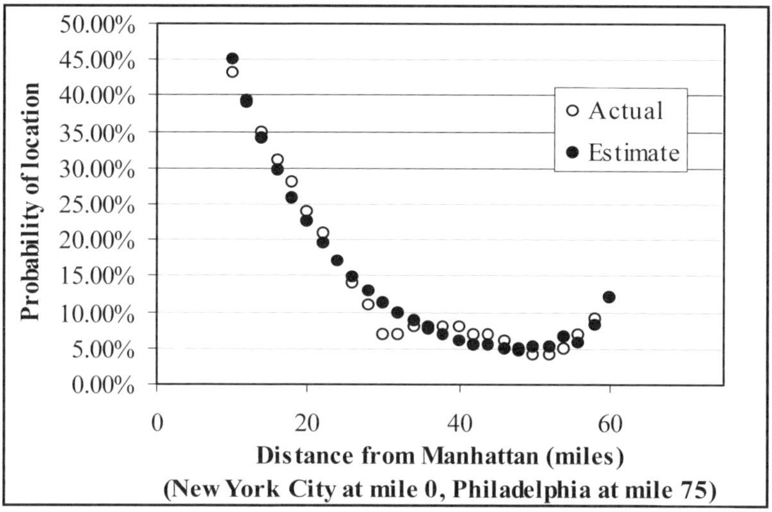

Figure 1. Actual and estimated values of business location model.

The traditional assumptions of economic rationality have been altered in the wake of September 11th. This in turn has produced some surprising phenomena. The World Trade Center events, far from leading to a tighter real estate market (to be expected after the sudden disappearance of 25% of downtown's office space), released a flood of sublet space (*New York Times*, 2002), which seems to be a reflection of the changes in business patterns and a weakening economy. Ultimately these changing business patterns are having a significant impact upon the processes of work-residential choice, as well as transportation mode and route choice. It is still too early to determine if these changes are transient or permanent. Those who survived the attacks and those who were forced to relocate because of changes in business patterns had to develop new commuting patterns.

The impacts of September 11th on travel behavior extend far beyond the New York City metropolitan area. As it was widely reported in the media, air transportation throughout the nation experienced a dramatic drop, while train alternatives such as the ones provided by Amtrak and the regional commuter lines had to deal with a significant upsurge in demand. Interestingly enough, the information provided to the authors by the technical staff of Greyhound and other regional bus operators indicated an across-the-board drop in demand, which they attributed to the fact that their customer base is formed primarily by low-income individuals who were directly affected by the weakening of the economy in the post-September 11th period.

The research reported here was conducted by faculty members of the Rensselaer Polytechnic Institute and the City College of New York as part of a project funded by the National Science Foundation to assess the changes on passenger travel behavior produced by the terrorist attacks of September 11, 2001. This paper specifically focuses on describing the behavioral changes in intercity travel. A separate component of this project—dealing with the impacts upon urban travel—could not be reported here because of the delays experienced in the corresponding data collection process. This delay was, in part, due to the inherent and understandable difficulties of coordinating work with transportation agencies directly affected by the events of September 11th and/or involved in the rescue and recovery operations in its aftermath. The regional planning organization, the New York Metropolitan Transportation Council, a research partner to this project and the lead agency in transportation data collection in the New York City metropolitan area, had its headquarters destroyed in the collapse of 1 World Trade Center, and is still in the process of reassembling the data sets lost in the attack.

Research Approach

This section provides a summary of the research approach implemented by the project team. This description, for the most part, focuses on the conceptual aspects of the work done, minimally describing the methodological formulations of the models used. This decision was taken in order to ensure that this paper is able to reach as wide an audience as possible.

The unique nature of this research project necessitated the implementation of a research approach able to capture the most significant changes in passenger travel behavior that have taken place as a consequence of the disaster of September 11th. To this effect, the research team decided to use behavioral models based on Random Utility Theory (RUT) to assess such behavioral changes. In this context, the random utility models provided the methodological framework for the assessment of behavioral changes, while the transportation surveys conducted provide the data to be used in the analyses and model estimation processes.

RUT is a behavioral/economic theory that postulates that decision makers choose the alternative that maximizes the utility derived from their choices. The distinguishing feature in RUT is the assumption that the utility function is composed of a systematic component, which depends upon the socio-economic characteristics of the decision maker and the alternatives' attributes, and a set of random terms that consider the fact that the analyst does not have full information about all relevant variables and decision processes. The latter translates into unobserved effects and measurement errors that are operationalized as observational randomness in RUT. The consideration of a random error term enables the formulation of random utility models based on probability principles. The origins of RUT can be traced to Thurstone (1927), though it was McFadden (1974, 1978) who provided the foundations for RUT and made it an operational theory that has become the standard tool for disaggregate modeling of transportation phenomena. The following section provides a brief overview of RUT. For the most part, the review follows Ben-Akiva and Lerman (2000), unless otherwise indicated.

RUT differs from the traditional (deterministic) utility theory in that the utility is assumed to have two components: (a) the systematic component, which is the one explained by the variables included in the model; and (b) the random component, which represents the unobservable factors and measurement errors that, as a rule, are not known to the analyst. The consideration of the random component enables the formulation of random utility models using probability concepts. In this context, the utility of alternative i for individual n, U_{in}, comprises the systematic (explained) component, V_{in} and a random term ε_i, as follows:

$$U_{in} = V_{in} + \varepsilon_i \tag{1}$$

As a result of the consideration of the random component, the choice process is formulated in terms of the probability that a given alternative is chosen, which is defined as the probability that its utility is higher than the maximum utility of the other alternatives. In mathematical terms:

$$P_n(i) = P[U_{in} > Max U_{jn}] = P[V_{in} + \varepsilon_i > Max(V_{jn} + \varepsilon_j)], \forall j \neq i \tag{2}$$

Different assumptions about the error terms lead to different discrete choice models. The model specification process, in short, consists of the specification of the systematic component of the utility function and the specification (assumption) of the distribution of error terms most appropriate to the problem at hand. The systematic component is, for the most part, assumed to be a linear-in-parameters combination (X_{in}) of socio-economic characteristics of the decision maker (S_n), the attributes of the alternative *i* as perceived by individual n (Z_{in}), and a vector of parameters β. Mathematically:

$$V_{in} = h(S_n, Z_{in}) = \beta' X_{in} \tag{3}$$

In general terms, the parameters of alternative models are estimated using maximum likelihood principles. In a typical application, the parameters and the model are tested for statistical significance. Depending upon the model structure, other tests such as the test for violations of the Independence of Irrelevant Alternatives may need to be conducted. In addition to the assessment of the statistical significance, the modeler is also required to analyze the conceptual validity of the model and its parameters.

Random utility models require, for calibration and forecasting purposes, disaggregate data that could represent (a) revealed preference data, i.e., actual choices; and (b) stated preference data, which is the data obtained by eliciting responses from the decision makers about choices and rankings in hypothetical choice situations. Using either revealed preference or stated preference data has advantages and disadvantages, though an increasing body of evidence indicates that discrete choice models estimated with stated preference data could be highly accurate, provided they are designed and conducted properly (Hensher, 1994). For that reason, the research team decided to rely on stated preference data to assess the changes in intercity travel behavior produced by the terrorist attacks of September 11th.

The Choice Experiment and the Survey Instrument

In order to provide a decision context for the respondents, the project team selected a choice situation that involved a compulsory trip, supposedly a business trip to another city. A business trip was selected because its compulsory nature eliminates one choice dimension, i.e., the decision to travel or not. This, in turn, presents a fairly clear choice situation that minimizes misunderstandings on the part of the respondents. Another benefit of using a compulsory trip in the choice situation is that the behavioral changes identified could be interpreted as lower bounds of the impacts, because non-compulsory trips (because of their inherent elasticity) are likely to be more impacted than compulsory trips.

Another relevant decision concerning the choice situation involved the trip distance. Since for long trip distances, air transportation may be the only practical alternative, focusing on long distances would have made it more difficult to assess behavioral changes in intercity travel because the dimension of mode choice would not have been present (which is the anticipated consequence of the decision makers' feeling "captive" of air transportation). For that reason, the project team decided to focus on the lower range of trip distances, for which the decision makers have different alternatives that effectively compete with each other. In this context, the behavioral changes would reveal themselves as components of the tradeoffs among alternatives captured by the systematic component of the utility functions.

The respondents were randomly assigned to three different trips: (a) New York–Washington, D.C., (b) New York–Boston, and (c) Boston–Washington, D.C. The percentages of respondents for each trip type were 43.47%, 28.80%, and 27.71% respectively. As shown in Table 1, the breakdown according to trip type correlates fairly well with the breakdown from the American Travel Survey (Bureau of Transportation Statistics, 1997). A similar situation happens with the mode split data. As shown in Table 2, the mode split in the sample seems to be in the appropriate order of magnitude (taking into account that the American Travel Survey includes all trip purposes). However, since there was no information available about the breakdown for business trips that could be used to further refine the sample, the project team decided to use the data as it came from the sample without using any correcting (weighting) factors. This decision was taken because, without the backing of solid statistics, using weighting factors would have introduced an unquantifiable amount of uncertainty in the estimation that did not seem justified in light of the results shown in Tables 1 and 2.

Table 1. Breakdown of number of trips (1000 trips/year) for selected trip interchanges (Bureau of Transportation Statistics, 1997).

Trip interchange	trips/year (1000)	Trip interchange	trips/year (1000)	Market breakdown	Sample breakdown
Boston-New York City	445	New York City-Boston	393	31.21%	28.80%
Boston-DC	253	DC-Boston	202	16.95%	27.71%
New York City-DC	676	DC-New York City	716	51.84%	43.47%
Totals	1,374	Totals	1,311	2,685	

Table 2. Mode split in sample and the American Travel Survey (Bureau of Transportation Statistics, 1997).

Mode split in sample				Mode split in ATS		
Mode	Trips	%		Mode	1,000 trips/year	%
Train A (Metroliner)	354	21.38%				
Train B (Acela)	202	12.20%		Other	242	18.29%
Commercial airplane	622	37.56%		Commercial airplane	396	29.93%
Personal use vehicle	478	28.86%		Personal use vehicle	685	51.78%
Total	1656	100.00%		Total	1323	100.00%

Approximately half of the respondents were told that their employer was paying for the trip, while the other half were told that they (the respondents) were paying for trip expenses. The respondents were provided with nine different choice scenarios involving four different transportation choices: two alternatives of rail service (Amtrak's Metroliner and Acela); one air transportation alternative; and a car alternative, and were asked to rank order the alternatives. Although rank order data was available, only the first choices were used in the analyses. A bus alternative was not included after consultations with the technical staff at the regional bus companies indicated that buses do not compete with these modes for business travel.

The alternatives in the choice set were characterized in terms of cost, travel time, and inspection/boarding time at the airport. Cost and time were broken down by segment of the trip (beginning of the journey, main trip, and end of the journey). Since the objective of the analysis is to assess behavioral changes, everything else being equal, the attributes of the different alternatives were—for the most part—assumed to be equal to the ones corresponding to the alternatives available in the market place. The only exceptions were the scenarios that included a "high speed" rail alternative (a variation on the current Acela service), not currently available in the market place. The attributes (factors) that were varied in the experiment were the inspection and boarding time at the airport that was assumed to have three factor levels (25, 60, and 120 minutes); and the departure and arrival times of the train alternatives and air (three factor levels each). Throughout the experiment the car alternative remained the constant option, i.e., with attributes that did not change values. The factor levels were combined in nine scenarios (treatment combinations). The scenarios were screened to eliminate those deemed to be not feasible from the technological or policy standpoint.

The questionnaire had five major sections, in addition to the stated preference section. The first section was intended to ascertain if participants had ever traveled to the target city, how frequently they travel there, the primary reason for traveling there, what mode they use and prefer, why they chose that particular mode, and the perceived level of quality of that mode. They are also asked to rate the mode of travel chosen on items such as cost, cleanliness, service, comfort, and safety. The second section contained a choice scenario in which the respondents were asked to indicate what their preference of four travel alternatives would have been before September 11th. This choice scenario was exactly the same as the base case in the stated preference section. The third section of the questionnaire included questions about the impacts of September 11th on the respondents: how much September 11th changed their travel choices on a 7-point scale (1=not at all; 7=significantly). In addition, participants indicated whether September 11th affected them in six different ways by checking the statements that apply to them: "I am more conscious of security," "I avoid traveling by plane," "I am more aware of people traveling with me," "I am more selective in choosing my travel mode," "I plan to change type of work," and "As much as I can, I avoid traveling altogether." The fourth section was aimed at gathering the socio-economic characteristics of the respondents to describe sample characteristics. Single items assessed age, gender, marital status, number of people in household, number of children in household, education, and income. The fifth section consisted of a set of four questions aimed at assessing perceived stress level. A four-item version of the Perceived Stress Scale (PSS4; Cohen and Williamson, 1988) was used to assess the degree to which respondents appraise their life as stressful. Respondents indicated how frequently they felt unable to control the important things in life, felt confident about handling personal problems, felt things were going right, and felt unable to overcome difficulties. Each of these items was rated on a 5-point scale ranging from 1 (never) to 5 (very often) with two reverse-scored items. A total stress score (PSS4) for each subject was calculated by summing item responses. The sixth section contained the stated preference scenarios described above.

The questionnaire was administered to an initial set of volunteers, graduate and undergraduate students at the City College of New York. The graduate students were asked to administer the questionnaire to three other individuals selected by them in order to maximize the variability in the socio-economic characteristics of the sample. The undergraduate students were only asked to respond to their questionnaires. Each volunteer filled out a consent form and one of four versions of the survey. The confidentiality of the responses was guaranteed, in accordance to National Science Foundation's human subject research guidelines. A total of 192 volunteers participated in the study. The questionnaires were administered between March 14, 2002, and April 4, 2002, about six months after the September 11th disaster.

It is important to highlight that the data collection and the experimental design process faced significant limitations due to the unique circumstances in which the data was collected. This translated into a rather unorthodox experimental design and data collection process. Faced with the decision either to wait for more resources to become available, or to spend resources and time in fine tuning the experimental design, the authors made a pragmatic decision regarding the scenarios to be included in the experiment and the main focus of the investigation, in order to avoid more delays in the data collection process that would have further dissipated transient behavioral effects. These decisions have proven to have advantages and disadvantages that are discussed throughout the paper.

Description of the Sample

The study sample comprised 192 participants, with 184 providing descriptive information (see Table 3). The majority were male, aged 20 to 25, and single with no children. The median income level reported was between $35,000 and $49,999, with two people in the household. Due to using a convenience sample, the majority of participants were college educated.

The sample differs from the general population of the five boroughs of New York City in a number of ways. Census data from 1990 on sociodemographic variables are included in Table 3. The sample is disproportionately male and single compared to the general population, which is 53% female, 41% married. In addition, the sample is younger, more educated, and wealthier than the general population. While household size seems to be similar to the general population, the majority of participants in this study have no children (60%), whereas only 46% in the population is without children.

Figure 2 shows the geographic distribution of the zip code of residence of the respondents (with triangles). As shown, the bulk of the respondents are residents of New York City, while there is a smaller, though significant, number of respondents that live in northern New Jersey and the rest of the New York City metropolitan area.

Results

This section is divided into two parts. The first section highlights the descriptive analyses and the second section describes the behavioral models. It is important to highlight that many of the results discussed in this section, specifically those pertaining to behavioral responses after September 11th, are highly dynamic in nature and, as a result, are likely to change with time as the respondents regain comfort in the routine nature of daily life. In this context, the results shown here are to be interpreted as a snapshot taken six months after September 11th. A second wave of panel data collected approximately a year after September 11th (not available at the moment of producing this document) is likely to provide more information about the dynamic transient behavioral effects.

The reader is asked to note that standard statistical abbreviations will be used. For example, SD is the standard deviation, α refers to Cronbach's alpha, a measure of the internal consistency of a scale (the higher the number, the better the scale's reliability), χ^2 is the chi-square statistic for differences in nonparametric data, and r is the Pearson's coefficient of correlation.

Table 3. Sample and population descriptives.

Variable	Sample* n	Sample* %	NYC Population **	NYC Population ** %
Gender (n=184)				
Male	112	61%		47%
Female	72	39%		53%
Age (n =183)				
<25	64	35%	17-24	15%
26-30	34	19%	25-29	12%
31-36	39	21%	30-34	12%
37-45	28	15%	35-44	19%
46-55	10	5%	45-54	14%
>55	8	4%	>55	28%
Size of Household (n = 162)				
1	40	25%		33%
2	46	28%		27%
3	35	22%		16%
4	21	13%		12%
>4	20	12%		12%
Number of Children (n =158)				
0	111	70%	0	54%
1	21	13%	1 or more	46%
2	17	11%		
>2	9	6%		

Variable	Sample* n	Sample* %	NYC Population **	NYC Population ** %
Marital Status (n =184)				
Single	107	58%		38%
Married	69	38%		41%
Divorced	8	4%	***	12%
Widowed	0	0%		9%
Income (n =182)				
< $24,999	42	23%		43%
$25-34,999	26	14%		14%
$35-49,999	24	13%		16%
$50-74,999	45	25%		15%
$75-99,999	21	12%		6%
$100,000+	24	13%		6%
Education (n = 179)				
			< High school	32%
High school	17	9%	High school	26%
College UG	67	37%	College UG	14%
College grad	68	38%	College grad	18%
Postgraduate	27	15%	Postgraduate	10%

Notes:
* Percentage columns may not add up to 100% due to rounding errors.
** 1990 Census data for the five boroughs of New York City.
*** Individuals reporting a divorce or separation are collapsed in this number.

Descriptive Analyses

A majority of the participants indicated that they have actually traveled from New York City to Washington, D.C. (70%) and from New York City to Boston (72%); however only 26% of those surveyed had ever made the Boston-to-Washington, D.C. trip. This makes sense given that the sample consisted of those whose primary residence was New York City and thus one would expect their trips to originate from there.

Of those who had traveled to the target cities in the past, approximately 55% made that trip over 12 months ago, 19% 6–12 months ago, 21% 1–6 months ago, and 5% made the trip less than a month ago. In addition, 83% said they visit the target city twice a year, 10% visit 3–5 times per year, 3% visit 6–8 times a year, and 2 participants reported that they visit the target city

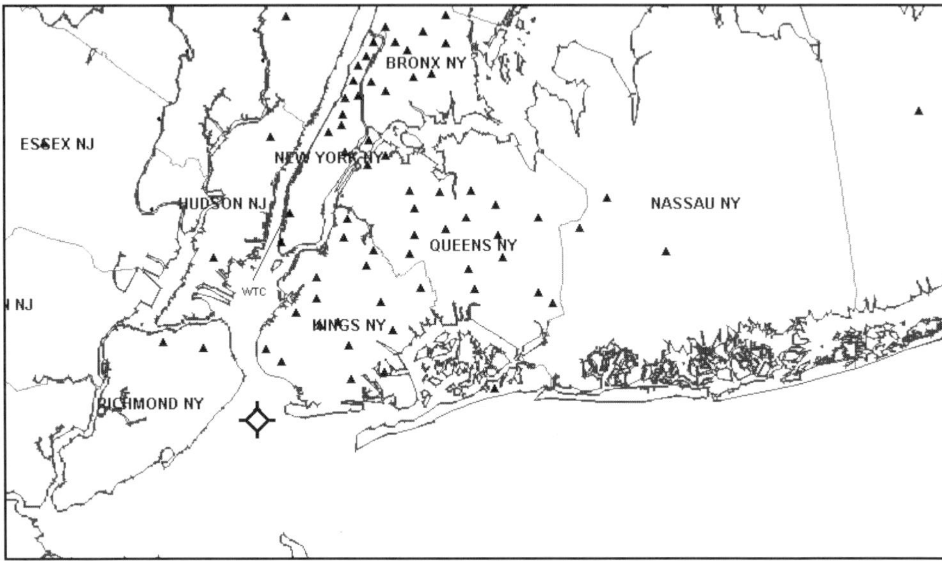

Figure 2. Geographic distribution of respondents.

more than 8 times per year (2 respondents did not answer this question). When asked to indicate the primary purpose of the trip, reasons were social (63%), work (15%), education (5%), and "other" (17%). Most traveled to the target city by car (63%), followed by air and train (14% each), and 8% said "other" (one person failed to answer this question). Participants were then asked to indicate all of the reasons that they chose the mode used on that last trip by checking off a list. Collapsing across modes, 21% marked reliability, 68% marked convenience, 53% cost, 17% safety, 12% security, 38% comfort, and 9% marked easy. On that last trip, the majority paid for the trip themselves (66%), which may explain why cost was important in mode selection. Of those who did not pay for the trip themselves, 12% had employers pay, 8% had family pay, and 9% had "other" pay for the trip.

Participants were also asked to rate the mode they selected on their last trip to the target city on various features including cost, cleanliness, service quality, reliability, security, comfort, and safety. Collapsing across modes, the mean for each feature ranged from 5.21 to 5.60 (SDs ranged from 1.3 to 1.4). A multivariate analysis of variance was conducted to assess if the mean ratings of each feature differed by mode (car, air, train, other), by trip (New York City to Washington, D.C., New York City to Boston, Boston to

Washington, D.C.), or by an interaction (mode x trip). The model indicated that there was a significant difference in ratings by mode (Wilks' lambda = 0.583, F = 1.98, $p < .01$). There was no significant effect for trip or the interaction term. Post hoc comparisons using the Bonferroni test (a method for adjusting for the increase in Type I error associated with running multiple unplanned comparisons) revealed which mode feature ratings were significantly different (Table 4 lists the means by mode).

Table 4. Mean values of quality ratings by mode.

	Feature						
	Cost	**Cleanliness**	**Service**	**Reliability**	**Comfort**	**Safety**	**Security**
Air	4.93	5.20	5.20	4.93	4.80	5.33	5.27
Train	4.33	5.13	4.73	5.47	4.73	5.27	5.00
Car	6.07	6.62	6.33	5.98	6.31	5.98	6.04
Other	4.89	4.55	4.11	4.44	4.33	3.88	4.33

Note: 1 = very bad, 4 = fair, and 7 = very good

For cost, cars were rated higher than air or train, but not different from "other." The only significant difference for cleanliness and service was between car and "other." Reliability was rated equally across modes. Air, train, and cars were rated equally for comfort; "other" was rated significantly lower. The only difference for safety was between car and "other," with cars rated higher. For security, cars were rated as more secure than train and "other;" air, train, and "other" were not significantly different.

Participants were then asked how much September 11th changed their choice of whether or not to travel. Scores ranged from 1 to 7, with an average change score of 3.39 (SD=2.0), which corresponds to "moderately." Analysis of variance tests revealed no significant differences on this change variable for gender, age, education, income, marital status, or number of children. However, there was a significant difference in change for household size (F = 2.11, df = 157, $p < .05$). Post-hoc comparisons revealed that those with two and four people in the household reported greater change than those in one-person households. No other comparisons were significant.

Participants were also asked to check off on a list how September 11th specifically affected their behavior: 74% indicated that they are now more conscious of security, 46% are more aware of other travelers, 33% are more selective in choosing their mode of travel, 22% avoid traveling by plane, 11% indicated that they now avoid traveling altogether, and 3% planned to change their jobs as a result of September 11th. Participants were asked if they would be willing to pay more to travel if those funds were used to increase security. Few participants said that they would be willing to pay more; the majority responded that they would not pay more (60%), while 23% were unsure.

Participants were then asked to indicate how often they felt overwhelmed and not in control per the perceived stress scale described above. Total stress (PSS4) scores could range from 4 to 20; the minimum and maximum total stress scores in this sample were 4 and 14, respectively. The mean was 9.56, with a standard deviation of 2.1. This corresponds to an item mean of 2.39 (SD = .5), or "almost never" having felt the way the item described. The reliability of the stress measure in this sample was acceptable (α = .60). These findings are highly similar to the published psychometrics for this scale. In a national area probability sample, the PSS4 had adequate reliability (α = .60), with a mean score (based on a 0–4 scale) of 4.49 (SD = 2.96) (Cohen and Williamson, 1988). This corresponds to "almost never" on the item response scale. Converting the current sample to a 0–4 scale yields a mean of 5.56 and a SD of 2.09. The item mean would then be 1.39, corresponding to "almost never" perceiving stress.

There were no significant differences in reported stress scores for any of the demographic variables (gender, marital status, education, income, number of people in household, number of children, age). These findings are different from the national sample on which the psychometrics for the scale were derived. In that sample, females reported greater perceived stress than males (Cohen and Williamson, 1988). Those who were divorced had greater perceived stress than those who were single, who had greater perceived stress than those who were married. In addition, stress scores decreased with age, income, and education; PSS4 scores increased with number of people in the household and number of children. The differences found in our sample may reflect the specific characteristics of the sample as described above.

Correlation analyses revealed a small but significant association between perceived stress and the degree to which respondents reported how much September 11th changed their choice of whether or not to travel ($r = .19$, $p<.02$). However, perceived stress scores were not significantly different for those who indicated having been impacted by September 11th in some particular fashion (i.e., becoming more conscious of security, more aware of others traveling, more selective in choosing travel mode, by planning to change their work, or by avoiding traveling by plane, or avoiding traveling altogether) than for those who did not report September 11th affecting them in any of these ways.

Figure 3 shows the geographic distribution of the perceived stress, while Figure 4 depicts the geographic distribution of the variable *Change*, which represents in a scale of 1 to 7 (1=not at all, 7=significantly) how much the September 11th events changed the respondents' travel choices. In both cases, the variables seem to be uniformly distributed across the geography of New York City. In the case of the variable *Change*, this seems to indicate that the September 11th events had similar impacts on the respondents, regardless of their proximity to the World Trade Center.

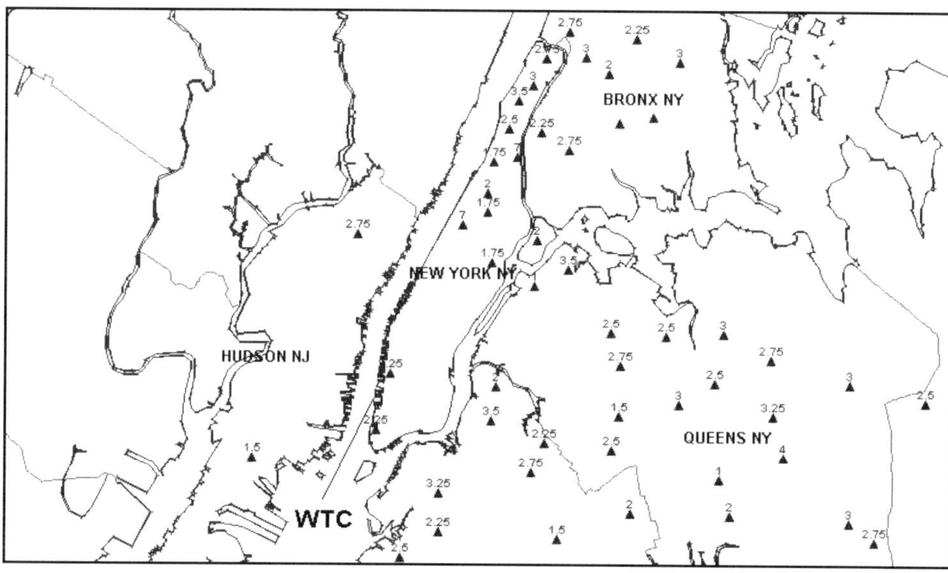

Figure 3. Geographic distribution of perceived stress.

Behavioral Analyses

This section describes the main results of the behavioral modeling conducted as part of this investigation. A number of different families of models were estimated. These families of models differ in the variables that were included in the models and in the specific type of discrete choice model used in the estimation. Two different types of discrete choice models were used: Nested Logit (NL) and Covariance-Heterogeneity Nested Logit (CHNL). The NL model is widely used in situations in which the analyst suspects a violation of the independent of irrelevant alternatives (IIA) property of the MNL model. This case arises when a subset of the alternatives is expected to share

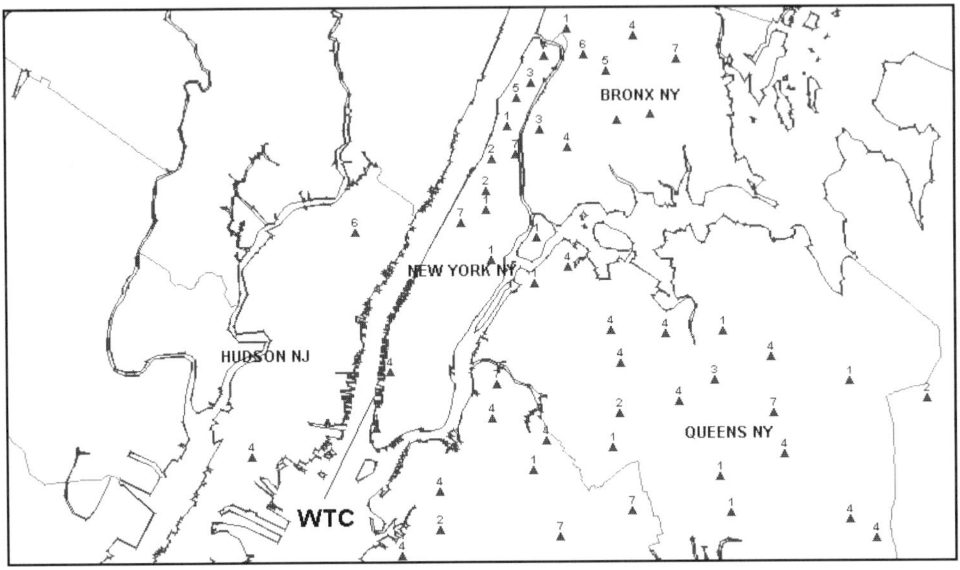

Figure 4. Geographic distribution of* Change *after September 11th.

unobserved attributes and/or measurement errors that cause the error terms to be correlated. Since the presence of the two rail alternatives (Metroliner and Acela) may introduce a violation of the IIA property for the reasons mentioned above, the NL model was selected for use in this investigation. The choice tree considered is shown in Figure 5.

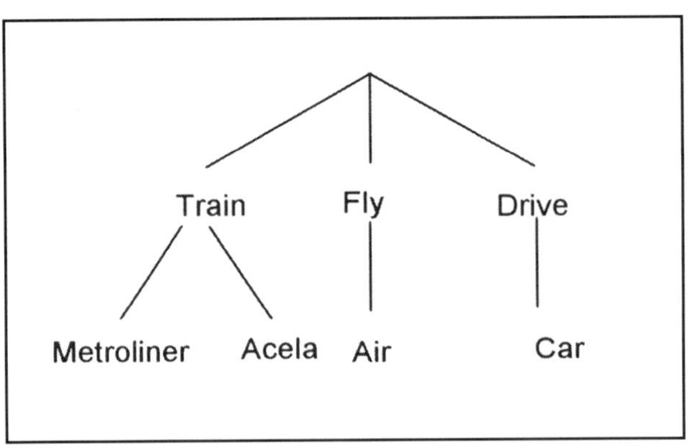

Figure 5. Choice tree.

Another issue that must be dealt with is the correlation introduced by the stated preference data. The different scenarios of stated preference are introduced in the model estimation process as different observations that share the values of the same socio-economic characteristics. The fact that the observations for the same individual are not statistically independent introduces error in the estimation process. This is usually dealt with by using discrete choice models that allow for specific consideration of population heterogeneity (e.g., Random Parameters Logit). As a compromise between ease of implementation and theoretical applicability, the project team decided to use the covariance-heterogeneity nested logit (CHNL) model available in LIMDEP (Greene, 1998), which is based on the model developed by Bhat (1997). The CHNL considers the case in which the parameters of the inclusive values exhibit a systematic relationship with some socio-economic characteristics of the decision makers. Since the coefficients of the inclusive values—related to the ratio of the scale parameters for the lower and upper levels—determine the sensitivity of choice between the alternatives in the nested branch and the others in the tree, the CHNL model enables the explicit consideration of the role that variables, such as income, may play in determining the cross elasticities of choice. Higher values of the coefficients of the inclusive value terms imply higher cross elasticities of choice. In this context, a direct relationship between the explanatory variable in the covariance heterogeneity term would lead to increased cross elasticities. A side benefit of using the CHNL model is that, since covariance heterogeneity is a particular form of population heterogeneity, using the CHNL model helps mitigate the repeated measurement problem introduced by the stated preference data.

In mathematical terms, the basic equations for the conditional and the marginal probabilities in the NL are (after Ben-Akiva and Lerman, 2000; Bhat, 1997; and Greene, 1998):

$$P_n(i/m) = \frac{\exp(\mu_m V_{in})}{\sum_{j \in C_m} \exp(\mu_m V_{jn})} = \frac{\exp(\mu_m \beta' X_{in})}{\sum_{j \in C_m} \exp(\mu_m \beta' X_{jn})} \tag{4}$$

$$P_n(m) = \frac{e^{\mu V_{mn}}}{\sum_{m'=1}^{M'} e^{\mu V_{m'n}}} = \frac{\exp(\frac{\mu}{\mu_m} I_{mn} + \alpha' X_{mn})}{\sum_{m'=1}^{M'} \exp(\frac{\mu}{\mu_m} I_{m'n} + \alpha' X_{m'n})} = \frac{\exp(\gamma I_{mn} + \alpha' X_{mn})}{\sum_{m'=1}^{M'} \exp(\gamma I_{m'n} + \alpha' X_{m'n})} \tag{5}$$

Where: μ and μ_m are the scale parameters for the upper (marginal) and lower (conditional) models; α , β are vectors of parameters; and X is the vector of attributes included in the utility functions.

In the CHNL model (Bhat, 1997; Greene, 1998), the parameter γ, which is the coefficient of the inclusive value I_{mn}, is allowed to vary across individuals as a function of a vector of parameters δ' and a set of variables Y_{mn} as shown in Equation 6.

$$\gamma_{mn} = \gamma_m \exp(\delta' Y_{mn}) \tag{6}$$

Where: $\gamma = \dfrac{\mu}{\mu_m}$ is the coefficient of the inclusive value I_{mn} ; δ is a vector of parameters; and Y represents the vectors of variables explaining the covariance-heterogeneity.

Both NL and CHNL were applied to two basic cases. The first case considered utility functions in which the inspection/boarding time and the rest of the total travel time were treated as separate variables. Among other things, this specification allows the analyst to specifically assess the role of inspection/boarding time at the airport as a factor in mode choice. The second case considered utility functions in which only the total travel time was considered. The families of models considered are represented schematically in Table 5.

Table 5. Families of models considered.

Type of model:	Inspection/boarding time and the rest of total time as two variables	Total travel time as one variable
Nested Logit (NL)	Family of models A	Family of models C
Covariance Heterogeneity Nested Logit (CHNL)	Family of models B	Family of models D

The best models from each of the families described above are discussed next. The model results are shown in a table containing the variables, coefficients, and t-statistics in the traditional format of discrete choice modeling. The models were estimated using the set of variables collected in the sample. The main emphasis of the modeling work was on the attitudinal variables related to September 11th impacts.

Variables Considered in the Models

The models considered alternative specific constants, usually for the air and car alternatives. Travel costs were considered by means of two variables: "*Company costs*" and "*User costs*" (in US dollars) that represent the actual charges incurred either by the traveler or the company (depending on who pay for the trip expenses). The role of travel time was considered using three different variables: "*Inspection/boarding time*," "*Main travel time,*" and "*Total travel time*," all of them in minutes. *Inspection/boarding time* refers to the time spent at the airport checking in and going through the security check points. *Main travel time* is the time spent in door-to-door travel excluding inspection/boarding time, i.e., total travel time minus inspection and boarding. *Total travel time* is the door-to-door travel time. "*Time (1 and 2) before meeting*" are two variables comprising a piece-wise linear approximation to non-linear effects in the utility functions, as shown in Figure 6. *Time 1 before meeting* represents the time up to the cutoff value of 30 minutes, while *Time 2 before meeting* represents the time in excess of 30 minutes. A similar approach was used with the variable *Age*, which was decomposed into three pieces: *Age 1* (less than 25 years old), *Age 2* (number of years in excess of 25, up to 50 years), and *Age 3* (number of years in excess of 50). These piece-wise approximations were intended to capture effects such as the one illustrated in Figure 6, in which time available in excess of the first 30 minutes has a negative utility (for more information, see Ben-Akiva and Lerman, 2000).

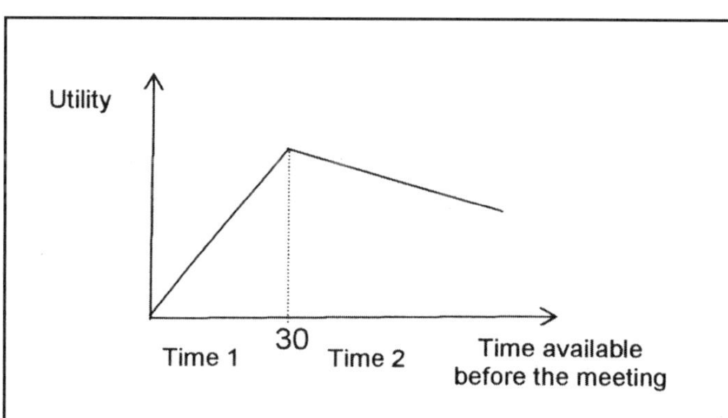

Figure 6. Piece-wise linear approximation to non-linear effects of time available before meeting.

The models included variables that measured both the stress level (*Stress*) and the stated impact produced by September 11th (*Change*). The variable *Change*, that captured their responses to the question of how much September 11th impacted them (1= not at all; 7=significantly), was used in interaction terms with travel time. It was assumed that the variables *Stress* and *Change* could be treated as if they were ratio scales (when in fact they are ordinal scales). This simplifying assumption was taken in order to expedite the model estimation process. Other socioeconomic variables that were found to be significant were income, education level (represented by a set of binary variables), and marital status (a set of binary variables).

It is also important to acknowledge the likely existence of endogeneity bias in some of the models discussed here. This would arise if the variable *Change* shares common unobserved attributes with the mode choice. Correcting for endogeneity bias, either by using instrumental variables (as in Holguín-Veras, 2002) or by explicitly modeling the econometric interactions between the error terms (as suggested in Train, 1986) could not be undertaken here because of the project constraints. This remains the subject of future research.

Models obtained using Inspection/boarding time and Main travel time as two separate variables—Tables 6 and 7 show both the NL and the CHNL versions of the models estimated treating inspection/ boarding time and main travel time as two separate variables. The results outlined in Table 6 illustrate a number of key results that are found in all models. The alternative specific constants for air and car are statistically significant and positive, which indicates bias toward the use of these modes.

In the tradition of discrete choice modeling, the role of the different explanatory variables in the choice process is assessed using the concept of marginal disutility, which represents the rate of change of disutility with respect to an explanatory variable, as shown in Equation 7. In general terms, if a linear in parameters specification is used and there are no interaction terms, the marginal utilities reduce to the coefficient of the variable. If interaction variables are used, the marginal utilities will have additional terms that correspond to the partial derivatives of the interaction variables.

$$U'_{x_{ink}} = \frac{\partial U}{\partial x_{ink}} = \beta_{ink} \tag{7}$$

The marginal disutility of company costs is approximately half the value of the marginal disutility of user costs (the reader should notice that since the coefficients are negative, these are disutilities). This indicates that when the company pays, users behave as having a valuation of travel time, which is double the valuation of their time when they are paying for the expenses. Table 6 also indicates the significance of inspection/boarding time, which has a marginal disutility higher than the marginal disutility for the main travel time (except for air).

As shown in Table 6, the main travel time interacts with the variable *Change*. For that reason, its marginal disutility must be computed taking into account this interaction. In mathematical terms:

$$U(Train\ \ A,B)^{'}_{t_m} = \frac{\partial U(Train\ \ A,B)}{\partial t_m} = -0.010983 - 0.000486\,Change \qquad (8)$$

$$U(Air)^{'}_{t_m} = \frac{\partial U(Air)}{\partial t_m} = -0.6435 - 0.003222\,Change \qquad (9)$$

$$U(Car)^{'}_{t_m} = \frac{\partial U(Car)}{\partial t_m} = -0.01517 - 0.001221\,Change \qquad (10)$$

Where: t_m is the main travel time (total travel time minus inspection/ boarding) in minutes.

Equations 8, 9, and 10 indicate how onerous (in utility terms) it is to travel by each of the transportation modes considered. Equations 8, 9, and 10 show that the marginal disutilities have two components (one attributed to the main travel time and another due to the interaction term between main travel time and *Change*). As shown, the marginal disutilities for the train alternatives are much lower than the ones corresponding to air and car. Furthermore, the marginal disutility for rail is much less affected by the impacts of September 11th on the users, captured through *Change*. This can be appreciated by noting that the coefficients of *Change* in Equation 8, i.e., 0.000486, is 2.5 times smaller than the coefficient of car (0.001221), and approximately 9 times smaller than the same coefficient in the utility function of air (0.003222). As shown in Equation 9, the marginal disutilities for air are much higher than the one corresponding to the other alternatives. Equation 9 also shows that, on a per-minute basis, the impacts of September 11th have had a more noticeable effect on the utility function of air than on the others (due to magnitude of the coefficient of *Change*, i.e., 0.003222).

This indicates the mechanisms by which air demand was affected by the September 11th events, and train demand increased. As shown, the psychological impact of September 11th measured by *Change* had virtually no effect in the decision to use train alternatives. The choices of air and car, in this sample, were significantly impacted by the September 11th events.

Table 6. Nested Logit (NL) version (Model A).

Variable	Rail alternatives: Utility function of Metroliner	Rail alternatives: Utility function of Acela	Fly: Utility function of Air	Drive: Utility function of Car
Alternative specific constants			85.00947	2.66552
			(3.812)	(2.128)
Company costs ($)	-0.012631	-0.012631	-0.012631	-0.012631
	-(4.532)	-(4.532)	-(4.532)	-(4.532)
User costs ($)	-0.028119	-0.028119	-0.028119	-0.028119
	-(7.275)	-(7.275)	-(7.275)	-(7.275)
Inspection/boarding time (mins)	-0.014219	-0.014219	-0.014219	-0.014219
	-(3.919)	-(3.919)	-(3.919)	-(3.919)
Main travel time (mins)	-0.010983	-0.010983	-0.643502	-0.015174
	-(2.999)	-(2.999)	-(3.703)	-(2.478)
(Main travel time) (Change)	-0.000486	-0.000486	-0.003222	-0.001221
	-(1.154)	-(1.154)	-(2.674)	-(2.354)
Time 1 before meeting (< 30 mins)	0.031467	0.031467	0.031467	0.031467
	(3.619)	(3.619)	(3.619)	(3.619)
Time 2 before meeting (> 30 mins)	-0.015489	-0.015489	-0.015489	-0.015489
	-(2.981)	-(2.981)	-(2.981)	-(2.981)

Variable		Utility function of RAIL	Utility function of FLY	Utility function of DRIVE
Income ($ 000/year)		-0.007105		
		-(4.156)		
College Undergraduate (binary)		-0.324041		
		-(2.022)		
College Graduate (binary)		-0.181456		
		-(1.149)		
Stress (PSS4)			-0.096745	
			-(3.460)	
Age (years)			0.009479	
			(1.667)	
Single (binary)				-0.261534
				-(1.887)
High School education (binary)				-0.400265
				-(1.615)
Inclusive value parameters		0.569605	0.563207	0.553665
		(4.512)	(5.844)	(3.682)
Log likelihood function	-1665.667		R-squared	0.1906
Restricted log likelihood	-2057.806		R-squared Adj	0.1866
Chi-squared	784.2782			
Degrees of freedom	23			
Significance level	0.00000			

The perceived stress was also found to negatively affect the choice of air alternatives. As shown in the second part of Table 6, the higher the *Stress*, the less likely the decision makers to choose air. Other socio-economic variables also play a role in the choice process, though for the sake of brevity their effects are not discussed here.

Table 7 shows the CHNL version of the model. As in the NL model shown before, the alternative specific constants for air and car are positive and statistically significant. It is also worthwhile to note that the alternative specific constant for air is smaller than the one estimated from the previous model, which is more in line with what one would reasonably expect. As before, the marginal disutilities of company cost (0.0085) are much smaller than that corresponding to user cost (0.0217), which highlights the respondents' differing valuations of time depending on who pays.

A peculiar feature of Model B is that it does not include the travel time in the utility functions for train, though interaction terms involving travel times and socioeconomic characteristics are statistically significant. The interaction variable (*Main travel time*)(*Change*) was not significant and was taken out of the model. As shown in Equations 11, 12, and 13, Model B is consistent with Model A in pointing out that traveling by air has higher disutilities on a per-minute basis than all the other modes, and that it has been impacted more severely by the post-September 11th events.

$$U(Train\ \ A,B)'_{t_m} = -0.017863\,\delta_{HS} - 0.003946\,\delta_{CU} - 0.010561\,\delta_{CG} \tag{11}$$

$$U(Air)'_{t_m} = -0.2645 - 0.001454\,Change - 0.0532\,\delta_{HS} - 0.006776\,\delta_{CU} - 0.0289\,\delta_{CG} \tag{12}$$

$$U(Car)'_{t_m} = -0.00395 - 0.000434\,Change - 0.01769\,\delta_{HS} - 0.00708\,\delta_{CG} \tag{13}$$

The covariance heterogeneity function considered only included *Income* as an explanatory variable. As shown in the second part of Table 6, this covariate was highly significant and positive. This indicates that individuals with higher income exhibit higher cross elasticities, which conceptually makes sense. This result is also consistent with Bhat (1997).

Models obtained using total travel time—As indicated previously, the choice situations did not contain the wide range of travel times that would have allowed proper estimation of parameters such as subjective travel time values (because in the choice situation, the travel times only changed with the corridor being considered). In order to mitigate this problem, the research

Table 7. Covariance Heterogeneity Nested Logit (CHNL) version (Model B).

Variable	Rail alternatives: Utility function of Metroliner	Rail alternatives: Utility function of Acela	Fly: Utility function of Air	Drive: Utility function of Car
Alternative specific constants			40.28663	2.59457
			(2.922)	(3.031)
Company costs ($)	-0.008547	-0.008547	-0.008547	-0.008547
	-(5.331)	-(5.331)	-(5.331)	-(5.331)
User costs ($)	-0.021749	-0.021749	-0.021749	-0.021749
	-(8.836)	-(8.836)	-(8.836)	-(8.836)
Inspection/boarding time (mins)	-0.011175	-0.011175	-0.011175	-0.011175
	-(4.028)	-(4.028)	-(4.028)	-(4.028)
Main travel time (mins)			-0.264585	-0.003951
			-(2.459)	-(1.692)
(Main travel time) (Change)			-0.001454	-0.000434
			-(3.397)	-(2.653)
Time 1 before meeting (< 30 mins)	0.028372	0.028372	0.028372	0.028372
	(3.250)	(3.250)	(3.250)	(3.250)
Time 2 before meeting (> 30 mins)	-0.010273	-0.010273	-0.010273	-0.010273
	-(2.294)	-(2.294)	-(2.294)	-(2.294)
(Main travel time) (High School)	-0.017863	-0.017863	-0.053292	-0.017691
	-(4.601)	-(4.601)	-(4.757)	-(3.917)
(Main travel time) (College Underg)	-0.003946	-0.003946	-0.006773	
	-(3.321)	-(3.321)	-(2.499)	
(Main travel time) (College Grad)	-0.010561	-0.010561	-0.028961	-0.007085
	-(4.203)	-(4.203)	-(4.725)	-(3.324)

Variable		Utility function of RAIL	Utility function of FLY	Utility function of DRIVE
(Main travel time) (Age 1)		0.000095		
		(1.607)		
(Main travel time) (Age 3)		-0.000231		
		-(3.903)		
Stress (PSS4)			-0.146633	
			-(4.772)	
Single (binary)				-0.402543
				-(2.702)
Inclusive value parameters		0.754331	0.611342	0.705494
		(5.375)	(6.507)	(4.511)
Coefficient of the Cov-Het term				
Income ($ 000/year)	0.002896			
	(4.179)			
Log likelihood function	-1623.847		R-squared	0.2109
Restricted log likelihood	-2057.806		R-squared Adj	0.2063
Chi-squared	867.918			
Degrees of freedom	27			
Significance level	0.00000			

team decided to estimate two families of models, similar to the ones discussed above, using total travel time instead of inspection/boarding time and the main travel time. This section reports the findings of these efforts.

Table 8 shows the statistics of Model C. As shown, the alternative specific constants reduced their values significantly to more realistic levels. In both cases, these constants are positive, indicating that, in equality of conditions, users would favor these transportation modes. As in the previous models, the marginal disutilities of user costs are much higher than the ones corresponding to company costs.

The marginal disutilities for total travel time vary by mode. The estimation results indicate that traveling by air has the higher disutility of time (0.009693) followed by car (0.008083) and train (0.02527). The marginal disutilities of travel time are increased by the interaction terms between total travel time and the variable *Change*. As shown, the marginal disutilities of (*Total travel time*)(*Change*) for the air alternative (0.001105) are approximately twice the value of that corresponding to car (0.000568), while this variable plays no role whatsoever in the utility of rail.

$$U(Train\ \ A,B)'_{t_T} = -0.002527 \tag{14}$$

$$U(Air)'_{t_T} = -0.009693 - 0.001105\,Change \tag{15}$$

$$U(Car)'_{t_T} = -0.008083 - 0.000568\,Change \tag{16}$$

Where: t_T is the total travel time (door to door) in minutes.

Stress, as in the previous models, was found to have a statistically significant negative impact on the choice of air. There were other socio-economic attributes that were also found to play a role in mode choice, among them level of education and age.

The CHNL version of this model is shown in Table 9. The parameters of the model are highly consistent with the parameters of the models discussed in the previous section, in that (1) the marginal disutilities of company costs (0.00693) are much smaller than the ones corresponding to user costs (0.022031); (2) the disutilities of travel time for air (0.009781) are higher than that corresponding to car (0.008294) and train (0.002615); (3) the interaction variable (*Total travel time*)(*Change*) has a more pronounced impact upon the choice of air (0.001017) than for any other mode; (4) the amount of free time before the meeting has a positive effect on mode choice as long as it is less

Table 8. Nested Logit (NL) version (Model C).

	Rail alternatives		Fly	Drive
Variable	Utility function of Metroliner	Utility function of Acela	Utility function of Air	Utility function of Car
Alternative specific constants			3.30624	2.42130
			(3.920)	(2.683)
Company costs ($)	-0.006787	-0.006787	-0.006787	-0.006787
	-(4.357)	-(4.357)	-(4.357)	-(4.357)
User costs ($)	-0.022127	-0.022127	-0.022127	-0.022127
	-(8.451)	-(8.451)	-(8.451)	-(8.451)
Total travel time (mins)	-0.002527	-0.002527	-0.009693	-0.008083
	-(2.180)	-(2.180)	-(3.142)	-(2.545)
(Total travel time) (Change)			-0.001105	-0.000568
			-(3.499)	-(2.879)
Time 1 before meeting (< 30 mins)	0.027873	0.027873	0.027873	0.027873
	(3.065)	(3.065)	(3.065)	(3.065)
Time 2 before meeting (> 30 mins)	-0.015641	-0.015641	-0.015641	-0.015641
	-(3.033)	-(3.033)	-(3.033)	-(3.033)

Variable		Utility function of RAIL	Utility function of FLY	Utility function of DRIVE
Income ($ 000/year)		-0.006718		
		-(4.021)		
College Undergraduate (binary)		-0.355810		
		-(2.268)		
College Graduate (binary)		-0.199004		
		-(1.291)		
Stress (PSS4)			-0.097166	
			-(3.435)	
Age (years)			0.009208	
			(1.606)	
Single (binary)				-0.243119
				-(1.772)
High School education (binary)				-0.300308
				-(1.211)
Inclusive value parameters		0.696508	0.627950	0.723297
		(4.922)	(5.907)	(4.181)
Log likelihood function	-1676.746		R-squared	0.1852
Restricted log likelihood	-2057.806		R-squared Adj	0.1815
Chi-squared	762.1207			
Degrees of freedom	21			
Significance level	0.00000			

Table 9. Covariance Heterogeneity Nested Logit (CHNL) version (Model D).

Variable	Rail alternatives: Utility function of Metroliner	Rail alternatives: Utility function of Acela	Fly: Utility function of Air	Drive: Utility function of Car
Alternative specific constants			4.60847	3.35028
			(4.976)	(3.195)
Company costs ($)	-0.006930	-0.006930	-0.006930	-0.006930
	-(4.259)	-(4.259)	-(4.259)	-(4.259)
User costs ($)	-0.022031	-0.022031	-0.022031	-0.022031
	-(8.399)	-(8.399)	-(8.399)	-(8.399)
Total travel time (mins)	-0.002615	-0.002615	-0.009781	-0.008294
	-(2.075)	-(2.075)	-(3.026)	-(2.524)
(Total travel time) (Change)			-0.001017	-0.000590
			-(3.287)	-(2.775)
Time 1 before meeting (< 30 mins)	0.029564	0.029564	0.029564	0.029564
	(3.323)	(3.323)	(3.323)	(3.323)
Time 2 before meeting (> 30 mins)	-0.015805	-0.015805	-0.015805	-0.015805
	-(3.119)	-(3.119)	-(3.119)	-(3.119)

Variable		Utility function of RAIL	Utility function of FLY	Utility function of DRIVE
College Undergraduate (binary)		-0.155024		
		-(1.120)		
Stress (PSS4)			-0.122360	
			-(4.392)	
Age (years)			0.007727	
			(1.205)	
Single (binary)				-0.367154
				-(2.629)
High School (binary)				-0.400432
				-(1.472)
Inclusive value parameters		0.584141	0.525658	0.581367
		(4.816)	(5.880)	(4.155)
Coefficient of the Cov-Het term				
Income ($ 000/year)	0.002582			
	(2.756)			
Log likelihood function	-1683.443		R-squared	0.1819
Restricted log likelihood	-2057.806		R-squared Adj	0.1784
Chi-squared	748.726			
Degrees of freedom	20			
Significance level	0.000000			

than 30 minutes, after this threshold, it has a negative impact; and (5) the higher the *Stress* level, the less likely the users are to choose air. The marginal disutilities for total travel time are shown in Equations 17, 18, and 19.

$$U(Train\ \ A,B)'_{t_T} = -0.002615 \tag{17}$$

$$U(Air)'_{t_T} = -0.009781 - 0.001017\,Change \tag{18}$$

$$U(Car)'_{t_T} = -0.008294 - 0.000590\,Change \tag{19}$$

The modeling results confirmed previously held assumptions about the factors determining intercity mode choice. Variables such as travel costs, time, income, gender, and the like were found to be statistically significant explanatory variables in the mode choice process. These results are in complete agreement with the intercity mode choice literature (e.g., Forinash and Kopppelman, 1993; Bhat, 1997).

As indicated by the model results, the impacts of extreme events on intercity passenger travel behavior consist of modifications of the utility functions for the different modes that translate into a departure from what is normally expected. The authors' conjecture is that these impacts are dynamic in nature and, for that reason, some of them are likely to evolve over time. This suggests that further research is needed to distinguish among the transient and permanent behavioral changes produced by September 11th on intercity travel.

Conclusions

This paper summarized the research conducted on the impacts of extreme events upon intercity travel behavior. This research relied on stated preference data provided by a convenience sample of residents of the New York City metropolitan area, collected six months after September 11th, and the use of modern econometric techniques based on Random Utility Theory to assess behavioral changes. The data collected was based on stated preference techniques, by which the respondents are asked to rank order the different alternatives (air, car, Metroliner, and Acela) as part of a hypothetical choice situation.

The findings may be limited in their generality given the nature of the sample. The participants in this study were mostly young, male, and single without children. They were also highly educated and reported higher incomes than the general population of the five boroughs of New York City. Despite these specific characteristics, a majority of the participants reported having made trips to the target cities, which indicates that the sample was an appropriate one to use and that these individuals do constitute a subset of the market demand in the corridors studied.

Participants reported that before September 11th they were most likely to choose transportation mode based on convenience and cost, and the mode of preference for most was car. Indeed, trips by car were rated better on cost than air or train, and were rated as more secure than trains. While there were no other significant differences among the primary modes (air, train, car) on any of the other features assessed (cleanliness, comfort, and safety), these features may not be as important as cost when choosing mode, especially when one is paying for the trip oneself, as was the case for the majority of participants.

On average, participants reported that September 11th affected travel change “moderately” but it is important to note that the full range of the scale was endorsed by participants. The most frequently reported specific changes were that people became more conscious of security and more aware of other travelers. Participants also had average perceived stress levels that correspond to “almost never” on the response scale, which is comparable to the data from a national probability sample (Cohen and Williamson, 1988). Despite low levels of perceived stress, the general change measure was significantly associated with stress levels, such that those who reported a greater September 11th impact on travel behavior also reported greater levels of perceived stress.

In terms of behavioral modeling, two different types of random utility models were estimated: Nested Logit and Covariance-Heterogeneity Nested Logit models. These two basic types were used in the estimation process using the variables gathered during the data collection process. The estimated models are highly consistent among themselves in highlighting a set of fundamental conclusions about travel behavior in the aftermath of an extreme event.

The modeling results confirmed previously held assumptions about the factors determining intercity mode choice. Variables such as travel costs, time, income, gender, and the like were found to be statistically significant explanatory variables in the mode choice process. These results are in complete agreement with the intercity mode choice literature (e.g., Forinash and Koppelman, 1993; Bhat, 1997).

The research was successful in finding statistically significant linkages between changes in travel behavior and the impact of an extreme event, in this case September 11th. These linkages revealed themselves as additional terms in the utility functions estimated using Random Utility Theory. In all models estimated, the variable that measured the impact of the September 11th events upon the individuals that participated in the survey, i.e., *Change*, and a psychometric scale of perceived stress level, i.e., *Stress*, were found to play a statistically significant role in the mode choice process. Interestingly, *Change* and *Stress* seem to have different mechanisms.

The variable *Change* was found to have significant interactions with the travel time. This, in turn, translates into the marginal disutility of travel time being modified by *Change*, as shown in Equations 8 to 19. The magnitude of this modification is clearly related to the transportation mode. In all cases, the contribution of the interaction term between *Change* and travel time is highest for air, followed by car and then train. This may be related to the fact that the terrorist attacks involved airplanes and the effect this had on the general public's perception about the safety of the air transportation system after September 11th. The modeling results clearly indicate that the utility functions of the train alternatives are minimally affected by the interaction terms between *Change* and travel time (in some cases, the interaction term drops out of the utility functions). The latter suggests that the users perceive the train alternatives as being less taxing to them, in utility terms, after an extreme event such as September 11th.

The perceived level of stress, i.e., *Stress*, was found to have a statistically significant impact in mode choice. However, the interpretation of the impact of *Stress* and its relation to September 11th is obscured by the fact that the psychometric measure used (PSS4; Cohen and Williamson, 1988) provides a measure of overall stress level, not of the stress specifically produced by September 11th. In any case, *Stress* specifically impacted the utility function of air without interacting with any other variable or utility function. In general terms, the higher the stress level, the less likely the decision makers to choose the air alternative.

In spite of the numerous and significant limitations faced in this research, the authors are confident in the ability of this research to provide insights into the effects that extreme events have upon intercity travel behavior. This modest success should not obscure the fact that this paper is nothing more than a first step in the long march towards a better understanding of the impacts of extreme events on travel behavior.

References

Ben-Akiva, M. and S. Lerman. 2000. *Discrete Choice Analysis*. Cambridge, MA: MIT Press.

Bhat, C.R. 1997. "Covariance Heterogeneity in Nested Logit Models: Econometric Structure and Application to Intercity Travel." *Transportation Research* 31B(1): 11–21.

Bureau of Transportation Statistics. 1997. *American Travel Survey*. Publication No. BTS/ATS95–US. Washington D.C.: U.S. Department of Transportation.

Cohen, S. and G.M. Williamson. 1988. "Perceived Stress in a Probability Sample of the United States." Pp. 31–67 in S. Spacapan and S. Oskamp, eds., *The Social Psychology of Health*. Newbury Park, CA: Sage.

Forinash, C. and F. Koppelman. 1993. "Application and Interpretation of Nested Logit Models of Intercity Mode Choice." Presentation at the 72nd Annual Meeting of the Transportation Research Board, Washington, D.C.

Greene, W. 1998. *LIMDEP: User's Manual*. Plainview, NY: Econometric Software, Inc.

Hensher, D. 1994. "Stated Preference Analysis of Travel Choices: The State of the Practice." *Transportation* 21: 107–133.

Holguín-Veras, J. 2002. "Revealed Preference Analysis of the Commercial Vehicle Choice Process." *Journal of Transportation Engineering* 128(4): 336–346.

Holguín-Veras, J., R. Paaswell, and E. Marin. 2002. "Business Location Modeling Working Paper." New York: University Transportation Research Center.

Holusha, J. 2002. "In Office Market, a Time of Uncertainty." *New York Times* (January 6): Real Estate Section 1,9.

McFadden, D. 1974. "Conditional Logit Analysis of Qualitative Choice Behavior." Pp. 105–142 in P. Zarembka, ed., *Frontiers in Econometrics*. New York: Academic Press.

McFadden, D. 1978. "The Theory and Practice of Disaggregate Demand Forecasting for Various Modes of Urban Transportation." Pp. 1–27 in W. Brown, ed., *Emerging Transportation Planning Methods*. Washington, D.C.: U.S. Department of Transportation.

Thurstone, L.L. 1927. "A Law of Comparative Judgement." *Psychological Review* 34: 273–286.

Train, K. 1986. *Qualitative Choice Analysis*. Cambridge, MA: MIT Press.

Acknowledgments

The research team expresses its gratitude and admiration for the staff of the New York Metropolitan Transportation Council and the Port Authority of New York and New Jersey, who despite having suffered terrible loses in September 11th had the energy, resiliency, and motivation to collaborate in this and other research projects. The research team is also grateful to the approximately 200 volunteers who generously agreed to participate in this important investigation. The investigators also recognize the collaboration of the staff of the Bureau of Transportation Statistics, particularly Mr. John Wells, Chief Economist, for providing data crucial to this investigation. Mr. Gary Grayley, from Greyhound, provided key insights into the nature of intercity mode competition. The extraordinary work done by Ms. Eva Marin, Graduate Research Assistant at the City College of New York, in putting together the data files and supporting the analyses is to be commended. The insights provided by Professor Chandra Bhat, University of Texas at Austin, into the Covariance Heterogeneity Nested Logit model are gratefully acknowledged. The contributions of all these individuals contributed to enhancing the quality of the analyses conducted here, but do not make any one of them responsible for the shortcomings of the research.

This research was partially supported by a grant from the National Science Foundation. This support is both acknowledged and appreciated.

Correspondence should be directed to José Holguín-Veras, Ph.D., P.E., Associate Professor, Department of Civil and Environmental Engineering, Rensselaer Polytechnic Institute, 4030 Jonsson Engineering Center, 110 Eighth Street, Troy, NY 12180-3590; e-mail: jhv@ti-mail.engr.ccny.cuny.edu or to Robert E. Paaswell, Ph.D., P.E., Distinguished Professor, City College of New York, Bldg. Y-220, New York, NY 10031; e-mail: paaswellcius@ccny.cuny.edu.

A Complex Organizational Adaptation to the World Trade Center Disaster: An Analysis of Faith-based Organizations

Jeannette Sutton
Department of Sociology
University of Colorado–Boulder

Introduction

The disaster on September 11, 2001, was massive in scope, affecting a major metropolitan area of world significance and all levels of social life. It was a complex, human-conceived disaster, resulting in the loss of more than 2,800 lives, enormous material destruction, and worldwide economic turmoil and political strife. The long-term effects remain unquantifiable in terms of levels of individual and community health, finance, and recovery.

Faith-based organizations have always responded to disasters, playing important roles in all phases of disaster relief. For instance, as a regular course of action, the Southern Baptist Men and the Salvation Army traditionally provide volunteers at feeding stations, local congregations routinely offer building space for emergency housing and meeting space, and Church World Service consistently provides a coordinating effort for financial and long-term resources for recovering communities.

The attacks on the World Trade Center created immense need throughout New York City and the tri-state metropolitan area, where an outpouring of aid was provided through governmental and non-governmental channels, including hundreds of congregations, faith-based community organizations, and disaster response agencies. Each of these organizations, whether they had existing disaster protocols or not, was challenged and stretched in task and structure as it addressed the immediate and long-term needs of disaster victims and the surrounding city.

Theoretical Problem

The mission of many faith-based organizations includes the provision of aid to persons in need. This activity informs the structures from which those organizations operate and the tasks in which they engage. In a disaster, many organizations must adapt to meet needs within their communities that may or may not be within the scope of their normal ranges of activity. Although faith-based organizations have historically participated in disaster response in numerous ways, little research has been conducted on the ways that they adapt in structure and task over the course of a disaster response. One question that is raised in particular is: how do latent functions (unintended consequences of which the participants are unaware) impact the adaptation of faith-based organizations after a disaster?

Research Problem

The purpose of this research is (1) to discover what impact, if any, latent functions affected the way faith-based organizations adapted to meet needs in the week after the World Trade Center attacks, and (2) to determine if the established typology depicting organizational adaptation applies to faith-based organizations in this situation, and if not, to suggest changes to that typology.

Literature Review

Organized behavioral responses in disasters have been the subject of extensive sociological research for several decades. Within the disaster literature, specific attention has been given to organizations that are traditional emergency responders, such as police and fire departments, as well as groups responding to emergent needs, such as search and rescue and other citizen groups. Of particular interest here is an analysis of faith-based organizations that exist before the disaster and continue to function within their communities as part of the disaster response. This analysis makes use of the typology of organizational adaptation in disaster developed by the Disaster Research Center, known as the "DRC typology," and the few developments made in analyzing established, expanding, and extending organizations and their disaster responses.

DRC Typology

Proposed as a model by which to analyze organizational adaptation in crisis (see Quarantelli, 1966; Brouilette and Quarantelli, 1971; Dynes, 1970), the DRC typology has provided a foundation from which to identify and analyze the changing structures and tasks of organized behaviors and has served as a springboard for further theoretical work. This typology clustered a wide range of organized behaviors observed after disaster into four simple categories based upon the organizational tasks and structures. Tasks were identified as being either routine or non-routine. Structures that were used to accomplish these tasks were identified as either old or new, that is, operating within the pre-disaster time frame, or emerging after its occurrence. These structures develop as a result of tasks—activities that are undertaken—resulting in social organization in disaster response (Bosworth and Kreps, 1986). When the two dimensions are cross tabulated, four analytic categories are produced, demonstrating four forms of organized behavior.

Type 1, **established organizations**, are established before the disaster and continue to function with routine tasks and old structures after a disaster. Police and fire departments typify a Type 1 organizational response. Type 2 organizations are **expanding organizations**, for which routine tasks continue but new structures are added after a disaster. The new structures may comprise an increase in volunteer bases, such as within the American Red Cross, or an increase in structural complexity, such as the reorganizing of employees within an organization to meet emergent needs. A Type 3 organization is an **extending organization**. This type continues to function within its old structure, but adapts by means of employing nonroutine tasks to meet needs. An example of this could be an elementary school that serves as a shelter for earthquake victims. A Type 4 organization is an **emergent organization**, where nonroutine tasks and new structures are developed to meet unmet needs. Local citizens, for example, may identify a long-term need that is not being met by existing structures, thereby forming a coalition to address these issues.

Although the existing DRC typology has served as a fruitful analytical tool for organizational adaptation in disaster, several researchers have argued that the typology is too limited (see Bardo, 1978; Stallings, 1978; and Quarantelli, 1996), resulting in elaborations by Bardo (1978), Quarantelli (1983 and 1996), and Drabek (1987). Each of these additions capitalizes on the existing typological categories and creates further distinctions and complexities.

Bardo's (1978) inclusion of latent and manifest disaster behavior within organizations develops the argument that pre-established organizations responding in disasters often make use of latent tasks in their response. These latent disaster behaviors are not engaged in under normal circumstances, but only become manifest during the post-disaster time phase. Bardo provides the example of a "disaster plan" for an established organization not normally maintaining a role in disaster, explaining that in a disaster, an organization can respond in a systematic, pre-specified manner by making manifest these normally latent activities (that is, invoking the disaster plan).

Upon further examination of empirical data, Quarantelli (1983) also provided an expansion of the DRC typology to account for emergent behaviors within established organizations. He found that within most of the established groups, there were considerable emergent phenomena. Three adaptations were noted: quasi-emergence, structural emergence, and task emergence. Quasi-emergence was the adaptation of established groups that underwent no major alterations in their structures or functions but exhibited some temporary or minor emergent qualities. Emergency organizations that temporarily improvise communication procedures when telephone connections are knocked out for a short period in a flood are one example of this type of emergence. Structural emergence was seen in organizations that carried out old functions or tasks and developed some new structure, but did not become an expanding group. One example of structural emergence is the temporary use of amateur radio groups when the traditional means for spreading public warnings is unavailable. Task emergence was seen in pre-existing groups where the structure was in no way altered or changed, but a major new task was assumed. Task emergence may be seen when a group carries out a temporary task such as setting up a public shelter when normal channels for this function are closed. In each of these examples group adaptation is temporary, and organizations return to their old form of functioning soon after implementation. The return to normal tasks and structures is what marks this new formulation, distinguishing the adaptation of established groups from the emergent groups described by the original DRC typology.

Faith-based Organizations in Disaster Research

Although past studies of disaster recovery have included organizations without predetermined disaster-oriented tasks and structures, few studies have focused specifically on the adaptation of faith-based organizations in disaster. Most work has been limited to descriptive accounts of church activities, pastoral leadership, and the activities of groups such as the Salvation Army in disaster (for example, Form and Nosow, 1958; Barton, 1969; Martin, 1976; Drabek and Key, 1984), or the impact of devotionalism on helping behavior (Nelson and Dynes, 1978). However, a small number of studies make use of the DRC typology in their analysis of faith-based organizations.

Ross and Smith (1974) provide an organizational analysis of emergent interfaith disaster recovery groups examining the later stage of emergence identified by Quarantelli as the "crystallization phase" (Quarantelli 1985), during which emergent groups solidify. Ross (1980) also examined interorganizational relationships between ecumenical disaster recovery groups focusing on the "post-crystallization conditions" (Quarantelli 1985), leading to institutionalization. At a congregational level, Smith (1978) examined local church responses to natural disasters, identifying organizational task extension based upon their pre-disaster demand-capability balance and disaster conditions.

One key trend identified by Smith (1978) was the expansion of the conceptualization of disaster organizations to include organizations that perform tasks and develop structures that are different from their pre-disaster tasks and structures. In other words, the definition of organizations responding in disaster expanded beyond those organizations that were developed with the purpose of providing disaster response, such as emergency management organizations, to those groups that play some role in disasters beyond their day-to-day operations, such as faith-based organizations. More importantly for the purposes of this paper, is the identification of the adaptive measures within faith-based organizations that are based upon those latent tasks that modify existing structure or help develop new ones. In contrast with Bardo's conception of latency (such as the creation of a disaster plan that is invoked in a disaster), here latency refers to the underlying mission, or function, of many faith-based organizations; that of providing some sort of "humanitarian aid" and "spiritual care." In the case of disaster response, many faith-based organizations make manifest these latent functions, meeting needs produced by the disaster.

Quarantelli's concept of adaptive behavior within existing groups is helpful because it shows that for some existing groups, there may be experiences of emergent phenomena in disaster response. Existing groups may not respond entirely within their predetermined domains, but instead make use of their resources in non-traditional ways. However, the emphasis on non-traditional behavioral adaptation necessarily conflicts with the idea of latent tasks made manifest. When latent tasks and inactive structures come into being, they are not newly emergent but simply put into practice. This paper examines several pre-existing faith-based organizations and their disaster response after the World Trade Center attacks, making use of the DRC typology with an emphasis on their latent functions.

Research Methods

This research is based in qualitative data collection and analysis approaches following Lofland and Lofland (1995), in which gathering, focusing, and analyzing data occurs concurrently throughout field research. The researcher as witness and observer also becomes a participant in the field site, leading to direct observation of and participation in the setting (Lofland and Lofland, 1995) while engaging in open-ended interviews with subjects. Different data collection techniques are appropriate during different stages of research (Weller and Romney 1988). For example, in the early stages, informal exploratory interviewing is necessary in order to obtain a general notion of the setting, activities, and organizational structures in which the activities are taking place. In the immediate post-disaster time frame, this same concept of exploratory interviewing and observation provides a framework from which to identify particular phenomena for consideration. These phenomena are then analyzed and interpreted with the aid of theory.

Due to the perishable nature of data immediately after a disaster, one of the first goals during this project was to enter the field as quickly as possible in order to observe and collect data that would have otherwise quickly become lost. In the days immediately after September 11, telephone contact was made with representatives from the Presbytery of New York City (PNYC), Church World Service (CWS) and the national Presbyterian Church's Disaster Assistance (PDA) office with the goal of identifying congregations in Manhattan and local faith-based organizations that were providing aid and services to victims within New York City. Although there are several hundred religiously based communities and faith-based organizations in New York City, the choice to contact the PNYC and PDA was based on this researcher's personal qualifications as a Presbyterian clergy-person. This insider status opened doors to access information about the response of the Presbyterian Church at the national level and the response of the PNYC, while adding credibility to requests for interviews with local clergy members.

The initial phone contacts made to the PNYC and members of CWS began a snowball sample of local spiritual care providers located in Manhattan and the boroughs of New York City. Contact was made with the Executive Directors of the Temple of Understanding (a broadly reaching interfaith educational center on the east side of Manhattan) and the Interfaith Center of New York (a center directly involved in local education and outreach of an interfaith nature). Within the PNYC, unstructured interviews were conducted with several persons working at an administrative level, and contact was made with several ministers in the surrounding boroughs of Manhattan. The PNYC also held an unscheduled consultative meeting for all ministers of the Presbytery, attended by many representatives from disaster response agencies (PDA, CWS disaster relief, and the American Red Cross (ARC)), who provided information and resources on short- and long-term recovery. As a result of interaction with these organizational representatives, interviews were scheduled with members of the ARC, including members of the ARC-affiliated disaster child care team, and disaster spiritual care team.

Over this week-long site visit, a total of 24 qualitative interviews were conducted with representatives of 10 faith-based organizations. These interviews include representatives from two interfaith ministries, four local Presbyterian congregations, administrators and ministers of the PNYC, representatives of CWS, and the two ARC-affiliated groups that provided disaster child care and spiritual care. These interviews focused on service delivery and organizational change in this particular disaster and included questions about normal organizational structure and function in contrast with the post-disaster tasks and structures, such as the types of needs assessed within the communities each organization traditionally served, additional emergent needs as a result of the disaster, and the services offered based upon the normal working structures and traditional activities of the organization.

Field observation of disaster activities took place at settings such as the Disaster Head-Quarters for the ARC, the Presbytery offices on the upper west side of Manhattan, and Pier 94 where the Family Assistance Center was established for this relief effort. These observations were noted in field notes, focusing on the setting of the activity and the types of activities in which organizational members were engaged.

Data Analysis

After all interviews and observations were transcribed, inductive methods were used for this analysis; in other words, descriptive material that had been collected was examined to identify themes and develop empirical generalizations. Analysis was based in the theoretical model established by the DRC on organizational adaptation, paying close attention to the structures made manifest and identifiable latent tasks and functions associated with the organizational change.

The next section describes the post-disaster structures and tasks of nine faith-based organizations: the Presbytery of New York (PNYC), four local Presbyterian congregations, two interfaith organizations, and two faith-based organizations affiliated with the American Red Cross response. It will show how they chose to meet the needs that emerged as a result of the World Trade Center attacks and provide an analysis of how each organization adapted as a result of latent functions.

Findings

The Presbytery of New York

The Presbytery of New York (PNYC) is responsible for the mission and government of local Presbyterian Church congregations throughout its geographic district. The PNYC provides governance to 99 churches in all five boroughs (the Bronx, Brooklyn, Queens, Manhattan, and Staten Island). Paid staff members provide executive leadership within the Presbytery offices, while elders and clergy serve as members on committees and commissions, providing oversight in designated areas.

Presbytery tasks include coordinating the work of its member churches; providing pastoral care and counseling to its congregational membership and pastors in forms such as supportive listening, physical assistance, and seasoned advice; and providing similar resources to member churches as needed (Book of Order, 1999, p. G-11.0103). On occasion, special commissions are established to deal with judicial matters or other issues. These commissions, which comprise elders and ministers, have the ability to make decisions on behalf of the larger membership of the Presbytery.

Disaster Activities

In response to the attacks, the PNYC immediately began making calls to congregational pastors to determine congregational and pastoral needs, fielded phone calls from congregations and Presbyteries across the country, and called an emergency consultative meeting of member congregations. The purpose of the meeting was broad and included offering a forum for local pastors to share their experiences of the attacks, to encourage one another in their ministry with members in their congregations, to share ideas for outreach to the neighborhoods around them, and to learn about the roles of governmental and non-governmental disaster relief efforts.

One of the questions that emerged through the forum and outreach calls related to the monetary donations that were coming in through the Presbyterian Church's national offices, the PDA offices, and the local PNYC. This question was coupled with a concern about how to plan for and respond to long-term congregational needs and the provision of internal pastoral care to the ministers of the Presbytery.

At the time of this meeting, preliminary needs assessments had been underway for several days. However, the extent of the impact on congregations was not yet known and decisions about financial assistance, pastoral training, and pastoral care could not immediately be made. In response to the need for an official decision-making task force and coordination of PNYC resources to the long-term disaster needs, the Presbytery executive body made the decision to establish an administrative commission that would attend to any ongoing and emergent issues.

Structure and Task

As a permanent faith-based organization in New York, the PNYC had an established structure with specific responsibilities designated through the Book of Order for the Presbyterian Church, USA. Additionally, this faith-based organization provided services to its member congregations, which are well within the boundaries of Presbytery responsibilities, where the activation of an administrative commission to address ongoing disaster needs is one service provided by the PNYC to its member congregations.

Although the structure of the commission is defined within the Book of Order of the Presbyterian Church (1999) the tasks for which it is responsible depend upon the emergent needs of the Presbytery and its member congregations. These needs, such as training in trauma counseling and crisis management, as well as the distribution of donated funds and material resources, can be identified as non-routine, requiring adaptive measures that were met by the expansion of organizational structures.

Expanded Organizational Structures

The administrative commission of the Presbyterian Church is one example of making manifest a suspended structure, whereby the organization expands to meet emergent needs. Although this structure does not actively function in the normal operations of the organization, a task force such as this is traditionally assembled when specific needs arise. The addition of such a commission increases the structural complexity of the organization in non-routine situations. However, the addition itself is built into the operations of the Presbytery.

The PNYC did not have any specific internal pre-disaster plans. Resources were made available through the PDA program and by connections to the Federal Emergency Management Agency (FEMA) and other disaster relief organizations. However, the organization found itself adapting to meet the long-term needs of member congregations. This need was met by the addition of the administrative commission, which created an expansion of the organizational structure and an increase in organizational complexity. Therefore, the Presbytery can be analyzed as an expanding organization within the DRC typology, where structures were activated in order to provide the means for the addition of a commission to address emergent needs.

Local Presbyterian Congregations

The first of the four congregations studied is a predominantly white, upper-middle-class congregation in New Jersey. The congregation experienced the death of one member, and is the home to 20 members who escaped from the towers on September 11. Those who escaped and survived were described as experiencing "survivor guilt" and their family members were having difficulty knowing how to provide support in the aftermath. In the first week, the church sanctuary was opened for prayer and meditation, and crisis counseling was offered to congregants and to neighborhood visitors who did not have a church home.

In the week after the attacks, the co-pastors made calls to every member of the church to assess needs and to offer counseling. Educational materials were produced addressing such issues as how to talk to children about death and the effects of posttraumatic stress disorder. By the second week, support groups for spouses had been developed and were underway.

The second congregation is a Hispanic congregation, led by a newly installed minister. No members of this congregation died in the disaster, but many expressed a great deal of fear and were questioning life, faith, and the meaning of community.

In the days after the attack, the church was open for prayer services and silent meditation. The pastor of this congregation reported that he noticed more visitors and persons in attendance who were turning to the church for answers about the meaning and purpose of their lives and the disaster events.

The third congregation is a small, racially diverse group whose most consistent attendees are numerous neighborhood children. The pastor of this congregation explained, "If I walk the four blocks around my church, I walk around the world," noting the international flavor of residents of this outlying borough in New York City. On September 11, the local children came to the church and asked to pray and to sing together. The children, in particular, expressed a need for a place to go to where they would be protected and provided assurance of safety. In the midst of this, the church continued to provide ongoing programs, modeling stability and consistency for the children.

The fourth congregation is a large Presbyterian church located within several miles of the World Trade Center. This area of Manhattan, below Canal Street, had restricted access for several weeks after September 11. Many persons in the neighborhood voluntarily evacuated and those who remained offered housing to friends who had fled the lower part of Manhattan. The church facility houses a church-run nursery school and the congregation experienced the loss of three members and several parents of nursery school children.

In the days after the attacks, the church conducted a memorial service, was open for nightly prayer, and collaborated with a local synagogue to provide an interfaith service. Two weeks after the attacks, the church was no longer holding special prayer services for the community, but the pastors were preparing for memorial services for their congregants who were killed.

Increase in Manifest Tasks

Each of these congregations had established structures and tasks before the disaster. Although they may not be considered "disaster-related," many of the services they offered did address the needs of the community and membership, while remaining well within their traditional realm of activities. Even with the emergence of new needs, these churches continued to function within their daily mission of congregational care and community outreach, operating as existing organizations under the DRC typology.

The maintenance of daily functions and the increase in resources aimed toward the congregation and community is important to note. Past research has focused on existing organizations and their operations in emergency management, rescue, and relief, but this examination has not generally included faith-based organizations or congregations as one of those existing structures involved in disaster relief. They have traditionally been identified with extending organizations, as they take on additional tasks that are beyond the scope of their daily operations. In this case, this sample of congregations maintained their pre-disaster structures and tasks while addressing disaster-needs and serving as community leaders.

Interfaith Groups

The Temple of Understanding is a small organization located in New York City that works globally to provide education about the beliefs and practices of multiple faith groups. It has the specific mission of developing partnerships with the United Nations and providing interfaith education in a variety of contexts. After September 11th, the Temple continued to provide education and outreach to churches and academic institutions in order to teach about the Muslim faith and Arab populations in New York City. They were particularly responsive to Christian church groups that requested a representative from the Muslim faith to provide education to their congregations. Additionally, the Temple was preparing to provide interfaith education to local and international universities with an emphasis on interpretations of the recent violence.

The Interfaith Center in New York City traditionally provides facility space and leadership for the provision of public educational, artistic, and cultural programs; creates opportunities for diverse groups to work together to meet specific common challenges and to raise mutual concerns in communities; and provides information, referrals, and consultation to religious organizations, community groups, and the media through ongoing working relationships.

Before the attacks, the Temple of Understanding and the Interfaith Center of New York assisted UNICEF in its preparations for the "Annual Interfaith Service of Commitment to the Work of the United Nations." This multi-faith service, designed to mark the opening of the 56th Session of the United Nations General Assembly, was subsequently dedicated to "all those whom we have lost and their loved ones in this week's tragedy" and was adapted to include memorial prayers. Held on September 13, it included representatives from such faith traditions as Judaism, Buddhism, Native Americans, Christianity, Islam, Hindi, Sikh, Jan, Tao, Zoroastrian, Shinto, and Yoruba. The service was representative of the memorial and prayer services held throughout the city in the weeks after the attacks that purposefully included multiple faith traditions.

As accounts of harassment were reported across the country, educational events and interfaith memorial services served as opportunities to speak publicly against acts of hate and violence. These incidents included verbal and physical intimidation, racial and ethnic slurs, vandalism, and physical assaults against Arabs, Muslims, and others who appeared to be of Middle Eastern heritage. In New York City, many Arab-looking persons were singled out and became the target of harassment. This "retaliation" against persons who were perceived by some to have a relationship to the group assumed to be responsible for the attacks evoked fear and concerns for personal safety. The interfaith faith-based organizations worked in partnership with multiple faith groups to present statements to the public and to serve as a religious and moral voice against the reported violence.

Increase in Manifest Tasks

These two interfaith faith-based organizations did not have pre-established disaster response plans. However, their organizational missions and structures prepared them to continue the provision of services with their messages extended into the realm of public statements as a moral and religious authority. Their everyday tasks were consistent with their post-disaster response; they were prepared to offer education, religious interpretation, and moral voice in response, altering their regular programs to accommodate the emergent needs.

Both interfaith faith-based organizations cited here can be seen as existing organizations whose manifest tasks were continued and increased according to the emergent needs that presented themselves after the disaster. The resources they offered were consistent with their daily functioning as they maintained their patterns of outreach and education within the community.

Faith-based Organizations affiliated with the American Red Cross: Child Care and Spiritual Care Organizations

The American Red Cross (ARC) defines itself as "a humanitarian organization, led by volunteers, that provides relief to victims of disasters and helps people prevent, prepare for, and respond to emergencies." As an independent entity with federally designated responsibilities, the ARC operates as a neutral organization, making services available to all persons regardless of race, religion, creed, national origin, gender, etc. The ARC is a non-sectarian organization and thus must access external resources to ensure the availability of coordinated, qualified, and appropriate responses.

At local, state, and national levels, the ARC works with government, business, labor, religious, and community organizations as well as other voluntary agencies to identify responsibilities and roles, share information, and find ways to coordinate an efficient response to disasters. Among the results of coordinated disaster response are the relationships established between the ARC and the Church of the Brethren to provide disaster child care and between the ARC and a set of professional chaplain cognate groups to provide disaster spiritual care in aviation disasters. These two groups represent faith-based organizations that are attached to the ARC through statements of understanding through which the ARC provides the administrative support and guidance to the mobilized teams in disaster response.

Title VII—Aviation Disaster Family Assistance Act of 1996

The Federal Aviation Re-authorization Act was signed into law in 1996, designating the National Transportation Safety Board (NSTB) as the federal agency responsible for assisting families of passengers involved in aviation disasters. In accordance with certain Victim Support Tasks described in the Act, the NTSB approached the ARC to coordinate emotional support for families. With the undertaking of this responsibility, the ARC formed the Aviation Incident Response (AIR) team. This team grew to include specially trained mental health, childcare, and spiritual care teams. The Childcare Aviation Incident Response (CAIR) team was coordinated through the Church of the Brethren. The Spiritual Care Aviation Incident Response (SAIR) team was coordinated through the ARC along with an advisory team of representatives from the professional chaplain cognate groups.

The CAIR and SAIR teams had been mobilized to multiple aviation disasters before September 11, functioning as an expansion of ARC disaster services. As a disaster relief organization, the ARC maintains an organizational structure that expands its volunteer base in a disaster. As the structure changes, growing in complexity, the tasks of each volunteer participant are pre-established in accordance with specific disaster response plans. Therefore, the organizational structure is new, but the tasks are routine.

The attacks on September 11 pressed the local and national offices of the ARC into service, requiring a quick mobilization of hundreds of disaster volunteers from across the country. The CAIR and SAIR teams were also mobilized. These teams were established in structure and task in accordance with the Aviation Family Disaster Assistance Act with the mandate that in an aviation disaster, the AIR team would be mobilized and all necessary resources would be made available. Although airliners were the precipitating agent of destruction on September 11, the disaster itself was not technically classified as an aviation disaster. Therefore the Act was not invoked, the

NTSB did not have jurisdiction for leading any investigation, and the ARC did not have the primary responsibility for victim support in the resultant response. The CAIR and SAIR teams were specifically designated for aviation disasters and were to be mobilized only when the Act is invoked and the NTSB and ARC provide primary support services. Therefore, with the mobilization of CAIR and SAIR in this disaster response, the ARC extended the domains of these teams, simultaneously adapting in multiple dimensions.

Childcare Aviation Incident Response

The Childcare Aviation Incident Response (CAIR) team is composed of six persons designated to provide administrative leadership and child care at an aviation disaster. It is part of the Church of the Brethren's Emergency Response Services Ministry program and has jurisdiction over the provision of aviation child care as authorized by the ARC under a statement of understanding created after the passage of the 1996 Act. Although not an ARC program, it is integrated into the services that are the responsibility of the ARC, receiving support and guidance from the Disaster Mental Health officer on the AIR team.

The Church of the Brethren designed these teams of specially trained disaster child care workers in 1980 when volunteers saw that families were spending hours in long lines for assistance. Emotionally charged children who were required to sit and wait for long periods of time were found to be in need of a safe space designed in accordance with child development and the probable psychological stress they were undergoing as a result of the disaster. The provision of aviation disaster child care is an offshoot of the Church of the Brethren's normal role in natural disasters.

CAIR volunteers receive training in child development and disaster response, serving as volunteers on "less traumatic" events such as natural disasters, before becoming trained and certified to work as a member of the AIR team. These volunteers are located nationwide and are available for mobilization within hours of a call received from the ARC through Church of the Brethren headquarters.

Disaster Activities

After the September 11 attacks, the CAIR team set up a secure child care area in the Family Assistance Center on Pier 94 where parents or guardians could drop off their children with volunteers while they attended to the paperwork and tasks of applying for victim assistance. The secure child care area was a large room with high walls and a locking door. For security reasons, each child was photographed with his or her guardian and remained in the facility until that guardian returned for the child.

As children played with the abundance of toys, read books, listened to music, or interacted with the therapy dogs, CAIR volunteers watched for signs of distress or need and intervened when necessary. Many of the activities available were designed to stimulate self-expression and be creative outlets for the feelings the children could be carrying with them. Although the CAIR volunteers do not counsel the children, they do refer them and their guardians to ARC disaster mental health services if needed.

By the third week, CAIR volunteers had seen a shift in the children who were coming in to the child care area; there were many more children who did not speak English and a realization was setting in that those who were missing were most likely deceased. Additionally, the CAIR volunteers were gearing up for what could prove to be particularly stressful for children and their guardians: the announcement that death certificates would be made available at the Family Assistance Center. CAIR coordinators were preparing for larger groups of children who would be reacting to greater levels of stress and would need extra volunteers to provide care.

This third week was also a period of staff transition for both leaders and workers on the CAIR team. A normal aviation disaster is "completed" within two weeks of the crash. For AIR team members, the meaning of completion varies, but in a general sense it means that the crash location has been excavated, remains have been recovered, and families have received notification of the death of their loved ones. In some instances, a memorial service has been prepared and conducted as well. However, after this disaster, CAIR team members rotated in for two- to three-week periods over the course of several months.

Spiritual Care Aviation Incident Response

The Spiritual Care Aviation Incident Response (SAIR) team is a seven-person team that provides management of the spiritual care component of the ARC AIR team. Like CAIR, SAIR is not an ARC program, but is integrated into the services that are the responsibility of the ARC, receiving administrative support and guidance from Disaster Mental Health.

Members of the SAIR team have ecclesiastical endorsement, are members of one of five certifying groups (the Association for Professional Chaplains, the Association for Clinical Pastoral Education, the National Association for Catholic Chaplains, the National Association for Jewish Chaplains, or the International Conference of Police Chaplains), and have attended the ARC-SAIR training program. Chaplains are on call for an aviation disaster for one month of the year, making themselves available to respond within four hours notice and to be away from their workplace for up to 10 days.

Chaplains who are members of the SAIR team are dedicated to three goals: ensuring that appropriate spiritual care is provided to those affected by aviation disasters; recruiting, training, and administering local spiritual care volunteers; and controlling who has access to victims and rescue workers. In addition, SAIR eventually provided chaplaincy services and supervision of local spiritual care volunteers at the temporary morgue at Ground Zero.

Disaster Activities

During their first few days in New York, SAIR team members trained over 500 local clergy who would provide ongoing spiritual care at the Disaster Assistance Service Center (DASC), the Family Assistance Center (FAC), the Respite Centers, and the morgue at Bellevue Hospital throughout the rescue and recovery efforts in New York City. The DASC offered resources to displaced families and workers; the FAC offered resources to families of the missing and deceased; the Respite Centers housed feeding and resting spaces for rescue workers at Ground Zero; the morgue, on the east side of the city, was a receiving area for all human remains found in the rubble.

Spiritual care represented different things at each site, but overall it consisted of a chaplain offering a presence or a listening ear for those who wished to talk. The ARC defines spirituality as "each person's way of finding meaning in his or her life experiences in light of a relationship to themselves, their community, and the transcendent." Therefore spiritual care is "a sustaining care that draws upon a person's own inner religious and/or spiritual resources . . . in its fundamental sense is service to others, including the religious/spiritual; emotional; social and, at times, physical care of those entrusted in the mist of crisis." Chaplains who work with the ARC are required to provide non-sectarian spiritual care, meaning they do not proselytize, they are available to persons of all faiths and persuasions, and they do not force themselves upon anyone. Instead, they make themselves available to those who are in need of spiritual support by "providing accompaniment, ritual, and prayer . . . and assisting others to regain and utilize their own spiritual resources."

ARC-affiliated chaplains were required to work without an agenda (such as the goal to convert a person to his or her own faith), following the lead of the persons to whom they were providing care in order to help them access their own spiritual resources. For some, this included offering prayer or other ritual, such as a blessing or anointing. For others, the simple accompaniment of a spiritual person through the process of gathering resources and dropping off DNA samples was considered supportive.

Controlling access to the different sites became an issue of security as well as assuring appropriateness of spiritual care. It was not unusual for ministers of differing faiths to arrive unannounced at the FAC, DASC, or morgue claiming that they had been "called by God" to provide care to survivors, grieving family members, and rescue workers. This was compounded by the fact that many New York City police were highly respectful of persons representing faith traditions, so persons wearing clerical collars were often given access to secured sites based on appearance alone. On the streets around Ground Zero and in local parks where impromptu shrines and memorials had sprung up, many persons who claimed to be religious professionals would often stop people on the streets and request to pray, hand out religious tracts, or proselytize. The SAIR team was charged with preventing this type of activity within the boundaries of each of the sites and worked diligently to screen for ministerial appropriateness, permitting access only to local volunteers who agreed to work in an interfaith manner and to abide by charter of the ARC.

During the second week after the disaster, hope remained that survivors would be found within the rubble of the fallen towers. Emergency personnel and construction crews at Ground Zero worked 24 hours a day with extreme urgency. At the FAC, family members came to provide descriptions and to post flyers with pictures of their missing loved ones. At the beginning of the third week, New York City Mayor Rudolph Giuliani announced that death certificates would be available to family members. This marked the end of the rescue operation and a beginning of recovery. Many family members interpreted the option of applying for death certificates as "giving up hope" that their loved one would be found, thereby creating an additional layer of tension at the FAC as families were torn between receiving assistance and holding out for a miraculous rescue.

The announcement of death certificate availability marked an increased need for additional spiritual care support staff at the FAC. This also marked the beginning of daily boat trips from Pier 94 to the World Trade Center area. ARC chaplains and Disaster Mental Health personnel accompanied family members as they traveled to see the space where the towers fell, and to leave flowers at a makeshift memorial.

Like the CAIR team, SAIR staff also made several transitions through these early weeks. For the first four days, spiritual care was coordinated out of the local chapter, advised by staff from the ARC National Headquarters in Falls Church, Virginia, and led by a national SAIR member who resided in New York City. Although SAIR members were on call and ready to fly in to New York, airports had been shut down and most team members were not able to fly in until the third day of operations. Many drove, and one group even flew in on a FEMA transport, accompanied by military fighter jets. It took several days to determine available office space for screening and training, to receive phone lines and computers, and to provide badges and ARC vests to local volunteers.

Many Jewish holy days fell within these three weeks, requiring rabbinical SAIR members to absent themselves from their positions and SAIR leadership to be temporarily restructured. Additionally, most SAIR team members made 10-day commitments to the operation. So just as the operation was changed from rescue to recovery, an entirely new team of SAIR volunteers was arriving and faced new challenges for coordination and response.

Latent Tasks Leading to Expansion and Extension

Quite obviously, the latent tasks and resources available for the highly tragic nature of an aviation disaster cross over to the setting of this particular disaster. Because this disaster was not considered aviation and the Act was not invoked, CAIR and SAIR became an extended resource for the ARC. The specialized training of CAIR members to work with children who are bereaved and undergoing stress was a naturally transferable skill for the FAC setting in New York City. Additionally, with the high numbers of casualties, and the massive amount of loss, the SAIR team was well equipped to provide administrative support to the local volunteers at all sites where spiritual care was offered through the ARC throughout the city.

However, the expansion of ARC disaster services coupled with an extension of these two teams into non-routine settings requires making use of the DRC typology in a new way. This was a multi-dimensional organizational adaptation in which expansion and extension occurred simultaneously. An increase in structural complexity and an expanded volunteer base coincided with the mobilization of specialized teams to non-traditional settings.

The latent tasks of CAIR and SAIR made this adaptation a rather simple expansion of services to a disaster of overwhelming proportions. The two groups continued to function in fairly similar ways to their pre-disaster design, but in a setting that was outside of the scope for which they had been initially developed. Although the responders have self-evaluated their activities as successful for the most part, the potential for inter- and intra-organizational conflict and turf battles created an added tension as the organizations adapted to their new domains. Latent functions helped the organizations carve out new territory as structures expanded to include massive numbers of local and national volunteers as well as new levels of administration.

Summary

The adaptive strategies of faith-based organizations described within this paper have been analyzed along the DRC typology with the additional factor of latent organizational tasks leading to activation of suspended structures or the expansion of structures to meet disaster response needs. By making note of those available resources that come into practice during a disaster, it has been seen that these organizations are able to draw upon structures and tasks not normally manifest in day-to-day operations.

Structural adaptation was seen in the PNYC as an administrative commission was mobilized to meet ongoing needs within congregations across the city. In the language of the DRC typology, the PNYC expanded in complexity due to provisions that were created through the denomination to be activated when needs arise that are beyond the scope of normal functioning. These pre-existing provisions were put into practice with the assessment of needs following the disaster.

The local congregations and interfaith organizations increased their daily activities to meet needs within the community, but largely operated within their normal tasks and structures. They can be framed within the quadrant of existing organizations in the DRC typology. The increase in activities, such as memorial services, prayer services, support groups and counseling, as well as education and public statements condemning violence were all within the normal scope of the existing organizations' structure and tasks. One important point related to the lack of adaptation seen within these organizations is the implication that disaster-related tasks extend beyond the rescue and recovery efforts of emergency managers into the local community as faith leaders seek measures of maintenance that include outreach and assessment of needs.

Multi-dimensional adaptation is shown by the response of the CAIR and SAIR teams. These teams expanded in volunteer base and structural complexity while extending their response to a new form of disaster for which the teams were not initially designed or formed. The latent tasks of these teams were ideally suited for this disaster response in light of the number of casualties and the traumatic nature of the event.

Conclusions

The analysis of faith-based organizations presented above leads to comments directed at the methods used for this study as well as the assessed outcomes of specific organizations. For instance, past studies of local congregations have provided analyses that identify congregations as extending organizations rather than existing organizations, where they have opened their doors to flood victims and served as shelters, or as expanding organizations where new volunteers increase the structural complexity of the organization.

This paper offers only a snapshot of relief efforts provided by congregations in the week after the attacks, but it provides an interesting contrast to the "normal" adaptive measures reported in other studies. This may be due to several things: (1) the sample of representatives interviewed was biased towards those congregations located outside the restricted zone of lower Manhattan. Had the research concentrated on congregations closer to Ground Zero, there may have been a different outreach to groups such as the homeless and other persons in need; (2) the nature of the metropolitan area of New York, with its abundance of resources, may have resulted in congregations' referring persons to other agencies that are normally prepared to address those needs; and (3) the congregations may have shown adaptive measures over a longer period of time. The data available from the short site visit prevents a longitudinal analysis that may provide differing conclusions. Nonetheless, this study does provide an account of the work of faith-based organizations in the response period immediately after the disaster and offers several insights to the application of the current DRC model.

Applications for Emergency Management

Organizations that concurrently expand their volunteer base while reaching into new territory will have to innovate and be creative in their work. This challenges organizational flexibility and resourcefulness as organizations navigate the terrain and work to maintain a sense of order. The benefit of activating pre-existent structures is that there will be some sense of regularity and cohesiveness in the midst of novel situations. Additionally, as organizations branch into new areas and establish themselves in new domains, they will develop structures to meet those needs that can be called upon in the future.

Understanding that multi-dimensional adaptation is occurring may help emergency responders to explain what appears to be disorganization in situations that are fraught with structural challenges. This may affect planning and mitigation measures as personnel evaluate and develop response initiatives for future events. Such developments may include the training and structuring of specialized response teams that are prepared to mobilize to multiple jurisdictions, taking more of an "all-hazards approach" to disaster response.

Additionally, a greater awareness of the functions that congregations, denominational agencies, and interfaith educational centers perform can expand the way that emergency response professionals view disaster response. For instance, as professional emergency managers provide rescue and relief efforts at the most severely affected disaster sites, ongoing work within the community to assess long-term needs and to provide support to victims and their families can be met and addressed by these local organizations, which will continue to operate long after the emergency response has concluded. This recognition of the work and leadership of congregations should lead to the inclusion of local faith-based organizations in community recovery plans for disasters of all types.

Theoretical Implications

Multi-dimensional adaptation of the SAIR and CAIR teams push the DRC typology into new directions. Past studies of organizational adaptation have focused on singular adaptive measures—organizations are situated in only one of the four quadrants of the typology and new behaviors take place within existing organizations. Multi-dimensional adaptation requires a shift in the way that organizations are analyzed and framed, away from single dimensions to a multi-faceted understanding. Just as Bardo (1978), Stallings (1978), and Quarantelli (1996) noted that the DRC typology is too limited, the existing two-by-two matrix does not allow for theoretical complexity. The existence of two-dimensional adaptation raises questions about the DRC model, begging the question of how the typology might be restructured to provide for an inclusion of these types of organizations and their adaptive measures. Further clarity might be gained by adding theories of organizational processes such as those by Kreps and Bosworth (1993, 1994) and emphasizing that structures are in the constant state of formation as tasks and activities affect social organization.

This multi-dimensional adaptation also raises questions about the type of disaster and organizational preparation for disaster response. Had the CAIR and SAIR teams been established as teams that were prepared to respond to any incident with mass casualties under the auspices of the ARC, regardless of the ARC jurisdiction, this type of adaptation may not have been seen. These implications result from definitions of disaster and the appropriate routes for organizational management and response. Future policy changes within CAIR, SAIR, and the ARC may precipitate new agreements through which these teams will become regular responders to mass casualty incidents along with their normal role in aviation disasters, thereby aligning their adaptations within the single quadrant of expanding organizations in the DRC typology.

This research also furthers the DRC concept of expansion to include growth of organizational complexity as well as the traditional understanding of an increase in volunteers. It also raises the idea that existing organizations are not just those groups with specified emergency response structures and functions, but includes faith-based organizations that continue to function in traditional ways, meeting emergent needs of their congregations and community. However, this is coupled with additional research questions about the use of manifest/latent tasks in organizational adaptation. For instance, in this analysis, latency has served as a concept to explain the conditions leading to the adaptation of faith-based organizations. How does latency function in other types of organizations that adapt in disaster response, such as for-profit businesses or non-profit community agencies? To what extent does the mission and function of the organization affect the response and relief efforts in meeting employee and community needs after a disaster?

In addition to the conditions leading to organizational adaptation, there remain questions regarding the consequences of these types of adaptive measures, including the implications of patterned change based on latent functions and the assessment of their success. For instance, are organizations that make use of latent tasks and functions in adaptation more or less successful in disaster response than organizations that take on nontraditional tasks and behaviors? What are the impacts of the behavioral change on the creation of new structures? And what are the long-term implications of change for these organizations? Is there a greater potential for institutionalization in contrast with emergent groups and those not making use of latency?

Finally, it is important to note the specific type of disaster invoked by the attacks on the World Trade Center as it relates to the adaptation of these faith-based organizations. Due to its complex nature, the resultant war efforts, and the assessment of religiously oriented causality, faith-based organizations played a special role in responding to moral questions and offering resources, critique, and interpretation to all those affected by the disaster. This made it especially important to focus on faith-based organizations as a particular group of responders while paying attention to each of them separately as individual organizations adapting to crisis and stress in a violently shocked community.

References

Bardo, John. 1978. "Organizational Response to Disaster: A Typology of Adaptation and Change." *Mass Emergencies* 3: 87–104.

Barton, Allan H. 1969. *Communities in Disaster: A Sociological Analysis of Collective Stress Situations.* Garden City, NY: Anchor Books.

Book of Order: The Constitution of the Presbyterian Church (U.S.A.) Part II. 1999. Louisville, KY: Office of the General Assembly.

Bosworth, Susan Lovegren, and Gary A. Kreps. 1986. "Structure as Process: Organization and Role." *American Sociological Review* 51: 699–716.

Brouillette, John R. and E.L. Quarantelli. 1971. "Types of Patterned Variation in Bureaucratic Adaptations to Organizational Stress." *Sociological Inquiry* 41: 36–46.

Dynes, Russell R. 1970. *Organized Behavior in Disaster*. Lexington, MA: D.C. Heath.

Drabek, Thomas E. 1987. " Emergent Structures." Pp. 259–290 in R.R. Dynes, B. DeMarchi, and C. Pelanda, eds., *Sociology of Disasters.* Milan, Italy: Franco Angeli Libri.

Drabek, Thomas E. and William H. Key. 1984. *Conquering Disaster: Family Recovery and Long-Term Consequences*. New York: Irvington Publishers.

Form, William H. and Sigmund Nosow. 1958. *Community in Disaster.* New York: Harper.

Kreps, Gary A. and Susan L. Bosworth. 1993. "Disaster, Organizing, and Role Enactment: A Structural Approach." *American Journal of Sociology* 99: 428–463.

Kreps, Gary A. and Susan L. Bosworth. 1994. *Organizing, Role Enactment and Disaster*. Newark: University of Delaware Press.

Lofland, John and Lyn Lofland.1995. *Analyzing Social Settings*. Belmont: Wadsworth.

Martin, Roy N. 1976. "Religious Organizations in Disaster: The United Presbyterian Church in the USA and Tropical Storm Agnes 1972." Ph.D. Dissertation. Madison, NJ: Drew University.

Nelson, L.D. and Russell R. Dynes. 1976. "The Impact of Devotionalism and Attendance on Ordinary and Emergency Helping Behavior." *Journal for the Scientific Study of Religion* 15(1): 47–59.

Quarantelli, E.L. 1966. "Organizations Under Stress" Pp. 3–19 in Robert Britson, ed., *Symposium Emergency Operations*. Santa Monica, CA: System Development Corp.

Quarantelli, E.L. 1983. "Emergent Behavior at the Emergency Time Periods of Disasters." Final project report. Newark, DE: Disaster Research Center, University of Delaware.

Quarantelli, E.L. 1985. "Emergent Citizen Groups in Disaster Preparedness and Recovery Activities." Final project report. Columbus, OH : Disaster Research Center, The Ohio State University.

Quarantelli, E.L. 1996. "Emergent Behaviors and Groups in the Crisis Time of Disasters." Pp. 47–68 in Kian M. Kwan, ed., *Individuality and Social Control: Essays in Honor of Tamotsu Shibutani*. Greenwich, CT: JAI Press.

Ross, G. Alexander. 1980. "The Emergence of Organizational Sets in Three Ecumenical Disaster Recovery Organizations." *Human Relations* 33: 23–39.

Ross, G. Alexander and Martin H. Smith. 1974. "The Emergence of an Organization and an Organization Set: A Study of an Inter-Faith Disaster Recovery Group." Columbus, OH: Disaster Research Center, The Ohio State University.

Smith, Martin H. 1978. "American Religious Organizations in Disaster: A Study of Congregational Response to Disaster." *Mass Emergencies* 3: 133–142.

Stallings, Robert A. 1978. "The Structural Patterns of Four Types of Organizations in Disaster." Pp. 87–103 in E.L. Quarantelli, ed., *Disasters: Theory and Research*. London, England: Sage.

Weller, Susan C. and A. Kimball Romney. 1988. *Systematic Data Collection*. London, England: Sage.

Acknowledgments

This project was supported by a grant from the Natural Hazards Research and Applications Information Center, University of Colorado, Boulder. The author wishes to thank Kathleen Tierney and Dennis Mileti for their advice, Jacquelyn Monday for her thoughtful assistance on early drafts of this paper, and two anonymous reviewers for their helpful comments. I am indebted to the generosity of many volunteers who shared with me their knowledge and experience during the early weeks of the response while continuing to serve those affected by the September 11th disaster.

Correspondence should be addressed to Jeannette Sutton, Department of Sociology, 327 UCB, University of Colorado, Boulder, CO 80309; e-mail: suttonj@mail.colorado.edu.

Part 5

Private Sector Roles and Activities

Business Responses to the World Trade Center Disaster: A Study of Corporate Roles, Functions, and Interaction with the Public Sector

David A. McEntire
Department of Public Administration
University of North Texas

Robie J. Robinson
Department of Public Administration
University of North Texas

Richard T. Weber
Department of Public Administration
University of North Texas

Introduction

One of the most prevalent trends in emergency management concerns increased partnerships between the public and private sectors. Initiatives such as the Federal Emergency Management Agency's "Project Impact"[1] and the Institute for Business and Home Safety's "Showcase Community" and "Disaster Recovery Business Alliance" programs are examples of this collaborative effort (Armstrong, 2000; IBHS, 2000). Ironically, research about the roles of the private sector and the functions it performs in emergency management is scarce. Nor is there sufficient information about the interaction of local jurisdictions with each of the organizations that participates in disaster response operations (Waugh, 2000; Erickson, 1999). In particular, there is a lack of findings about the collaboration that takes place among businesses and government agencies to solve mutual disaster problems (Webb et. al., 2000; Mileti, 1999).

Taking into account these trends in the field and gaps in the disaster studies literature, and utilizing the attacks on the World Trade Center on September 11, 2001, as a case study, this paper seeks to answer three important questions. First, what roles and functions do businesses play or perform in emergency management? Second, how does the private sector interact with government agencies during times of disaster? Finally, what are the implications of business involvement in emergency management for both scholars and practitioners? Before proceeding with this analysis, the methods used to gather information for this study will be discussed.

Methodology

Information about private sector roles, functions, and interaction with the government has been obtained from a variety of sources and through various methods. Comments in the next section (regarding the roles of the private sector in emergency management) are based on the combined 18 years of academic and professional experience of the authors of this paper. Each of the authors teaches in the Emergency Administration and Planning program in the Department of Public Administration at the University of North Texas, has interest and specialization in emergency management, and has taught various courses including Emergency Preparedness, Disaster Response, Disaster Recovery, and Private Sector Issues in Emergency Management. One of the authors was previously employed as a County Emergency Management Coordinator and as a Regional Fire Coordinator for the Forest Service in the State of Texas, obtaining first-hand experience in collaborating with both public and private organizations in disaster situations. Another has worked for large corporations in the retail and petroleum industries in California and Colorado, and has observed the behavior of private sector entities in disasters while performing case worker functions for the Denver Branch of the Mile High Chapter of the American Red Cross. This author has arranged several internships with businesses for students interested in pursuing careers in the private sector. The final author served previously as an intern working on Project Impact initiatives with Region IV of the Federal Emergency Management Agency (FEMA) in Chicago, Illinois. He has also spent time in the occupational and safety area with Heritage Environmental (a hazardous materials spill response and remediation company), and interacts frequently with businesses involved in emergency management due to his position as the Professional Development Coordinator in the Center for Public Management at the University of North Texas. These distinct backgrounds have enabled the authors to understand the roles of businesses in emergency management.

Information about the private and public response to the World Trade Center terrorist attack was obtained through a Quick Response grant provided by the National Science Foundation and the Natural Hazards Research and

Applications Information Center at the University of Colorado at Boulder. Immediately after the attacks, the authors identified and contacted potential informants from the public and private sectors who were involved in the response to this disaster. A week later, the authors traveled to the scene and spent several days interviewing people involved in the operations or affected by the disaster. During these interviews, informants were asked a series of questions about the functions they were performing, the successes and failures of coordination across the public/private sectors, and lessons gained about the private sector and multi-organizational collaboration. Where possible and when required, additional open-ended questions were asked to elicit more information or clarify the responses given. At the close of these interviews, the practice of "snowball sampling" was utilized to uncover additional informants and these individuals and agencies were subsequently contacted. Among others, the informants (N=47) included representatives of businesses that were affected, employees and volunteers of corporations that responded to the incident, and public officials from various departments and agencies (e.g., the Mayor's Office of Emergency Management, Public Works, FEMA, the Office of the Inspector General, etc.).

After conducting this field research, the authors obtained further information and arrived at additional findings by attending emergency management conferences pertaining to the September 11th disasters, and through follow-up e-mail correspondence and phone conversations with the prior interviewees. The authors also relied on internet articles, media coverage, and news clippings to support findings where needed. After reviewing field notes, transcribing the most salient interviews, and discussing the information gathered in a series of meetings, the authors then generated an outline and began writing their findings. Once the initial draft was completed, it was circulated to various informants to verify content and obtain additional insights. The paper was revised several times based on the responders' feedback and authors' review of the research gathered. Finally, the paper was peer-reviewed before publication in this volume.

Roles of the Private Sector

Research has typically underscored the problems of the private sector in emergency management. Studies reveal that businesses often contribute to disaster or amplify their adverse effects in a variety of ways. For instance, corporate practices may make people, communities, and nations more vulnerable to natural hazards (Blaikie et. al., 1994). Industry increasingly develops complex systems and relies on high-risk technologies in the production process (Perrow, 1999). And businesses often embrace preparedness measures in a superficial manner and may even intentionally neglect the need to plan their responses to potential crises and other unforeseen events (Webb et. al., 2000). In spite of these obvious and disturbing weaknesses, it should also be recognized that corporations do play important and diverse roles in emergency management (Table 1). The private sector is involved in volunteer and donation activities, insurance provision, occupational health and safety, risk management, planning for and preventing transportation disasters, emergency medical care, hotel disaster preparedness and sheltering, reporting and information dissemination, business continuity, and the vending of goods and services for emergency management.[2] Each of these areas will be discussed in turn.

Table 1. Roles of the private sector.

Volunteering and donations	Insurance coverage
Occupational health and safety	Risk management
Transportation-accident planning and prevention	Emergency medical care
Hotel preparedness and sheltering	Reporting and information dissemination
Business continuity	Vending of goods and services

In addition to the concerned citizens, charitable organizations, and government officials/agencies that respond to disasters, the private sector also participates in the typical emergence and convergence process. Employees often serve as volunteers during response operations and companies frequently donate needed supplies and services to disaster-affected communities. For instance, after the 1981 Hyatt Regency skywalk collapse, hotel staff kept spectators away from dangerous debris, removed the wounded from the rubble, and set up a first aid station (Waugh, 1988). In other cases, restaurants donate food to emergency workers while soft drink distributors bottle water for communities with severed water lines. It is also common that manufacturers and discount retailers send diapers, baby formula, clothing, or other necessities of life to affected areas. In addition, home improvement stores such as the Home Depot give victims of disaster lumber, plastic sheeting, and other construction equipment and supplies for temporary or permanent repairs to damaged homes.

Although the government provides some insurance coverage (through the National Flood Insurance Program), it is the private sector that writes the vast majority of policies for fire, wind, hail, earthquake and other hazards. This includes the coverage of residential and commercial properties, personal vehicles, and the fleets of major transportation firms and carriers. Insurance and reinsurance companies not only help individuals, families, and other corporations recover after disaster by covering losses, they also play a vital role in mitigation by assigning a dollar value to risk; spreading the costs of disaster among a larger population; and reducing vulnerability through education, training, and the safe location of buildings and personal property. State Farm encourages mitigation with its "Good Neighbor House" in Deerfield Beach, Florida. This showcase residence is built with the latest in disaster prevention construction materials and techniques (e.g., impact-resistance glass, lightning protection, smoke detectors, fire sprinklers, high wind shudders, water damage detection equipment, and hurricane straps for the roof).

Since the industrial revolution governments have mandated that businesses address occupational health and safety concerns. Laws have traditionally focused on the length of the employees' work day and the need for periodic breaks, but attention has also been given to the prevention of slips, trips, falls, back injuries, and industrial accidents involving machinery or heavy equipment. With increasing government regulations and fines for violations, manufacturing firms and related companies have been forced to become more interested in maintaining a safe work environment by means of employee education and training, clean and well-organized factories, safety audits, first aid stations, and access to emergency response equipment.[3] Preventive and planning measures for hazardous materials fires, industrial explosions, and chemical spills are major concerns for occupational health and safety, especially since the 1984 Union Carbide disaster in Bhopal, India. Under SARA Title III (the Superfund Amendment and Reauthorization Act), industries with reportable quantities of lethal chemicals must file Tier II reports to notify fire departments, Local Emergency Planning Committees, and the state agency in charge of environmental protection of the types, locations, and amount of hazardous materials in the facility. Businesses including Texas Instruments, Lockheed Martin, and oil/gas companies also have their own emergency teams that respond to industrial disasters before and in conjunction with public emergency service personnel. Industry is now giving attention to workplace security/violence and possible terrorist attacks on petroleum refineries and other plants that process, manufacture, store, or use hazardous substances.

Many businesses have employees that perform risk management functions. Risk management includes activities directed towards the goal of protecting the monetary and other interests of the company. Risk managers familiarize themselves with the hazards associated with the workplace and are aware of the potential for lawsuits resulting from the operation of the business. They also purchase insurance policies to limit the financial liability of the company. Consequently, risk management is closely related to the insurance industry and the goals of occupational health and safety. In some ways, risk managers are similar to emergency managers in that they attempt to prevent and prepare for crises and disasters. Risk managers may even represent their companies as members of the jurisdiction's Local Emergency Planning Committee.

Transportation firms have always played emergency management roles. Companies that operate ocean vessels attempt to steer clear of adverse weather, and have often provided assistance to ships in distress (e.g., the *Carpathia* rescued those that survived the sinking of the *Titanic*). In the transportation sector, railroad companies are required by law to complete a "consist" (shipping manifest) that acknowledges the amount and type of hazardous materials above the reportable quantity. They may also have their own teams of employees that respond to derailments to ensure a quick recovery of normal operations. Similarly, trucking companies such as SAIA Motor Freight often have their own personnel that investigate vehicle accidents and clean up hazardous materials spills according to state and federal environmental and transportation policies. Aviation firms promote the safety of passengers through the maintenance of planes and the training of pilots and crews. Airlines are also required under the Aviation Family Disaster Assistance Act of 1996 to plan and prepare for aviation crashes. Responsibilities include information dissemination, body identification, and psychological counseling.

There are numerous companies involved in emergency medical health care as it relates to disaster. Before an event, hospital administration and staff meet with community leaders to plan and prepare for earthquakes, hazardous materials spills, terrorist incidents,[4] and other disasters that may involve large numbers of victims. When an emergency or disaster occurs, ambulance companies—many of them privatized (e.g., Rural Metro in Scottsdale, Arizona)—dispatch emergency medical technicians to practice triage, care for the wounded, and transport victims to nearby hospitals. Hospitals, in turn, must treat the large numbers of patients that are self-referred or arrive by ambulance or with friends or family. At times, hospitals must also protect or evacuate patients if their facilities have been directly affected by the disaster agent(s).

The tourist industry is becoming more involved in emergency management activities (Drabek, 1994). Fires such as the one that occurred at the MGM Hotel in Las Vegas in 1980 have compelled hotels to better prepare for emergency situations. For example, Marriott has its own Crisis Management and Business Continuity division. Hotels also have policies and procedures relating to warning employees or guests of potential disasters, and even serve as places of refuge for those who have evacuated due to unfolding disasters and other emergencies. These preparedness measures are necessary since hotels are often located in vulnerable areas (e.g., along the coast) and because the size and occupancy of many hotels have risen dramatically over the past few decades. Hotels often establish agreements with the American Red Cross to shelter victims in times of individual, family, or community disaster.

Media organizations, such as newspaper, radio, television, and cable companies, are also heavily involved after disaster strikes. These organizations send reporters en masse to the scene of incident to obtain interviews as well as audio and/or video footage. Once this information is compiled, organized, and edited, it is distributed in print, via the internet, over the air waves, or on the screen. Much of this news reporting and commentary will focus on what happened, why it occurred, and what the effects were (Scanlon et. al., 1985). In many cases, the information provided will be incomplete or inaccurate (Payne, 1994). At other times, media representatives may get in the way of first responders (Scanlon et. al., 1985) and the coverage will reinforce widely held disaster myths (Fischer, III, 1998). Regardless of these problems, the media is a valuable conduit for government officials to spread important information about what citizens in the affected community can do to protect themselves or where they can go to receive disaster assistance.

As a result of the at-times alarming number of businesses that fold after disaster,[5] corporations are becoming increasingly involved in mitigation, preparedness, response, and recovery activities. For instance, a brewery in California spent an impressive amount of money to shore up equipment to prevent it from tipping over in the event of an earthquake. It is estimated that these measures saved the company millions of dollars when the Northridge earthquake struck a short time later. Businesses are also increasingly involved in continuity planning. Business continuity planning includes the identification of vital operations, the potential negative impacts of disaster, and methods to help the corporation run in spite of fires, floods, or other catastrophes. In many cases, computer experts, technology, and plans are relied upon to back up files, maintain communications, restore the operation of advanced industrial equipment, and resume normal business operations. The private sector therefore appears to be reacting positively to the rising quantity and toll of disasters and is expected to embrace these efforts even more closely in the future. However, many smaller businesses do not participate in these endeavors because they lack human and material resources.

The private sector is also a major provider of goods and services for emergency managers and other businesses, organizations, or communities in need of equipment and technical expertise. For instance, Halff Associates and Dewberry & Davis are corporations made up of engineers, architects, planners, and others that consult with governments on ways to incorporate mitigation into major development, infrastructure, and transportation projects. Federal Flood maps floodplains in conjunction with the National Flood Insurance Program to determine appropriate insurance rates. Simpson Strong-Tie Co., Inc. manufactures connectors that are used to strengthen wall, floor, and roofing joints against severe wind hazards. ENPRO is a distributing company that sells window film products to make glass more resistant to strong winds, terrorist bombings, and other hazards. The Institute of Business and Home Safety helps families and corporations prevent accidents and prepare for disasters. SAIC works with governments and businesses to assess risks and write plans to mitigate or respond more effectively. High Sierra Electronics, HI-GO, and American Communications manufacture and sell weather warning stations, dam and reservoir monitoring systems, weather alert radios, tornado warning sirens, and communications equipment in first-responder vehicles. Other vendors sell or rent sand bags, personal protective equipment, generators, computer aided decision support systems, and other supplies for first responders and those working in emergency operations centers. Emergency & Disaster Management, Inc. trains officials and staff at airports to deal with major aviation incidents. Cura Emergency Services and Hulcher Services are companies that respond to hazardous materials spills on highways and railways respectively. DRC, Inc. provides logistical support during response by providing labor, workforce housing, potable water, and other items. Phillips and Jordan is a company that contracts with government agencies to remove, burn, and dispose of debris and animal carcasses left in the aftermath of disaster. Verizon, ConEd, and other utility providers restore phone, electric, gas, and water infrastructure systems that have been rendered inoperable due to the powerful forces associated with hazard agents. BMS Catastrophe is well known for its ability to restore buildings and office equipment after major floods and fires. Parsons Brinkerhoff works with FEMA to verify structural damage resulting from disasters and estimate the likely costs for repairs. Numerous contractors and builders also descend on disaster-affected areas to restore damaged buildings or rebuild entire communities.[6] Thus, it is evident that the private sector plays varied and vital roles in emergency management.

Private Sector Involvement in the September 11th Disaster

The September 11th disaster at the World Trade Center required significant private sector involvement as well as close coordination with officials and agencies in the public sector. Functions performed by the private sector and in conjunction with public officials included warning and evacuation, Emergency Operations Center relocation and management, emergency operations at Ground Zero, mitigation of additional terrorist attacks, perimeter control and law enforcement, security and medical staffing, logistical support of urban search and rescue teams, information dissemination, communications, and infrastructure repair. Other functions included building restoration, sanitation services, business relocation and resumption, disaster assistance and insurance coverage, mass fatality management, debris removal, transportation assistance, donation management, and equipment repair and replacement.

Warning and Evacuation

Although there was no specific and credible warning that terrorists would hijack a plane and fly it into a building, there was concern that a similar event could recur at the south tower of the World Trade Center after the north tower was hit. For this reason, occupants of the south tower were advised to leave the building while others left voluntarily. However, other announcements inside the south tower instructed people that it was safe to return to their offices because the attack only occurred at the north tower. Although not everyone obeyed this latter announcement, many people remained in the building or returned to work when the second tower was hit. When the south tower collapsed at 10:05 a.m., the fire chiefs decided that the north tower would also be in jeopardy of structural failure. Fire officials therefore worked with businesses and employees to evacuate it. In both of the 110-story towers, virtually all of those working on or above the floors impacted by the airplanes were unable to evacuate. These, and other people in or near the buildings, were killed by fire or subsequent collapses. As of June 24, 2002, 2,823 people (including 403 emergency workers) lost their lives, were still missing, or had been issued death certificates. Nonetheless, the design and construction of the buildings, in addition to the adequacy of well-lighted stairways and prior evacuation training exercises involving businesses in the World Trade Center complex, allowed thousands of workers to exit the buildings safely. During the 1993 World Trade Center bombing, complete evacuation took 11 hours as stairwells were dark, carry-chairs were not available for the physically disabled, and coordination between emergency responders and tenants was less than perfect. Many of those issues were addressed after the 1993 incident, and undoubtedly increased the speed and effectiveness of the evacuation on September 11th.

EOC Relocation and Management

The public and private sectors also interacted closely in the Emergency Operations Center (EOC). As the devastating impact of the terrorist attacks became apparent, the city emergency management staff decided to evacuate the EOC (which was located in World Trade Center building 7). This proved to be a wise decision because the collapse of the north tower damaged building 7 and resulted in the building's being gutted by fire. Within a surprisingly short time, the city obtained office space at Pier 92 on the

Hudson River to establish a new EOC. Manufacturers donated necessary electronic office equipment (including computers, printers, and fax machines) and utility companies were sought to establish sufficient phone lines for the new facility. Once the EOC was operational, volunteers from the private sector arrived to help in any way they could (e.g., providing food or running errands). Public officials and representatives from various corporations (including those from the World Trade Center) met periodically to coordinate response priorities and operations. When personal meetings could not be arranged, phone calls took place between the EOC staff and company leaders. It is generally felt that this coordination was effective. Prior meetings between the EOC staff and local business continuity planning groups were credited for the successful improvisation and management of the EOC.

Ground Zero Operations

Various functions had to be performed at Ground Zero, including damage assessment, search and rescue, and evidence collection. To facilitate these operations, a geographic information system (GIS) database was established and the affected area was divided into 75-foot quadrants. Emergency personnel were then assigned to individual grids and briefed before they were put to work. This training provided an update on the situation with reference to secondary hazards such as hanging debris and unsafe wreckage. LIDAR (light detection and ranging) was utilized to detect ongoing fires as well as voids and potential shifts in the debris pile. Experts from MAPINFO arrived in New York to assist with the GIS. The data entry required a significant amount of input from the private sector. In addition, E-TEAM software was used to provide situation updates on the location and assignment of resources (e.g., staging areas, food, ice, water, restrooms, and personnel). Because over 200 organizations were involved in the response, frequent updates on the resources being deployed by the private sector were required.

Mitigation of Potential Future Attacks

Almost immediately after the unfortunate events unfolded on the East Coast of the United States, the Federal Aviation Administration (FAA) ordered all planes to be grounded to avert the occurrence of similar terrorist attacks elsewhere. This precautionary step required close collaboration between the government, airports, and individual airlines. At the same time, the Federal Bureau of Investigation (FBI) contacted the airlines involved in the attacks in order to obtain thousands of manifests (passenger lists) to assist in the identification of the terrorists involved and gain information about other possible hijackings. Later on, the FAA worked with airports and airlines to resume flights and implement new security measures. Both the cancellation and resumption of flights posed incredible logistical challenges in that planes were diverted away from their destinations and flight crews were unavailable (e.g., some rented cars and drove home to be with their families while others were reluctant to return to work). Periodic conference calls between the FAA and airline officials took place as they slowly brought the system back to full operation. The implementation of new security measures was also troublesome. Airlines had difficulty absorbing and communicating new security policies to employees, as the FAA issued new policies almost daily. Nonetheless, it is believed that the FAA and airlines worked well together in spite of the difficult situation with which they were presented.

As these activities were taking place, federal, state, and local officials thought it necessary to increase security in New York City. Police presence was increased in the subways, on the streets, in the harbor, and at government buildings. A major concern was the vulnerability of government buildings to vehicle-delivered bombs. While the local government had an existing program and schedule for installing fixed, retractable, and removable bollards (metal and concrete barriers) in front of buildings, there was a desire to speed up the process. Local officials contacted Secure USA within two weeks after the incident to increase orders and accelerate the installation. While the coordination between the public and private sectors was adequate, budgeting issues got in the way and slowed down the process. Nonetheless, the public sector relied heavily upon this business to increase security after the September 11th disasters. The private sector therefore performed important security functions and collaborated closely with the government to implement security policies before and after the events of September 11th.

Perimeter Control and Law Enforcement

One of the major challenges after the collapse of the World Trade Center buildings was to control access to affected area. In the immediate aftermath of the disaster, police escorted business owners into the affected area and allowed them to survey damage, collect needed documents or goods, and start processing insurance claims. However, emergency management officials desired to keep all non-essential persons out of the area for health and safety reasons. In addition, it was also believed that the presence of people at the disaster scene would slow down important response and recovery functions, and pose a potential security threat (as terrorists could blend into the crowds and attack emergency personnel). Consequently, a perimeter was established a few blocks away from and surrounding the impacted area. Fences were acquired from National Rent-a-Fence. A security check-in station was set up and a policy was established that detailed who would be allowed into the area and for what purposes. It was noted that these measures posed a few problems as some employees from various businesses (e.g., those affected and others involved in recovery activities) had valid reasons for getting in to the restricted area. Therefore, exceptions had to be made so that the response and recovery operations could proceed. What is more, the process of checking people in was slow and cumbersome at times. Within a few days, executives in the private sector contacted the EOC directly to ask for bulk credentialing. This sped up the security check-in process and improved the coordination among those at the check-in point, the EOC, and businesses.

The terrorist disasters in New York created an atmosphere conducive to potential and actual looting, disaster assistance fraud, and other criminal behavior. The collapse of the World Trade Center towers and other nearby buildings resulted in a situation in which classified documents, precious metals, and even weapons were accessible to those working at Ground Zero. In addition, the stores in the malls under the World Trade Center could not be secured due to damaged doors and shattered windows. For this reason, private security guards were hired to protect the interests of companies and businesses in the vicinity. Representatives from the FBI and Office of the Inspector General arrived at the scene to deter potential criminal activity. In some cases, this meant that government law enforcement personnel coordinated with the private sector to patrol Ground Zero, nearby businesses, and the malls underneath the World Trade Center. While research has consistently reiterated that deviant behavior is infrequent in the overwhelming majority of disasters, there was some looting during recovery operations at Ground Zero. According to a representative of the Office of the Inspector General, some of the steel (valued at hundreds of thousands of dollars) from the World Trade Center towers had been diverted in the debris removal process to be resold on the black market. The Office of the Inspector General therefore worked with debris removal contractors to install global positioning systems on heavy equipment to monitor the mileage, location, and timing of trucks hauling debris from the site of the disaster. Law enforcement officials also worked at other locations (e.g., the debris collection point to prevent theft of victim belongings) and with the private sector (to deal with fraudulent practices relating to the disaster assistance process).

Security and Medical Staffing

Finding an adequate number of personnel to fulfill certain critical functions, such as site security and medical care, was a significant challenge in the disaster. After the incident, police worked 12-hour shifts (or longer) each day to control access to the affected area. Their goals were to keep curious citizens away from dangerous debris and prevent their interference with response operations. Because the restricted zone extended several blocks in every direction from the World Trade Center complex, a sizable force was needed. Law enforcement officers from across the state were brought to New York City to augment the local police contingent. Additionally, officers from other states as far away as Florida and Texas were allowed to work side-by-side with New York City police. All of this kept police from performing other more routine tasks. The Governor of New York therefore called out the National Guard on September 22, which included scores of reservists. This affected a number of businesses in the New York area, most of which were more than willing to see Guard personnel answer the call to serve.

Hospitals also needed help in dealing with the large numbers of patients needing specialized care, such as burn victims. Like police officers and other emergency responders, medical personnel worked long hours to tend to their patients. In some cases, shelters were set up to provide housing for medical personnel who did not have time to go home between long shifts. The Salvation Army operated one such shelter for the employees of St. Vincent's Hospital. When a hospital's staff could not fill the demand for specialists, they made requests through the emergency management network and specialists from other jurisdictions were brought in to work for up to two weeks at a time. The facilities from which these employees came were willing to absorb the overtime and scheduling hassles to fill in for the personnel who went to New York. Unfortunately, the expected influx of patients never materialized; most of the disaster victims were killed immediately when the towers collapsed, and very few were found alive and in need of medical care.

Logistical Support of USAR Teams

Even though urban search and rescue (USAR) teams are intended to be self-sufficient for the first 72 hours, they must be assisted in various ways when disaster strikes. During the September 11th response, USAR teams were housed at the Jacob K. Javits Center in Manhattan. The Javits Center is a convention center that normally hosts a wide variety of events that were simply rescheduled or canceled in the weeks immediately following September 11th. The facilities of the Javits Center proved to be adequate for the needs of the teams it served. Nonetheless, the provision of food and other supplies for the many USAR teams was quite a burden and a wildland fire hotshot crew was called up to assist with the task. Although the Manhattan landscape was quite different from the mountains of the western United States, good coordination among all involved ensured that the tasks were accomplished. The American Red Cross, the Salvation Army, and other volunteer agencies assisted with the meals. Some local restaurants also provided bulk orders of food to be picked up by response personnel. In some cases, customers waited to be served until the large orders were filled for USAR personnel (as was the case at Starbuck's Coffee). What is more, many local restaurants refused to accept any payment from emergency personnel who ate at their locations. Others donated thousands of meals to be served to the responders at facilities around the area, including a cruise ship docked at Battery Park. The staff for food preparation and delivery was largely made up of volunteers from the food service industry who wanted to do something to help with the response.

Information Dissemination

Disseminating information to the community and affected persons (such as employees and renters) was vital during the response to the September 11th disaster. Government officials needed to let the public know how to respond to the events, what they could or should not do to help victims and responders, where the restricted zones were on any given day, how to find out about lost loved ones, and where to go for disaster assistance. Coordination between the public and private sectors was critical. For instance, reporters from the print, radio, television, and cable media met with public officials frequently and at various locations to obtain updates on the situation. Although there were significant discrepancies in terms of the reported number of lives lost, emergency management officials thought that the role of the media in the performance of this function was crucial. In fact, one official observed that the government could not handle public relations issues without the private media.

Employees affected by the incident, as well as their families, looked for up-to-date information regarding the status of the businesses, employee welfare concerns, paycheck information, insurance coverage, and many other issues. Many of these companies relayed this information to employees via recorded telephone messages. Once the contact number was communicated to all employees, the employer needed only to change the recording that callers would hear in order to keep all employees abreast of the latest developments. Some companies appeared to have had information lines available before the incident, while others used lines that were previously assigned for different purposes. Each of the private sector businesses surveyed handled this function on its own.

As already mentioned, site security was a major concern and a wide security perimeter was set up. In addition, the air was filled with hazardous particles that hindered breathing. These factors led to the shutting down of the housing stock near the World Trade Center. According to a few of the apartment dwellers who were interviewed, there was virtually no attempt by the public sector to communicate with tenants in lower Manhattan apartment buildings. They indicated that the only parties with whom the public sector agencies communicated were building superintendents and owners. Therefore, many tenants did not have information about the status of their housing in the immediate aftermath of the disaster. This lack of information probably resulted from a failure of landlords to communicate with tenants, rather than government officials failing to communicate with building owners.

Communications and Infrastructure Repair

The World Trade Center disaster resulted in a massive failure in communications and a loss of important utilities. The World Trade Center provided cellular telephone antennas and other communications infrastructure for the downtown area. When these buildings collapsed, 10 cell sites were destroyed (Moss and Townsend, 2001). The World Trade Center also lost hundreds—if not thousands—of hard lines (perhaps a number of switches equal to that used by a city the size of Cincinnati) (Moss and Townsend, 2001). Street-front businesses near Ground Zero were unable to accept credit cards, as phone service was still not working several weeks after the event. A manager mentioned that the inability to accept this form of payment was detrimental to business. In light of these problems, Verizon repair vehicles could be seen throughout the lower Manhattan area. In addition, a communications company provided cell phones for emergency workers, government officials, disaster victims, and anyone else who needed to make a call. The company also brought in charging units and established several portable cell towers to meet the demand near Ground Zero. This was arranged in conjunction and with the input of government leaders.

Getting the infrastructure restored was a massive and critical function after the disaster. The restoration of water, electric, and gas service required the participation of numerous public and private organizations. Many of these projects were extremely large and labor intensive. For instance, ConEd installed approximately 20 miles of shunting for electrical service (Berkowitz, 2001). Water and gas restoration proved to be equally challenging. The restoration of these and other utilities required close coordination between public organizations and the utility companies themselves. Access, timing, engineering concerns, and traffic control were only a few of the factors that had to be addressed so that utilities could be restored in the most expeditious and effective manner possible.

Building Restoration

The fires in the towers, and the collapse of these and other buildings, did not only affect the World Trade Center. Instead, the resulting debris and dust cloud impacted a large number of buildings in lower Manhattan. For instance, ash was several inches deep inside the buildings nearest Ground Zero. Even buildings blocks away had dust particles in their elevator shafts or heating, ventilation, and air conditioning systems. For this reason, facility restoration became a top priority for many public agencies and private corporations. Almost immediately after the incident, restoration companies began converging at Ground Zero. Executives from BMS Catastrophe—perhaps the nation's largest and most respected restoration company—arrived in New York on September 12. Over the next several weeks, as many as 800 employees worked for BMS Catastrophe to remove the contaminants from scores of buildings on Wall Street and in and near the World Trade Center complex. Many of the facilities requiring restoration assistance had existing contracts with BMS Catastrophe. However, other work was initiated by BMS Catastrophe as knowledge of its services spread by word of mouth to tenants or building owners. While most of BMS Catastrophe's involvement was directed towards the private sector, its personnel did interact and coordinate with government agencies and officials. BMS Catastrophe obtained approval from the FAA to fly a private jet to New York on September 12 even though the President had grounded all aircraft. Before BMS Catastrophe could work, it also needed permission to enter the restricted areas in and surrounding Ground Zero. It therefore provided a list of employees to the Office of Emergency Management and the Department of Design and Construction (which were in charge of site security). Workers were thus allowed access when they checked in to begin the cleanup. In addition, the company worked closely with the Department of Sanitation. Because electricity was lost in a vast number of buildings in lower Manhattan, food was spoiling in many refrigerators. The Sanitation Department asked BMS Catastrophe to remove the food that was posing a health threat. The Chief of the Sanitation Department then asked BMS Catastrophe executives to meet him at 6:00 a.m. one day so that he could personally escort the employees through the checkpoints to the work area. Scott BaVier, Vice President of BMS Catastrophe, commented that the Sanitation Chief "was very cooperative." This close contact between the public and private sectors proved invaluable for facility restoration.

Sanitation Services

In light of the destroyed utility infrastructure and because of the massive influx of emergency workers and other responders into lower Manhattan, there was a large need for portable sanitation units. The Mayor's Office of Emergency Management, American Red Cross, ConEd, FEMA, and others contracted with Mr. John to provide 750 toilets at Ground Zero, staging areas, bridges and tunnels, the landfill, and at other locations (e.g., the company provided a number of sanitation units free of charge for the ongoing funerals for fire fighters and police officers). In addition to working with public officials to arrange the terms of the contracts, the company coordinated with the Sheriffs Department to escort the sanitation units in from New Jersey. The company also worked with government agencies to identify the locations that needed the units and to obtain vehicle permits that would allow the company to enter the area. One major problem that arose was the dynamic nature of the road closures. In many cases, Mr. John would attempt to drop off a unit only to find that access into the area had not been approved, and would have to wait until the proper permits could be granted. In another case, no one at Ground Zero knew where the units were to be placed. Regardless, of Mr. John representatives reported that everyone was cordial and helpful, and did the best they could to resolve the situation for the company.

Business Resumption and Relocation

Business continuity was a top priority after the disaster. Issues for corporations included the inability to operate, employee relations, expense concerns, city ordinance enforcement, facility relocation, and record retrieval. Because offices were destroyed, damaged, or dirty, and since roads were closed and security was tight, many corporations could not reopen for business in a timely manner. Consequently, many employees were not able to work during this time and suffered the resulting loss of income. At other times, businesses did not know when to tell employees to return to work because of the astounding degree of disruption caused by the disasters. Alternatively, some businesses needed additional employees or required their staff to work overtime in order to clean or relocate their facilities. This was also problematic in that many businesses had little or no revenue coming in to pay for these expenses.

Another important issue after September 11th pertained to the enforcement of emergency ordinances intended to limit the number of sightseers clogging the streets in lower Manhattan. The goal of such ordinances was to aid both response efforts and recovering businesses, and to show sensitivity to the victims of the disaster and their families. None of the businesses interviewed indicated that they had input into the ordinances. However, almost all supported them and wanted to see them strongly enforced. The businesses wanted the crowds on the sidewalks to be dissipated or forced to move on even though they also needed customers in their shops and offices. Nonetheless, many respondents felt there was a lack of enforcement of ordinances. In addition, those that were enforced were done so in an inconsistent manner with no apparent method for determining when they should be enforced (e.g., picture taking was not permitted at first but was allowed later).

Relocation proved to be another major concern for businesses. Approximately 20 million square feet of office space was taken from the downtown area when the towers collapsed. This left an impacted area roughly the size of Atlanta's central business district. Therefore, obtaining space for businesses located in and near the World Trade Center area was an important part of the recovery process. Many firms had to find new office space, as their previous facilities no longer existed. In these cases, there was much variation. One business person stated that her company rented 102,000 square feet of office space in midtown Manhattan on September 17th, just six days after the event. Others struggled with relocation for several weeks after the disaster. Many of these relocating businesses were absorbed into different areas in the New York metropolitan area. Midtown, Jersey City, Stanford, and Westchester were all areas that received displaced companies. In several instances, even competitors opened their offices and shared space to help speed the recovery process. Other businesses moved to temporary offices created in hotel rooms throughout the city. It is unclear how much coordination occurred between the public and private sectors regarding this relocation. Several informants said they were sure that the leaders of their company must have talked to some public sector agencies, but they had no real knowledge of this type of coordination occurring. Thus, businesses often took care of themselves by relying on the market forces of supply and demand, or the good will and generosity of others in the private sector.

An additional challenge for businesses involved the availability of vital records. Some companies were able to rebound immediately as their corporate data was backed up at off-site facilities around the nation. In these cases, only the work in progress was lost when the towers collapsed. Other companies were not so fortunate. They either did not back up their information, or the alternate storage facilities were located in nearby World Trade Center buildings. These companies are struggling to recover to this day. The retrieval of data did not, to our knowledge, involve the services of the public sector.

Disaster Assistance and Insurance Coverage

The private sector was involved in the disaster assistance process. For example, the airlines affected by the hijackings provided disaster assistance to the families of deceased passengers, including information, psychological counseling, and other forms of aid. Nonetheless, most of the disaster assistance came from FEMA and other government entities such as the Small Business Administration. This assistance was directed toward citizens or the private sector and was provided to help repair physical damage and/or provide working capital. As of June 20, 2002, the SBA had provided 3,550 loans in Manhattan for a total of $390,642,300 (375 were to residential condominiums and co-ops, and 3,175 were for other private companies). While much of this assistance was closely coordinated, other aid was not. Some businesses affected by the disaster noted that they received flyers from Congressional representatives announcing a meeting to be held at a nearby university to outline the requirements of disaster assistance programs. Federal representatives stated that a number of outlets were used to disseminate information about loans and aid programs. Mayoral announcements, six disaster recovery centers, assistance from the Empire State Development Corporation, television spots, press conferences, publications, newspapers, and outreach programs (door-to-door) all were used to get information to businesses and residents. However, some of the businesses interviewed did not hear from the government or about any assistance programs; they did not know where to turn for help.

In order to facilitate recovery, the Lower Manhattan Development Corporation (LMDC) was created (as a subsidiary of the Empire State Development Corporation) to assist in redeveloping the downtown area. The U.S. Department of Housing and Urban Development provided LMDC with a $2 billion community development block grant. LMDC offers financial assistance and incentives in order to retain individuals who currently live in lower Manhattan and to encourage others to move to the area. The LMDC also provides job training assistance to help those preparing for employment in lower Manhattan. The organization therefore played important roles in helping businesses return to normalcy after the disaster.

Congress also approved an aid package involving millions of dollars of grants for the airlines affected by the September 11th incidents. The airlines insisted that they could not survive the economic impact of the terrorist attacks, lower passenger numbers, and possible post-September 11th lawsuits without federal assistance. A victims' compensation fund was therefore established and intended to make cash payments to victims and families at the expense of the federal government. In order to qualify for these federal payments, victims and families had to waive all rights to sue in federal or state courts. The overall success of this federal effort to help and protect the airlines remains to be seen as few families have signed the necessary waivers and a few lawsuits have begun to emerge.

Insurance companies similarly played a major role after the disaster. Companies sent scores of adjusters into the affected areas to deal with the huge numbers of claims. In some instances, insurance adjusters completed their jobs without interacting with officials from the public sector. However, the State of New York had set up a Disaster Coalition six months earlier. Modeled after the Institute for Business and Home Safety coalition during Hurricane Andrew, it included the New York Department of Insurance, the New York Department of Emergency Management, the New York Insurance Association, several insurance companies, and FEMA (Ryland, 2002). The coalition held a tabletop exercise during the summer of 2001, which is credited for facilitating the collaborative and successful public/private response to the World Trade Center disaster.

A few minor problems emerged during recovery, however. Insurance companies had to access the disaster scene after damaged buildings were inspected by engineers. This proved to be difficult at times due to the tight security. Another major challenge that insurance companies had was determining if the disaster would be covered or if it would fall under the "act of war" exclusion. Some companies covered losses while others are still determining what should be done, if anything, for their clients. It is likely that this will result in a number of prolonged lawsuits in the future. Another problem surrounded insurance coverage for those working at Ground Zero. In this case, the private sector was hesitant to provide coverage in light of the potential for additional terrorist attacks and danger involved in the response and recovery operations. The government did allow the contractors to begin operations before the normally required insurance policies were in place.

Mass Fatality Management

The World Trade Center disaster was one of the largest mass fatality incidents in the history of the United States. Not only did the plane crashes and resulting structural collapses kill thousands of people, but the process of body removal and identification was difficult in that there was simply no trace of many victims due to the intense heat from the burning jet fuel and nature of the building collapses. Consequently, the confirmed number of dead is low and is still being revised to this day. The private sector was a valuable asset to the government in the management of the large number of fatalities. For example, public officials communicated with the airlines involved in the terrorist attack, businesses in the World Trade Center, and local hospitals (the Greater New York Hospital Association) to develop lists of the missing and presumed dead. Private corporations and mortuaries also participated in body identification and fatality management. Conversely, the government was also considered to be a valuable asset to private corporations. Airlines referred the families of victims to the New York City website to inform them about the process of acquiring death certificates. Most of the activities surrounding fatality management witnessed collaboration across sectors. For instance, the Associated Press, the *New York Times*, CNN, and Kinko's helped to disseminate information about victim identification. However, fire department personnel confronted police when private contractors were allowed to bring in heavy equipment to speed up the debris removal process; they were concerned that the remains of their comrades would be removed in an insensitive manner.

Debris Removal

As already mentioned, the use of airplanes as weapons resulted in the collapse of both the north and south towers of the World Trade Center. However, the twin towers were not the only buildings destroyed in the incident. As they collapsed, steel and other building materials fell on top of or into nearby buildings. This created additional structural failures and spread fire across the World Trade Center complex. As many as 10 major buildings were destroyed or damaged, leaving behind estimated 1.2 million tons of debris. With this enormous amount of rubble before them, government officials designated the Design and Construction Department as the lead agency for debris removal. This government department then divided the 16-acre World Trade Center site into quadrants and signed agreements with four contractors. In turn, scores, if not hundreds, of subcontractors were utilized to assess debris stability and voids, monitor safety, cut steel beams, remove and load debris, and haul it away for further processing including investigation, disposal, and recycling. Such a massive undertaking required the close collaboration of the public and private sectors. For instance, heavy equipment such as grapplers and dump trucks had to be acquired from businesses around the nation. Some corporations donated the use of 750-ton cranes for the operation. Moreover, ingress and egress routes had to be determined by public officials and communicated to the companies involved in the debris removal. The military and police also searched all vehicles involved in debris removal to ensure that bombs would not be delivered to Ground Zero. Trucks had to be hosed down before leaving the area to limit the transport of dust to other areas. Although all of this proved to be a major logistical nightmare, it is believed that the coordination of this function was exceptional. In fact, the debris was removed at a much faster pace than was originally anticipated (by May 2002).

Transportation

All forms of transportation in the Manhattan area participated in the response or were affected by the terrorist events. When the terrorist attack occurred, thousands of people evacuated Manhattan by ferry. Others utilized the subway system and taxis to leave Ground Zero. Later on, USAR teams relied upon large buses to move from the Jacob Javits Center to Ground Zero and back. As the response continued, the airline industry became heavily involved. To assist with the efforts in New York City, several carriers provided reduced fares to workers and volunteers who responded. Additionally, airlines provided flights to the families of victims to assist them in obtaining death certificates and receive psychological counseling. Transportation in and around Manhattan was severely affected by the disaster and ongoing response operations. Initially, the city government closed the downtown area south of 14th Street. Roads adjacent to the World Trade Center were also shut down and vehicle and pedestrian traffic was altered to expedite the removal of debris from the area. Local government conveyed this information to the private sector periodically so transportation companies would be aware of the street closures. The private sector therefore provided transportation services during the response and the public sector collaborated with businesses to coordinate traffic detours after the disaster.

Donations Management

Donations management is a constant concern for the agencies involved in disaster response. Because of the altruistic nature of the American society, citizens will respond by sending goods and supplies to the scene of disaster. This was certainly the case with September 11th as an unbelievable outpouring of relief arrived in New York City. Many of the donations were provided by the private sector. For example, a manager of a sporting goods store mentioned how first responders used his store for protection when the buildings collapsed. The manager then provided swimming goggles and socks to help the responders equip themselves in order to continue their emergency response. The U.S. Forest Service received containers of coffee from Starbuck's for personnel at the USAR staging area. Likewise, the personnel at Ground Zero dawned overalls and other protective equipment provided by various manufacturers. Respirators and mask cartridges were given to responders to alleviate breathing problems created by fire, smoke, and unknown particulate matter. Gloves, batteries, and other supplies were sent by private companies. In many cases, the donations were closely coordinated with officials in the public sector. There were other instances in which unneeded supplies were given, however. For instance, Veterinary Medical Assistance Teams were sent dog food that could not be given to the animals participating in search and rescue operations. This lack of coordination was especially apparent in the area of financial donations. Citizens and corporations alike sent hundreds of thousands of dollars to the American Red Cross to help victims and fund the agency's response. Because this pool of money was so large, the Red Cross decided against using all of the money on the September 11th victims in order to have sufficient reserves for future disasters. This resulted in an outcry from the public and a Congressional review of the use of donations by this non-profit organization. The Red Cross has since undergone a change of leadership and revised its policy regarding the use of financial donations by citizens and corporations.

Equipment Repair and Replacement

The September 11th disaster resulted in a massive loss of equipment owned and operated by the Fire Department of New York (FDNY). It is estimated that the emergency vehicles destroyed in the incident included at least 10 ambulances, two Emergency Medical Service Suburbans, 24 sedans used by staff chiefs, 17 Suburbans used by batallion chiefs, two heavy rescue units, one tactical support rescue unit, two high-rise units, four hazardous materials vans, one self-contained breathing apparatus unit, two road-side emergency trucks, 18 ladders, and at least 28 pumpers. In light of these losses, private and public organizations worked diligently to meet the needs of the FDNY. For instance, Seagrave Fire Apparatus sent employees to New York to work with city mechanics to repair 76 damaged fire engines. Fire departments from around the country asked that their orders be delayed to speed up the delivery of emergency vehicles to New York. Many companies that manufacture fire equipment donated emergency vehicles to the FDNY. As an example, Pierce-Kenworth worked with 70 other manufactures to donate an air and light support rescue vehicle to New York. A similar gift was provided by Emergency One and other donors. Seagrave received a $25 million contract to build 54 units for the FDNY, however. It therefore requested the support of the labor union, mayor, and citizens of Clintonville, Wisconsin, to speed up production of this equipment. The company communicated with the FDNY to receive the finished vehicles and continues to produce those that remain to be built. The September 11th disaster therefore showed that the private sector is a major supplier of emergency vehicles to the public sector.

Discussion and Conclusion

This study of private sector involvement in the September 11th disaster provides several important lessons and implications for researchers and practitioners alike. While these findings should be regarded as preliminary in nature, they may have significant impact upon the future of emergency management theory and its application.

- It is apparent that the private sector plays both vital and varied roles in emergency management. In fact, it is not an exaggeration to state that the contributions of businesses in mitigation, preparedness, response, and recovery activities have been woefully underestimated.

- The private sector interacts frequently with the public sector to fulfill necessary community disaster functions. Therefore, the lines between the public and private sectors appear to be blurring, disappearing, or perhaps even artificial.

- The public sector relies heavily upon the goods and services provided by the private sector. Many functions, such as public information, debris removal, and emergency medical care, could not be adequately performed without the assistance of the private sector.

- Coordination issues surrounding site security proved to be a major challenge during the response to the September 11th disaster in New York City. Check-in procedures for contractors should be streamlined (especially at terrorist incidents which require stringent site security).

- Numerous factors facilitate coordination among the private and public sectors. Previous disaster experience, planning meetings, mock exercises, communications equipment, and a willingness to cooperate were mentioned as variables that promoted close collaboration after the terrorist attacks.

- Much more needs to be known about businesses in disasters. For example, what other roles did businesses play in the response to the September 11th disaster? Were there additional functions performed by corporations than those revealed through this project? Do other types or locations of disaster exhibit similar patterns of private sector involvement?

- Additional research on coordination will be required. Future scholarship should focus specifically on the interaction of the public and private sectors in emergency management.

- Methods of educating and involving businesses in emergency management must be promoted. Public officials and agencies should include, where possible, corporations in all types of disaster prevention and planning activities.

- Practitioners must continue to emphasize networking and partnering. The performance of emergency management is increasingly a result of the successful collaboration taking place among government agencies and corporations.

- The factors that hinder and foster coordination must be explored by academia. Practitioners should also familiarize themselves with the lessons provided by scholars in order to augment future emergency management capabilities.

In conclusion, it is apparent that the private sector plays important roles in emergency management and interacts frequently with government officials and agencies to perform important disaster functions. It is hoped that this paper will foster more discussion about the involvement of businesses in disasters and their coordination with the public sector. There has never been such a great need to overcome the problems associated with businesses in disasters while harnessing the potential and actual contributions of the private sector in emergency management. The authors therefore invite and encourage others to study these topics in order to add to the knowledge base of disasters, and improve our ability to prevent and respond to their adverse consequences.

Notes

1. Project Impact is now known as the Pre-Disaster Mitigation Program. It is modeled after the Hazard Mitigation Program and is supported by the Section 103 of the Disaster Mitigation Act of 2000. The Pre-Disaster Mitigation Program was implemented in 2002 and will provide $25 million in grants to local and state governments.

2. Other disaster-related roles of the private sector include loss prevention, site security, and legal services.

3. Unfortunately, many businesses still overlook the importance of occupational health and safety as they downplay the potential for accidents and disasters, and ignore prevention and preparedness activities in order to maintain a higher profit margin.

4. Terrorism preparedness is often conducted with public health agencies to build the Metropolitan Medical Response System.

5. For a variety of reasons, it is extremely difficult to determine the number or percentage of businesses that fail in the wake of disaster (Webb et. al., 2000, p. 86). However, it is generally believed that there will be at least some business mortality after a disaster based on the magnitude of the hazard, the steps taken for mitigation, the strength of the company, and the condition of the economy. It is estimated that nearly 2,000 small businesses went under in the months after the September 11th disaster (Shah, 2002).

6. There have been instances of dishonest business practices in the aftermath of disaster (e.g., price gouging, contractor fraud, etc.).

References

Armstrong, Michael. 2000. "Back to the Future: Charting the Course for Project Impact." *Natural Hazards Review* 1(3): 138–144.

Berkowitz, Michael. 2001. "Learning from Urban Disasters." Presentation at the National Science Foundation Workshop on Response and Opportunities for Future Research. New York: Institute for Civil Infrastructure Systems, Robert F. Wagner Graduate School of Public Service, New York University.

Blaikie, Piers, Terry Cannon, Ian Davis, and Ben Wisner. 1994. *At Risk: Natural Hazards, People's Vulnerability and Disasters*. New York: Routledge.

Drabek, Thomas E. 1994. *Disaster Evacuation and the Tourist Industry*. Environment and Behavior Monograph No. 57. Boulder, CO: Institute of Behavioral Science, University of Colorado.

Erickson, Paul A. 1999. *Emergency Response Planning for Corporate and Municipal Managers*. New York: Academic Press.

Fischer, III, Henry W. 1998. *Response to Disaster: Fact Versus Fiction and Its Perpetuation*. Lanham, MD: University Press of America.

Institute for Business and Home Safety. 2000. "What is a Disaster Recovery Business Alliance?" http://www.ibhs.net/html/businesses/drba.htm.

Mileti, Dennis S. 1999. *Disasters by Design: A Reassessment of Natural Hazards in the United States*. Washington, D.C.: Joseph Henry Press.

Moss, Mitchell and Anthony Townsend. 2001. "Learning from Urban Disasters." Presentation at the National Science Foundation Workshop on Response and Opportunities for Future Research. New York: Institute for Civil Infrastructure Systems, Robert F. Wagner Graduate School of Public Service, New York University.

Payne, Christopher F. 1994. "Handling the Press." *Disaster Prevention and Management* 3(1): 24–32.

Perrow, Charles. 1999. *Normal Accidents: Living with High-Risk Technologies*. Princeton, NJ: Princeton University Press.

Ryland, Harvey. 2002. "Insurance and Disasters: Andrew's Lessons Served Responders Well in New York." *Natural Hazards Observer* (March): 1. http://www.colorado.edu/hazards/o/maro02/maro02a.htm.

Scanlon, Joseph, Suzane Alldred, Al Farrell, and Angela Prawzick. 1985. "Coping with the Media in Disasters: Some Predictable Problems." *Public Administration Review* (Special Issue): 123–133.

Shah, Angela. 2002. "The Silent Losers of Wall Street." *Dallas Morning News* (September 3): 1A.

Waugh, William L., Jr. 1988. "The Hyatt Skywalk Disaster." Pp. 115–129 in Michael T. Charles and John Choon K. Kim, eds., *Crisis Management: A Casebook*. Springfield, IL: Charles C. Thomas Publisher.

Waugh, William L., Jr. 2000. *Living with Hazards, Dealing with Disasters: An Introduction to Emergency Management*. Armonk, NY: M.E. Sharpe.

U.S. Fire Administration. 2002. "The World Trade Center Bombing: Report and Analysis."
http://www.usfa.fema.gov/applications/publications/tr076.cfm.

Webb, Gary R., Kathleen J. Tierney, and James M. Dahlhammer. 2000. "Business and Disasters: Empirical Patterns and Unanswered Questions." *Natural Hazards Review* 1 (3): 83–90.

Acknowledgments

This article is adapted from Weber, Richard T., David A. McEntire, and Robie J. Robinson. 2002. *Public/Private Collaboration in Disaster: Implications from the World Trade Center Terrorist Attacks*. Quick Response Report #155. Boulder, CO: Natural Hazards Research and Applications Information Center, University of Colorado. The authors would like to thank the National Science Foundation and the Natural Hazards Research and Applications Information Center, for a Quick Response grant that made this research possible. The authors are also grateful to the many individuals and organizations in the private and public sectors that provided information for this study. We finally express thanks to the anonymous reviewers for their valuable recommendations to improve the content in and presentation of this report.

Correspondence should be directed to David A. McEntire, Emergency Administration and Planning, Department of Public Administration, University of North Texas, P.O. Box 310617, Denton, TX 76203-0617; e-mail: mcentire@unt.edu; Robie J. Robinson, Emergency Administration and Planning, Department of Public Administration, University of North Texas, P.O. Box 310617, Denton, TX 76203-0617; e-mail: robinson@unt.edu; or to Richard T. Weber, Center for Public Management, Department of Public Administration, University of North Texas, P.O. Box 310617, Denton, TX 76203-0617; e-mail:weberr@unt.edu.

Information Technology Firms Respond to the Immediate Aftermath of the September 11th Terrorist Attacks

Sarah Michaels
School of Planning
University of Waterloo

Introduction

If communication is the lifeblood of disaster recovery, information technology firms provide the arteries through which communication flows. Relief coordination is increasingly underpinned by information technology (McEntire, 1998) that speeds the delivery of known knowledge (Bukowitz and Williams, 1999). How quickly decision makers know what is happening and help can be summoned and organized bears directly and significantly on the success of recovery.

The synchronized explosions of September 11, 2001, caused the greatest-ever loss of human life from terrorism on U.S. soil. People throughout the country and the world recognized the symbolism of attacking the Pentagon and the World Trade Center. The economic consequences reverberate internationally.

In our knowledge-based economy, a key concern is the disruption in information flows. Arguably, at the time of the September 11th explosions, the World Trade Center housed the densest fiber-optic network in the world (Hall and Mearian, 2001). The immediate aftermath of September 11th highlighted the vital role of information technology.

This paper reports on exploratory, quick response research that sought to understand how selected information technology firms contributed to recovery immediately after the attacks of September 11th. How did they contribute to recovery from the disaster? How did firms decide what to do?

The next section identifies the gaps in the hazards literature that this research begins to fill by addressing the above questions. The research approach is described next. After that, the findings are discussed under three themes; disaster relief assistance provided, how firms decided what to do, and how offers of assistance were made known to potential users. Based on the findings, the nature of corporate disaster assistance is considered and suggestions are made for future research to inform practice. In closing, the contribution of this research is considered.

Related Hazards Literature

Traditionally, research on organizations providing disaster recovery assistance has focused on the responses, roles, and contributions of government agencies, such as the Federal Emergency Management Agency (FEMA), (U.S. General Accounting Office, 1993) and nonprofit groups, such as the Red Cross (Cohn et al., 2000), that have primary responsibility for providing emergency assistance.

In the past, when hazards researchers have investigated businesses, it has been in terms of vulnerability, preparedness, or how businesses impacted by disasters have recovered or failed to recover. Research has been done on business vulnerability to disasters (Tierney, 1994), and on business disruption, preparedness, and recovery (Alesch et al., 2001;Webb et al., 2000; Tierney and Dahlhamer, 1998; Alesch and Holly, 1997). Guidebooks have been written on business continuity planning and recovery (Fulmer, 1999; Bell, 2000; Stringfield, 2000).

Recently, corporate social responsibility in disaster reduction has emerged as an important theme (Twigg, 2001; Ariyabandu and Hulangamuwa, 2002; Sharma et al., 2002). While this body of literature emphasizes mitigation, it does consider the contribution of business post-disaster. Weber et al. (2002) describe various roles corporations play in emergency management and suggest there is a scarcity of research on private sector performance in this field.

The impact and increasing ramifications of information technology on disaster response have been acknowledged in the hazards literature (Stephenson and Anderson, 1997; Chartrand and Punaro, 1985; Fischer, 1999; National Research Council, 1996; Quarantelli, 1999). For example, Grant (1996) examined the impact of information technology on organizational relationships within the first ten days of flooding in 1996 in Allegheny County, Pennsylvania. She focused on onsite assistance by traditional emergency responders, such as the Red Cross, the Salvation Army, and FEMA. Other than the media, she did not single out participants in the for-profit sector in the list of organizations participating in the response to the floods. The same pattern of spotlighting the work of the public and nonprofit sectors can be seen outside the United States. For example, Comfort (2000), when she studied the use of information technology in the on-site interorganizational response to the August 17, 1999, Marmara, Turkey, earthquake, interviewed government officials and representatives of nonprofit groups, such as the Turkish Radio Amateur Club and AKUT, a volunteer search and rescue team. The decisions and actions of firms that normally provide information technology and related services have not been considered extensively in the immediate post-disaster phase.

The centrality of information technology to business recovery has been recognized (Toigo, 1989), as has the role of networks, telecommunications, and data communications (Bates, 1992). Ensuring business continuity and viability is absolutely critical. Yet little attention has been paid in the hazards literature to the role of information technology firms in providing disaster assistance to make this happen.

Research Approach

Recurring features of qualitative research true of this project are that the researcher attempts to gain an overview of the context under study and to capture the perception of insiders. The primary task is to understand how people in particular circumstances take action and manage specific situations (Miles and Huberman, 1994). While many actors were engaged in numerous situations immediately after the September 11th terrorist attacks, this research focuses solely on the contribution to recovery made by a few information technology firms.

Qualitative interviewing was desirable because it generated information that was not feasibly available in any other form (Mason, 1996). Qualitative interviews enabled the researcher to capture the thinking behind the genesis and evolution of decision making that business executives are unlikely to document. It was important to find out how decisions were made while the experiences were fresh in people's minds. Speaking to interviewees soon after September 11th increased the likelihood that they could remember in detail what prompted their firms' actions and why they chose to respond the way they did.

The firms contacted were either ones with which the researcher had pre-existing contacts and/or were identified in news reports or advertisements from *Computerworld*, *Silicon Alley Daily*, and the *New York Times*. Electronic media reports were an essential means of identifying firms that are response innovators. Since electronic media reports are not necessarily well archived, it was essential to review electronic publications as they were posted.

While there was urgency in collecting the needed information, there was no need to go to the physically impacted sites. The web sites needed for review are equally accessible from anywhere with an internet connection. The in-depth, elite interviews were conducted by telephone since one respondent was based in Europe while the others were based across the United States. Interviews were conducted as guided conversations (Rubin and Rubin, 1995).

Ten individuals from nine firms were interviewed between September 18 and October 4, 2001. Seven interviewees were in management positions. Two had responsibilities related to technology, three to marketing, and two to corporate social responsibility, including humanitarian relief. To varying degrees, all were involved in deciding how their firms would respond to the immediate aftermath of the September 11th terrorist attacks. Three interviewees worked in communications and routinely acted as spokespeople for their large firms. Immediately after the World Trade Center collapse, two of them had been charged with liaising with the media about their companies' response activities.

Nine interviewees talked about the immediate post-event contributions of their firms. The firms contacted are engaged in different aspects of information technology. The majority of interviewees who talked about their firms insisted on anonymity for themselves and for their firms. They were concerned about potential breaches of security and about their competitive positions within their industry. These interviewees made it clear that they did not want the nature of their businesses portrayed with any degree of specificity, lest a reader be able to identify particular firms. Some firms provide communication services, such as web hosting, web design, voice mail, e-mail, long distance voice, data, and video. Others manage specialized information or provide infrastructure, technology, and software for telecommunications companies.

Karl F. Rauscher, Director of Network Reliability at Lucent Technologies, was interviewed not about his firm's response, but about the Wireless Emergency Response Team (WERT) he founded on the evening of September 11th. WERT's mission was to provide coordinated wireless industry mutual aid support for search and rescue efforts at the World Trade Center. In reviewing its endeavor, in October 2001, WERT concluded that its efforts had provided value in the following ways.

- WERT helped keep rescue teams from being exposed unnecessarily to physical danger by quickly discrediting false reports that there might be survivors with cell phones in the rubble.
- WERT was able to confirm that individuals thought to be missing were safe.
- The team was able to help family members achieve closure on the death of loved ones by dispelling rumors that cell phone calls had been made when they hadn't.
- WERT assured the public both domestically and internationally that all known technological approaches were being used to listen for any cellular or pager communication emanating from the collapsed World Trade Center towers.

- To enhance capability to respond to future events, WERT documented lessons learned and made recommendations in a final report made available on the web (Wireless Emergency Response Team (WERT) 2001). WERT's intent was to operate as an ad hoc group in future emergencies (Wirbel, 2001).

No survivors were found as a result of WERT's efforts. Thirty-three organizations participated (see Appendix A). Over 250 industry subject matter experts participated in WERT's efforts. One hundred and twenty reports of cell phone or pager use from the rubble were investigated. The WERT Public Call Center received 5,039 calls. The Center was set up to record information from people who had received contact from a missing person via an electronic device, from people with electronic devices trapped in the collapse of the World Trade Center, and from people who were aware of any electronic devices associated with a missing person at the World Trade Center. Approximately 500 Bell South employees volunteered as operators at the public call center (WERT, 2001).

Findings

The results of the interviews are organized under three complementary themes. The first involves what and how firms provided disaster assistance. The second is how firms decided what to do, to whom to provide assistance, and for how long. The third is how offers of assistance were made known to potential beneficiaries.

Disaster Assistance Provided

Firms provided help to businesses as well as giving aid to support rescue and community response. Much of what firms provided was either part of normal everyday business or an extension of it. Some firms provided technical services free to those directly impacted by the events of September 11th. These included normally fee-based services, such as web and telephone hosting. Assisting firms increased the amount of client support they supplied and undertook activities that clients normally performed. Donating firms provided telecommunication networking expertise, rerouted phone systems, provided answering services, undertook data center recovery and call center recovery, and contributed logistical support.

The flexibility of communication systems and the ability to scale up means that firms providing communication services have the capacity to take on new business at short notice. At least initially, these firms were not limited in what they could provide by available infrastructure.

Firms did provide assistance close to Ground Zero at the World Trade Center. Corporate disaster relief teams were activated and provided humanitarian assistance. The services they performed were a function of what they were asked to do and what they volunteered in light of the needs they saw. Free telephone access was made available temporarily at the impact site and at airports. Prepaid telephone cards were distributed at some locations. WERT aided search and rescue efforts by monitoring cellular networks and detecting signals from wireless devices carried by those caught in the collapse of the towers. Firms provided access to their facilities. One respondent described how one of her firm's buildings, which was near the World Trade Center, was used as a staging area for rescue operations. The catering services run for employees in that building were made available to relief workers.

In addition to services, firms donated supplies they either had in-house or to which they had ready access. Supplies that were donated included telecommunication devices, such as cell phones and two-way pagers, as well as general hazard response supplies, such as suits, gloves, and eyewash used in hazardous materials incidents. One firm provided approximately 7,000 cell phones. Another firm supplied a self-contained, fully functional portable telecommunications site that provided high quality cellular transmission.

Financial assistance for humanitarian relief efforts included corporate donations and facilitating employee donations. The latter included, at least for one firm, the corporation's matching the employee contributions, up to a set maximum. The money was given to established nonprofit groups, such as the Red Cross, or to newly created funds dedicated to disbursing assistance to those directly impacted, such as the Twin Towers Fund set up by Mayor Rudolph W. Giuliani to help the families of police and fire personnel killed or injured responding to the World Trade Center explosions.

Employee contributions were not only monetary. In at least one firm, staff donated blood, and employees could use eight hours of work for volunteer activities.

Employees expressed enormous pride in their companies' responses. Unprompted, three respondents said they were proud to be part of their firms, given how their firms had responded to the events of September 11th. Each noted the speed with which the right things got done.

How Assistance was Provided

Firms provided help by supporting the work of traditional emergency responders and by assisting impacted businesses in recovery. Direct assistance to business was more prevalent than working through industry associations or groups formed in the immediate aftermath of September 11th.

Contributing to traditional emergency responders—Firms used pre-established, formal channels to offer assistance. For example, more than one firm volunteered help to the Red Cross through existing connections with that nonprofit. Firms already had links to the Red Cross either because the Red Cross was a client, through the firm's charitable foundation, and/or through its disaster response team. In addition, individuals were familiar with the Red Cross. The Red Cross was seen as the lead nonprofit responder and as such the initial organization to contribute resources, particularly financial aid.

Two respondents expressed qualified frustration in dealing with traditional emergency response organizations. The concerns focused on the lack of prompt adaptiveness. One respondent described how his firm offered to fly federal agency personnel for free on planes the firm had chartered. By the time the offer was processed through the bureaucracy, the planes had flown without government personnel on board. The individual had found it difficult to work through public channels to identify a decision maker who could expedite consideration of his firm's offer. Another respondent described how nonprofit organizations were unable to act as brokers for businesses offering assistance. The respondent thought there was a potential role for a nonprofit known for its response work to take on this function. Her rationale was that it would occur to those in need to seek help through such an organization. She asked if there was a way for nonprofits to determine how business could fit into their approach to disaster response.

Providing business-to-business assistance—The primary form of business-to-business assistance was meeting the needs of existing clients. All other forms of help were in addition to this. As one interviewee explained, customers are not sacrificed, the number one priority is to focus on customers and ensure that they have adequate resources. Service subscribers are taken care of first. As one interviewee explained, the reason firms subscribe to contingency services is to guarantee availability of resources. Non-customers are helped within reason.

Firms continued to provide the same services after the disaster that they had provided to clients before September 11th. Firms delivered on post-disaster services contracted for before the disaster, such as providing back-up data support centers, or supplementary assistance. For a number of firms, what they did after the event built upon the disaster recovery planning they had done for their clients. Assisting firms took on functions clients were not able to perform, such as systems monitoring or providing call center capability.

Some firms made available one or more of their services for free temporarily, such as web hosting or phone service. Such offers were extended to those who had not been clients of the firms.

Providing assistance through existing industry associations and outlets—Industry associations and outlets were not the primary vehicles through which firms provided post-disaster assistance. An interviewee described how there was a national gathering in Washington, D.C. on September 11th, of one of the main industry associations to which his firm belongs. The association coordinated getting people back home from Washington, D.C. While his firm offered assistance to the association in doing so, its help was not needed. One interviewee spoke about being in contact with industry associations in the New York City area to keep them informed about his firm's activities and to learn what other firms were doing. WERT was created as a new organization drawing on the contacts established through existing associations.

Providing assistance through groups created in the aftermath of the disaster—WERT was the only example interviewees provided of a new group that was a focal point for participatory, technical disaster assistance. One interviewee described an example of opportunistic post-disaster assistance provided by her firm. On learning of a September 21 telethon, her firm volunteered to use its telemarketing capability and call centers, enabling employee volunteers to answer the phones. Financial contributions were made to funds created after the event to help those most affected by the disaster.

How Firms Decided What to Do

Interviewees described how they wanted to do something to aid in the recovery from the terrorist attacks. Watching events on television was a powerful prompt to action. On September 11th, the founder of WERT saw on television how cell phones had been used by individuals in the World Trade Center rubble. He realized that his industry colleagues understood the technology of how to monitor signals from cell phones and pagers. That evening he got on the phone to them and began to organize the Wireless Emergency Response Team (WERT).

For firms that did not have a disaster response orientation before September 11th, a powerful motivation was the sense of kinship and affinity with New Yorkers. People knew people in New York who were relatives, friends, and colleagues.

Interviewees described how initiatives to assist others were taken at different levels and in different divisions within the company. One respondent explained how disasters create opportunities for those not in executive positions to demonstrate leadership. Another respondent was careful to note that offers of assistance, within her firm, were undertaken in consultation with supervisors and management. One interviewee described how the firm pulled together its top talent to address the question of how they could help those in need.

Non-routine assistance to other businesses and to charities was the result of thought-through business decisions. One respondent described how his firm did the numbers and calculated how long they could provide a resource-intensive service free without really hurting the business. As a for-profit business, the firm gave what it could afford to and consequently, "were not going to provide everything for free." The firm felt that was the best they could offer. The same interviewee explained that donating money only goes so far, providing a service is providing what those in the firm know best and so that is why they provided it. Another respondent described how the process of deciding what to contribute on September 11th was infused with the urgency of the circumstances. She described how there was a sense that what they, as business leaders, might do may not be perfect, that they may not have thought it through completely, but that the firm had to get something out there or time would pass.

In contrast to developing a new initiative, one interviewee described how providing technical assistance was straightforward—where systems were not working, technicians and equipment were sent in. He regarded it as part of customer support and relations. That work gets done and then you see who pays for it.

Precedence for Action

How those in the corporate sphere understood and responded to the events of September 11th was shaped by their previous exposure and participation in responding to disasters.

Two respondents with experience of disasters worldwide put September 11th in a broad context. They talked about disasters that had claimed more lives than the September 11th terrorist attacks and described situations where people had fewer pre- and post-event resources than those directly affected by the September 11th event. These two respondents pointed out that private enterprise was often at the disaster site sooner than government or international relief providers. They described how private firms with their technical expertise and prompt arrival at the disaster site rebuilt the infrastructure governments and international donors then used to provide their assistance.

Four of the respondents explained there was no precedence for their actions. Six respondents acknowledged that experience in previous disasters influenced their firms' decision making. Experiences mentioned included recent earthquakes in Washington State and India, hurricanes in Florida, and the Oklahoma City bombing. Of those who acknowledged precedents for action, four respondents emphasized that the scale of this event and its novelty required their firms to do things they had not done before, such as making arrangements to get key personnel home from off-site locations. One respondent stressed that developing and promoting a culture of service is the only answer for providing appropriate and timely response.

Respondents noted that an important way in which this disaster differed from previous disasters in the United States was loss of access by air, not only to the impact site but also to places not directly affected. The assumption that a firm could immediately fly staff and materials wherever they were needed did not hold for this event.

Disaster Recovery Plan in Place

All of the respondents described how their firms had plans in place for what to do if struck by a disaster. A number emphasized that it was essential because they were in the business of protecting their clients' information assets. One respondent explained that it is normal for telecommunication firms to want to keep networks up so that they do have disaster recovery procedures that get activated readily when networks fail. Another respondent described how in his firm's line of work system failure was not an option. In firms with well-developed disaster response plans not all actions taken stemmed from the plans.

Disaster Assistance Plan in Place

Although all respondents described how their firms had corporate disaster recovery plans in place, not all had a plan for helping others impacted by a disaster. Five respondents explained how the assistance they provided to others after September 11th was developed after the synchronized explosions. For example, before this event there was no plan for coordinated emergency wireless response. The two firms with disaster assistance teams and another firm that had a long history of providing disaster assistance had pre-existing strategies for helping. Firms with pre-existing strategies complemented what they had in place with activities tailor-made to the unique circumstances stemming from September 11th.

Deciding to Whom to Provide Assistance

Aiding current customers was the number-one priority in providing business-to-business assistance. Businesses checked with current customers to find out if they needed help. One respondent described how the firm inventoried customer needs, prioritized them, and went from there. Firms that provide highly specialized infrastructure and support were not likely to be approached by non-customers for technical assistance.

Firms that made known through the media what services they had to offer were willing to provide those services to whomever responded. For example,

WERT tried to track down all the leads they received about signals from wireless devices.

Respondents declined to name or describe those to whom they had provided business assistance. Firms that offered one or more of their usual services for free to those directly impacted by September 11th were unwilling to say how many firms or individuals had used those services. The reasons given for not divulging details of beneficiaries were either proprietary or security related or because firms hadn't tracked information about recipients of the assistance they provided.

Firms used different screens in deciding which charitable organizations would be given money. Firms considered which group was providing the most fundamental assistance, who would benefit the most from what a firm could provide, or which groups were perceived to be not receiving their fair share, or the areas of interest to the donating firm.

Corporate disaster response teams initially offered their services to whomever was in charge. By doing so, they tapped into a well-defined interjurisdictional incident command structure.

Anticipated Length of Providing Assistance

The nature of the assistance determined whether respondents knew of an end date and, if so, when that date would be. For example, while it was not clear at the time of the interview how long free local telephone service in the affected area would go on, the provision of free airport phone service ended September 21, 2001.

Technical support was to be provided for as long as needed. One respondent explained that by the time of the interview, his firm was no longer providing extra customer support. Other respondents noted that technical support would not be needed for long since the peak usage of phone service occurred immediately after the synchronized explosions on September 11th.

One respondent didn't know when his firm would stop providing free technical service, another said that based on market research his firm would provide such help for three months, and a third explained it depended on individual situations. The third firm initially offered assistance through the end of September, then extended it to the end of October.

One respondent explained, "You make a charitable donation for x amount and then stop." Financial contributions are capped.

One respondent explained that corporate programs of social responsibility must build up credibility by coming when asked and working closely with others in all situations. If firms get into the business of providing post-disaster humanitarian assistance they can't stop simply because they have had enough. If they do, they lose more than they gain by exiting the scene. Another respondent estimated that his firm's humanitarian assistance might well go on for a year or two with a reduced intensity after the massive first response.

How Offers of Assistance Were Made Known to Potential Beneficiaries

Making potential users aware of what is available is a critical step in getting offered services used. In the immediate aftermath of September 11th, firms used their pre-existing connections, the contact network of employees, and media releases to make people aware of what they were offering.

Individuals within firms used their own network of contacts in other firms and in nonprofits to offer their services. Employees in sales and marketing, engineering, and security were instrumental in making direct contact through e-mail and by phone.

Media press releases were important means for letting potential users know about services being offered. One firm ran an ad in a daily electronic news journal geared to the information technology industry. WERT issued press releases to explain where people should call to provide phone or page numbers of individuals missing in the World Trade Center.

For those firms that had not previously offered post-disaster assistance, the biggest challenge was getting businesses and individuals to know what was being offered. More than one respondent described surfing the web to try to locate intermediaries to publicize services offered.

One respondent emphasized that most of his efforts to let firms know about disaster response and mitigation options came through the conversations he had with companies, cities, and agencies before September 11th. He had met with decision makers, such as executives or elected officials, to discuss risk by facility or organization since his firm provides preparedness planning and can be hired on retainer for disaster response.

Discussion—the Nature of Corporate Disaster Assistance

One way to think about the nature of disaster recovery assistance from business is on a continuum from ordinary activities to extraordinary activities.

Ordinary activities refer to the set of actions firms take as part of doing business. This involves providing a particular service, such as web hosting. It also involves activities to which firms have an ongoing commitment based on corporate social responsibility, such as supporting disaster assistance teams. At the ordinary-activities end of the continuum are those actions that firms undertake regularly. Extraordinary activities refer to those actions that are not routine to a firm, such as making travel arrangements for government officials. At the extraordinary-activities end of the continuum are actions that are novel and exceptional. Closer to the middle may be activities the firm does on a less-routine basis. These are logical extensions of normal activities, such as making available for free services that companies usually charge for, such as phone calls.

One measure of a firm's resilience is the ease with which it can move along the continuum. Locating a firm's actions on the continuum provides a snapshot of a particular time frame. An extraordinary activity for a firm at one time may become part of the firm's suite of ordinary activities.

Another way to think specifically about business-to-business assistance after a disaster is according to beneficiaries targeted and services provided (Table 1). This perspective highlights the extent to which providing disaster assistance is part of doing business. Making sure that current clients have functional operations is vital.

Table 1. Providing existing and new services to existing and prospective clients.

	Existing services	New services
Existing clients	Customer support	Development, marketing
Prospective clients	Outreach marketing	Development, outreach marketing

Disasters may lead to an increase in business. Existing clients may opt to purchase new services, new clients may opt to purchase existing services. Existing and prospective clients may purchase new services. On September 19, 2001, one firm launched a new service that in light of September 11th would be appealing to existing and prospective clients.

Firms do not only provide assistance by meeting the needs of existing and prospective clients. Firms contribute monetarily to traditional disaster assistance providers. In addition, humanitarian assistance is provided directly and on an extended basis through disaster response teams. One simple way to consider the full range of types of business disaster assistance is by disaster recovery phase (Table 2).

Table 2. Business contribution to phases of response and economic recovery.

Phase	Business contribution
Rescue	Disaster response teams, financial contributions
Taking care of life essential needs of impacted people	Disaster response teams, financial contributions
Restoring shared infrastructure	Ordinary activity, extraordinary activity
Economic rehabilitation	Ordinary activity, extraordinary activity
Economic advancement	Ordinary activity

Evolution of Corporate Disaster Assistance as a Function of Technology

How the information technology component of the private sector provides post-disaster assistance is evolving as a direct function of improving and expanding technology. For example, WERT's contribution to search and rescue deployed advanced network monitoring techniques to track signals from wireless devices that have become widely used only in the last five years.

Technology enabled firms to provide spatially unrestrained assistance. Firms were not tied to equipment in one location. Remote sites were used to handle back up data. New York–based firms were provided with Los Angeles– based phone numbers by a company physically based in a third city.

Two way paging turned out to be one of the most effective, reliable, and secure mechanisms for contact among emergency response personnel, including those in the Mayor's office and Governor's office. The system had few users compared to cellular phones. Particularly popular were the BlackBerry Wireless Email Solution devices that firms put into the hands of response personnel. These wireless e-mail devices are sufficiently intuitive that novice users working in stressful situations were readily able to communicate using them.

One respondent explained that the potential of emerging technology for disaster recovery has yet to be tapped. He said that since September 11th it will be easier to promote the development and application of these technologies for disaster relief than it was before, when no one could see the business case for applying emerging technology in non-commercial settings. The respondent provided two illustrations of the use of emerging technology in disaster settings. First, cellular broadcasting from a central switch to all users in an area could be deployed. In a particular area, cell phone users could be asked to turn off these devices, if they are making nonessential calls, to enable emergency workers to get through. Second, through cellular systems, mobile positioning systems could be used to locate signals from phones and pagers. He suggested this would not be difficult to develop. WERT is the first generation of sophisticated initiatives to attempt to do so to enhance rescue operations. It demonstrates some of the potential of coordinated wireless industry mutual aid. Consequently, it is not unreasonable to consider the emergence of other sophisticated initiatives either during crises or, ideally, in advance of them.

Recommendations for Research to Inform Practice

This exploratory research suggests the value of investigating systematically how and why the for-profit domain participates in providing assistance after a disaster. It would be particularly valuable to investigate thoroughly the assistance offered by information technology firms. Such a study would provide a picture of how an economic sector that has redefined modern life shapes such a time-sensitive and communication-dependent function as disaster recovery.

Exploring the interface between businesses offering technical disaster assistance for the first time and traditional governmental and not-for-profit disaster responders would be valuable. A related need is to identify constructive avenues for businesses offering assistance to let potential beneficiaries know what help is available. Such efforts would help groups such as the American Red Cross that are trying to devise ways to better harness offers of volunteer help from individuals with highly specialized skills (Clizbe, 2002).

It is important to learn more about how being in a particular sector—be it for-profit, nonprofit, or government—shapes the nature of the assistance provided, how it is provided, and the spin-offs garnered from providing assistance. How applicable to the for-profit sector are the models of disaster assistance developed from examining governmental and nonprofit operations? Preliminary evidence suggests that this question would be worth pursuing. McEntire (1998) developed a model of coordination among nonprofit groups.

He noted that pre-disaster links among non-governmental humanitarian groups increase coordination during disaster relief operations. Before September 11th, linkages among businesses made the quick formation and activation of WERT possible.

Contribution

This research focuses on information technology firms as assistance providers in the immediate aftermath of a disaster. Thus it extends our understanding of the network of disaster response beyond the traditional organizational emphasis on the public and nonprofit sectors. At the same time, it expands the research agenda on the private sector during response to include the decision making of intact firms. In the past there has been an understandable preoccupation with the recovery of impacted firms.

By focusing on the decision making of firms normally engaged in information technology and allied services in the immediate aftermath of a disaster, this research provides a different perspective on the impacts and ramifications of information technology in disaster response. Past investigations have emphasized the use of information technology by traditional emergency responders. This work highlights how firms adapt and extend their normal business practices to extraordinary circumstances.

Examining the business contribution to disaster assistance, specifically from the information technology sector, provides new insights into what constitutes disaster assistance and how it is delivered. Firms can provide expertise and technical resources not necessarily available in the public and nonprofit sectors. At the same time, individual businesses can respond nimbly with tailor-made contributions.

This research highlights that disaster assistance is not necessarily delivered at ground zero and that providers and recipients do not need to be in close physical proximity. Non-spatially restricted assistance is a function of the information technology revolution. Firms providing some forms of telecommunications relief, such as web hosting, could do so without having employees travel to the disaster site. This was particularly important given the ban on commercial and general aviation precisely when the need for telecommunications was at its peak—right after the attacks of September 11th. For example, AT&T Wireless reported its highest calling volume ever on that day (Barnes, 2001).

Since information technology firms transacted assistance from around the country, the author did not conduct the investigation in Washington, D.C. or New York City. In the past, it has been a given that research into response decision making requires investigators to be physically present at the disaster site. Of course, for the majority of quick response research on decision making in the aftermath of an event, the need to travel to the disaster site will remain. Still, as the nature of urgent post-disaster assistance evolves, how quick response research is conducted will reflect such change.

Conclusion

In providing assistance in the immediate aftermath of the September 11th terrorist attack, information technology firms undertook ordinary and extraordinary activities. Firms combined prior disaster experience, pre-existing disaster response plans, and post-event ingenuity to deliver previously contracted services, to provide new business related services, and to donate humanitarian aid. Considering the contribution of information technology firms both reinforces the value of traditional emergency response and expands what constitutes post-disaster assistance and how it is delivered.

References

Alesch, D.J. and J.N. Holly. 1997. "Small Business Failure, Survival, and Recovery: Lessons from the January 1994 Northridge Earthquake." *CUREE Conference Proceedings*. Green Bay, WI: University of Wisconsin.

Alesch, D.J., J.N. Holly, E. Mittler, and R. Nagy. 2001. *Organizations at Risk: What Happens When Small Business and Not-For-Profits Encounter Natural Disasters*. Fairfax, VA: Public Entity Risk Institute.

Ariyabandu, M.M. and P. Hulangamuwa. 2002. *Corporate Social Responsibility and Natural Disaster Reduction in Sri Lanka*. Colombo, Sri Lanka: ITDG–South Asia.

Barnes, J.E. 2001. "For Cellphone Holdouts, Worry Closes the Sale." *New York Times*. http://www.nytimes.com/2001/09/19/technology/19CELL.html?ex=1001909019.

Bates Jr., R.J. 1992. *Disaster Recovery Planning: Networks, Telecommunications, and Data Communications*. New York: McGraw-Hill.

Bell, J.K. 2000. *Disaster Survival Planning: A Practical Guide for Businesses*. Port Hueneme, CA: Disaster Survival Planning, Inc.

Bukowitz, W.R. and R.L. Williams. 1999. *The Knowledge Management Fieldbook*. Harlow, England: Financial Times Prentice Hall.

Chartrand, R.L. and T. Punaro. 1985. *Information Technology Utilization in Emergency Management*. Washington, D.C.: U.S. Library of Congress, Congressional Research Service.

Clizbe, J. 2002. "The Red Cross after September 11, 2001." The International Emergency Management Society, Ninth Annual International Conference, May 15. Waterloo, Ontario, Canada: University of Waterloo.

Cohn, R., K.M. Carley, J.R. Harrald, and W.A. Wallace. 2000. "Emotions in Crisis Management: An Analysis of the Organizational Response of Two Natural Disasters." *International Journal of Technology Management* 19 (3/4/5): 313–335.

Comfort, L.K. 2000. *Information Technology and Efficiency in Disaster Response: The Marmara, Turkey Earthquake, 17 August 1999*. Quick Response Report #130. Boulder, CO: Natural Hazards Research Applications and Information Center, University of Colorado.

Fischer, H.W. 1999. Using Cyberspace to Enhance Disaster Mitigation, Planning and Response: Opportunities and Limitations. *Australian Journal of Emergency Management* 14(3): 60–64.

Fulmer, K.L. 1999. *Business Continuity Planning: A Step-by-Step Guide*. Brookfield, CT: Rothstein Associates.

Grant, D.B. 1996. *The Potential Impact of Information Technology on the Structure of Interorganizational Relationships During Crisis Response: The Pennsylvania Floods of 1996.* Quick Response Report #85. Boulder, CO: Natural Hazards Research Applications and Information Center, University of Colorado.

Hall, M. and L. Mearian. 2001. "Recovery." *Computerworld.* http://www.computerworld.com/cwi/story/0,1199,NAV47_ST063723,00.html.

Mason, J. 1996. *Qualitative Researching*. Thousand Oaks, CA: Sage.

McEntire, D.A. 1998. *Towards A Theory of Coordination: Umbrella Organization and Disaster Relief in the 19974,3398 Peruvian El Nino*. Quick Response Report #105. Boulder, CO: Natural Hazards Research Applications and Information Center, University of Colorado.

Miles, M.B. and A.M. Huberman. 1994. *Qualitative Data Analysis*. Thousand Oaks, CA: Sage.

National Research Council. 1996. *Computing and Communications in the Extreme: Research for Crisis Management and Other Applications*. Washington, D.C.: National Academy Press.

Quarantelli, E.L. 1999. The Computer Based Information/Communication Revolution: A Dozen Problematical Issues and Questions They Raise for Disaster Planning and Managing. Article # 340. Newark, NJ: Disaster Research Center, University of Delaware.

Rubin, H.J. and I.S. Rubin. 1995. *Qualitative Interviewing*. Thousand Oaks, CA: Sage.

Sharma, A., M. Gupta, and K. Gupta. 2002. Corporate Social Responsibility and Disaster Reduction: An Indian Overview. Seeds, India.

Stephenson, R. and P.S. Anderson. 1997. "Disasters and the Information Technology Revolution." *Disasters* 21(4): 305–334.

Stringfield, W.H. 2000. Emergency Planning and Management: Ensuring Your Company's Survival in the Event of a Disaster. 2nd edition. Rockville, MD: Government Institutes, Inc.

Tierney, K.J. 1994. "Business Vulnerability and Disruption Data from the 1993 Midwest Floods." Preliminary Paper No. 213. Newark, DE: Disaster Research Center, University of Delaware.

Tierney, K.J. and J.M. Dahlhamer. 1998. *Business Disruption, Preparedness and Recovery: Lessons from the Northridge Earthquake*. Newark, DE: Disaster Research Center, University of Delaware.

Toigo, J.W. 1989. *Disaster Recovery Planning: Managing Risk and Catastrophe in Information Systems*. Englewood Cliffs, N.J.: Computing Series.

Twigg, J. 2001. *Corporate Social Responsibility and Disaster Reduction: A Global Overview.* London: Benfield Greig Hazard Research Centre, University College London.

U.S. General Accounting Office. 1993. *Disaster Management: Recent Disasters Demonstrate the Need to Improve the Nation's Response Strategy*. Washington, D.C.: U.S. General Accounting Office.

Webb, G.R., K.J. Tierney, and J.M. Dahlhamer. 2000. "Businesses and Disasters: Empirical Patterns and Unanswered Questions." *Natural Hazards Review* 1(2): 83–90.

Weber, R.T., D.A. McEntire, and R.J. Robinson. 2002. *Public/Private Collaboration in Disaster: Implications from the World Trade Center Terrorist Attacks*. Quick Response Report #155. Boulder, CO: Natural Hazards Research Applications and Information Center, University of Colorado.

Wirbel, L. 2001. "Emergency Wireless Team Issues Call for Help." *EE Times*. http://www.commsdesign.com/news/tech_beat/OEG20011129S0052.

Wireless Emergency Response Team (WERT). 2001. Final Report for the September 11, 2001, New York City World Trade Center Terrorist Attack.

Acknowledgments

The author would like to acknowledge the Natural Hazards Research and Applications Information Center, University of Colorado–Boulder and the National Science Foundation for their support through a Quick Response grant. Thanks to all the respondents who provided generously of their time and insights. The names of individuals and firms are not provided in accordance with the wishes of a number of interviewees. Thanks also to Lori Peek, Natural Hazards Center, University of Colorado, for her initiative in identifying some of the contacts and to Wanda Headley, also of the Natural Hazards Center, University of Colorado, for her help with bibliographic references. I am grateful for the constructive recommendations of two anonymous reviewers.

This article is adapted from Michaels, Sarah. 2001. *How and Why Selected Information Technology Firms Contributed to Recovery Immediately after the September 11, 2001 Terrorist Attacks*. Quick Response Report No. 141. Boulder, CO: Natural Hazards Research and Information Applications Center, University of Colorado. A shorter version of this paper, entitled "Disaster Assistance from Information Technology Firms," was presented December 12, 2001 at the workshop, Learning from Disasters: National Science Foundation Response and Opportunities for Future Research Workshop, Institute for Civil Infrastructure Systems, New York University, New York.

Correspondence should be directed to Sarah Michaels, School of Planning, University of Waterloo, Waterloo, Ontario, Canada N2L 3G1; (519) 888-4567, x6863; fax: (519) 725-2827; e-mail: michaels@fes.uwaterloo.ca

Appendix A

Organizations Participating in the Wireless Emergency Response Team
(WERT 2001, p. 2)

Arch Wireless
Argonne National Laboratory
AT&T
AT&T Wireless
BellSouth
Cellular Telecommunications and Internet Association (CTIA)
Cingular Interactive
EDO Corporation
Ericsson
Federal Communications Commission (FCC)
Lucent Technologies
Metrocall
Motorola
National Communications System (NCS)
National Coordinating Center for Telecommunications (NCC)
Network Reliability Steering Committee (NRSC)
Network Reliability and Interoperability Council (NRIC) V
Nextel
New York City Police Department (NYPD)
New York City Mayor's Office
Nortel Networks
Personal Communications Industry Association (PCIA)
U.S. Secret Service
SkyTel
Sprint PCS
Telcordia Technologies
TruePosition
U.S. Department of Energy
U.S. Marshals Service, Electronic Surveillance Unit
Verizon
Verizon Wireless
VoiceStream
Wheat International

Part 6

Public Policy and Political Context

Global Justice Protesters Respond to the September 11th Terrorist Attacks: The Impact of an Intentional Disaster on Demonstrations in Washington, D.C.

Patrick F. Gillham
Department of Sociology
University of Colorado

Bob Edwards
Department of Sociology
East Carolina University

Introduction

This paper examines the September 11, 2001, terrorist attacks on the United States from a perspective that provides insight into the future of social movements and political dissent in a context of heightened concern over threats to national security. Specifically, we examine the way a range of social movement organizations responded to the dramatically changed political climate in the three weeks following the terrorist attacks. The organizations examined here had been planning a series of global justice demonstrations against the World Bank and the International Monetary Fund to be held in Washington, D.C., during the last week of September 2001. Based on field observations, interviews, and systematic review of e-mail list-servs and websites, we chronicle the organizational field of protesters and the roster of protest events as they changed during the three weeks after the terrorist attacks. The crisis precipitated by the September 11th disaster dramatically heightened concern over national security and the public rallied around the President with nearly unanimous approval of his handling of the situation. During such crises the contrast between groups and events like those examined here and broad public sentiment appears stark. The political and social costs of dissent are heightened and its public display is either curtailed by dissenters themselves, as in the case of most of the groups discussed here, or at times actively quashed. Yet such periods of unanimity never seem to last very long and a social and political climate more favorable for dissent returns rather quickly, as the 15 months since the events analyzed here have shown.

Background

For months a diverse coalition of national and international advocacy groups, local church and community organizations, and a smaller loosely affiliated contingent of anarchists had been organizing a series of protest events in Washington, D.C. The series of protests and related events called the Mobilization for Global Justice (MGJ)* was to coincide with the meetings of the World Bank and the International Monetary Fund (IMF) scheduled for the last week of September 2001 (see the glossary for list of acronyms and description of organizations and events). The MGJ would be the latest in a series of high-profile, mass demonstrations in world policy and financial centers since the "Battle in Seattle" had nearly brought the November 1999 meetings of the World Trade Organization (WTO) to a grinding halt. Having experienced a similar series of disruptive actions in April 2000, the Washington, D.C., police were expecting the MGJ to be an even larger and more-difficult-to-manage series of events. Management of the protests would be especially complicated since the World Bank and IMF meetings would be held in a mixed business and residential district, making it difficult to control access to the area. World Bank and IMF officials were just as concerned as police that the protests might prevent their meetings from occurring. Their concerns intensified throughout the summer as the MGJ Coalition and the roster of planned protest events grew. On July 11, 2001, World Bank and IMF officials decided to relocate the meetings to their central buildings in downtown Washington, D.C. (Fernandez and Blustein, 2001). Within a month they had shortened the meetings to two days and reduced the number of participants. It was hoped that these decisions would make it easier for the police to defend the space and control protesters (Santana and Blustein, 2001).

Through the first week of September 2001, planning for the MGJ progressed smoothly although the overall coalition had subdivided into three distinct branches. The smallest and most confrontational branch included the Anti-Capitalist Convergence (ACC) and other anarchist groups who were secretly planning confrontational and illegal actions aimed at halting the World Bank and IMF meetings. A second branch intended to disrupt the meetings and communicate opposition to the global trade establishment by engaging in non-violent acts of civil disobedience like sit-ins, obstructing

*This and other organizational acronyms are explained more fully in a Glossary that appears at the end of this paper.

traffic, or blockading buildings. The third and largest branch included a wide range of prominent advocacy organizations, church groups, environmental groups, and labor unions that had obtained permits for legal marches, rallies, and teach-ins. By the morning of September 11th planning for the permitted events had advanced so well that their sponsors had scheduled a major press conference to publicly articulate the MGJ's four demands against the World Bank and IMF, and to announce final plans for the mass rally and march set for September 30, 2001. Present would be AFL-CIO President John Sweeney, Friends of the Earth President Brent Blackwelder, as well as Feminist Majority President Eleanor Smeal (Cavanaugh, 2001; AFL-CIO, 2001a). The press conference was scheduled for September 11, 2001, at 11 a.m., EDT, outside the Methodist Building one block east of the Capitol and across the street from the Supreme Court. The news conference never happened.

Tragically, at about 8:45 that morning the first jetliner was piloted into the north tower of the World Trade Center in New York City. By 10:37 a.m. authorities had reported that other planes had crashed into the second tower, the Pentagon, and a Pennsylvania field (Murray, 2001). In a state of shock, the organizers of the press conference released a perfunctory cancellation notice (AFL-CIO, 2001b), which received no local press coverage. Members of the press along with everyone else in Washington, D.C., had, for the time being, lost all interest in the upcoming demonstrations. The terrorist attacks of September 11th instantly transformed the American political context, leading each organization involved to reappraise its mission and activities. The World Bank and IMF cancelled their meetings. Protesters cancelled certain events and planned new ones considered more appropriate for the current circumstances. The roster of organizations involved changed dramatically as well, with some dropping out as previously uninvolved groups joined.

Originally, the MGJ had planned a 10-day episode to occur from September 23—October 4, 2001 (MGJ, 2001a) outside the Woodley Park Hotel, the meeting place used by the World Bank and IMF for the last 20 years (Fernandez and Blustein, 2001). However, September 11th profoundly changed the MGJ and related events. By September 14 a significant exodus occurred among larger organizations who had been planning the legally permitted events. A few days later most of the MGJ Coalition was dissolved. Within another two weeks, only a few groups had maintained the same level of involvement that they had planned before September 11th happened, a few others had altered their involvement, and a number of new groups had entered the organizational field. Pre-MGJ events were either cancelled or significantly changed. The majority of MGJ events were cancelled, especially those involving public outdoor gatherings. Only one activity continued as planned. New events and previously planned but altered events focused on providing sympathy for those killed in the attacks, showing solidarity with those threatened by racist retaliation for the attacks, discussing the causes of terrorism, and promoting the nascent peace movement. The next section describes our research methods and data collection and clarifies the specific questions to be addressed in the remainder of this paper.

Research Methods, Questions, and Data

This research capitalizes on a "natural experiment," which provided the unique opportunity to examine how a specific coalition of social movement organizations was affected by, and responded to, the suddenly imposed transformation of the political climate after the terrorist attacks.[1] The coalition planning the protests, the Mobilization for Global Justice Coalition (MGJ Coalition), represented a diverse range of advocacy groups, churches, unions, and community organizations. The volatile atmosphere after September 11th reshaped the prevailing context of political opportunity within which the MGJ Coalition and its constituent organizations had oriented their actions (McAdam, 1996). In the span of a few hours each group involved, whether reformist or radical, risked losing legitimacy in the eyes of its core constituents. More mainstream groups, most of which had only recently joined the ranks of active, public MGJ participants, risked irreparable loss of credibility with policy makers and segments of the public who supported their positions on a range of non-MGJ issues, creating a powerful incentive to withdraw. Conversely, the most radical groups involved felt an equally strong incentive to forge ahead with their intended confrontational and disruptive actions lest their core supporters think they had been co-opted. The politically charged nature of deciding, in that situation, whether or not to continue with planned protests, divert resources to relief efforts, or engage in other displays of solidarity and symbolic support had become particularly acute for protest organizers (Vanderslice, 2001a; Grusky, 2001).

In this paper we focus on two broad questions facing these groups: should they remain involved, and if so, what would they do? Specifically, we consider changes in the roster of organizations involved and changes in the types events undertaken. Given their differences in organizational form and constituency the response of coalition members may vary significantly. The research addresses these questions:

(1) Changes in the Organizational Field—What groups were involved in the protest planning before September 11th? Which groups remained involved? Which groups dropped out? And which previously uninvolved groups joined in after the terrorists attacks?

(2) Changes in Events—What events were originally planned before September 11th? Which events were continued? What new events were organized after the attacks?

The organizations in this study had been identified during previous research as those actively planning this "episode of contention" against the World Bank and IMF (Gillham, 2003; McAdam et al., 2001). The episode, known as MGJ by its organizers, was scheduled for the week of September 23–30, 2001 and was to include a series of protest events culminating in a day of mass rallies, marches, and direct action. Eleven organizations officially sponsored the MGJ Coalition and numerous other organizations participated in working groups that were formed to carry out the goals of the MGJ. Together, these organizations formed a diverse coalition consisting of labor unions, religious organizations, environmental, human and civil rights groups, as well as anarchists and other loosely tied anti-capitalist grassroots organizations.

Data were collected through in-depth interviews, by observing strategy meetings and protest events, and by monitoring websites and e-mail list-servs. Interviews were conducted during the two weeks immediately after September 11th. Six leaders from four Washington, D.C., advocacy groups were interviewed for 30–60 minutes each. All respondents had been involved in planning for the MGJ: one was with the national AFL-CIO office; another with the AFL-CIO Washington, D.C., Metropolitan Labor Council (MLC); two represented the Jubilee USA Network; another was affiliated with Saint Stephens Episcopal Church; and the last directed the Washington, D.C., chapter of the American Friends Service Committee (AFSC).[2] During the same time period, the researchers observed three different decision-making meetings, an AFL-CIO "disaster relief blitz," two activist gatherings at the Anti-Capitalist Convergence Center, as well as numerous protest events (a teach-in, People's Summit, an interfaith prayer gathering, two legally sanctioned sets of rallies and marches, and one illegal march). Numerous documents were collected at the observation sites. Finally, electronic information was gathered from the websites and e-lists of advocacy organizations during the two weeks preceding and following the September 11th disaster.

From "The Battle in Seattle" to the Mobilization for Global Justice

In this section we set the stage for mass protests planned for September 23–30, 2001 in Washington, D.C. First, we provide background information about the global justice movement and its activities and the coalition of trade policy protesters converging on Washington, D.C., in September 2001. Second, we describe how the coalition of organizations changed after September 11th.

The organizations under analysis were involved in coordinating a week-long series of events challenging the policies of the World Bank and IMF.

The World Bank and IMF were meeting in Washington, D.C., for their annual fall convention, which typically brings together several thousand staff persons from around the world, as well as foreign and U.S. dignitaries. The protests were meant to coincide with the World Bank and IMF's annual conference rescheduled for September 29–30, 2001.

This was the third contentious episode held in the United States against institutions associated with the promotion of economic globalization and trade liberalization. The Battle in Seattle was the first, in which 50,000–100,000 activists successfully shut down the WTO meetings in Seattle during the fall of 1999 (Gillham and Marx, 2000). A second episode targeting the policies of the World Bank and IMF brought upwards of 20,000 protesters into the streets of Washington, D.C., in April 2000 (Santana and Fernandez, 2001).

In all episodes, activists singled out the WTO, IMF, and World Bank as the driving forces behind the global expansion of neo-liberal economic policies. Activists blamed these policies for everything from global environmental degradation and human rights violations, to the undermining of public education and health care, to the worsening of fiscal problems in developing nations, to the growth of the AIDS epidemic in Africa (Danaher, 1994; Danaher and Burbach, 2000; Jubilee USA Network, 2002; Mander and Goldsmith, 1996; MGJ, 2001a; Thomas, 2000; Welton and Wolf, 2001).[3] The fall 2001 meetings provided advocacy organizations yet another opportunity to express their dissent before an international audience and the IMF and World Bank elites themselves.

Some of the groups planning this episode of protest had been actively involved in the global justice movement since the 1999 Battle in Seattle brought wider public attention to policies of the WTO, World Bank, and IMF. Each round of recent protests against the global trade and international lending institutions had been planned, in part, by a coalition using similar organizational structures, though a shifting membership. The first episode involved a broad network of coalitions that shared information with each other, but also carried out distinctly separate plans. Thus, in Seattle for example, mainstream environmental groups, the religious community, and organized labor worked together to organize a number of marches and rallies. Each of these groups also worked with Public Citizen, the advocacy organization started by Ralph Nader, to organize teach-ins, marches, and rallies. The Direct Action Network (DAN), a more radical group credited for orchestrating the civil disobedience that shut down the first day of WTO trade talks, also coordinated activities with Public Citizen, although mainstream labor and environmental organizations distanced themselves from the DAN activists. The overlapping, yet often discrete, threads of the Seattle coalitions suggest a latticed networking of groups that shared overarching critiques of neo-liberal economic policies, but often disagreed about appropriate strategies for social change and protest tactics.

As in Seattle, each subsequent episode of contention in the emerging global justice movement was collaboratively organized by a loosely knit and diverse array of social movement organizations. Many of the organizations were involved at some level in the original Seattle protests. As the movement evolved, the network shrank and increased in density. This occurred as the number of these coordinating coalitions decreased while their organizational make-up broadened to include more established and mainstream groups, while at the same time pushing to the margins those who would not renounce the use of vandalism.

Thus, in the earlier protests, groups like the AFL-CIO and the Sierra Club had kept a degree of distance from organizations involved in the direct action. While the larger and more mainstream organizations continued to distance themselves from confrontational tactics, the 2001 MGJ was the first episode where prominent national social movement and public interest advocacy organizations had agreed to work in a more formal manner with the medium-sized national and international advocacy organizations, and smaller local groups. Groups entered the coalition knowing that some MGJ members planned to use non-violent direct action, but assumed that violence and vandalism would not be directly associated with the mobilization.

Terrorist Attacks Reconfigure the Global Justice Coalition

Mobilization for Global Justice Planners before September 11th

The MGJ Coalition had formed, in large part, around the task of planning and coordinating the series of inter-related MGJ protest events against the World Bank and IMF. The MGJ Coalition generally functioned "as a host for all groups . . . organizing and planning actions" that week (MGJ, 2001b). Specifically, the MGJ had hoped to provide the infrastructure for the protests by setting up temporary housing, preparing legal and medical training, establishing a convergence training and welcoming space, and coordinating a series of events in which local and visiting activists could partake (MGJ, 2001b). The MGJ explicitly demanded that the World Bank and IMF (1) open all World Bank and IMF meetings to the media and public; (2) cancel all impoverished country debt to the World Bank and IMF using the institutions' own resources; (3) end all World Bank and IMF policies that hinder people's access to food, clean water, shelter, health care, education, and right to organize; and (4) stop all World Bank support for socially and environmentally destructive projects (MGJ, 2001c).

Twenty-nine U.S. organizations officially sponsored events that would occur during the contentious episode (Table 1, column 1). Some were national labor organizations like the AFL-CIO, while others were national human rights advocacy groups like 50 Years is Enough and Jubilee USA Network. There were also local advocacy organizations such as the

Table 1. Organizations involved in Washington, D.C., protests before and after September 11th.

INVOLVED BEFORE SEPT 11TH [1]	INVOLVED AFTER SEPT 11TH	DIFFERENCE [2]
Anti-Capitalist Convergence (ACC)	ACC	ACC
Essential Action	Essential Action	Essential Action
Global Exchange	Global Exchange	Global Exchange
Homes Not Jails	Homes Not Jails	Homes Not Jails
Religious Working Group[P-11]	Religious Working Group	Religious Working Group
Alliance for Responsible Trade		~~Alliance for Responsible Trade~~
AFL-CIO [P-11]		~~AFL-CIO~~
American Lands Alliance		~~American Lands Alliance~~
Center for Economic Justice		~~Center for Economic Justice~~
Coalition of Immokalee Workers		~~Coalition of Immokalee Workers~~
Feminist Majority [P-11]		~~Feminist Majority~~
Friends of the Earth [P-11]		~~Friends of the Earth~~
Institute for Public Policy		~~Institute for Public Policy~~
International Rivers Network		~~International Rivers Network~~
Jobs with Justice [P-11]		~~Jobs with Justice~~
Mexico Solidarity Network		~~Mexico Solidarity Network~~
Mobilization for Global Justice (MGJ) [P-11]		~~MGJ~~
Oxfam America [P-11]		~~Oxfam America~~
Rainbow Coalition [P-11]		~~Rainbow Coalition~~
Rainforest Action Network		~~Rainforest Action Network~~
Union of Needletrades, Industrial and Textile Employees (UNITE)		~~UNITE~~
United Students Against Sweatshops (USAS) [P-11]		~~USAS~~
World Bank Boycott		~~World Bank Boycott~~

50 Years Is Enough [P-11]	50 Years Is Enough	*50 Years Is Enough*
American Friends Service Committee (AFSC)	AFSC	*AFSC*
Jubilee USA Network [P-11]	Jubilee USA Network	*Jubilee USA Network*
Ruckus Society	Ruckus Society	*Ruckus Society*
	International Action Center (IAC) Washington Peace Center	**IAC** **WPC**
Interfaith Service 57 Groups (e.g., Bread for the World, Center of Concern, Church of Brethren, Friends of the Earth, 50 Years Is Enough)	Interfaith 57 + American Muslim Council, Fellowship of Reconciliation, Global Justice, Home for Peace & Justice, Insight Mediation Community of DC, Jubilee Colorado, Jubilee Oregon, National Coalition Building Institute, Salvadoran American National Network, School of International Service of American University, Shalom Center/EPH, Alliance for Jewish Renewal, Sikh Kharma	Interfaith 57 **+ American Muslim Council, Fellowship of Reconciliation, Global Justice, Home for Peace & Justice, Insight Mediation Community of DC, Jubilee Colorado, Jubilee Oregon, National Coalition Building Institute, Salvadoran American National Network, School of International Service of American U., Shalom Center/EPH, Alliance for Jewish Renewal, Sikh Kharma**

[1] P-11 superscript denotes one of 11 groups sponsoring some of the largest legal permitted MGJ events.
[2] Level of Involvement: Plain = Same; *Italics* = Decreased level of involvement; ~~Strike-Through~~ = Withdrawal; **Bold** = Newly involved after September 11.

People's Repo and the ACC. Many of the groups' memberships consist primarily of other organizations. Some are mixed, relying on both organizations and individuals for support. A few, such as the ACC, rely primarily on the support of individuals and only come together for specific events such as the protests in Washington, D.C. (ACC, 2001a, 2001b).[4] Most MGJ sponsors concentrate on national or international issues, while some groups focus on both. The ACC addresses local issues within the national and international context of global capitalism. The People's Repo is specifically interested in local issues and concentrates on returning under-utilized and abandoned housing to homeless people in Washington, D.C. (Homes Not Jails, 2001).

Planning for the MGJ began long before September 2001 and proceeded through an increasingly formalized and "on-site" (Washington-based) process as the time of the protests approached. On March16, 2001, the coalition issued its first call to action meant to recruit other organizations to join in planning (Essential Action, 2001a). The call originated on the Essential Action e-list and quickly was disseminated to other advocacy lists around the United States and the world.[5] While it is unclear how many people and organizations received the original call, given the exponential nature of e-mail forwarding, it was likely in the several thousands. Planning for the protests occurred primarily in the Washington, D.C., metropolitan. During the early MGJ meetings a number of working groups were formed in order to facilitate organizing the protest.[6]

Spokes-council and Affinity Groups

As in previous global protests, the MGJ Coalition made broader decisions using a spokes-council/affinity group structure and used working groups for accomplishing tasks. The spokes-council provides a forum in which a facilitator manages discussions among affinity groups, aimed at reaching consensus on issues at hand. Each affinity group consists "of 5-20 people who work together [yet autonomously from the larger group] on direct action or other projects" (MGJ, 2001e). This decision-making model is organic in nature and challenges the hierarchy of top-down decision making (MGJ, 2001e). Each local affinity group and MGJ sponsor sent a representative or a "spoke" to a weekly spokes-council meeting. While these meetings were open and attended by many different people, only spokes could speak during the session. The spokes-council engaged in two types of decision making. Agenda setting entailed making decisions about what would be covered during a specific meeting. Once the agenda had been established, the spokes-council dealt with more instrumental questions about how to proceed with the mobilization. The council then delegated responsibility for developing such plans to working groups, which became affinity groups to the larger spokes-council. Each working group therefore sent a representative to report back and make decisions during the spokes-council meeting.

Working Groups

Immediately after the MGJ Coalition came together, 14 working groups were formed to pursue specific tasks. In contrast to the formally representative structure of the spokes-council, the working groups were open to anyone who wanted to participate and decisions were made on the basis of consensus among all participants. Moreover, not all organizations actively involved in working groups were official sponsors of the larger mobilization. Some dealt with generic responsibilities like media relations, fundraising, outreach, and logistics (MGJ, 2001f). Other tasks were more political in nature, including crafting the message and disseminating it through working groups focused on public education, website development, arts, and action. Still other working groups planned permitted rallies and marches, a concert for the largest permitted rally, training activities, and a civil disobedience scenario meant to close down the IMF and World Bank meetings. Some working groups had subgroups; for example, outreach was subdivided into student, labor, and neighborhood working groups (MGJ, 2001g).

The MGJ relied in part on list-servs for planning. Moderated and un-moderated MGJ discussion lists were started within the first month of planning, and un-moderated working group lists were formed whenever a specific group was started. Working groups also added a "members only" feature to their lists to help screen those who could join and review their e-mail archives.[7] Should a member prove unhelpful in the planning or be suspected of being an undercover police officer, the group could remove the person from the list. The lists allowed for running feedback across the MGJ Coalition and within working groups on the evolving plans for the September 2001 protests. They were also used to announce meeting times and locations, and to post breaking news such as changes in dates of the World Bank and IMF meetings. The un-moderated list provided an open forum for anyone on the planet with access to e-mail. Discussion there focused on a broader range of movement-related issues, provided general information regarding the planning and coordination of these efforts, and gathered feedback on the emerging plans from a far-flung constituency.

The permitted-scenario working group (P-11) organized a number of legal protest events throughout the week for which it had obtained official permits from local authorities. These events were to conclude with a large rally and march on September 30, 2001. Activists called this group the P-11 because it consisted of 11 sponsoring organizations including the AFL-CIO, Jubilee USA Network, 50 Years is Enough, and Jobs With Justice, which were only officially involved in organizing and engaging in legally permitted activities. Some of these organizations were also general sponsors of the MGJ. Besides the 11 sponsoring organizations, the P-11 working group consisted of other Washington, D.C., advocacy organizations, including prominent social movement organizations like the Sierra Club, Friends of the Earth, Bread for the World, and Oxfam America. These informal members did not formally endorse either the broader array of MGJ events or the more limited list of legally permitted events organized by the P-11 working group, in part perhaps because of concerns that some MGJ events would include illegal and confrontational actions. Most organizations in the P-11 had cooperated together for years on legislative advocacy issues of mutual interest (Cavanaugh, 2001). Meeting behind closed doors at the AFL-CIO headquarters, these groups shared their plans and concerns about the larger mobilization. Decisions about permitted events and concerns were then reported back at MGJ spokes-council meetings. Other working groups operated similarly, though the P-11 group is the only one that consisted primarily of large national and international social movement organizations.

The specific form of the MGJ and the exact roster of events it would encompass emerged gradually out of this back-and-forth process between the spokes-council and the various working groups, and through input from the larger constituency of advocacy groups and individuals accessing the e-lists from around the world. The variety of events created the potential for conflict within the coalition. To allow for internal disagreement, MGJ Coalition members could agree not to sign on to events that they disliked. For example, while the AFL-CIO was a central sponsor for the permitted mass rally and march scheduled for September 30, it did not sign on for the non-violent direct action scheduled for the same day.

Organizational Attrition and New Arrivals

Like everyone else in Washington, D.C., activists were significantly affected by the attacks of September 11th. They witnessed the nonstop news footage of the destruction that occurred at the Pentagon, the World Trade Center, and in Pennsylvania. Some advocacy groups had constituencies directly affected by the attacks, such as the AFL-CIO, which reported that 634 union members were killed (AFL-CIO, 2002). Moreover, many Washington, D.C., advocacy groups were affected by evacuations that occurred on the day of the attacks. For example, members of the Jubilee USA Network were evacuated from their office situated three blocks east of the Capitol building (Vanderslice, 2001a). Others experienced sorrow and anger not only because of the attacks, but also because they felt that the attacks could have been avoided had politicians heeded concerns raised by advocacy organizations over the years (Hoover, 2001). The grief and anger extended to concerns about an imminent war and further loss of life (Vanderslice, 2001a; Hoover, 2001).

P-11 Withdrawal

One way that the MGJ changed after September 11th involved the withdrawal of central organizations. Organizations involved in the permitted events withdrew first. Of these, the AFL-CIO was the first to publicly announce its removal of support. According to Mike Cavanaugh, Deputy Director in the AFL-CIO's Department of Field Mobilization, the decision to withdraw was made during an emergency meeting with AFL-CIO national staff on September 12.[8]

The next day, during the P-11 working group's weekly meeting, the AFL-CIO told those present that it would redirect all material and human resources to disaster relief (Cavanaugh, 2001). At the same meeting representatives from other national and international organizations also voiced that they were considering withdrawing from the protests (Cavanaugh, 2001). Some groups, like Jubilee USA Network, wanted to wait until after an emergency spokes-meeting of the MGJ Coalition (scheduled for September 15) before deciding (Vanderslice, 2001b). It was agreed not to publicize decisions to withdraw until Monday, September 17. This delay would allow both the MGJ Coalition to hold its emergency spokes-council meeting, and for P-11 organizations to make final decisions and write press releases. All organizations officially withdrawing from the broader MGJ and specific P-11 sponsored events could then release the news on the same day in a show of cooperation and solidarity.

The next day (September 14), however, AFL-CIO President John Sweeny released a press statement officially withdrawing the confederation of unions from the protests (AFL-CIO, 2001c). Other national and international social movement organizations, like Friends of the Earth, Bread for the World, Oxfam, the Sierra Club, and Greenpeace, quickly followed, withdrawing from the protests before the spokes-council meeting could be held. Some organizations informally involved in the P-11 did not officially withdraw because they were never officially involved. Instead, these organizations either dropped out quietly, or released press statements expressing sympathy for those who lost family and friends in the attacks. Rather than follow suit, Jubilee USA Network and Essential Action sent representatives to the September 15 MGJ emergency meeting.

MGJ Coalition Disbands and Peace Organizations Step Forward

An emergency spokes-council meeting was quickly arranged using the pre-existing moderated MGJ and working group list-servs that had been used to coordinate planning throughout the mobilization. At first it was scheduled to occur at All Souls Unitarian Church in the Columbia Heights neighborhood, but was relocated to a large room at St. Aloysius Catholic Church, near Union Station. A temporary and quickly scrawled sign placed on the door at All Souls, which directed people to the new location, revealed the urgency of the meeting.

Approximately 150 people attended the three-hour meeting, though members of the ACC, and many of the established labor and environmental groups involved in the P-11 working group were noticeably absent. An ad hoc committee had developed a tentative agenda focused primarily on whether and how to continue with the MGJ. A long-time Washington, D.C., activist and director of the local AFSC and member of the Washington Peace Center (WPC) facilitated the meeting. Each agenda item was briefly outlined and amended in the larger group. Then, those in attendance separated into their original working groups (e.g., media and un-permitted scenario) for a 15-minute "breakout" session to discuss the agenda items.

By the end of the evening it was decided that MGJ Coalition would withdraw support from the permitted and non-permitted demonstrations scheduled for September 30, but would continue to endorse the People's Summit and September 29 Prayer Vigil. The general agreement among those at the meeting was that the MGJ could neither continue as planned nor could the coalition formally endorse the peace marches that were now being scheduled by the WPC, the International Action Committee (IAC), and the ACC. This was because organizations and people from around the world who had sponsored the MGJ (through money donations and endorsements) could not be contacted quickly enough to confirm such a shift in focus. Nevertheless, several of the MGJ organizations, many with roots in the peace movement, agreed to informally encourage local and visiting participants to attend a peace march being sponsored by the WPC on September 30. At the same time, most organizations decided not to support the two September 29 anti-war marches independently organized by the ACC and IAC.

Support for the ACC was rejected because the group was seen as too confrontational for the post-September 11th political climate. The IAC march was rejected because the MGJ Coalition considered the IAC to be both too autocratic and undemocratic in its decision making and too controversial in its goals (Vanderslice, 2001a). Groups involved in previous demonstrations with the IAC had noted its inclination to make unilateral decisions without consulting other coalition members. Moreover, there was great concern that the IAC had taken a pro-Saddam Hussein position during the Gulf War, when it engaged in overly controversial actions like featuring a speaker favorable to Saddam Hussein's regime at a Gulf War protest in 1991. The vast majority within the MGJ Coalition and the broader global justice movement opposed such IAC positions vigorously, as did the broader peace movement during the Gulf War.[9]

The final decision of the night was that the media working group would release a statement by Monday, September 17, cancelling the mobilization. After the spokes-meeting, organizers expressed disappointment that the MGJ events they had worked so hard on would be cancelled or altered. This disappointment was evident in that most of the 150 participants quickly left the building when the meeting recessed and only a handful of activists joined together on the following day to forge the press release. On Sunday,

September 16, a day ahead of schedule, the MGJ media working group released a terse press statement officially cancelling the protests (MGJ, 2001h).

In all, most organizations involved in the MGJ Coalition withdrew their support. The major advocacy groups in the P-11 were the first to withdraw. Organizations less integrated into Washington-based policy advocacy withdrew some of their support, though a few stayed in for the People's Summit, Globalization Teach-in, and Prayer Meeting. The most radical groups (ACC and Homes Not Jails) or those with their own foundational support (such as Essential Action or Global Exchange) stayed involved in the same capacity, and two new groups joined in after the attacks (see Table 1, columns 2 and 3). Organizations like Essential Action and Global Exchange are not radical and thus did not remain involved to prove their credibility to a small, radical constituency. Rather, these groups could remain involved for at least two reasons. First, they have rather secure sources of funding. Essential Action's resources come primarily, if not exclusively, from the Public Citizen Foundation, while Global Exchange raises the largest portion of its revenue from proceeds on its own eco- and other-issue educational travel tours and the sale of merchandise. In addition, Global Exchange had received a substantial donation from a patron a year earlier. Second, since both organizations have strong ties to the Green Party, neither would have felt pressure to drop out of the MGJ in order to maintain ties or legitimacy with Democratic politicians. Public Citizen founder Ralph Nader ran for President, and Global Exchange Co-Director Medea Benjamin ran for the U.S. Senate in California in 2000 as Green Party nominees critical of the Democratic Party.

Not only did the groups that participated shift substantially in the immediate aftermath of September 11th, but the roster of events changed as well (Table 2). Clearly, with the MGJ Coalition disbanding and withdrawing its official sponsorship from all previously scheduled events, the roster was profoundly altered in response to the terrorist attacks.

What Was Being Planned and What Actually Happened

In this section we shift our attention from the changing cast of organizations to the roster of events originally planned and how they were changed in response to the terrorist attacks. First we describe the protest events initially planned to occur during the week of September 23–30, 2001, then discuss how this roster of events was changed in response to September 11th, and finally describe the events that were actually undertaken.

Events Planned before September 11th

The roster of MGJ events planned for the week of contention included activist training in civil disobedience, art and puppet making and first aid; legal

Table 2: Time-table of events planned before the attacks, occurring after, and changed between September 11 and October 10, 2001.

DATE	EVENTS PLANNED	EVENTS OCCURRED	DIFFERENCE [1]
9/11	P-11 Press Conference		~~P-11 Press Conference~~
9/11		MGJ Emergency "Spokes" Meeting	**MGJ Emergency "Spokes" Meeting**
9/12		1st of Daily Peace Vigils	**1st of Daily Peace Vigils**
9/12		AFL-CIO Emergency Meeting	**AFL-CIO Emergency Meeting**
9/13		AFL-CIO Memorial Service	**AFL-CIO Memorial Service**
9/13	P-11 Weekly Meeting	P-11 Weekly Meeting	P-11 Weekly Meeting
9/14		AFL-CIO Withdrawal from MGJ & Toleration Press Release	**AFL-CIO Withdrawal from MGJ & Toleration Press Release**
9/14		Silent Processional through DC streets	**Silent Processional through DC streets**
9/14–9/18	Ruckus Action Camp		~~Ruckus Action Camp~~
9/15–9/16	AFL-CIO/MGJ Protest Blitz		~~AFL-CIO/MGJ Protest Blitz~~
		AFL-CIO Relief Blitz	*AFL-CIO Relief Blitz*
9/15	MGJ Weekly "Spokes" Meeting	MGJ Emergency "Spokes" Meeting	**MGJ Emergency "Spokes" Meeting**
9/16		MGJ Withdrawal Press Release	**MGJ Withdrawal Press Release**
9/24		Rally for Restraint	**Rally for Restraint**
9/24–10/2	MGJ Welcome & Convergence Center		~~MGJ Welcome & Convergence Center~~

9/24–10/1	Anti-Capitalist Convergence Welcome & Convergence Center	ACC Welcome & Convergence Center	ACC Welcome & Convergence Center
9/25	Immigrant Rights Rally		~~Immigrant Rights Rally~~
9/25	Taco Bell Protest		~~Taco Bell Protest~~
9/26	Citi-Corp Protest		~~Citi-Corp Protest~~
9/27		Local People's March for Peace Announced by AFSC & WPC	*Local People's March for Peace Announced by AFSC & WPC*
9/27–9/28	People's Summit	People's Summit	*People's Summit*
9/27–9/29	Ending Global Apartheid Teach-In	Ending Global Apartheid Teach-In	*Ending Global Apartheid Teach-In*
9/28	Sweatshop Retailers Action		~~Sweatshop Retailers Action~~
9/28	Clean Energy March		~~Clean Energy March~~
9/29	ACC Direct Action	ACC March Against Hate, Cap & War	***ACC March Against Hate, Cap & War***
9/29		International Action Center ANSWER Rally & March	**International Action Center ANSWER Rally & March**
9/29–9/30	Prayer Vigil, March & Intervention	Prayer Vigil	*Prayer Vigil*, ~~March & Intervention~~
9/30	MGJ P-11 Rally & March		~~MGJ P-11 Rally & March~~
9/30	MGJ Un-Permitted Direct Action		~~MGJ Un-Permitted Direct Action~~
9/30		Washington Peace Center – Local People's March for Peace	**Washington Peace Center – Local People's March for Peace**

[1] Plain text = An unchanged event ~~Strike-through~~ = An event that was cancelled
Italics = An event that was Changed **Bold** = An event added after September 11th

permitted public dissent events such as rallies outside of corporate businesses targeted for abusing labor rights and the environment; educational and networking events such as teach-ins and workshops; solidarity actions like an interfaith prayer vigil; and illegal public assemblies or "un-permitted" events meant to disrupt the World Bank and IMF meetings on September 30.[10] Two important activities planned in advance of the MGJ included a nonviolent civil disobedience training camp to prepare for the un-permitted action, and a joint AFL-CIO/MGJ Blitz, where small groups of union and MGJ activists would visit several thousand metropolitan Washington, D.C., union households, recruiting participants to attend the legal events for which demonstration permits had been secured from local authorities (see Table 2, column 1).

Pre-MGJ Activities

Two pre-MGJ events were of particular importance for training activists and recruiting participants. The Ruckus Society had organized a four-day "Global Justice Action Camp" to be held September 15–18 in Middlebury, Virginia, just outside Washington, D.C. (MGJ, 2001i). Cosponsoring the training were the Washington, D.C.- based advocacy groups Institute for Policy Studies and Jobs With Justice, along with the Bay Area's Global Exchange. The camp was directed towards activist leaders and was to include teach-ins on issues regarding the World Bank and IMF, as well as workshops for training in nonviolent direct action and in how to conduct high visibility tactics. The camp also wanted to advance networking between national and international activists (MGJ, 2001i). Much of the training would prepare activists to engage in more confrontational tactics during the MGJ week. Just as important, these newly trained activists would lead training at the MGJ Convergence center in the days before the mass civil disobedience.

Another important pre-MGJ activity was a joint AFL-CIO/MGJ Blitz in which union volunteers would team up with MGJ activists to visit 10,000 union member households. The Blitz, sponsored by the national office of the AFL-CIO and the Washington, D.C. Metropolitan Labor Council, was scheduled for the weekend of September 15–16. Organizers expected 200 union volunteers and activists to make home visits at which they would explain the upcoming MGJ events, outline the criticisms leveled against the World Bank and IMF, note reasons for the AFL-CIO's involvement, and identify ways that the rank and file could get involved (Cavanaugh, 2001).

MGJ Week

The MGJ week itself would include activist training, large spoke-council decision-making meetings, educational and networking activities, a prayer vigil, and numerous legal and illegal demonstrations.

Activities focused on training and democratic decision making were central. The MGJ had planned to open a convergence and welcome center from September 24 to October 1. It would be located in a warehouse within walking distance of the protests or near a metro line. The convergence space and welcome center would provide a place for activists to network, share meals, construct and store large props and puppets, share child care, and receive training in an assortment of skills. Training would focus on nonviolent civil disobedience, puppet making and street theater, talking with the media, dealing with legal and medical issues, and engaging in affinity group and spokes-council decision making. The affinity group and spokes-council training was important since the MGJ spokes-council would now shift its focus from building infrastructure for the broader protest episode (including teach-ins and permitted events), to carrying out the mass non-violent direct action on September 30. The spokes-council would therefore need to integrate hundreds of additional affinity groups coming from out of town. The better trained these groups were in consensus decision making, the more efficiently the potentially cumbersome spokes-council meetings would function. The ACC planned to mirror the MGJ convergence, and would concentrate on welcoming and training a relatively small but zealous contingent of anarchists coming for the protests.

Educating, networking, and strategizing—The MGJ also emphasized education, networking, and strategy events. Educational events included teach-ins that critiqued IMF and World Bank policies, while networking and strategizing occurred through interactive workshops.

The Ending Global Apartheid Teach-in was the largest educational event planned. It was scheduled to fall right before the weekend of the mass demonstrations, with the intent to educate the large number of activists who would be arriving from around the globe. Sponsored by 50 Years is Enough, Essential Action, International Rivers Network, and Jubilee USA Network, the teach-in would include over 20 speakers from around the world (Essential Action, 2001b). Some of the sponsoring organizations provided transportation and lodging for the international speakers. Teach-in topics included "Democratizing Development" and "Corporate Globalization and Indigenous Rights" (Essential Action, 2001b). Since no single building secured by organizers was large enough to accommodate all the people they hoped to draw, teach-ins were dispersed among different churches throughout the Columbia Heights neighborhood (Essential Action, 2001b). Inside each church, literature tables would be set up, both to help organizations allied with the global justice movement recruit new members and to encourage networking among activists.

The People's Summit combined education, networking, and strategizing into a single event where cadre activists could join with activist leaders and experts from around the world in more intimate workshop settings. The summit, which overlapped with the Ending Global Apartheid teach-ins, was scheduled for September 27–28 at the Luther Place Memorial Church. Many of the international leaders would be shuttled from the global apartheid gathering in time to participate in the workshops. Each workshop would include at least one speaker to lead discussions on issues such as "Dialogue on Diversifying the Movement" and "Privatization: Making the Local-Global Connection" (DC Indymedia, 2001; Essential Action, 2002).

Solidarity and action—In order to encourage solidarity and provide the faith community a unique means to educate themselves and express their concerns about the IMF and World Bank, the Religious Working Group on the IMF and World Bank and Jubilee USA Network scheduled an interfaith prayer vigil.

Jubilee USA Network leaders and the Religious Working Group had organized the prayer vigil for Saturday, September 29 to kick off the mass demonstrations.[11] The vigil was to begin at 7 p.m. and last until early the next morning. It would include times of singing and both public and contemplative prayer. These activities would be structured around the testimonies of visitors from developing nations describing the negative effects of international debt, and the policies and programs of IMF and World Bank on the lives of those in their countries. Many of these same speakers would speak during teach-ins and workshops earlier that week. In the early morning, vigil participants would then march to the un-permitted direct action, which would be taking shape outside the World Bank and IMF buildings in downtown Washington, D.C.

As noted by one organizer (Vanderslice, 2001b) the decision that vigil attendees would march to the direct action was based on concerns voiced during earlier MGJ spokes-meetings that violence might erupt between police and protesters. Some of the anarchists had suggested they would show less restraint than they had during the April 2000 protests, and police were worried that vandalism associated with other global protests would be repeated in Washington, D.C. (Santana and Fernandez, 2001).[12] In response to these concerns, organizers of the prayer vigil hoped that their presence would defuse possible confrontations between police and direct action activists.

Permitted public dissent—The P-11 working group and other organizations organized a number of permitted activities dedicated to expressing public dissent. These events included an immigrants' rights rally as well as rallies outside branches of multinational fast food restaurants, banks, and retail stores.[13] In planning these events the P-11 cooperated with local authorities by using the institutionalized protest permitting process or "Public Order Management Systems" (POMS) that has grown up in Washington, D.C., since the early 1970s (see McCarthy, et al., 1999).[14] This is a marked contrast to the policy of "open secrecy" used by other groups to plan large-scale acts of civil disobedience discussed below.

Opening the week on September 24 was to be an immigrants' rights rally to be held outside of the Capitol. It was organized by the Mexico Solidarity Network and hoped to bring attention to the poor working conditions of immigrant workers in the United States, as well as IMF and World Bank policies that forced their migration in the first place. In another permitted event scheduled for September 25th, the Coalition of Immokalee Workers had planned to hold a rally outside a Taco Bell restaurant demanding that the food chain ensure that their contractors pay living wages to North American tomato pickers. Until the demands were met, the organizers would encourage customers to join a national boycott against the giant fast food chain (CorpWatch, 2001; Coalition of Immokalee Workers, 2002).

Labor activists from the Union of Needletrades, Industrial and Textile Employees (UNITE), United Students Against Sweatshops, and Behindthelabel.org planned another protest. They would hold a march and short rallies outside of downtown Washington retail stores selling clothes made by people believed to be working under sweatshop conditions.

Environmental concerns were also to be addressed through legal public dissent. In one case, the Rainforest Action Network of San Francisco, and the American Land's Alliance of Washington, D.C., had planned a demonstration for September 26 outside a branch office of Citi-Corp. The advocacy groups would demand that the corporation stop funding logging projects in old growth forests (World Bank Bond Boycott, 2001). Likewise, the Institute for Policy Studies organized a clean air march for September 28, demanding that the World Bank promote public transportation and renewable energy projects in developing nations.

The week was to conclude with a large permitted MGJ rally and march scheduled by the P-11 working group for September30th at the Ellipse, immediately south of the White House. Organizers, including the AFL-CIO, hoped to draw tens of thousands of participants to listen to activist speakers and musicians passionately describe the international ills generated by the World Bank and IMF. After the mass rally, protesters would then march through a corridor of union marshals along a preplanned route, past the fence and barricades surrounding the World Bank and IMF, and then back to the Ellipse. Using the POMS, police and organizers worked out details in advance so that marshals and police would keep the permitted march from combining with the civil disobedience occurring with the other major MGJ event.[15]

Un-permitted public dissent—By contrast, a working group not directly connected to larger organizations planned the un-permitted event. The direct action scheduled for September 30 hoped to force the cancellation of the World Bank and IMF meetings. Based on a belief that the policies of the World Bank and IMF were morally indefensible, participants in the direct action would engage in intentional lawbreaking to shut the meetings down. Affinity groups from around the world would take on the task of shutting down or delaying the meetings. In order to coordinate the mass direct action and provide participating affinity groups an idea of what to expect from each other, spokes-council meetings would be held at the MGJ Convergence and Welcoming Center the four nights before the mass civil disobedience.

Organizers for the un-permitted mass direct action refused to cooperate with police in pre-protest planning and relied on a structure of "open-secrecy" in their planning. The mass direct action was intentionally organized without input from or cooperation with police. While all spokes-meetings were open to the public both before and during the MGJ week, only general plans were discussed regarding the un-permitted scenario, such as the plan to have affinity groups blockade intersections around the IMF and World Bank meetings. Final planning on how these blockades would occur was made within affinity groups among people who were well acquainted with each other, minimizing what police could actually know in advance. This reliance on structured secrecy allowed affinity groups to cooperate in general terms while also maintaining an element of surprise on the day of the direct action, and preventing police from undermining the plans through infiltration and co-optation.

While all the plans were not settled as of September 11th, the direct action would have likely included activities similar to demonstrations previously held in Seattle and Washington, D.C. If so, large numbers of affinity groups, some using mechanical devices such as lock boxes, would blockade intersections to prevent IMF and World Bank delegates from entering the World Bank Headquarters where the meetings were scheduled to occur. Some activists were planning to participate under the ACC banner and intended to engage in even greater confrontational actions.

The Reconfigured Roster of Events

Because the composition of groups involved in the mobilization changed significantly, events were also affected. While the MGJ itself was cancelled, a few events continued, others were changed, and a few newly added.

Pre-MGJ Events

The pre-MGJ events were affected by travel delays and occupational solidarity with those affected by the attacks. Furthermore, new events arose out of the grassroots Washington, D.C., peace community. After September 11th, all commercial airlines were grounded for several days. As a result, the three-day Ruckus Society training camp scheduled for September 15–18 was scrapped because the San Francisco-based trainers could not travel to Washington, D.C., in time to set up their camp (Ruckus, 2002).

The AFL-CIO Blitz dramatically changed its focus to disaster relief. Since many of those injured and killed in the attacks were union members, the AFL-

CIO felt an obligation to assist with mutual aid. Relying on the infrastructure of the already-scheduled Blitz, organizers changed the focus from disseminating information about the MGJ to recruiting unionists to give blood, donate money for mutual aid funds, and to gather contact information for volunteers should another disaster occur. The MGJ activists who had planned to team up with the 200 union volunteers withdrew, perhaps because their wishes to recruit unionists for anti-war protests were met with little support from AFL-CIO organizers (Miller, 2001).

Peace and solidarity activities occurred immediately after September 11th as some organizations within the MGJ returned to their peace activism roots. The two major activities included the organizing of daily silent vigils at Dupont Circle, and a 1,000-person candlelight peace and solidarity march through Washington, D.C., neighborhoods the Friday after the attacks (Hoover, 2001).

MGJ and Anti-War Events

Mobilization for Global Justice training and decision-making events were cancelled, while the educational and networking events and the prayer service continued, with significant modifications. All MGJ permitted and un-permitted protest events were also cancelled. Many of the main MGJ permitted events were replaced with anti-war events. In some cases permits granted for a cancelled MGJ event were transferred to the organizers of a new peace event.

Training and Decision Making

Since it was decided at the emergency spokes-meeting to cancel all activist training and decision-making events, there was no reason to open the MGJ Convergence and Welcome Center. On the other hand, the smaller ACC opened its convergence center as originally planned. In the two weeks after September 11th, ACC activists argued that the attacks could be traced to violence perpetuated by global capitalism, including policies advanced by the World Bank and IMF. While unable to find a space near downtown, the ACC rented a community center in the Columbia Heights neighborhood.[16] This site was in close proximately to a subway station, allowing relatively easy access between the ACC Convergence Center and the newly scheduled peace and anti-war protests downtown. The ACC continued as planned with some scaling back of training, perhaps due in part to fewer people coming to town for the new protests than had promised to come for the MGJ.

Educational and Networking Events

At the MGJ emergency spokes-meeting it was agreed that greater emphasis should be placed on educational events. Some organizers hoped that the tragedies of September 11th might open up discussions around the causes of such animosity towards the United States. Thus, no educational and networking events were cancelled, although all were substantially changed to reflect issues of causality behind the attacks, to oppose war in Afghanistan, and to note that economic policies of the World Bank and IMF kill people just as efficiently as do terrorist attacks.

The WPC had joined as a central organizer for the Ending Global Apartheid Teach-ins. Its involvement marked a shift towards rooting the causes of the attacks in U.S. militarism and growing concern among activists about a possible war in Afghanistan. The teach-in continued although sessions were added addressing these concerns. Progressive authors like Noam Chomsky and Howard Zinn would lead discussions for the new sessions. U.S. and international speakers previously scheduled for the teach-in redirected their comments towards addressing causes of terrorism and in noting how their countries had experienced economic terrorism at the hands of the World Bank and IMF.

The People's Summit grew in significance as well and focused on describing the connections between the World Bank and IMF violence and terrorism. For example, a workshop titled "Anti-Arab Racism and Anti-Semitism in Europe and the US," was added to the schedule (MGJ, 2001k). Another workshop called "Where do we go from here?" addressed concerns among activists that the movement would be significantly reshaped by the events of September 11th.

Interfaith Service

The prayer vigil and interfaith service continued with some changes. It was decided to limit the service to the evening rather than an all-night vigil, to cancel the public march and peacekeeping, and to show solidarity with people of Middle-Eastern descent by denouncing reactionary racist attacks occurring across the country. The prayer service focused on articulating the linkages between U.S. foreign policy and international poverty. Approximately 400 people attended the service, which began with a local choir singing hymns and peace songs. This was followed by a processional and candle-lighting ritual inside the church. Later, several citizens from developing nations spoke about the ways that economic globalization promoted by U.S., World Bank, and IMF policies contributed to the violence of poverty and war experienced each day in their countries. Organizers and speakers also contended that such policies benefit the interests of elites at the expense of the world's poor, creating desperation and animosity that gives rise to terrorism. Many of the speakers gently noted that the 4,000 deaths attributed to the terrorist attacks at the time, while tragic, were nothing new in nations from the global south. One South African spokeswoman noted that millions of Africans have died from AIDS after being denied access to lifesaving drugs. Her final analysis was that global capitalism favors large corporations through establishment of trade agreements meant to protect patents at the expense of human beings (Rowden, 2001).

Permitted and Un-permitted Events

Permitted Events

All MGJ permitted events were cancelled, while two new permitted anti-war events were scheduled for the September 29–30 weekend. Cancelled events included the immigrant rights rally, the clean air march, Taco-Bell and sweatshop product retailer actions, and the mass MGJ rally and march originally planned by the P-11 working group.

In contrast to the cancelled MGJ events, the IAC scheduled its permitted protest for Saturday, September 29 at Freedom Plaza several blocks east of the White House. It was called the Act Now to Stop War & End Racism (ANSWER) rally and march. The demonstrations focused dissent on the potential war against Afghanistan, which organizers labeled as a "racist war." The event began with a two-hour rally and concluded with a march up Pennsylvania Avenue to the Capitol. Speakers in the rally included an emergency medical technician from New York City who had lost friends in the collapse of the World Trade Center. Several thousand people participated in the march and rally. Many of the participants had come from cities on the East Coast.

Likewise, the WPC sponsored another new event, the "People's March for Peace," on September 30. Unlike the IAC protest from the day before, organizers relied on local networks and drew many Washington area activists. The People's March for Peace began at Malcolm X Park in Colombia Heights. After a rally emphasizing empathy for those affected by the attacks and suggesting the need to avoid further bloodshed by not going to war, protesters marched along a pre-arranged route through Dupont Circle and down Embassy Row. The march stopped briefly at Sheridan Circle to recognize and mourn violence against immigrants. The park was the site where a Kurdish man had been killed a year earlier, apparently because of his Middle-Eastern appearance. Marchers then retraced their steps back to Malcolm X Park in the late afternoon.

Un-permitted Events

No previously planned, un-permitted MGJ events occurred as planned. The ACC did, however, change their plans for direct action into a new anti-war rally featuring a flag burning followed by an illegal march to World Bank headquarters. At 9 a.m. September 29, approximately 1,000 activists, many dressed in black and wearing bandanas over their faces, converged in a park near Union Station. Nearly as many police dressed in full riot gear were waiting to meet them. After activists ceremoniously burned an American flag, police told the organizers to disband or risk arrest for unlawful assembly. After a short stand-off, police negotiating through intermediaries from the Lawyers Guild agreed to allow an un-permitted march to Edward R. Morrow Park just outside the World Bank. Protesters flanked by columns of body-armored police made the two-mile march together. Police vehicles led the unusual procession of black-clad youth, while cross streets at every intersection were blocked with police cars. This provided a backup should the crowd break through the corridor of police on foot and ensured that the march would continue moving towards the agreed-upon destination.

Scuffles between police and ACC activists slowed the march at times. Eventually, protesters arrived outside the World Bank, where they found hundreds of Metro Police, Secret Service, National Park, and other law enforcement officials surrounding a perimeter of steel barricades. Authorities forced the marchers into the large human corral where they detained them for over an hour. The crowd was eventually herded several blocks to Freedom Plaza where the IAC was holding its permitted rally. While there were rumors that members of the ACC and IAC harbored hostilities towards each other, there were no incidents at the rally or during the IAC march back towards the Capitol. In all, the crowd numbered somewhere near 10,000. By 4 p.m. the independently organized ACC and IAC events ended at the same park where the ACC march had begun earlier that morning. The crowd slowly dispersed as activists walked to the Union Station subway to board trains back to the ACC Welcome Center, home, or temporary lodging.

Discussion

The MGJ protests were significantly affected by the September 11th terrorist attacks. As noted, most of the original protest events were cancelled. This was in part due to the withdrawal of sponsorships and endorsements for the MGJ, including the withholding of important resources by core supporters. Strains were also created within the coalition, especially between groups that at times experience conflicting interests in other settings, such as unions and environmental groups. Moreover, tensions emerged around the more radical groups, like ACC, because of its decision to continue with the flag burning and an un-permitted march to the World Bank and IMF Headquarters on September 29. The events of September 11th also generated new activities and the re-distribution of resources to such activities as disaster relief, solidarity actions, and three additional protest events.

As discussed above, a number of organizations who had worked for months on the MGJ events quickly distanced themselves from the MGJ or dropped out entirely after September 11th. Organizations that withdrew from the planned protests most quickly or became notably less involved tended to be those whose primary or core goals were not directly related to those of the MGJ. For example, labor unions and environmental groups withdrew quickly, while groups like 50 Years Is Enough, World Bank Boycott, Essential Action, and ACC remained actively involved although in sometimes different capacities. The AFL-CIO and Sierra Club were core sponsors and organizers of MGJ before September 11th. Neither organization had been similarly involved in Seattle in 1999 or in subsequent global justice events (Gillham, 2003). For each organization, the MGJ represented a relatively recent extension of its issue repertoire and relatively new public affiliations with some P-11 organizers and sponsors. The MGJ represented new issues for both groups and their involvement had also been carefully crafted as a way to educate their large and somewhat diverse constituencies about globalization and its relationship to long-standing concerns of their core constituents. These goals could not have been accomplished in the post-September 11th environment. Continuing to try would have created internal confusion and controversy among their members.

Such groups also withdrew to preserve their credibility with the general public by avoiding de-legitimizing affiliations. The P-11 organizations had taken great care to clearly differentiate themselves and the permitted events they were sponsoring from more radical groups and un-permitted and controversial actions. In the immediate aftermath of September 11th it became quite apparent that such distinctions carefully and painstakingly crafted over months of planning would be completely lost on the press, the general public, and perhaps most importantly upon their own members. Without the attacks they could have successfully distanced themselves from the confrontational actions of the ACC. But because of September 11th, mainstream labor and environmental organizations would likely have been lumped together with the ACC and IAC, risking substantial loss of legitimacy with the general public.

Groups like the AFL-CIO and Sierra Club withdrew, in part, because the goals they had for MGJ involvement could no longer be accomplished, and to avoid de-legitimizing affiliations that would have undercut their public support and offended large segments of their members. More radical and confrontational groups like the ACC remained involved and continued with their planned actions for similar reasons. Their ability to accomplish their goal of disrupting the World Bank and IMF meetings and symbolically communicating their utter antipathy for global capitalist institutions could both still be accomplished in the post-attack situation. In fact, the post-September 11th environment may have been even more conducive. Moreover, following through with their planned actions offered an even greater opportunity to confirm their radical stance to their constituents and members. Small, radical groups like the ACC maintain legitimacy with their own members precisely by demonstrating their antipathy for mainstream economic, political, and cultural institutions. If the ACC had withdrawn after September 11th, it would have been widely criticized from within for "selling out" and many of its own members would have questioned the ACC's legitimacy and perhaps ended their affiliation with the group. Thus, confrontational anti-war and anti-capitalist groups were endeared to their narrow, radical constituencies by protesting in the immediate aftermath of the attacks. Such actions preserved their legitimacy just as withdrawal did for groups like the AFL-CIO and the Sierra Club.

Conclusion

The peace movement, whose initial re-emergence is captured above, has since gained momentum with over 120 rallies or demonstrations held nationwide on December 9, 2002, in conjunction with World Human Rights Day (Carr, 2003), and a series of mass anti-war actions in Washington, D.C., San Francisco, and other cities over the Martin Luther King Day holiday weekend in 2003. At the time of this writing, peace groups were coordinating "penny purchase" actions at gas stations around the country in conjunction with the upcoming State of the Union Address to demand that the U.S. not attack Iraq to gain control over its oil. Similarly, the global justice movement, while slowed, has remained active, planning for the next round of World Bank and IMF meetings. Moreover, advocates on a range of other issues continue to protest and demonstrate, despite continued national security concern and large-scale military deployments to the Middle East, as recent demonstrations on the anniversary of the *Roe v. Wade* decision by both pro-choice and pro-life advocates makes clear. The events chronicled above clearly altered one episode of contention in the ongoing global justice movement, yet the issues of concern persist and the movement continues to mobilize.

For decades disaster preparedness and response activities have been relatively non-politicized. Yet, on the eve of the formation of the cabinet-level Department of Homeland Security, it remains unclear how extensively the disaster community will become involved in preparing for and responding to the kinds of national security threats posed by terrorist attacks. What seems clearer is that in the future disaster preparedness and response may no longer enjoy relatively a-politicized operations. To the extent that disaster and emergency response agencies become associated with perceived curtailments of civil rights and government intrusions on privacy, they may well find that their policies and actions become the object of the sorts of public protest or political contention examined here.

Notes

1. One author had been researching the evolving coalitions behind a series of contentious anti-globalization protests since before their emergence to public attention in Seattle in 1999. His research into the series of protest events being planned and organized in Washington, D.C., in September 2001 was well underway when the terrorist attacks occurred on September 11th.

2. AFL-CIO is the largest federation of labor unions in the United States. During the last 10 years the federation has begun to reemphasize workplace organizing and political advocacy in response to economic globalization and global free trade agreements. Jubilee USA Network is a coalition of religious and non-religious organizations working for international debt cancellation. Saint Stephens is a Washington, D.C., congregation long known for its involvement in social justice causes since the Civil Rights movement. Activism on progressive causes by the AFSC reaches back to Abolitionist support for the Underground Railroad and has extended consistently through the 20th century.

3. The United States with its allies formed these institutions near the end of World War II. Each was to perform a specific task related to rebuilding war-torn Europe. The Bank was to lend capital for rebuilding national infrastructures. The IMF loaned money for restoring banks and improved fiscal management of European nations. The General Agreement of Trade and Tariffs (GATT), from which the WTO was later formed, provided oversight to encourage free trade across Western Europe. It was thought that the creation of these institutions would prevent future war because it was assumed that nations who traded together would not go to war against each other. After Europe was rebuilt, rather than be dismantled, the institutions redirected their focus to poverty reduction in developing nations, primarily through the provision of loans and the management of infrastructure projects like building dams and roadways.

4. The ACC is a network of anarchist activists formed primarily for the week of protests. It differs from a coalition in that those involved in the ACC regularly share information, without actually maintaining a formalized organizational structure. Instead, the ACC is a loose affiliation of individuals and smaller collectives from around the world. Most of those who endorsed the ACC were from the eastern United States. The ACC is linked electronically through websites and discussion lists with other anarchist organizations including those involved in protests in Seattle. While there are no formal leaders in the ACC, the individuals and collectives that create and maintain the network's web pages and list-servs are essential actors and appear to be leaders by default. There was much organizational overlap between the ACC and MGJ, though some observers wanted to make a distinction between the two. For some, the ACC represented revolution, whereas the MGJ worked for reform of the World Bank and IMF. According to the MGJ website, this distinction was artificial and promoted in part by police efforts to divide the movement. The MGJ and ACC intentionally played

down their differences to provide a unified front (MGJ, 2001d). For clarity, we treat the MGJ as the primary coalition organizing the protests and the ACC as a more radical and relatively small member of this coalition.

5. Essential Action, a Washington, D.C., clearinghouse for trade and international debt information, is part of Public Citizen, a Washington, D.C.-based public interest advocacy organization founded by Ralph Nader.

6. The ACC issued a similar call directed primarily towards radical activists in the Baltimore/Washington metropolitan area. They participated in the MGJ planning and established their own parallel spokes-council and working groups in order to conduct independent actions during the MGJ week (DC Indymedia, 2002; ACC, 2001c).

7. "Members only lists" require a moderator to add potential members to the list. This allows the moderator to screen potential members, adding an element of control over who may join the list. Whether or not such screening actually occurs or acts as a deterrent is unclear. Regardless, organizers claim that no "covert" discussion occurred over e-mail. This openness reflects a central value for many in the global justice movement and is the basis for their critique of the World Bank and IMF's closed meetings (MGJ, 2001c).

8. This meeting occurred immediately after a silent vigil held outside of the AFL-CIO headquarters for the families and friends of those killed in the attacks, which included 634 AFL-CIO members (Cavanaugh, 2001).

9. In 1991 during the peace movement's mobilization against the U.S. military buildup in the Persian Gulf, two "warring" coalitions formed to oppose U.S. intervention, culminating in rival "national" demonstrations on successive weekends in Washington, D.C. The IAC was a leading group in the much smaller of the two coalitions, which took a highly controversial pro-Saddam Hussein position. By contrast, the much larger and broader-based peace movement coalition refused to cooperate with the IAC and publicly denounced its position (Marullo and Edwards, 1997). We note that at the time of this writing (January 2003), this same split appears to be developing in the re-emergent peace movement as a larger and broader-based contingent of groups opposing a U.S. invasion of Iraq moves to disassociate itself from the more controversial positions of the IAC and ANSWER.

10. The strategy to interrupt the meetings was successfully used in Seattle in 1999, and to a lesser extent in Washington, D.C., in April 2000. In response, police planned to install a 9-foot-high fence around the World Bank, IMF, and White House (Fernandez, 2001).

11. Their decision to hold the vigil the night before the protests was based in part on the success of the Jubilee Northwest gathering the night before the protests

against the WTO in Seattle in 1999. Upwards of 10,000 people participated in that event.

12. The possibility for confrontation may have been further increased since police had decided to construct a security fence around the World Bank and IMF buildings. MGJ organizers believed the fence would create additional frustration and anger among some demonstrators regarding perceived violations of the First Amendment, and might also generate a siege mentality among police themselves (MGJ, 2001j).

13. This same strategy was used during the WTO protests in Seattle. Not only did activists target the WTO meetings, they also held public rallies outside of Weyerhaeuser and Monsanto offices demanding that the corporations end unsustainable logging practices and the development of genetically modified organisms, respectively. According to Pellow (2001), target expansion to non-government entities like corporations is a relatively new phenomenon.

14. Such "permitted events" are completely legal and usually entirely non-confrontational. Even when acts of civil disobedience are performed, their timing and location are often coordinated with law enforcement. Many argue that the POMS produces positive outcomes for all involved. Protesters are able to express their dissent, often for a media audience, without concern that police will end their show early. Police on the other hand, know what to expect from protesters and are thus able to manage protests in ways that reduce the risk of having to use force, or expending large amounts of money in police overtime and other expenses.

15. Police reports suggest that one of the main failures in Seattle resulted in not keeping permitted and un-permitted protesters apart (McCarthy and Associates, 2000).

16. This location was used the previous year by the MGJ after its original convergence zone was closed down in a pre-emptive police raid.

References

Anti-Capitalist Convergence. 2001a. "Donate to the ACC." http://www.abolishthebank.org/en/donate.html.

Anti-Capitalist Convergence. 2001b. "General Information Packet." http://www.abolishthebank.org.

Anti-Capitalist Convergence. 2001c. "Anti-Capitalist Convergence." http://www.abolishthebank.org/en/index.htm.

American Federation of Labor–Congress of Industrial Organizations. 2001a. "Broad Coalition to Release Joint Principles and Plans for Sept. 30 March and Rally for Global Justice." Press release September 7. Common Dreams News Center. http://www.commondreams.org/news2001/0907-07.htm.

American Federation of Labor–Congress of Industrial Organizations. 2001b. "Cancellation of Press Conference Today." Press release September 11. Common Dreams News Center. http://www.commondreams.org/news2001/0911-01.htm.

American Federation of Labor–Congress of Industrial Organizations. 2001c. "Statement by John J. Sweeney On September IMF/World Bank." Press release September 14. http://www.aflcio.org/publ/press2001/pr0914.htm.

American Federation of Labor–Congress of Industrial Organizations. 2002. "Sept 11—Remember, Reflect and Recommit." http://www.aflcio.org/sept_11/index.htm.

Carr, Jen. 2003. Interview with Jen Carr of Peace Action. Conducted January 26.

Cavanaugh, Mike. 2001. Interview with Mike Cavanaugh, AFL-CIO. Conducted September 16.

Coalition of Immokalee Workers. 2002. "Boycott the Bell." http://ciw-online.org/tz_site-revision/home/home.html.

CorpWatch. 2001. "S-25: Show Taco Bell What Democracy Looks Like!" Bulletin board post July 25. http://www.corpwatch.org/bulletins/PBD.jsp?articleid=410.

Danaher, Kevin, ed. 1994. *50 Years is Enough: The Case Against the World Bank and the International Monetary Fund.* Boston: South End Press.

Danaher, Kevin and Roger Burbach, eds. 2000. *Globalize This! The Battle Against the World Trade Organization and Corporate Rule.* Monroe, ME: Common Courage Press.

DC Indymedia. 2001. "Media Care Package Sept. 21—Oct. 1, 2001."

DC Indymedia. 2002. "Ocotber RACB Intiative [sic]: Revolutionaries: Organize and Converge on Washington DC!" Discussion list correspondence dated May 6, 2001. http://dc.indymedia.org/display.php3?article_id=8684&group=webcast.

Essential Action. 2001a. "Actions at IMF/WB Annual Meetings, Sept. 28—Oct. 4 2001—Washington." Discussion list correspondence dated March 16, 2001. http://lists.essential.org/pipermail/stop-imf/2001q1/000382.html.

Essential Action. 2001b. "Ending Global Apartheid Calendar." Ending Global Apartheid Organizing Packet distributed before September 11th.

Essential Action. 2002. "Tonight at 7PM—MGJ Education meeting!" http://lists.mutualaid.org/pipermail/mgj-discuss/2002-July/000804.html.

Fernandez, Manny. 2001. "Globalization's Diverse Foes: Wide Range of Protesters Unites Against IMF, World Bank." *Washington Post* (September 5): B01.

Fernandez, Manny and Paul Blustein. 2001. "IMF, World Bank Move Meeting Site." *Washington Post* (July 11): B01.

Gillham, Patrick F. 2003. "Social Movement Organization Structure and Action: Three Contentious Episodes in the Global Justice Movement." Ph.D. dissertation. Boulder, CO: University of Colorado.

Gillham, Patrick F. and Gary T. Marx. 2000. "Complexity and Irony in Policing and Protesting: The World Trade Organization in Seattle." *Social Justice* 27(2): 212–236.

Grusky, Sarah. 2001. Phone interview with Sarah Grusky. Conducted September 12.

Homes Not Jails. 2001. "Homes Not Jails DC." http://www.homesnotjails.org/dc/index.htm.

Hoover, Betty. 2001. Interview with Betty Hoover, AFSC, Washington, D.C.. Conducted September 18.

Jubilee USA Network. 2002. "Jubilee USA Network Front Page." http://www.jubileeusa.org.

Mander, Jerry and Edward Goldsmith, eds. 1996. *The Case Against the Global Economy: And for a Turn Toward the Local.* San Francisco: Sierra Club Books.

Marullo, Sam and Bob Edwards. 1997. "Persevering for Peace: Organizational Survival and the Transformation of the Peace Movement, 1988–1992." Pp. 181–202 in Thomas R. Rochon and David S. Meyer, eds., *Coalitions and Political Movements: The Lessons of the Nuclear Freeze.* Boulder and London: Lynne-Rienner Press.

McAdam, Doug. 1996. "Political Opportunities: Conceptual Origins, Current Problems, Future Directions." Pp. 23–40 in Doug McAdam, John D. McCarthy and Mayer N. Zald, eds., *Comparative Perspectives on Social Movements: Political Opportunities, Mobilizing Structures, and Cultural Framings.* New York: Cambridge University Press.

McAdam, Doug, Sidney Tarrow, and Charles Tilly. 2001. *Dynamic of Contention.* New York: Cambridge University Press.

McCarthy, John D., Clark McPhail, and John Crist. 1999. "The Diffusion and Adoption of Public Order Management Systems." Pp. 71–94 in Donatella della Porta, Hanspeter Kriesi, and Dieter Rucht, eds., *Social Movements in a Globalizing World.* New York: St. Martin's Press, Inc.

McCarthy, Ronald M., and Associates. 2000. "An Independent Review of the World Trade Organization Conference Disruptions in Seattle, Washington November 29—December 3, 1999." Final Report.

Mobilization for Global Justice Coalition. 2001a. "Call to Action!" http://www.september30.org/s30/news.cfm?ID=76.

Mobilization for Global Justice Coalition. 2001b. "Frequently Asked Questions: Who is the Mobilization for Global Justice?" http://www.september30.org/s30/knowledgebase.cfm?section=About.

Mobilization for Global Justice Coalition. 2001c. "Frequently Asked Questions: What Does the Mobilization for Global Justice Stand For?" http://www.september30.org/s30/knowledgebase.cfm?section=About.

Mobilization for Global Justice Coalition. 2001d. "Frequently Asked Questions: What's Your Relationship to the Anti-Capitalist Convergence?" http://www.september30.org/s30/knowledgebase.cfm?section=About.

Mobilization for Global Justice Coalition. 2001e. "Affinity Group Information and Resources." http://www.september30.org/s30/feature.cfm?ID=153.

Mobilization for Global Justice Coalition. 2001f. "Working Group Contacts." http://www.september30.org/s30/feature.cfm?ID=9.

Mobilization for Global Justice Coalition. 2001g. "Outreach Working Group." http://www.september30.org/s30/feature.cfm?ID=39.

Mobilization for Global Justice Coalition. 2001h. "Mobilization for Global Justice Cancels its Call for Street Demonstrations Against World Bank/IMF at End of September." Press release September 16. http://www.september30.org/s30/feature.cfm?ID=208.

Mobilization for Global Justice Coalition. 2001i. "Announcing Global Justice Action Camp! 9/14 – 9/18." http://www.september30.org/s30/news.cfm?ID=160.

Mobilization for Global Justice Coalition. 2001j. "8/17 Press Statement – Fence Represents Symbol of Repression and Division." Press release, August 17. http://www.september30.org/s30/news.cfm?ID=97.

Mobilization for Global Justice Coalition. 2001k. "People's Summit: To Globalize Justice and Peace." http://www.september30.org/s30/feature.cfm?ID=218.

Miller, Rachael. 2001. Interview with Rachael Miller of Metropolitan Washington, D.C. Labor Council. Conducted September 16.

Murray, Frank J. 2001. "INFAMY; Hijacked Planes Destroy World Trade Center, Ram Pentagon." *The Washington Post* (September 12): A1.

Pellow, David. 2001. "Environmental Justice and the Political Process: Movements, Corporations, and the State." *Sociological Quarterly* 42(1): 47–67.

Rowden, Rick. 2001. "Prayer Vigil video compiled by Rick Rowden."

Ruckus. 2002. Personal telephone correspondence with Nisha Anand of Ruckus Society. August 5.

Santana, Arthur and Paul Blustein. 2001. "IMF Trims DC Session To 2 Days; Police to Seek $38 Million in Federal Aid for Security." *The Washington Post* (August 11): A01.

Santana, Arthur and Manny Fernandez. 2001. "Police Await 100,000 Protesters." *Washington Post* (August 18): B01.

Thomas, Janet. 2000. *The Battle in Seattle: The Story Behind and Beyond the WTO Demonstrations.* Golden, CO: Fulcrum Publishing.

Vanderslice, Mara. 2001a. Interview with Mara Vanderslice of Jubilee USA Network. Conducted September 15.

Vanderslice, Mara. 2001b. Interview with Mara Vanderslice of Jubilee USA Network. Conducted September 29.

Welton, Neva and Linda Wolf. 2001. *Global Uprising: Confronting the Tyrannies of the 21st Century.* Gabriola Island, BC: New Society Publishers.

World Bank Bond Boycott. 2001. "RALLY AGAINST CITIGROUP: The World's Largest and Most Destructive Financial Institution." E-list announcement made September 10. bank-boycott@yahoogroups.com.

Acknowledgments

We thank two anonymous reviewers for helpful comments on an earlier draft of this paper. We also thank the University of Colorado's Natural Hazards Center and the National Science Foundation for support through the Quick Response grant program.

Correspondence should be directed to Patrick Gillham, Department of Sociology, 327 UCB, University of Colorado, Boulder, CO 80309-0327; e-mail: Gillham@colorado.edu.

Glossary of Organizational Acronyms and Names

Acronym	Name of Organization
ACC	Anti-Capitalist Convergence
AFL-CIO	American Federation of Labor–Congress of Industrial Organizations
AFSC	American Friends Service Committee
ANSWER	Act Now to Stop War & End Racism march emerged after September 11th attacks and sponsored by IAC.
DAN	Direct Action Network, the activists responsible for organizing much of the direct action that paralyzed Seattle during the WTO protests. Planning for the September 2001 protests was modeled on the success of DAN.
Global Justice Movement	Refers to a broad, international social movement opposed to neo-liberal trade policies and their consequences
IAC	The International Action Committee organized the ANSWER Rally and March after September 11th attacks.
IMF	International Monetary Fund
MGJ	Mobilization for Global Justice was one contentious episode within the broader global justice movement, which was originally scheduled for September 23—October 4, 2001, in Washington, D.C.
MGJ Coalition	Mobilization for Global Justice Coalition was a diverse coalition of social movement organizations that had formed temporarily to sponsor and plan the MGJ.
MGJ Coalition Spokes-Council	A democratic, deliberative body comprising working groups and formal organizations involved in the MGJ Coalition. Meetings are open to all, but speaking and decision making are limited to members.
MLC	Metropolitan Labor Council is a Washington, D.C., chapter of AFL-CIO
P-11	"Permitted Eleven" refers to 11 organizations that had received demonstration permits for a series of legal MGJ related events.

POMS	Public Order Management Systems
SMO	A social movement organization is any named group or organization that pursues the social change goals of a social movement. SMOs include conventional issue advocacy organizations, community-based groups, church-related organizations, as well as radical groups using confrontational, direct action tactics.
UNITE	Union of Needletrades, Industrial and Textile Employees
WPC	Washington Peace Center
WTO	World Trade Organization

Representation of the September 11th Terrorist Attacks in the Online Edition of the *Los Angeles Times*

Christine M. Rodrigue
Department of Geography
California State University, Long Beach

Introduction

This paper evaluates the representation of the terrorist attacks on the World Trade Center and the Pentagon on September 11, 2001, by one prominent daily American newspaper, the *Los Angeles Times,* during the first 12 weeks after the disaster, using literature content analytic methods. These attacks were unprecedented experiences for Americans in terms of their scale, the geographical reach and extreme venom of a non-state enemy, and the deliberate targeting of non-combatant civilians. Since 1900, Americans have experienced incidents sharing one or two attributes with the September 11th attacks: several wars overseas with huge casualties, a stunning attack at Pearl Harbor by a state-based enemy nation, natural disasters that have killed thousands of people in one incident (e.g., the Galveston Hurricane of 1900) or destroyed tens of billions of dollars in property and economic activity at once (e.g., the Northridge earthquake of 1994 and Hurricane Andrew in 1992), and the daily backdrop of the unusually high violent crime rate in the United States. There is, thus, a body of media analytic literature on which to draw for guidance on evaluating media coverage of an incident that uniquely combines disaster, war, and crime.

Review of Media-analytic Literature

Media affect the individual perception and social understanding of hazards, crimes, and wars, as well as nearly every other subject, in which public perceptions affect political decision making. Their shaping of Americans' understanding of and responses to the horrific events of September 11th bears attention by the hazards research community, and this paper is meant to contribute to that discussion.

The role of the media in shaping perception and policy in general and in the area of hazards in particular has been inspected in a large body of literature scattered across a variety of disciplines. Themes in this scholarship include risk amplification (sensationalism) and attenuation, emergency mass communication, biases in coverage, and agenda-setting.

A common criticism is of the sensationalism many media bring to hazard and crime stories, which can amplify public concern about minimal risks or even hamper efforts to respond to a disaster (Bennett, 2002; Elliott, 1989; Fishman, 1978; Friedman, 1994; Kasperson and Kasperson, 1991; Kasperson et al., 1988; Mazur, 1998; Scanlon, 1989; Smith, 1992). Alternatively, by not focusing on an important hazard, media can attenuate the development of public concern and pressure on decision makers to deal with a risky situation (Kasperson and Kasperson, 1991; Rodrigue, 2001a, 2001b).

Some efforts have been made to quantify sensationalism by comparing media coverage with objective measures of damage or danger (Rovai and Rodrigue, 1998; Rodrigue et al., 1997; Sandman, 1994; Singer and Endreny, 1994; Smith, 1992). Others argue that such comparisons are not fair: the media are, in this view, not there faithfully to reproduce in print, radio, or images the exact probabilities or estimates approved by experts. Rather, they are to provide helpful information and education for people to evaluate and reduce their risk (Mazur, 1998; Quarantelli, 1989). Others concerned with the mass media functions argue that they are to report on possible breakdowns in institutional protections for people, and, most importantly, they are to provide a public forum or arena for debate on issues that might not be well encompassed by official statistics (Peters, 1994).

Until now only sporadically linked to hazards literature is a large body of generic media criticism mostly targeted to an educated lay audience with progressive political sympathies (e.g., Bagdikian, 1997; Cohen and Solomon, 1995; Cohen, 1993; Gans, 1989; Herman and Chomsky, 1988; Lee and Solomon, 1991; McChesney, 1997; Schechter et al., 1997; Stevens, 1998). This work identifies a variety of filters purported to bias media selection of newsworthy items from the chaos of daily events, of which the most often cited are capital concentration in media and media dependence on advertising revenue.

Approximately 20 corporations control over half the revenue in the global media industry, including television, radio, daily newspapers, magazines, movies, and books (Compaine and Gomery, 2000; Herman and McChesney, 1998). A common concern raised about this centralization is that the narrowness of ownership structure may produce a narrowing of the political spectrum featured for citizens to bring to bear on a given situation of importance to society. If so, narrowing of the range of views and information available could diminish the informed consent on which effective democratic oversight of policy depends (Bagdikian, 1997; Herman and Chomsky, 1988; Herman and McChesney, 1998; Lee and Solomon, 1991). A counterpoint to this concern is that, while narrowing of views may affect traditional media on which citizens depend for the bulk of their news, new companies and new media do arise and can destabilize the situation (Compaine and Gomery, 2000; Rodrigue, 2001c).

Media generate profits through advertising revenue, and this can lead to a skewing of coverage in a disaster towards the interests of the usually more prosperous "desirable demographics" the advertisers are trying to reach or towards editorial perceptions of the needs of the advertisers themselves (Herman and Chomsky, 1988; Maher, 1997). If such skewing occurs in the coverage of a disaster, it can lead to media marginalization of the needs of poorer victims or geographical areas, which can affect disaster recovery (Rodrigue et al., 1997; Rovai and Rodrigue, 1998; Rovai, 1994).

Media, while thus arguably prone to biases in coverage and to sensationalism, remain critical to the operations of a democratic society. They influence public opinion by directing audience attention to the particular issues that they emphasize (Brewer and McCombs, 1996; Dearing and Rogers, 1996; Gans, 1989; Iyengar et al., 1982; McCombs and Shaw, 1972). The media industry may not, as Bernard Cohen once famously put it, "be successful much of the time in telling people what to think, but it is stunningly successful in telling its readers what to think about" (1963, p. 13). This effect may not be a simple and measurable correspondence between an issue's salience in the media to its salience in public opinion. Rather, it may work through "priming," or activating clusters of associated ideas and values in the readers' or viewers' minds. Media presentation can then make the selected issues particularly "available" for viewers' and readers' reflection and association with other things on their minds, with a variety of often unpredictable effects on public opinion about a given issue or politician's performance (Bennett, 2002; Iyengar et al., 1982; Mazur, 2001).

The selection of news themes by editors itself may reflect a similar process of priming. Mark Fishman, in discussing crime waves, comments that "the chances that an event or incident will be reported increase once it has been associated with a current theme in the news" (1978, p. 535). Editors seek to reduce the chaos of everyday occurrences and reports to a manageable number of story themes to pursue, and this selection process is unavoidably ideological. As Fishman puts it, "This procedure requires that an incident be stripped of the actual context of its occurrence so that it may be relocated in a new symbolic context: the news theme" (1978, p. 536).

Media agenda-setting, while effective in getting items into individual attention and public agendas, is not linearly connected with them. If an issue does not grab the attention of individuals, they will not read, view, or listen to stories about it; if an issue is important to them but they are confident that they already understand it, they may do little more than scan related stories once in a while. In neither of these situations can media govern the construction of public agendas perfectly. If, however, an issue in the media does tie in with individuals' interests and emotions and they feel uninformed about it, they may well become hungry for stories on such a salient issue; both the content of their concerns and what they think about the issue may be strongly determined by the media's decisions to follow a given theme, which can become a mutually reinforcing dynamic that encourages sensationalism and risk amplification, biased coverage, and poor contextualization (McCombs, 2002).

Issues placed on the public agenda with varying degrees of efficacy may then result in public pressure on elected decision makers. Depending on the presence of issues competing for their attention and the influences brought to bear on them, public pressure can set the agenda for policy makers' debates (Dearing and Rogers, 1996; Brewer and McCombs, 1996). This "pressurization" of policy debate affects all issues, including those of risk management (Birkland, 1996; Rodrigue, 2001a, 2001c).

Research Questions

The foregoing discussion of media sensationalism, biases, and agenda-setting roles establishes a basis for concern about the coverage of the events of September 11th, the effects it may have on readers', viewers', and listeners' understanding, and the pressure of that public understanding on elected officials' debates about the proper assessment of and management of the terrorism risk. It is outside the scope of this study to elicit audience perceptions and politicians' sense of the drift of public opinion in their constituencies. Rather, this study will focus on the post-event coverage itself, evaluating whether the coverage of one prominent daily newspaper's online edition in the 12 weeks after September 11th exhibits three features derived from media-analytic literature in its coverage of these attacks. The front-screen coverage of the *Los Angeles Times* from September 12, 2001, through December 4, 2001, should feature (1) sensationalism over context, (2) a narrow political and ideological framework for public understanding and debate, and (3) socio-economic bias in coverage of victims.

The attacks of September 11th were inherently and intensely sensational, so it is difficult to imagine how media could sensationalize something like this further, given the spectacular nature of the events themselves. The most sober and responsible of media would depict the second plane striking the building, the smoking towers, people falling or jumping out of them, the collapse of the towers, and the devastated population fleeing the impacted areas. Sensationalism may, in such a sensational event, be usefully explored as inadequate attention to the contexts generating the event. This study, then, determines the prominence of context in the *Los Angeles Times'* post-event coverage to operationalize the concept of sensationalism in coverage of a sensational tragedy.

There is quite a range of opinion in the world today about the roots of Muslim fundamentalism, the role of the United States in directly or indirectly fostering its development, the impacts of economic globalization and military intervention in fomenting resentment towards the United States in particular and the West in general, and ways of reacting to September 11th and managing terrorism. This range of views around the world, part of the mainstream public debates in other countries, is accessible here to the more determined consumer of news information through a sampling of books, speakers at special events, foreign media, some university curricula, and public broadcasting. It is argued here that, as far as the needs of the more time-stressed or casual parts of the "citizen-audience," American commercial media will peripheralize those parts of the debate that fall beyond the current rather short ideological range dividing the two dominant political parties. The *Los Angeles Times* is, thus, not expected to cover a spectrum much wider than this range in its showcased stories, i.e., those on its front page or home screen.

The people working in the World Trade Center and Pentagon and most of the victims of the four plane attacks were a mix— predominantly middle-class and often professionals, largely but by no means exclusively American. The buildings struck in the attacks, however, also had an extensive service staff and small service businesses that employed more blue-collar workers and immigrants. The *Los Angeles Times'* front page might be expected to underplay coverage of the latter. Also, the attacks will have ramifications far beyond the World Trade Center and Pentagon buildings themselves, in economic ripple effects that will strike both corporations and the people doing menial, skilled, technical, professional, and managerial work for them and the many small businesses directly affected by the attacks or vulnerable through their dependencies on larger corporations that were. If the *Los Angeles Times'* coverage resembles media coverage of disasters in general, corporations should be emphasized over their employees and smaller businesses.

Data and Methods

I utilized the online *Los Angeles Times* for the 12-week duration of the project. This is a nationally prominent newspaper, which maintains bureaus in Washington, D.C., and New York, as well as overseas. Its audience, moreover, is geographically removed from the two cities struck by the terrorists, so its coverage could be expected to concentrate on those stories with national appeal.

I used the online edition because of the ease of data collection that it enables. Copying and pasting from the web reduced the need for manual retyping of headlines and lead sentences into the spreadsheet used to record them for further analysis. There were eventually 558 front-screen stories in the database, of which 380 were directly related to the events of September 11th.

The research proceeded through the inductive and iterative development of a dual-level classification system for the central concerns of the articles on the front screen and then summary and analysis of the resulting counts. The dual-level classification resulted when examination of the main concerns of the articles showed that these categories could themselves be further grouped into three more general categories. Similarly, each article was classified by the range of views to which the story exposed the reader as either centrist or alternative views. Also, any article on impacts was classified as focusing on workers and small businesses or focusing on large corporations. This is a literature content analytic methodology, a form of qualitative analysis that yields quantitative characterizations of textual and other material.

Content analysis entails close reading of a text (or, in some variants, images or sounds) and development of categories for the coding of content (Krippendorf, 1980; Weber, 1990). Content may be coded as falling into these categories at a variety of levels: word by word, sentence by sentence, paragraph by paragraph, article by article, or image by image. Development of the categories may proceed inductively or deductively and is inherently subjective. While less subjective, coding of text into existing categories, too, occasionally entails difficult judgment calls about particular cases, which can vary among coders. Because of the subjectivity of the classification development process and the possibility of discrepant coding into the classification system, the validity of the results obtained through this method has long been questioned (Berelson, 1952).

In attempts to improve the descriptive and interpretive validity (Maxwell, 1992) of content analysis, some applications have entailed multiple independent coders trained similarly with the calculation of reliability statistics for the discrepancies in their coding (e.g., Neuendorf, 2002). Other variants involve coding independently and meeting to decide democratically on any discrepancies among codings (e.g., Carey et al., 1996).

Given that the funding source for this project, the Quick Response Research Program, specifically precludes wages and salaries for assistants, I opted for a single-coder system. Single coder content analysis may suffer from reduced reliability of coding of each article because it lacks a mechanism for soliciting possibly divergent interpretations of that article's chief concern and then deciding among such interpretations or reporting on their divergence. On the other hand, single-coder content analysis does facilitate consistency in coding from one article to the next as one person applies a single approach to all of them rather than training others to apply that approach with varying levels of motivation and understanding.

Coding of Front-page Articles

This project concentrated on the 558 articles appearing on the front screen (the home page or front page) of the *Los Angeles Times* for the first 12 weeks after the disaster. The study period, then, ran from September 12, 2001, through December 4, 2001. These are, thus, the stories that the editors of the paper deemed significant enough to showcase on the front screen.

The home page presents the reader with a visually intricate screen, including navigation buttons to other sections of the paper, teaser lines leading to other articles in the paper, advertisements, photographs, and, often, thumbnail images leading to stories elsewhere in the paper or to multimedia presentations (such as video/audio clips from television broadcasts). To be included in this study, the article had to include both a headline and at least one lead sentence on the home page, from which it would be possible to code the major concerns or themes of the article. This normally yielded six to eight articles each day, divided into a top-of-screen section (roughly the above-the-fold part of a paper newspaper) with from three to five articles and, usually, a local or state section, generally with three articles.

The headline and then the lead sentences were highlighted, copied, and then pasted into a spreadsheet (though some shorter ones were simply manually typed or longer words abbreviated). Besides these two variables, other items recorded included the date and key theme. As a backup measure, a hard copy was printed of the home page and the nine main sections of the paper each day. Unfortunately, I did not have access to the internet for two days during these 12 weeks (one day each during the ninth and tenth weeks), importing a small error into the analysis reported here (about 3% of the potential articles are missing).

To develop the main themes, I initially read the headlines and lead sentences and recorded a word that summarized the central concern of the article. I then read through the articles to confirm my categorization. After doing this for a week's coverage as a pilot sample, I sorted the spreadsheet by the key theme category and noted instances where I had used similar but not identical words. I settled on a consistent expression for such cases and then went back through the database, reclassifying and again sorting. After a few iterations of this process, I was consistently naming the main themes. I repeated the process for a couple more weeks' coverage until I was confident I was classifying the stories consistently and then used this system to complete the database for the full 12-week study period. This system yielded 17 key themes, but only 10 of these ultimately included more than five stories each. These are briefly described below.

Ten Key Themes in Front-page Articles

The 10 categories that emerged from this iterative coding process each covered anywhere from six to 103 individual stories.

- **Context** included stories about the cultural and geopolitical background to the terrorist attacks, including Islam, the history of American involvement in the economies and politics of the Middle East, anti-Western and anti-American sentiments around the world that could be organized to support terrorism, the suppression of women in Afghanistan and its tardy valuation as a rationale for confronting the Taliban, the workings of the Al-Qaeda network, and Osama bin-Laden's personal history. Examples of headlines that fell in the "context" category are "Jihad Fervor now Resentment" (December 2), "Sorting out the Terrorists" (October 4), and "Terror's World a Local One" (September 22).
- **Diplomacy** had stories about American negotiations with the Taliban; American efforts to build a multinational coalition against Al-Qaeda and the Taliban; and the varying interests of Pakistan, India, Uzbekistan, and Iran that shaped their responses to American entreaties and pressure. Sample headlines coded under "diplomacy" are "Afghanistan Offers to Free Aid Workers If US Stops Threats" (October 6), "Bush urges UN Members to Back War" (November 10), "Shoring up Saudi Support" (October 3), and "Evidence Impresses Pakistan" (October 4).
- **Impact** included articles about the human toll in lives and business relationships lost; the economic dislocations caused by the attacks in the aviation, entertainment, sports, tourism, and other industries, in governance, and in lower Manhattan. Examples of headlines falling into this category are "Trade Center toll Rises to 6,300" (September 20), "Controlled Chaos Seizes Day at LAX" (September 12), "Hawaii takes a Big Hit as it Waits for Tourists" (October 17), and "State's Economy Seems Unlikely to Dodge Fallout" (October 7).
- **Investigation** took in articles on the progress of the criminal investigations identifying suspects; the history of the suspects in the United States; and the genetic identity of the anthrax strains found in letters. Sample headlines coded as "investigation" are "How did the Hijackers get past Airport Security?" (September 23), "FBI Investigating Florida Anthrax Cases" (October 8), "Pentagon 'Black Boxes' Found" (September 14), "Terrorists' Checklist" (September 28), and "75 Questioned, 4 Held in Widening Probe" (September 18).
- **Military** included stories reporting on airstrikes in Afghanistan, deployment of special forces, deaths of military and intelligence

personnel in combat or while providing support for combat operations, and collateral damage to civilians. Some illustrative headlines in the "military" classification are "US Jets Open 2nd Week of Strikes" (October 15), "Errant Bomb Hits Housing Row in Kabul" (October 14), "Bush Takes CEO Role in Waging War" (September 23), "Aid Workers Airlifted to Freedom" (November 14), "Secretary of Defense Stirs and Delivers" (November 10), and "Both Sides Report Heavy Fighting near Key City" (November 8).

- **Mitigation** refers to stories about the proposal and implementation of heightened security measures in the wake of the September 11th attacks. Examples of "mitigation" stories include "House OKs Airport Security Bill" (October 12), " FAA Plans to Increase Security of Cockpit Doors" (September 18), and "New Push to Centralize Food Inspections" (November 23).
- **Reactions** include stories focussing on grief of victims' families and friends and the nation and world as a whole, commemorations of the victims, such emotions as fear of flying and anger, and attacks on Muslims and people who "look" Muslim. Articles coded under the heading of "reactions" include "Remembering Those Lost" (September 14), "Fear That LA Would Be next Grips Many" (September 12), "Turbans Make Sikhs Innocent Targets" (September 20), "US Keen to Avenge Attacks" (September 16), and "Anti-war Protest" (September 29).
- **Reconstruction** includes coverage of plans for rebuilding the destroyed buildings and the economy. Sample headlines deemed related to "reconstruction" include "Bush Seeks $60 Billion in Tax Cuts" (October 5), and "Free to Plant Opium" (November 22). There were no front-screen stories on rebuilding the World Trade Center or the Pentagon in the first 12 weeks.
- **Response** described stories that dealt with search and rescue operations, evacuations, and workers being sent home by employers. Examples of headlines falling under the "response" category are "How to Help" (September 12), "Rain Hampers NY Rescuers" (September 14), and "Anthrax Terrorism Spreads" (October 16).
- **Restoration** took in stories about the reopening of airports, resumption of sports events, workers returning to their jobs, and restoration of utilities. Headlines in the "restoration category" include "US Aviation Resumes" (September 13), "Residents Warily Return to Lives" (September 13), "Grim Work Continues in NYC" (October 1), and "People Take a Break from Catastrophe" (September 17).

Not all stories about September 11th fell tidily into these 10 areas. There were seven categories that had too few stories to warrant listing separately, so these were grouped together as "other related stories" (n=14). An example would be two stories reporting that Kofi Annan and the UN received the Nobel Peace Prize for "tackling challenges from poverty to terrorism." Another single article was coded as "politics," dealing with an incident in which President Bush embarrassed members of Congress by revealing leaks to the press traceable to some of them.

In addition, as time wore on, other unrelated stories began to return to the front screen of the *Los Angeles Times*. These were recorded as reflecting restoration of more ordinary concerns, but they were not specifically classified beyond "unrelated story." Some of these involved difficult judgment calls. For example, the ongoing Israeli and Palestinian confrontations returned to the front screen late in the fifth week and eventually included 24 headlines out of the 558. While the history of Israeli and Western presence in the Islamic holy places of the Middle East is mentioned by Osama bin-Laden and others as part of what allegedly drove them to terrorism, if these stories of present-day conflicts did not link the incidents they reported with the history of terrorism and the development of the September 11th attacks, they were classified as "unrelated." Examples of these stories include such headlines as "Israeli Cabinet Minister Slain," "Israelis Kill Hamas Leader," "Hamas Supporters Vow Revenge," and "Blaming Arafat, Israel Starts Counterattack." With these qualifications, there were eventually 179 unrelated front-screen stories in the first 12 weeks of *Los Angeles Times* coverage.

The Three Metastories on the Front Page

The 10 categories and the stories in them began themselves to converge into three overarching narratives or metastories: there were stories of the **disaster**, response to it, and recovery from it; there were stories about the **crime** and its investigation; and there were the **war** stories of diplomacy, deployment, airstrikes, and the fall of the Taliban. Each theme was eventually assigned to one of these metastories and then the other related story category was gone through, story by story, with individual stories assigned to one of the three larger narratives. There is a tacit fourth metastory implicit in all the other stories that were unrelated to the disaster, crime, and war. Their presence tells the stories of a nation returning to its more ordinary concerns, and they serve as a harbinger of at least partial recovery.

Range of Ideological Exposure on the Front Page

All 379 stories related to September 11th were further classified by the kinds of views to which they exposed the reader. Stories were coded as

centrist if they reported views that fall within the range of political opinion normally expressed in the two dominant American political parties, Republican and Democrat; the developing international consensus about the appropriateness of American military response to Al-Qaeda; or simply reported incidents without analysis. Examples of such headlines and lead sentences are

- “Evidence impresses Pakistan—For the first time, the only country keeping ties with the Taliban regime has joined the US in directly blaming bin Laden” (October 4).
- Crop-dusting Delay Threatens Agriculture—Officials worry that pests, weeds and disease could get a foothold. Aircraft group members are asked about training inquiries” (September 25).
- “Anthrax Patient Dies; Investigation Continues—Health officials have emphasized that no other cases have been reported and there is no evidence of a terrorist attack” (October 5).
- “President Reasserts Vow to Uproot Terror—As world’s attention focuses on the military deployment around Afghanistan, Bush uses radio address to remind nation ‘this will be a different kind of war’” (September 29).
- “Thousands Grieve at Memorial—People gathered to mourn those killed in the Sept 11 WTC attack. Above, a NYC firefighter places a family photo on a makeshift memorial” (October 28).

Stories were coded as **alternative viewpoints** if they covered viewpoints that expressed criticism of the United States’ foreign policy or domestic reactions to September 11th; represented immigrant, counter-cultural, or oppositional voices within the United States; or emphasized opinions in other countries that are highly divergent from American foreign policy and the international consensus about terrorism and the Taliban. Some headlines and lead sentences falling in this category are

- “Terror’s World a Local One—Thousands die around the globe each year in home-grown violence. But US war isn’t likely to target these killers” (September 22).
- “Japanese-Americans’ Wounds Reopened—Citing WWII internment camps, groups plan candlelight vigil in support of Muslims” (September 26).
- “Bombing alters Afghans’ Views—The US, once viewed as a savior, is increasingly being seen as the enemy—and the Taliban as a victim” (November 4).

- "Sharon draws US Rebuke—Rift over claims that Washington is ready to sell out Israel points up the intricacies of the war on terrorism" (October 6).
- "Anti-war Protest—Activists and anarchists chanted "no war" as they took to the streets in Washington today, their anti-globalization cause transformed by the terrorist attacks" (September 29).

Corporations, Employees, and Small Businesses

Forty-seven of the stories addressed business and employees in some manner related to September 11th. Some of the articles dealt only with workers and others only with businesses, while others mentioned both. Within the business-focused stories, it was possible to differentiate larger corporate concerns from small businesses as the focus, while in still others it was not possible to differentiate them or the article covered both scales of enterprise. As a result, six different categories emerged, which are listed below with illustrative headlines and lead sentences.

Employee focused—"Call to Ease LAX Security is Rejected—Officials cite continuing threat in rebuffing pleas from hundreds who say rules have cost them their jobs" (September 26); "Slowdown's Silent Victims—Few laid-off hospitality workers qualify for government aid, and those who do—because they hold temporary work permits—won't apply" (October 21); and "Unemployment Rate Leaps to 5.4%—The biggest one-month jump in more than 21 years is result of 400,000 jobs being lost" (November 2).

Large business focused—"United CEO Forced out as New Turmoil Rocks Airline—A warning that the carrier could 'perish' spread anxiety in an ailing industry" October 28); "Disney Results Expected to be Bleakest since 1993—Bruised by fallout from last month's terrorist attacks, the company is expected to report a major drop in income from its FL operations" (October 27); and "Another Brutal Day on Wall St.—Market suffers its fourth big decline in five days. Dow experiences its worst weekly point drop in history" (September 21).

Small business focused—"Private lots near LAX feel Crunch of Rules—The owners say the public isn't aware they're open and that their shuttles can go to the airport" (September 28); "Weaving a Ribbon of Unity—A local store tries to keep up with demand for the red, white, and blue symbols. Other scenes reveal our changing landscape" (September 16); and "Few Hate Crimes, but Police Act Quickly—In the most serious attack, gunmen shoot at a convenience store owned by a Syrian-American" (September 15).

Undifferentiated business focused— "Local Firms Brace for a Slowdown—Cancellation of the Emmys and sports events is affecting revenues at hotels and restaurants" (September 14), and "Retail Sales leap Record 7.1%—October sales, driven by heavy discounting, were the highest ever recorded" (November 14).

Large business and employee focused—"Guards Needed, but Low Pay a Problem—The nation's two largest security firms anticipate hiring a combined minimum of 10,000 to 15,000 new guards" (September 29), and "LA may see Boost in TV, Movie Work—Studios are rethinking plans as actors, executives voice desire to avoid travel" (September 24).

Undifferentiated business and employee focused—"Anaheim Feeling Pinch—With a travel-wary public, once-popular spots face financial devastation. Above a worker at the Anaheim Convention Center cleans in the lobby" (October 4), and "Stimulus Bill facing Change—The $100 billion package passed by the House must go to the Senate, where greater support exists for additional spending. Above, thousands at the 'Twin Towers Job Expo' in NY today" (October 25).

Coding of Front-page Photographs

Additionally, I examined the front-screen photographs, recording a one-sentence description of the scene and then classifying it by the metastory it illustrated: disaster, war, crime, or unrelated. Visuals can reinforce, amplify, or contradict impressions created by textual information and sometimes are the sole impression taken away by the casual or time-pressed reader.

Coding of September 11th Relevance of Other Major Sections of the Paper

Finally, I went through the other main sections of the paper for the 12 weeks, except for the Travel section, counting up the total number of stories (9,551) and the number that were related to the events of September 11th (2,736). The Travel section, though listed among the main sections, generally consists of one or two stories repeated each day for a week or so and resembles the narrow-theme minor sections (e.g., Health, Arts & Entertainment, and Books) more than it does the paper's main sections (e.g., Nation, California, Business, Editorials).

Findings

In this section, the analysis of front-screen coverage will be broken down by the 10 main themes and the three metastories of disaster, crime, and war, into which the 558 stories fell. The relative abundance of each theme and metastory will be discussed for the 12-week study period as a whole and then broken out by week. Following the findings for the front-page coverage, the balance between September 11th-related and unrelated stories among the main sections of the paper will also be traced by week.

Overview of the First Twelve Weeks on the Front Page

This section considers the 12-week study period as a whole. The 10 themes and then the three metastories are discussed in terms of dominant themes and metastories over the study period. A later section will follow key themes and metastories week by week.

The Ten Main Themes

Three themes were emphasized in the *Los Angeles Times'* front-page coverage of September 11th in the sense that they each received about 10% or more of the first 12 weeks' coverage. Two other themes were drastically de-emphasized, receiving less than 2% of front-page coverage during the study period.

Far and away the dominant concern through the first weeks viewed together was the military category with 103, or 18.5% of the 558 stories (Figure 1). Secondary themes more weakly emphasized were those of the crime investigation (n=57, or 10.2%) and of reactions to the disaster itself (n=55, or 9.9%).

Dramatically de-emphasized were two other themes: context and reconstruction. The context of the events of September 11th received the least front-screen coverage, with just six stories, or 1.1%. The context is an important part of this story and key for Americans to understand and prepare for what is to them a new hazard. In this sense, the attacks of September 11th generated needs for contextualized information similar to those for any disaster, and the weak showing of the context on the front screen here fits with prior work on how media perform in other hazards and disasters (e.g., Singer and Endreny, 1994). Moreover, the context of this disaster was deeply and unavoidably political and ideological at core. Its disappointing treatment here is consistent with the ideological narrowing that some media critics worry comes with corporate concentration in media, but it is also consistent with the long-established media need for drama and simplicity, which contextual information rarely satisfies.

If the 24 stories about the incidents in the ongoing Israeli-Palestinian conflict were included in "context," for the sake of argument, still only 30 stories could then be seen as addressing the context. This remains a paltry 5.4% of coverage.

Also receiving very little coverage was reconstruction (n=9, or 1.6%). Given that this study focused on just the first 12 weeks after the attacks, this is not too surprising: This disaster was still at a very early stage in the conventional post-event timeline of response-restoration-reconstruction (Haas et al., 1977).

The tacit story of recovery is seen in the eventual appearance of large numbers of stories that were unrelated to the events of September 11th and their aftermath. By the end of the 12-week study period, fully 179 or 32.1% of the front-screen stories in the *Los Angeles Times* fell in the unrelated category.

The Three Metastories

Collapsed into the three metastories of disaster, crime, and war, the dominant narrative was the war story, with 168 of the 558 stories, or 30.1% (Figure 2). The disaster was the second most prominent metastory for the 12-week study period, almost a co-dominant at 152 stories or 27.2%. The crime story was the least covered of the grand narratives, with 59 stories or 10.6%.

Images of the Three Metastories

Of the 82 photographs, fully 48 or 58.5%, represented scenes tied to the war story (Figure 3). Examples included a scene of protesters in Pakistan burning an American flag on the second day of airstrikes, a Defense Department image of U.S. troops at an undisclosed location, a head shot of the chair of the Joint Chiefs of Staff at a Pentagon press conference discussing the role of special forces in southern Afghanistan, and two Afghan women in burkas with two children on their way to Pakistan as refugees. Another 22 photographs or 26.8% depicted scenes related to the disasters themselves, such as firefighters hosing down Building 7 of the World Trade Center, the President and First Lady at a National Cathedral commemorative service, Bruce Springsteen at a fund-raising telethon, and a crowd shot of people at the "Twin Towers Job Expo." Only six photographs, or 7.3%, dealt with the crime investigation, such as a head shot of the President thanking investigators at the FBI, and head shot of the attorney general holding up a hijacker's suicide note or checklist, and three boys in Pakistan, sons of a suspect being detained by the United States on immigration charges, being held by the detainee's elderly brother. Images related to September 11th completely dominated the front-page photographs of the *Los Angeles Times* throughout the 12-week study period, comprising 76 photographs or 92.7%.

Range of Ideological Exposure on the Front Page

Of the 379 front-page articles that were related to the events of September 11th, only 48 (12.7%) were classified as exposing readers to alternative viewpoints as defined above. Fully 331 (87.3%) were centrist. This affirms the expectation that American media coverage would report a relatively narrow range, not often reaching outside the ideological spectrum that dominates American political discourse.

Corporations, Employees, and Small Business

Of the 47 front-screen stories that mentioned businesses and employees, 19 focused on employees and 20 on business, while another eight gave attention to both. Among the business-focused articles, 11 concentrated on

Key Themes		CONTEXT	DIPLOMACY	IMPACT	INVESTIGATION	MILITARY	MITIGATION	REACTIONS	RECONSTRUCTION	RESPONSE	RESTORATION	OTHER RELATED	UNRELATED	SUMS
Week 1	#	0	1	5	11	9	1	10	0	5	8	3	0	53
	%	*0.0*	*1.9*	*9.4*	*20.8*	*17.0*	*1.9*	*18.9*	*0.0*	*9.4*	*15.1*	*5.7*	*0.0*	*100.0*
Week 2	#	1	2	5	8	12	1	16	1	4	1	1	0	52
	%	*1.9*	*3.8*	*9.6*	*15.4*	*23.1*	*1.9*	*30.8*	*1.9*	*7.7*	*1.9*	*1.9*	*0.0*	*100.0*
Week 3	#	0	6	4	3	5	8	6	0	3	7	0	4	46
	%	*0.0*	*13.0*	*8.7*	*6.5*	*10.9*	*17.4*	*13.0*	*0.0*	*6.5*	*15.2*	*0.0*	*8.7*	*100.0*
Week 4	#	2	10	2	3	8	1	1	2	0	1	4	13	48
	%	*4.2*	*20.8*	*6.3*	*6.3*	*16.7*	*2.1*	*2.1*	*4.2*	*0.0*	*2.1*	*8.3*	*27.1*	*100.0*
Week 5	#	1	2	1	5	10	2	0	0	1	0	4	20	46
	%	*2.2*	*4.3*	*2.2*	*10.9*	*21.7*	*4.3*	*0.0*	*0.0*	*2.2*	*0.0*	*8.7*	*43.5*	*100.0*
Week 6	#	0	2	2	7	8	0	0	0	1	0	3	19	43
	%	*0.0*	*4.7*	*7.0*	*16.3*	*18.6*	*0.0*	*0.0*	*0.0*	*2.3*	*0.0*	*7.0*	*44.2*	*100.0*

Week 7													
#	0	0	6	5	7	3	2	1	1	0	0	20	45
%	*0.0*	*0.0*	*13.3*	*11.1*	*15.6*	*6.7*	*4.4*	*2.2*	*2.2*	*0.0*	*0.0*	*44.4*	*100.0*
Week 8													
#	0	2	2	5	7	1	9	1	2	1	0	19	49
%	*0.0*	*4.1*	*4.1*	*10.2*	*14.3*	*2.0*	*18.4*	*2.0*	*4.1*	*2.0*	*0.0*	*38.8*	*100.0*
Week 9													
#	0	5	0	2	10	1	2	0	0	0	2	17	39
%	*0.0*	*12.8*	*0.0*	*5.1*	*25.6*	*2.6*	*5.1*	*0.0*	*0.0*	*0.0*	*5.1*	*43.6*	*100.0*
Week 10													
#	0	6	0	2	10	3	4	1	2	0	0	15	43
%	*0.0*	*14.0*	*0.0*	*4.7*	*23.3*	*7.0*	*9.3*	*2.3*	*4.7*	*0.0*	*0.0*	*34.9*	*100.0*
Week 11													
#	1	0	2	3	9	1	0	2	1	0	0	28	47
%	*2.1*	*0.0*	*4.3*	*6.4*	*19.1*	*2.1*	*0.0*	*4.3*	*2.1*	*0.0*	*0.0*	*59.6*	*100.0*
Week 12													
#	1	2	4	3	8	2	2	1	0	0	0	24	47
%	*2.1*	*4.3*	*8.5*	*6.4*	*17.0*	*4.3*	*4.3*	*2.1*	*0.0*	*0.0*	*0.0*	*51.1*	*100.0*
All 12													
Weeks #	6	38	35	57	103	24	52	9	20	18	17	179	558
%	*1.1*	*6.8*	*6.3*	*10.2*	*18.5*	*4.3*	*9.9*	*1.6*	*3.6*	*3.2*	*3.0*	*32.1*	*100.0*

Highlighted cells are those with 10% dominance or more.

Data compiled by C.M. Rodrigue, 2002

Figure 1. Key themes by week.

Metastories		week 1	week 2	week 3	week 4	week 5	week 6	week 7	week 8	week 9	week 10	week 11	week 12	1st 12 weeks
War story	#	11	15	19	21	15	10	10	10	16	19	10	12	168
	%	20.8	28.8	41.3	43.8	32.6	23.3	22.2	22.2	41.8	44.2	21.3	25.5	30.1
Crime story	#	11	8	3	4	5	7	5	5	3	2	3	3	59
	%	20.8	15.4	6.5	8.3	10.9	16.3	11.1	10.2	7.7	4.7	6.4	6.4	10.6
Disaster story	#	31	29	20	10	6	7	10	15	3	7	6	8	152
	%	58.5	55.8	43.5	20.8	13	16.3	22.2	30.6	7.7	16.3	12.8	17	27.2
Unrelated story	#	0	0	4	13	20	19	20	19	17	15	28	24	179
	%	0	0	8.7	27.1	43.5	44.2	44.4	38.8	43.6	34.9	59.6	51.1	32.1
SUMS	#	53	52	46	48	46	43	45	49	39	43	47	47	558
	%	100	100	100	100	100	100	100	100	100	100	100	100	100

Highlighted cells indicate dominant or co-dominant metastories for that week

Data compiled by C.M. Rodrigue, 2002

Figure 2. Metastories by week.

Metastories		week 1	week 2	week 3	week 4	week 5	week 6	week 7	week 8	week 9	week 10	week 11	week 12	1st 12 weeks
War story	#	1	3	4	4	4	5	5	6	3	5	6	2	48
	%	14.3	42.9	57.1	57.1	57.1	71.4	71.4	85.7	50.0	83.3	85.7	28.6	58.5
Crime story	#	0	1	1	0	1	1	0	0	2	0	0	0	6
	%	0.0	14.3	14.3	0.0	14.3	14.3	0.0	0.0	33.3	0.0	0.0	0.0	7.3
Disaster story	#	6	3	2	3	1	0	2	1	1	0	1	2	22
	%	85.7	42.9	28.6	42.9	14.3	0.0	28.6	14.3	16.7	0.0	14.3	28.6	26.8
Unrelated story	#	0	0	0	0	1	1	0	0	0	1	0	3	6
	%	0.0	0.0	0.0	0.0	14.3	14.3	0.0	0.0	0.0	16.7	0.0	42.9	7.3
SUMS	#	7	7	7	7	7	7	7	7	6	6	7	7	82
	%	100.0	100.0	100.0	100.0	100.0	100.0	100.0	100.0	100.0	100.0	100.0	100.0	100.0

Highlighted cells indicate dominant metastories in that week's front page photographs.

Data compiled by C.M. Rodrigue, 2002

Figure 3. Front-page photographs classified into metastories, by week.

large corporations, four on small businesses, and another four on business in general or large and small businesses together. Of those balancing employers and employees, four dealt with large businesses and another four with both large and small employers. The coverage of workers and businesses, then, was almost perfectly balanced, offering no support for the expectation that the *Los Angeles Times'* coverage would be biased towards business.

Changes in Front-page Coverage over Time

Broken out week by week, there were interesting and revealing shifts and constancies in attention. Only one of the main themes remained salient throughout the study period. More commonly, a few main themes became prominent at one time during the study period and then submerged at other times. In terms of the metastories, the two co-dominant larger narratives switched dominance between the first three weeks and the last nine weeks.

The Ten Main Themes over Time

Only one of the 10 major concerns consistently garnered at least 10% of each week's coverage throughout the study period: military. Its share of weekly coverage ranged from 10.9% to 25.6%, averaging 18.5% .

Investigation was the second most common dominant of each week's coverage. It appeared among the list of categories winning at least 10% of a week's coverage in six weeks out of the 12. Not too surprisingly, the crime investigation was especially salient in the two weeks right after the attacks. It appeared again as a prominent story from the fifth through the eighth weeks, as the anthrax attacks reached their peak and as controversy developed over FBI warnings. It dropped off during the last four weeks of the study period. Figure 4 gives a reference timeline of events.

Two more themes appeared among the salient stories for still shorter periods of four weeks each: reactions and diplomacy. Reactions understandably dominated the first three weeks after the disasters. This theme became prominent again during the eighth week, when the first bin-Laden videotape was released, data confirming the economy had entered recession were published, and the governor of California instituted controversial precautionary measures to protect bridges on the basis of an FBI warning that was later retracted. Diplomacy became salient during the third and fourth weeks as the United States began to launch the air war in Afghanistan, and again during the ninth and tenth weeks (when the Taliban began clearly to fall apart, and the missionary aid workers held by the Taliban were rescued). Restoration was prominent during two weeks of the first month after the disaster (week one and week three) as New York and Washington, D.C., struggled to restore the lifelines and activities on which their safety and livelihoods depend. Two more categories flashed briefly into salience for one

Events that Week	week 1	week 2	week 3	week 4	week 5	week 6	week 7	week 8	week 9	week 10	week 11	week 12
WTC/Pentagon plane-bombings Airports reopen Stock markets reopen	1											
First hate killing in Los Angeles First peace protests First celebrity fund-raiser		2										
Pakistan allies with United States First unrelated story on front page First report of Special Forces in Afghanistan			3									
Anthrax in Florida First airstrikes Russian jet struck by missle				4								
Anthrax in New York First errant bomb kills non-combatants Anthrax in Washington, D.C. (Daschle)					5							
Anthrax in New Jersey First ground force military engagement Shift of bombing from cities to battlefields						6						
Anthrax spreading Air strikes and ground involvement increase FBI warns of new threats							7					
Bin Laden videotape Economic downturn FBI: Warning not credible after CA gov's measures								8				
US diplomacy on several fronts Plane crashes in Brooklyn Taliban defeats									9			
Anthrax seems to be domestic terrorism Hostage aid workers rescued Problems distributing aid to 9/11 victims										10		
Taliban collapsing International terrorism: Madrid, Indonesia Taliban prison revolt											11	
Recession deepening First American to die in combat American Taliban												12

*Data from **L.A. Times** compiled by C.M. Rodrigue, 2002*

Figure 4. Reference timeline of news events.

week each: impact during week seven, when anthrax seemed to be spreading through the mail enough to alter business mailroom practices, and mitigation during week three as more permanent security measures were being proposed and debated, especially for air travel.

The Three Metastories over Time

Grouping all these into the three larger narratives of disaster, crime, and war, it had seemed that disaster and war were nearly matched in their dominance of the coverage during the 12-week study period (see Figure 2). Broken out week by week, however, a strong shift in emphasis was seen. The disaster story completely dominated the first three weeks of coverage, from

58.5% of the stories in the first week to 43.5% by the third week. Dominance switched drastically in the fourth week to the war story. As airstrikes began in Afghanistan, the war story now won 43.5% of the coverage. Meanwhile, the disaster story dropped to 20.8%. The war story dominated from the fourth week through the sixth, co-dominated with disaster during the seventh week, and resumed dominance from the ninth through the twelfth weeks. The disaster story experienced a flurry of salience in the seventh week, when it matched the war story, and the eighth week when it actually once more dominated coverage. (The seventh and eighth weeks experienced increased numbers of anthrax cases in several places and the first economic data on the downturn associated with the attacks, both of which events briefly refocused attention on the disaster itself.)

The stories unrelated to September 11th first showed up on the third week after the attacks, making up 8.7% of the third week's front-screen stories. This percentage climbed dramatically over the next two weeks, to 27.1% on the fourth week and to 43.5% by the fifth week. It fluctuated from roughly 35% to 44% for the next few weeks, from the sixth through the tenth week. By the eleventh week, this had shot up to 59.6% and, by the twelfth week, to 51.1%. This is not too surprising a trajectory, as the paper's readership became accustomed to the disaster and its aftermath and gradually became interested in other matters going on in the world.

Images of the Three Metastories over Time

The photographs on the front page remained dominated by September 11th-related images throughout the 12 weeks. Images related to September 11th accounted for 100% of the front-page photographs for fully nine of the 12 weeks and, in each of the remaining three weeks, only one photograph was unrelated to September 11th. Images of the World Trade Center disaster dominated for only the first two weeks, a period even shorter than that seen when considering only the textual material on the front page. Unrelated images rose to dominance over the three metastories only in the very last week, compared with the last seven weeks of the textual material. The graphics of the *Los Angeles Times*, then, remained dominated by September 11th long after the front-page stories did, and the war metastory takes over sooner among the three metastories.

September 11th and the Main Sections of the Paper

The 9,551 stories in the Home and other main sections included 2,736 articles somehow related to September 11th, or 28.6% (Figure 5). Not too surprisingly, coverage of the disaster dominated the paper during the first week, with 356 stories related to it out of the 619 articles in the main sections of the paper, or 57.5%. This plummets permanently below dominance by the second week, with 332 of 820 articles (40.5%) related to the events of September 11th. Coverage drops below a third by the third week, dwindling unsteadily to 19.4% by the last week (Figure 6). This negative exponential trend in coverage is characteristic of coverage in other disasters (e.g., the Northridge earthquake) as media and audience attention shifts to other concerns more typical of the pre-disaster framework.

The prevalence of September 11th-related stories over the total study period varied considerably among the main sections of the paper from a low of 5.4% (Sports) to a high of 90.4% (Nation). Their rate of replacement by non-related articles, however, suggests the recovery of various sectors of society from the September 11th disaster.

The first major section of the paper to recover was Sports (see Figures 5 and 6). Far and away the dominant major section of the *Los Angeles Times* was Sports, with 2,940 stories out of the 9,551 (30.8%). Of the 2,736 September 11th-related stories, though, only 158 (5.4%) fell in the Sports section (Figure 7). Even so, Sports coverage the first week was dominated by the disaster (74 out of 145 stories, or 51.0%), dropping to 11.9% by the second week and then remaining below 5% for the next five weeks and generally below 2% for the remaining five weeks. The first week's Sports coverage was cut roughly in half by the disaster when compared, for example, against the 254 sports stories seen on average in weeks two through twelve. The remaining sports stories that first week consisted mostly of announcements of canceled events, discussion of the deaths of two Kings scouts killed in the hijacked flights, and non-sports activities of sports figures. By the second week, events were being rescheduled, and the Sports reporters finally had something more on their customary beat to report.

After Sports, the second most important major section of the *Los Angeles Times* in terms of the number of stories was Business. Business contained 2,227 of the 9,551 stories in the study period, or 23.8% (see Figure 5). The share of Business stories related to September 11th was about proportional to its share of all stories: 597 of the 2,736 September 11th-related stories, or 21.8%, fell in the Business section (Figure 7). Business coverage devoted to September 11th-related themes declined from 66.4% to 42.8% from the first to the second week but then declined in a steady, nearly linear fashion to 8.6% of stories in the last week of the study period (see Figure 5).

The front page was, obviously, constrained to relatively few stories (generally around 45 stories per week), but these stories were the ones the editors wished to showcase, the ones most likely to tempt the reader's attention. Only 558 stories of the 9,551 fell onto the front screen, or 5.8% (see Figure 5). Of the 2,736 main section stories that were related to September 11th, however, 378 or 13.8% were on the front screen (Figure 7). The events of September 11th and stories related to them dominated the home page for 11

Week	1	2	3	4	5	6	7	8	9	10	11	12	Total 9/11	ALL
Section	%	%	%	%	%	%	%	%	%	%	%	%	%	#
Home	100	100	94	72	53	53	57	59	54	65	51	43	68	*558*
World	12	5	8	4	5	1	12	4	7	9	40	53	14	*859*
Nation	100	100	95	94	95	97	99	92	89	91	66	68	90	*899*
Calif	13	42	29	27	19	29	16	20	9	20	10	14	21	*887*
Business	66	43	43	36	33	27	20	20	14	13	11	9	26	*2271*
Politics	64	50	44	50	21	69	85	42	18	67	40	18	45	*204*
Sports	51	12	3	4	1	3	5	2	1	0	2	0	5	*2940*
Tech	79	42	18	20	18	22	3	7	5	5	6	0	14	*438*
Op-ed	88	83	74	82	69	71	63	70	60	68	44	43	67	*495*
Total 9/11-related %	58	40	32	30	26	27	26	22	21	22	26	19	29	*9551*
Total of all stories #	*619*	*820*	*853*	*896*	*859*	*858*	*799*	*888*	*721*	*784*	*613*	*841*		

Data compiled by C.M. Rodrigue, 2002

***Figure 5. Percentage of coverage, by week, in the nine major sections of the* Los Angeles Times.**

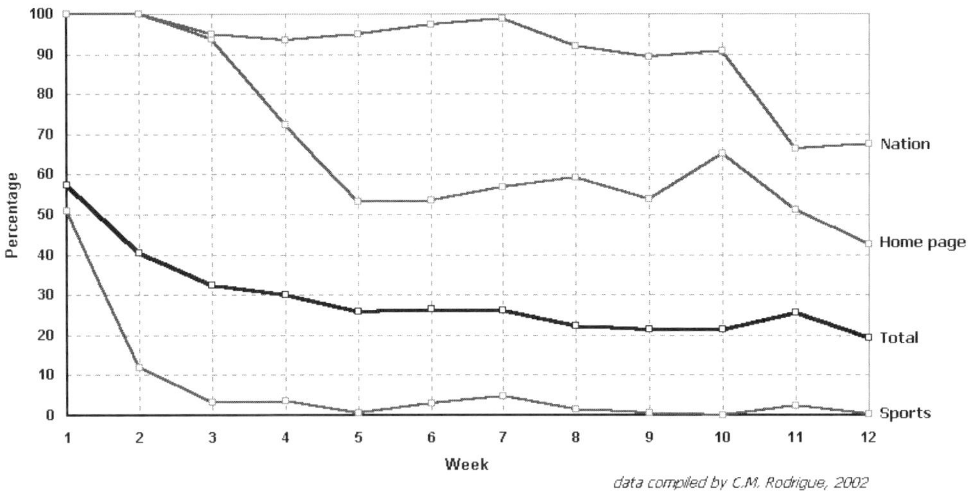

Figure 6. Percentage of* Los Angeles Times *coverage related to September 11th.

of the 12 weeks, only dropping below 50% in the last week of the study period (see Figures 5 and 6).

Op-Ed, the section printing editorials and regular commentaries, resembled the front screen in terms of September 11th coverage. Another small section of the paper with 495 of the 9,551 stories (5.2%), Op-Ed garnered 12.2% of all September 11th-related coverage in the main sections of the *Los Angeles Times* (see Figure 7). The events of September 11th declined in importance in an irregular but linear trend from 88.0% of stories in the first week to 43.2% in the last week (see Figure 5).

The last section of the paper to begin recovering was Nation. Nation normally occupies a modest part of the paper, steadily running about 11 articles per day in the online edition. Of the 899 articles in the Nation section over the course of the study period, fully 813 or 90.4% of the stories had to do with September 11th (see Figures 5 and 6). As a result, the Nation section was the likeliest venue for September 11th-related coverage, with 29.7% of all September 11th coverage appearing in this section (see Figure 7). All Nation stories for the first two weeks after the disaster focused on the attacks. September 11th accounted for 90+% of Nation stories for some eight weeks after that, dwindling to about two-thirds in the last two weeks of the study period (see Figures 5 and 6).

	Week 1			Week 2			Week 3			Week 4			Week 5			Week 6		
	Week Total			Week Total			Week Total			Week Total			Week Total			Week Total		
Section	Sum	9/11	%	Sum	9/11	%	Sum	9/11	%	Sum	9/11	%	Sum	9/11	%	Sum	9/11	%
Home	53	53	100.0	52	52	100.0	47	44	93.6	47	34	72.3	47	25	53.2	43	23	53.5
World	77	9	11.7	77	4	5.2	77	6	7.8	77	3	3.9	77	4	5.2	77	1	1.3
Nation	77	77	100.0	77	77	100.0	77	73	94.8	77	72	93.5	77	73	94.8	77	75	97.4
Calif	77	10	13.0	77	32	41.6	77	22	28.6	77	21	27.3	77	15	19.5	77	22	28.6
Business	137	91	66.4	180	77	42.8	157	68	43.3	198	71	35.9	192	63	32.8	202	54	26.7
Politics	14	9	64.3	16	8	50.0	25	11	44.0	24	12	50.0	19	4	21.1	13	9	69.2
Sports	145	74	51.0	269	32	11.9	306	10	3.3	306	11	3.6	291	2	0.7	295	9	3.1
Tech	14	11	78.6	24	10	41.7	40	7	17.5	46	9	19.6	34	6	17.6	36	8	22.2
Op-ed	25	22	88.0	48	40	83.3	47	35	74.5	44	36	81.8	45	31	68.9	38	27	71.1
TOTALS	619	356	57.5	820	332	40.5	853	276	32.4	896	269	30.0	859	223	26.0	858	228	26.6
	Week 7			Week 8			Week 9			Week 10			Week 11			Week 12		
	Week Total			Week Total			Week Total			Week Total			Week Total			Week Total		
Section	Sum	9/11	%	Sum	9/11	%	Sum	9/11	%	Sum	9/11	%	Sum	9/11	%	Sum	9/11	%
Home	44	25	56.8	49	29	59.2	39	21	53.8	43	28	65.1	47	24	51.1	47	20	42.6
World	77	9	11.7	68	3	4.4	44	3	6.8	54	5	9.3	77	31	40.3	77	41	53.2
Nation	77	76	98.7	74	68	91.9	66	59	89.4	66	60	90.9	77	51	66.2	77	52	67.5
Calif	77	12	15.6	76	15	19.7	66	6	9.1	66	13	19.7	77	8	10.4	63	9	14.3
Business	201	40	19.9	204	40	19.6	217	30	13.8	194	26	13.4	180	19	10.6	209	18	8.6
Politics	13	11	84.6	19	8	42.1	22	4	18.2	12	8	66.7	10	4	40.0	17	3	17.6
Sports	230	11	4.8	313	5	1.6	180	1	0.6	253	0	0.0	89	2	2.2	263	1	0.4
Tech	40	1	2.5	45	3	6.7	40	2	5.0	58	3	5.2	17	1	5.9	44	0	0.0
Op-ed	40	25	62.5	40	28	70.0	47	28	59.6	38	26	68.4	39	17	43.6	44	19	43.2
TOTALS	799	210	26.3	888	199	22.4	721	154	21.4	784	169	21.6	613	157	25.6	841	163	19.4

	All 12 Weeks			All 12 Weeks	
	9/11 as % of *			Section as % of *	
Section	Sum	9/11	%	Sum	9/11
Home	558	378	67.7	5.8	13.8
World	859	119	13.9	9.0	4.3
Nation	899	813	90.4	9.4	29.7
Calif	887	185	20.9	9.3	6.8
Business	2271	597	26.3	23.8	21.8
Politics	204	91	44.6	2.1	3.3
Sports	2940	158	5.4	30.8	5.8
Tech	438	61	13.9	4.6	2.2
Op-ed	495	334	67.5	5.2	12.2
TOTALS	9551	2736	28.6	100.0	100.0

Data compiled by C.M. Rodrigue, 2002

Figure 7. Number and percentage of stories, by week.

These examples illustrate the recovery of more ordinary concerns in the *Los Angeles Times* and the variability in this normalization among the main sections of the paper and, implicitly, among the social interests related to those sections. Sports recovered almost completely and that very quickly once crowd-drawing events were allowed to resume. The Nation coverage had not recovered anything resembling its pre-disaster character even after 12 weeks, reflecting the ongoing and agonizing decisions to be made by Americans in facing this new (to them) kind of hazard and the opportunity costs in civil liberties to be exacted by zealous application of the precautionary principle. This uneasy persistence of concern about September 11th was seen in other sections of the paper that also provided forums for social debate about managing this new risk, e.g., Op-Ed and the front page.

Discussion

Looking at these findings a little more closely, the rapid shift of attention to the war in Afghanistan, the recovery of non-September-11th-related coverage, and the varying rates of news movement from September 11th nearly concealed another story: the disaster story still affecting the memories and daily lives of New Yorkers and Washingtonians throughout the study period. As seen in Figure 2, the metastory of disaster was displaced by the war story on the front page by the end of the third week after the attacks and in front-page photographs no later than the second week (see Figure 3). The military-related coverage took up 30.1% of the 558 front-screen stories in the first 12 weeks of overall coverage (see Figure 2) and 58.5% of the front-page photographs (see Figure 3). Disaster-related stories (27.2% of all 558 articles) did dominate for the first three weeks of coverage, but they gave way to the war story for most of the last nine weeks (see Figure 2), and their images (26.8% of the 82 photographs) held a dominant position for only the first two weeks (see Figure 3). While the events of September 11th were a stupendous disaster, an audacious crime, and an act of war, the rapid movement from disaster-focused coverage to war-dominated coverage deprioritized the needs of New Yorkers and Washingtonians in recovering from these horrible events.

The September 11th attacks were quite possibly the greatest disaster to befall the United States in the last half century, due to the thousands of deaths, the huge property losses, and the economic and personal disruptions they caused, directly and indirectly. Estimates are that 3,041 people died because of the four hijackings, 2,650 in the World Trade Center, 125 in the Pentagon, and 266 on the four planes themselves (Wikipedia, 2002). Swiss Reinsurance estimates that insured property losses will reach $10-12 billion (Zanetti et al., 2002, p. 8), and business interruption insurance claims have been estimated at $3.5–10 billion (Zanetti et al., 2002, p. 8; Hartwig, 2001). The Congressional Budget Office estimates that insurance payouts for economic losses of all types will surpass these figures, amounting to at least $30 billion, and could top $50 billion (Torregrosa et al., 2002, p. ix). Uninsured losses add an unknown amount to these huge insured losses (Zanetti et al., 2002; Hartwig, 2001). By comparison, estimates of those killed by the Northridge earthquake range from 57 (e.g., U.S. Geological Survey, 1996, p. 2) to 72 people (e.g., Federal Emergency Management Agency, 1998). The temblor caused $16.6 billion (1999 dollars) in insured property losses (Torregrosa et al., 2002, p. 9), $9.5 billion in federal financial assistance (Torregrosa et al., 2002, pp. 15–16), $6.5 billion (1994 dollars) in business interruption costs (Gordon et al., 1998, p. 21, 28), and $20 billion in hidden uninsured losses, including deductibles paid by insured homeowners, repairs paid out of pocket, and damage to uninsured buildings (Platt, 2000, p. 29). Hurricane Andrew is estimated to have killed 65 people (Rappaport, 1993), produced insured losses of $17.1 billion (1999 dollars) (Torregrosa et al., 2002, p. 10), and $4.4 billion in federal financial assistance (Torregrosa et al., 2002, p. 15) . Given that the September 11th attacks produced a disaster far exceeding the loss of life in the two greatest recent natural disasters in the United States and at least comparable to them in economic costs, the rapid submersion of the September 11th disaster story under the war story seems abrupt.

The change in emphasis from the disaster itself to the war to exact vengeance and dismantle terrorist networks seems disproportionate. This war does not compare to others, from World War II on, in terms of American casualties and other losses. For example, the United States lost 405,399 people in WWII, 36,568 in the Korean Conflict, 58,203 in the Vietnam War (U.S. Department of Defense, 2000, p. 80), and 363 in the Gulf War (American War Library, 2002). By contrast, American casualties in the Afghanistan conflict reached 57 (25 in combat) on April 25, 2003 (Rodrigue, 2003). Even so, the war held most of the media focus after the second to third week, perhaps expressing American anxiety about getting into a potentially endless war.

The crime story is a relatively minor strand in front-screen coverage (10.6% of the 558 stories and 7.3% of the front-page photographs), though it may be the greatest crime ever committed in this country. This relative quiescence of the *Los Angeles Times* may simply reflect the necessary secrecy in which an investigation of this sort must be conducted, so there may not have been much information to report during the study period. Recent FBI controversies after the study period, however, with sensational allegations of blunders and internecine strife, do have elements that draw media attention. The crime investigation story did, thus, become a topic of media interest in May and June of 2002 (e.g., Rowley, 2002; Van Natta and Johnston, 2002).

In short, the September 11th attacks were unique in their combination of crime, war, and disaster. The relative coverage of these events could not be predicted ahead of time due to the unusual combination of these three elements. The underplaying of the crime aspect is understandable in terms of the secrecy of investigation and the study period ending early in that process. Less understandable and predictable is the magnitude of the shift from the disaster story to the war narrative, given the relative significance of the disaster aspect of the events compared to other disasters of the last half century and of their war aspect compared to other wars of the last 50 years.

In terms of the three expectations formulated from extant literature as the study began, there was support for the expectations that the less sensational context would be little covered and that the ideological exposure would be narrow. There was, however, no support for the expectation that the *Los Angeles Times'* coverage of those affected by the events of September 11th would be disproportionately skewed towards business interests and away from those of workers.

Context *vs.* Sensation

As often seen in the coverage of any disaster, the context of the events of September 11th was poorly drawn out. Only six stories appeared on the front screen of the *Los Angeles Times* about the geopolitical background that produced such murderous men so indifferent to their own and others' lives. Normally, a category with a mere six stories in it would have been lumped into the "other related" category, but the context of a disaster is central to the disaster research and policy communities. Preserving it as a freestanding category with its six paltry entries thus highlighted a serious gap in front-page disaster coverage, in this catastrophe as in so many others.

Sensationalism, a common criticism leveled at the media during disasters, could be argued based on the obsessively repetitive imagery of the plane striking the south tower on television and on the front-page graphics of newspapers, including the *Los Angeles Times*. The online format of this edition of the *Los Angeles Times* enabled readers to access the video imagery from the broadcast media. Given the extremely sensational nature of these events, however, such sensational coverage may have been entirely appropriate and commensurate. The anthrax-laden letters mailed to Senators Daschle and Leahy, NBC news anchor Tom Brokaw, the *New York Post,* and the supermarket weekly tabloid, *The Sun*, on the other hand, were sensationalized in the traditional manner often critiqued in analyses of disaster coverage. That is, the coverage of these bioterrorist incidents amplified public concern far above the actual numbers of people exposed, sickened, and killed by mailed anthrax, leading to pressure on physicians for prescriptions of Cipro.

Narrow Ideological Range

The *Los Angeles Times'* coverage of the events of September 11th, as expected, offered a relatively narrow ideological range, rarely taking readers outside the rather circumscribed American political discourse. Whether this was a function of the concentration of ownership in the American media cannot be answered in this study as it would be impossible to construct how the media would have covered something like this in the days when every town and city sported two dailies and local ownership of the media was the norm. Perhaps the news would have been even more parochial under past conditions. The point of examining the political range in coverage of this story, then, was to draw attention to the narrowness of the perspectives offered on the front screen, no matter whence it derived, because of its possible agenda-setting ramifications. For most stories most of the time, local and national self-interest may be well enough served by a certain parochialism of media content. In a stunningly new kind of geopolitical crisis, however, more high-priority media attention to context and to the ideological range outside the United States might help Americans better understand the nature of the risks they face from terrorism and the long-term consequences of the actions the country might take in mitigating this hazard. The agenda needs to be broadened to include all facets of this hazard, not narrowed when so much is at stake in dealing with an unprecedented new danger.

Bias toward Employers over Employees

Despite the expectations of media criticism literature and to its credit, the *Los Angeles Times* covered impacts on corporations and impacts on workers in roughly equal numbers in front-screen stories during the study period. Of the 47 front-page stories that dealt with job losses or business impacts, 19 focused on workers, 20 on businesses, and eight on both. Among the business stories, small businesses were the concern of four articles and larger corporations of 11, with another four addressing impacts on both large and small businesses. The paper was careful to draw out the job losses cascading from these events, as well as the impacts on a variety of businesses from large international corporations to small local businesses. This example shows that it is possible for a media entity to rise above the various filters that have been argued to bias coverage in an emergency of this scope and do an exemplary and fair job.

Conclusion

For people involved in clearing the rubble, restoring the full functionality of New York and the Pentagon, reactivating and restructuring the national transportation system, rebuilding the World Trade Center and the Pentagon, and trying to mitigate the risk of any similar event striking their regions, the disaster stories were soon submerged under the stories of war. While the war took more and more media attention away from the needs of the victimized people and places, their needs did not just gradually and proportionally fade away with the coverage.

Those responsible for recovery from any extreme event have to work to get the media to focus on the needs still lingering after a catastrophe, including this unprecedented paramilitary strike, criminal action, and disaster. The following recommendations follow from other work on media and disasters, in which I have been involved (Rodrigue et al., 1997; Rovai and Rodrigue, 1998; Rodrigue, 2001b; Rodrigue, 2001c).

First, government agencies and non-governmental organizations can actively cultivate personal relationships between particular reporters and particular representatives of their organizations. A good example of this is the relationship of national media with seismologists Kate Hutton and Lucy Jones of CalTech and the U.S. Geological Survey, respectively. Journalists face extreme time pressure during stories like these and work in a competitive environment. Anything that can help them use their time more effectively and promote accuracy in their final product is appreciated. One of the most helpful things for them in doing their jobs quickly and accurately is simply knowing who the peer-respected experts are ahead of time. Many reporters develop particular "beats" and are open to deepening their knowledge of those areas. Agencies and non-governmental organizations with responsibilities for disaster preparation or response will not be wasting their valuable time by sharing it with interested individual reporters on a regular basis.

Second, those in non-governmental and victim advocacy organizations are in a position to play to the media's need for human drama by generating "newsworthy" events, including demonstrations. A common framework for a story is to communicate information through an emotionally engaging human interest angle. The conflict and drama of a demonstration may generate the coverage needed to set issues of victim needs on the political agenda.

Third, the internet can be used to generate public interest in and support for victims' needs and the organizations trying to meet them. Other work has shown the efficacy of internet organizing in public risk debates (Rodrigue, 2001a; Rodrigue, 2001c). While the World Wide Web is all the rage, it is e-mail, listservers, and news groups that proved the most effective channels. These channels enable exponential expansion of a message to reach an audience of a size and geographical scope once the domain exclusively of national media conglomerates. Getting a message onto listservers or news groups can lead to readers' forwarding it to all of their internet friends and contacts, who forward it to theirs, and so forth. The organization of protests around the meetings of the World Trade Organization is just one example of how effective these new media can be. Disaster managers and victim advocates would benefit from studying internet organizing around a variety of political issues and working out similar techniques that might be appropriate to their own mandates.

References

American War Library. 2002. *Numbers of Americans Killed/Wounded, by Action.* http://members.aol.com/usregistry/allwars.htm.

Bagdikian, Ben Haig. 1997. *The Media Monopoly,* 7th edition. Boston, MA: Beacon Press.

Bennett, Simon A. 2002. "Lock and Load? Explaining Different Policies for Delivering Safety and Security in the Air." *International Journal of Mass Emergencies and Disasters* 20(2): 141–169.

Berelson, Bernard. 1952. *Content Analysis in Communication Research.* Glencoe, IL: Free Press.

Birkland, Thomas A. 1996. "Natural Disasters as Focusing Events: Policy Communities and Political Response." *International Journal of Mass Emergencies and Disasters* 14(2): 221–243.

Brewer, Marcus and Maxwell McCombs. 1996. "Setting the Community Agenda." *Journalism and Mass Communication Quarterly* 73(1): 7–16.

Carey, James W., Mark Morgan, and Margaret J. Oxtoby. 1996. "Intercoder Agreement in Analysis of Responses to Open-Ended Interview Questions: Examples from Tuberculosis Research." *Cultural Anthropology Methods* 8(3): 1–5.

Cohen, Bernard. 1963. *The Press and Foreign Policy.* Princeton, NJ: Princeton University Press

Cohen, Jeff, and Norman Solomon. 1995. *Through the Media Looking Glass: Decoding Bias and Blather in the News.* Monroe, ME: Common Courage Press.

Cohen, Jeff. 1993. *Adventures in Medialand : Behind the News, Beyond the Pundits.* Monroe, ME: Common Courage Press.

Compaine, Benjamin M. and Douglas Gomery. 2000. *Who Owns the Media? Competition and Concentration in the Mass Media Industry.* Mahwah, NJ: Lawrence Erlbaum Associates.

Dearing, James W. and Everett M. Rodgers. 1996. *Agenda Setting.* Newbury Park, CA: Sage Publications.

Elliott, Deni. 1989. "Tales from the Darkside: Ethical Implications of Disaster Coverage." Pp. 161–170 in Lynne Masel Walters, Lee Wilkins, and Tim Walters, eds., *Bad Tidings: Communication and Catastrophe.* Hillsdale, NJ: Lawrence Erlbaum Associates.

Federal Emergency Management Agency. 1998. "Northridge Earthquake." http://www.app1.fema.gov/NR/nr_0106.htm.

Fishman, Mark. 1978. "Crime Waves as Ideology." *Social Problems* 25: 531–543.

Friedman, Sharon M. 1994. "The Media, Risk Assessment and Numbers: They Don't Add Up." *Risk: Health, Safety, and Environment* 5(3): 203. http://www.piercelaw.edu/risk/vol5/summer/friedman.htm

Gans, Herbert J. 1989. *Deciding What's News : A Study of CBS Evening News, NBC Nightly News, Newsweek, and Time*. New York: Random House.

Gordon, Peter, Harry W. Richardson, and Bill Davis. 1998. "Transport-related Impacts of the Northridge Earthquake." *Journal of Transportation and Statistics* 1(2): 21–36.

Haas, J. Eugene, Robert W. Kates, and Martyn J. Bowden. 1977. *Reconstruction Following Disaster.* Cambridge, MA: MIT Press.

Hartwig, Robert P. 2001. "First Nine Months Results." Insurance Information Institute Media, Industry Financials and Outlook Section. December 19. http://www.iii.org/media/industry/financials/2001firstninemonths/.

Herman, Edward S. and Noam Chomsky. 1988. *Manufacturing Consent: The Political Economy of the Mass Media.* New York: Pantheon Books.

Herman, Edward S. and Robert W. McChesney. 1998. *The Global Media: The New Missionaries of Corporate Capitalism*. London: Cassell Academic.

Iyengar, S., M.D. Peters, and D.R. Kinder. 1982. "Experimental Demonstrations of the 'Not-So-Minimal' Consequences of Television News Programs." *The American Political Science Review* 76: 848–858.

Kasperson, Roger E. and Jeanne X. Kasperson. 1991. "Hidden Hazards." Pp. 9–28 in Deborah G. Mayo and Rachelle D. Hollander, eds., *Acceptable Evidence: Science and Values in Risk Management.* New York and Oxford: Oxford University Press.

Kasperson, Roger E., Ortwin Renn, Paul Slovic, Halina S. Brown, Jacque Emel, Robert Goble, Jeanne X. Kasperson, and Samuel Ratik. 1988. "The Social Amplification of Risk: A Conceptual Framework." *Risk Analysis* 8: 177–187.

Krippendorf, K. 1980. *Content Analysis: An Introduction to Its Methodology.* Beverly Hills, CA: Sage Publications.

Lee, Martin A. and Norman Solomon. 1991. *Unreliable Sources: A Guide to Detecting Bias in News Media*. New York: Lyle Stuart.

McChesney, Robert W. 1997. *Corporate Media and the Threat to Democracy.* Open Media Pamphlet Series. New York: Seven Stories Press.

McCombs, Maxwell. 2002. "The Agenda-Setting Role of the Mass Media in the Shaping of Public Opinion." Presentation to Mass Media Economics 2002 Conference, London School of Economics. http://sticerd.lse.ac.uk/dps/extra/McCombs.pdf.

McCombs, Maxwell and Donald Shaw. 1972. "The Agenda-Setting Function of Mass Media." *Public Opinion Quarterly* 36(2): 176–185.

Maher, T. Michael. 1997. "How and Why Journalists Avoid the Population-Environment Connection." *Population and Environment* 18(4): 339–372.

Maxwell, Joseph A. 1992. "Understanding and Validity in Qualitative Research." *Harvard Educational Review* 62(3): 279–300.

Mazur, Alan. 2001. "Looking Back at Fluoridation." *Risk: Health, Safety and Environment* 12(½): 59–65.

Mazur, Alan. 1998. *A Hazardous Inquiry: The Rashomon Effect at Love Canal.* Cambridge, MA, and London: Harvard University Press.

Neuendorf, Kimberly A. 2002. *The Content Analysis Guidebook.* Thousand Oaks, CA: Sage.

Peters, Hans Peter. 1994. "Mass Media as an Information Channel and Public Arena." *Risk: Health, Safety, and Environment* 5(3): 241. http://www.piercelaw.edu/risk/vol5/summer/peters.htm.

Platt, Rutherford H. 2000. "Natural Disasters, Politics, and Property Rights in the U.S.: Priorities in Conflict." *EarthMatters* (Spring): 29.

Quarantelli, E.L. 1989. "A Study of Disasters and Mass Communication." Pp. 1–19 in Lynne Masel Walters, Lee Wilkins, and Tim Walters, eds., *Bad Tidings: Communication and Catastrophe*. Hillsdale, NJ: Lawrence Erlbaum Associates.

Rappaport, Ed. 1993. "Hurricane Andrew, 16–28 August, 1992." Preliminary Report. National Oceanic and Atmospheric Administration. http://www.publicaffairs.noaa.gov/andrew92.html.

Rodrigue, Christine M. 2001a. *Construction of Hazard Perception and Activism on the Internet: Amplifying Trivial Risks and Obfuscating Serious Ones.* Working Paper #106. Boulder, CO: Natural Hazards Research and Applications Information Center, University of Colorado. http://www.colorado.EDU/hazards/wp/wp106/wp106.html.

Rodrigue, Christine M.. 2001b. "Impact of Internet Media in Risk Debates: The Controversies over the Cassini-Huygens Mission and the Anaheim Hills, California, Landslide." *The Australian Journal of Emergency Management* 16(1): 53–61. http://www.ema.gov.au/5virtuallibrary/pdfs/vol16no1/rodrigue.pdf.

Rodrigue, Christine M. 2001c. "Internet Media in Technological Risk Amplification: Plutonium on Board the Cassini-Huygens Spacecraft." *Risk: Health, Safety, and Environment* 12 (3/4): 221–254. http://www.csulb.edu/~rodrigue/cassini/risk01.html.

Rodrigue, Christine M. 2003. "U.S. Casualties during the Afghan War (Enduring Freedom)." http://www.csulb.edu/~rodrigue/afghanwar/.

Rodrigue, Christine M., Eugenie Rovai, and Susan E. Place, 1997. "Construction of the 'Northridge' Earthquake in Los Angeles' English and Spanish Print Media: Damage, Attention, and Skewed Recovery." Presentation to the Southern California. Environment and History Conference, Northridge, California. http://www.csuchico.edu/geop/chr/scehc97.html.

Rovai, Eugenie L. 1994. "The Social Geography of Disaster Recovery: Differential Response to the North Coast Earthquakes." *Yearbook of the Association of Pacific Coast Geographers* 56: 49–74.

Rovai, Eugenie and Christine M. Rodrigue. 1998. "The 'Northridge' and 'Ferndale' Earthquakes: Spatial Inequities in Media Attention and Recovery." *National Social Science Journal* 11(2): 109–120.

Rowley, Coleen. 2002. Statement by Coleen M. Rowley, FBI Special Agent and Minneapolis Chief Division Counsel. http://www.fbi.gov/short/sacrstate.htm.

Sandman, Peter M. 1994. "Mass Media and Environmental Risk: Seven Principles." *Risk: Health, Safety, and Environment* 5(3): 251. http://www.piercelaw.edu/risk/vol5/summer/sandman.htm

Scanlon, Joseph. 1989. "The Hostage Taker, the Terrorist, the Media: Partners in Public Crime." Pp. 115–130 in Lynne Masel Walters, Lee Wilkins, and Tim Walters, eds., *Bad Tidings: Communication and Catastrophe*. Hillsdale, NJ: Lawrence Erlbaum Associates.

Schechter, Danny, Browne Jackson, and Robert W. McChesney. 1997. *The More You Watch, the Less You Know: News Wars/(Sub)Merged Hopes/Media Adventures*. New York: Seven Stories Press.

Singer, Eleanor and Phyllis M. Endreny. 1994. "Reporting on Risk: How the Mass Media Portray Accidents, Diseases, Disasters and Other Hazards." *Risk: Health, Safety, and Environment* 5(3): 261. http://www.piercelaw.edu/risk/vol5/summer/singer.htm

Smith, Conrad. 1992. "Media and Apocalypse: News Coverage of the Yellowstone Forest Fires, Exxon Valdez Oil Spill, and Loma Prieta Earthquake." *Contributions to the Study of Mass Media and Communications* 36. Westport, CT, and London: Greenwood Press.

Stevens, Elizabeth Lesley. 1998. "Mouse-ke-fear." *Brill's Content* (November): 94–103.

Torregrosa, David, et al. 2002. *Federal Reinsurance for Disasters.* Washington, D.C.: Congressional Budget Office.

U.S. Department of Defense. 2000. *Department of Defense Selected Manpower Statistics Fiscal Year 2000.* Washington, D.C.: Department of Defense. http://web1.whs.osd.mil/mmid/m01/fy00/m01fy00.pdf.

U.S. Geological Survey. 1996. *USGS Response to an Urban Earthquake, Northridge '94.* Open-File Report 96-263. Reston, VA: USGS.

Van Natta, Jr., Don and David Johnston. 2002. "Wary of Risk, Slow to Adapt, FBI Stumbles in Terror War." *New York Times* (June 2): National section, p. 2.

Weber, Robert Philip. 1990. *Basic Content Analysis*, 2nd edition. Newbury Park, CA: Sage.

Wikipedia. 2002. "September 11, 2001 Terrorist Attacks/Casualties." *Wikipedia: The Free Encyclopedia.* http://www.wikipedia.org/wiki/September_11_2001_Terrorist_Attack/Casualties.

Zanetti, Aurelia, Enz Rudolf, and Werner Schaad. 2002. "Natural Catastrophes and Man-Made Disasters in 2001: Man-Made Losses Take on a New Dimension." *Sigma Report* 1/2002 (Swiss Reinsurance). http://www.swissre.com/INTERNET/pwsfilpr.nsf/vwFilebyIDKEYLu/SHOR-586KCU/$FILE/sigma1_2002_e.pdf.

Acknowledgments

The author wishes to acknowledge the National Science Foundation-funded Quick Response grant form the Natural Hazards Research and Applications Information Center at the University of Colorado–Boulder, which defrayed costs associated with this research. She would also like to express her gratitude to the NSF for paying for her travel to the Learning from Urban Disasters Workshop hosted by the New York University's Institute for Civil Infrastructure Systems in December 2001 to bring together social scientists, engineers, and public officials early in the research process.

The database for this analysis is available from the author by e-mail. That and other correspondence should be addressed to Christine M. Rodrigue, Professor and Chair, Department of Geography, California State University, Long Beach, CA 90840-1101, (562) 985-4895; e-mail: rodrigue@csulb.edu; http://www.csulb.edu/~rodrigue.

Natural Hazard Mitigation in Oregon: A Case Study

Andre LeDuc
Oregon Natural Hazards Workgroup, Community Service Center
University of Oregon

Robert Parker
Community Planning Workshop, Community Service Center
University of Oregon

Kathy Lynn
Oregon Natural Hazards Workgroup, Community Service Center
University of Oregon

Introduction

Professional planners, hazard mitigation specialists, and emergency response officials have long recognized that better planning can significantly reduce the cost of coping with the aftermath of natural disasters. Effective disaster mitigation requires the coordination of a variety of agencies at different levels of government with many levels of responsibility. This can be achieved by establishing mechanisms for interagency communication with strong working relationships between the involved agencies. Effective interagency communication and partnerships, however, can rarely be established once a disaster occurs. Therefore, it is critical that partnerships be established before disasters happen if damage is to be minimized and response and recovery efforts are to be effective. A recent article published in *Planning*, the magazine of the American Planning Association, highlights research showing that communities making the greatest improvements in safety are located in states that require hazards elements in local plans (Steinberg and Burby, 2002).

The disaster of September 11, 2001, illustrated America's vulnerability to terrorist attacks. In response to September 11th, federal, state, and local governments around the nation implemented their emergency response plans and communication processes which, in some cases, illustrated deficiencies and challenges facing interagency communication. Understanding the gravity of the terrorism threat, and reflecting on the many natural and technological hazards to which Oregon communities are vulnerable, the Oregon Emergency Management (OEM) Agency expressed an interest in evaluating how effective state agency communication is in responding in emergency response and disaster recovery situations. Working with OEM, the Oregon Natural Hazards Workgroup[1] (ONHW) at the University of Oregon received a Quick Response Grant through the Natural Hazards Research and Application Center at the University of Colorado at Boulder to conduct a survey of communication and coordination in the post-disaster environment through the Oregon Emergency Response System.

Oregon has worked to develop coordinated efforts and partnerships and is one of a few states that makes planning for natural hazards an integral element of a statewide land use planning program. Oregon's statewide planning program has required cities and counties to develop and adopt comprehensive land use plans since 1973. Moreover, land use plans must include an element that addresses development in areas of "known natural hazards" (State of Oregon, n.d.).

Well before September 11th, Oregon recognized the urgency of natural hazards planning after the damage and losses suffered statewide from severe winter storms that struck the state in February and November of 1996. These storms triggered heavy flooding and numerous landslides, resulting in property destruction and loss of life in several regions of the state. Immediately after these events, Oregon Governor John Kitzhaber requested that several state agencies review their programs and identify ways of reducing future risks from natural hazards.[2] The Governor specifically directed the Land Conservation and Development Commission to review Statewide Planning Goal 7, Areas Subject to Natural Disasters and Hazards (Community Planning Workshop, 1998). Goal 7 requires incorporated cities and counties to inventory natural hazards and to adopt "appropriate safeguards" to mitigate development in hazardous areas.

The evaluation of Statewide Planning Goal 7 in 1998 marked the beginning of a series of efforts between state agencies, Oregon communities, and the ONHW related to natural hazard planning and mitigation in Oregon. These efforts resulted in the development of resource materials and collaborative partnerships that have been sustained since that time. While the events of September 11th were tragic, they also provided an opportunity for government, businesses, and citizens alike to assess their readiness for natural disasters. Our research examines intergovernmental communication and coordination in this post-disaster environment.[3] We begin with a description of the development and organization of hazard mitigation activities in Oregon, showing how these activities have promoted mitigation planning and coordination and greater communication among governmental agencies. We then present data from a survey of state agency representatives conducted by ONHW to assess the nature of coordination and planning in the post-disaster environment after September 11th. Finally, we discuss implications that Oregon's experience may have for other states that hope to build a framework for sustainable collaboration for mitigation.

Natural Hazard Mitigation in Oregon, 1996–2002

Oregon communities are vulnerable to earthquakes, wildfires, floods, and other natural disasters with the potential to cause extensive loss of life and property and severe disruption to essential human services and the economy. Oregon is ranked third nationally for potential earthquake losses—projected to exceed $12 billion given a major event in the Cascadia Region Subduction Zone. In the past decade, major floods, earthquakes, drought, and severe windstorms, as well as other events, have resulted in over ten statewide Presidential Disaster Declarations (Federal Emergency Management Agency, 2002). With the impacts communities have suffered from disasters there is growing recognition about the importance for and benefits of long-term planning strategies to reduce risk from natural disasters.

In 1996, Oregon began investigating ways in which the state could minimize losses from future hazard events. Raging floods and landslides brought the devastating consequences of natural disasters to public attention, and Oregon state government and communities began to engage in a wide range of activities intended to mitigate the impact of future disasters. Federal and state agencies as well as Oregon universities provided research and technical assistance to local governments, the Oregon Legislature passed key regulatory legislation, and perhaps most important, a strong partnership of government and private agencies dedicated to mitigating natural hazards emerged.

The Catalyst for Mitigation Activities in Oregon: The 1996 Floods

The floods of 1996 and 1997 marked a turning point in statewide hazard mitigation efforts. Between October 1995 and January 1996, northwest Oregon had rainfall that was about 125% above average, producing an unusually wet winter that saturated soils and raised reservoirs to abnormally high levels. Up to this point, little snow had fallen, but beginning in mid-January, both the Cascades and the Coast Range experienced high snowfall in middle and high elevations. In just two weeks by the end of January 1996, average snowpack had reached a level of 112% of average (Oregon Cascades West Council of Governments, 1996, p. 16).

A period of intense cold at the end of January followed by warm rain and air temperatures soon melted the snowpack in the Cascade and Coastal ranges, causing enormous amounts of water to be released into those watersheds—a classic, though unusually powerful "rain-on-snow event." Consequently, stream levels rose, and many very quickly reached flood stage, creating vast amounts of flooding. The flood impacts were compounded by landslides throughout the flooded areas, caused by the overly saturated soils. Several people lost their lives and the region's damage was estimated at more than $280 million (Interagency Hazard Mitigation Team, 1996, p. 12).

The February flood of 1996 followed by flooding in November and December heightened the awareness of officials and the general public about the dangers of such natural disasters. The 1996 floods also suggested the need for cooperative planning within communities and across jurisdictional boundaries to prepare for future disasters. Since the floods of 1996 and 1997, Oregon has engaged in a series of coordinated activities around natural hazard preparedness and mitigation to address this need. Figure 1 provides a timeline of key events following the 1996–1997 floods. The events can be thought of as involving three interrelated lines of activity: (1) research and technical assistance, (2) legislation, and (3) coordination, all of which are described in detail in the following sections.

Figure 1. Timeline of events.

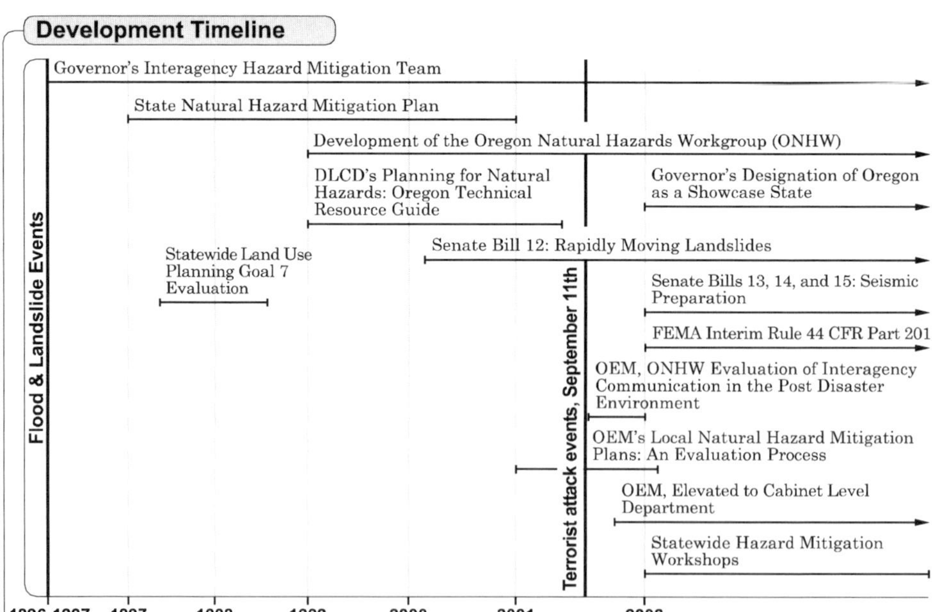

Helping Local Governments: Research and Technical Assistance

As a result of the floods of 1996 and 1997, and at the direction of the Governor, the Oregon Department of Land Conservation and Development commissioned the University of Oregon's Community Planning Workshop to conduct an evaluation of Statewide Planning Goal 7, Natural Hazards and Disasters (Community Planning Workshop, 1998). This evaluation established strong evidence of need for a more coordinated and comprehensive approach to providing technical assistance to local governments in Oregon. Creating effective inventories, policies, and ordinances to guide development in areas where known natural hazards exist requires large amounts of time, energy, labor, resources, and technical expertise to a degree that many local governments simply do not have (Community Planning Workshop, 1998). As a result, communities often experience difficulties in the areas of communication, information access, technical assistance, and general resources available to conduct the large and intricate process of hazard mitigation. Our report concluded that

> The wide range of hazards addressed in local jurisdictions, as well as the inherent complexity of the forces that cause hazardous conditions, stresses the ability of local jurisdictions to effectively plan for and evaluate development in hazardous areas and to ensure that appropriate safeguards are in place. . . . Incomplete or inadequate communication between Department of Land Conservation and Development, hazard specialists in other agencies and organizations, and local planners and decision-makers results in incomplete inventories and confusions concerning proper application of inventories and hazard ordinances. Planners are not always aware of the types of resources that are available to them, how they may access those materials, or whom they may contact for assistance in using the materials (Community Planning Workshop, 1998, p. 43).

Based on these findings we recommended that increased technical assistance be provided to local governments to enhance their (1) knowledge of agencies, specialists, organizations, communities, and other sources that could provide information and guidelines for working through the process; (2) ability to obtain these types of information and guidelines; (3) knowledge of agencies that could offer technical assistance in the process; and (4) ability to communicate with and gain technical assistance from agencies, specialists, organizations, communities and other resources that could assist in the process (Community Planning Workshop, 1998, pp. 11–50).

In response to these findings and recommendations, the state Department of Land Conservation and Development in partnership with ONHW produced the *Planning for Natural Hazards: Oregon Technical Resource Guide* (Oregon Natural Hazards Workgroup, 2001). ONHW developed as a sister program to Community Planning Workshop in 1999 and was founded with a mission to assist communities in reducing their risk to natural disasters. *Planning for Natural Hazards: Oregon Technical Resource Guide* provides tools that Oregon communities can use to plan for, and limit the effects of, threats posed by natural hazards. It describes the requirements of the Statewide Planning Goal 7, methods and data sources for developing accurate inventories, and model policies and ordinances to implement the planning provisions of Goal 7. Development of this guide was a first step in addressing risk reduction and providing education to planners and policy makers in Oregon communities.

The Department of Land Conservation and Development is not the only state agency addressing technical assistance needs of local governments in Oregon. Following on the successful development of the *Planning for Natural Hazards: Oregon Technical Resource Guide,* and in anticipation of the federal rule (the Federal Emergency Management Agency's Interim Final Rule, 44 *CFR*, Part 201) requiring states and jurisdictions therein to develop natural hazard mitigation plans, OEM asked the ONHW to develop a guide to assist in the evaluation of local natural hazard mitigation plans. This document, *Local Natural Hazard Mitigation Plans: An Evaluation Process*, provides guidance on evaluating hazard mitigation plans and a synthesis of standards and approaches developed by state and federal agencies and organizations to assist communities in achieving risk reduction. The criteria outlined in this document address preliminary federal criteria from 44 *CFR* Part 201, as well as other Federal Emergency Management Agency (FEMA) programs, including the National Flood Insurance Program's Community Rating System, the Flood Mitigation Assistance Program, and the Hazard Mitigation Grant Program. While the *Technical Resource Guide* is a broader resource document, the *Evaluation Process* is a tool that can be used to help communities define the planning process and develop strategies for public participation and activity identification during the development of natural hazard mitigation plans.

In addition to development of technical resources, we recognized the need to test these resources by helping communities develop natural hazard mitigation plans, as well as to promote the tools and resources by providing training and outreach opportunities for local communities. Since 2000, ONHW has assisted four cities and three counties in developing natural hazard mitigation plans. These cities and counties requested assistance through their participation in the Flood Mitigation Assistance Program and Hazard Mitigation Grant Program.

In 2002, OEM, in partnership with ONHW and the Department of Geology and Mineral Industries, coordinated Hazard Mitigation and Public Assistance Workshops across the state. Materials presented at the workshop included mitigation resources, programmatic information about the "Partners for Disaster Resistance and Resilience: Oregon Showcase State program," and information on technical assistance for mitigation plan development and implementation.

Over the past several years, Oregon communities have benefitted from increased technical assistance, resources, and public education and outreach programs focused on mitigation. The coordinated partnership between state agencies and ONHW increased the availability and dissemination of these resources and programs to Oregon communities. The real values and opportunities for replication are best illustrated by the time and resources dedicated to developing these programs and the capacity building that occurs at the local level. An organization such as ONHW can assist in developing and maintaining an organizational structure that facilitates long-term planning and implementation of mitigation strategies.

Enhancing the Regulatory Framework: Key Legislation

Since the flood and landslide events of 1996–1997, the Oregon State Legislature has enacted several key pieces of legislation addressing natural hazards. Two key bills have addressed risks from landslides. Senate Bill 1211, which was passed in 1997, addresses risks from rapidly moving landslides around steep forestlands. Senate Bill 12, which was passed in 1999, directs state and local governments to protect people from rapidly moving landslides. The bill has three major components affecting local governments: detailed mapping of areas potentially prone to debris flows, local government regulating authority, and funding for a model ordinance. The direction of these bills, developed through a collaborative process with the Governor's Interagency Hazard Mitigation Team, will result in planning resources such as maps and technical reports based on areas subject to rapidly moving landslides. Coordinating legislation with the development of technical resources can have a direct impact on education and development patterns among communities.

Three bills, all passed in the 2001 legislative session, relate to hazards from potential seismic activity. Senate Bill 13 requires each state and local agency and persons employing 250 or more full-time employees to develop seismic preparation procedures and inform their employees about the procedures and conduct drills in accordance with OEM guidelines. Senate Bill 14 requires the State Board of Higher Education to provide for seismic safety surveys of buildings that have a capacity of 250 or more persons and are routinely used for student activities by public institutions or departments under the control of the board. Finally, Senate Bill 15 requires the Health Division to provide for seismic safety surveys of hospital buildings that contain an acute inpatient care facility and seismic surveys be conducted on fire stations, police stations, sheriffs' offices, and similar facilities. Senate Bills 13, 14, and 15 again illustrate that coordination at the highest level of government can lead to commitment of essential resources and facilitate risk reduction activities among state, regional, and local stakeholders.

The value of these bills derives from the coordination that resulted in their development, the broad-based collective support, and in their geographic reach. Statewide legislation for natural hazards also provide a baseline of activity across communities, even when local legislation fails to pass emergency management-related measures. For example, in the November 5, 2002, election, Oregon citizens passed two seismic rehabilitation Senate Bills for schools and critical facilities, while a local municipality failed to approve a public safety measure.

Leveraging Resources through Partnerships: Coordination

One key outcome of the 1996 and 1997 disasters in Oregon was the development of strong partnerships that have become institutionalized and are involved in hazard mitigation and planning. These include the formation of the Governor's Interagency Hazard Mitigation Team, the development of the State Natural Hazards Mitigation Plan, and the development and implementation of Partners for Disaster Resistance: Oregon Showcase State program.

Governor Kitzhaber declared a State of Emergency for 18 counties in Oregon on February 8, 1996, due to disastrous flooding and landslides. The Governor then directed his administration to identify ways state and local government can minimize loss of life and damages to property from future events. The development of the Governor's Interagency Hazard Mitigation Team in 1997 was a direct outcome from that review. The Hazard Mitigation Team comprises 18 state agencies, including OEM, the Department of Geology and Mineral Industries, and other agencies representing natural resources and the environment, risk management, economic development, utilities, and transportation. The Hazard Mitigation Team's primary goal is to assist in guiding government action related to natural hazard mitigation activities in the state.

The Hazard Mitigation Team played an integral role in providing information and technical expertise during the development of the State Natural Hazard Mitigation Plan. Representatives from state agencies such as the departments of Geology and Mineral Industries, Forestry, Land Conservation and Development, among others, provided technical expertise and experience to participate in inventorying hazard-related issues and resources, and identifying appropriate action items.

On December 12, 2000, the Governor signed an Executive Order designating Oregon a "Showcase State for Natural Disaster Resistance and Resilience" (Executive Order No. 00-31). This Executive Order follows a model developed and tested in Rhode Island by the Institute for Business &

Home Safety (IBHS), an initiative of the insurance industry to reduce deaths, injuries, property damage, economic loss, and human suffering caused by natural disasters. The Executive Order builds upon the foundation established by the state's land use planning laws, building code requirements, emergency preparedness planning, hazards assessment, and other policies and programs. The Showcase State program provides a comprehensive framework for government and the private sector to prepare for and minimize risk and impact of natural hazards. Specifically, the mission of the Showcase State initiative is to prevent injuries and deaths, protect public and private property, and create a disaster-ready statewide economy through public and private partnerships.

The Showcase State program provides a unique opportunity to demonstrate how state and local governments, the insurance industry, and academia can work together to promote awareness of natural hazard risks and associated risk reduction strategies. ONHW is serving as the statewide coordinator and has taken a lead in further developing Oregon's Partners for Disaster Resistance program. The primary objective is to create and strengthen private/public partnerships to enhance disaster safety and preparedness statewide. In its coordinating role, ONHW facilitates and implements activities to motivate behavioral change among communities, individuals and businesses. Figure 2 illustrates the coordination and collaborative roles among communities, the private sector, public agencies, and academia, all partnering together to help realize program objectives. Partners for Disaster Resistance strives to create mutually beneficial

Figure 2. Partners for Disaster Resistance stakeholders.

relationships between the partners. One example of this kind of partnership is illustrated by the City of Tillamook, which experienced floods in 1996, 1997, and 2000. At the state level, OEM provided funds for the city to develop a flood mitigation plan. Making use of the state resources, the ONHW developed the plan with strong collaboration among federal, state, and local government, as well as local businesses and community members.

Organized around 14 interdependent elements (Figure 3), the Showcase State model provides an integrated, cost-effective and systematic approach for all levels of government and the private sector by bringing together resources—both human and financial—to prepare for and minimize natural disaster impacts. These elements, developed by IBHS, are measurable activities that serve both to institutionalize disaster protection into long-range policies, procedures, programs, designs, and plans and to take immediate action to begin to reduce costs associated with disasters. Figure 3 illustrates each of the fourteen elements, which range from building codes to incentive

Showcase State Model – 14 Elements

	Element	Description
1	**Formal commitment and strategic plan**	Obtain Governor-level executive order to formalize partnership. Create 5-year strategy with 1-year action plans.
2	**Statewide hazard and risk assessment**	Identify hazards and what's at risk statewide to help prioritize disaster-resistant actions.
3	**Business recovery alliances**	Develop partnerships with businesses for coordinated mitigation, preparedness, response and recovery.
4	**Enforceable building code**	Adopt and enforce a statewide model code that incorporates hazard-resistant design.
5	**Landuse plans**	Address relevant hazards in state-level landuse decisions. Encourage adoption of local plans that incorporate hazards and mitigation strategies.
6	**Response and recovery plans**	Maintain a state emergency response plan. Develop a state post-disaster recovery plan coordinated with local post-disaster plans.
7	**Rating and regulatory systems**	Improve compliance and participation in natural hazard-related rating and regulatory systems (e.g. the National Flood Insurance Program, Community Rating System, Fire Suppression Rating, Building Code Effectiveness Grading Schedule, etc.)
8	**Lifeline protection**	Incorporate disaster protection measures into public and private lifeline utilities, infrastructure and critical facilities.
9	**Community-level disaster resistance**	Encourage the development of disaster resistant communities within the state and coordinate at local and regional levels.
10	**Public awareness and outreach**	Develop programs to increase the public's awareness of natural hazards and how to reduce or prevent damage.
11	**School curricula**	Incorporate natural hazard awareness and reduction programs into grade-school and higher education curricula.
12	**Protection of childcare centers**	Support IBHS and its partners in the nonstructural retrofit of nonprofit childcare centers.
13	**Professional training**	Conduct mitigation training for building design and construction professionals and others to incorporate disaster resistance into policy and practice.
14	**Incentives and disincentives**	Identify existing incentives and disincentives for hazard loss reduction action. Develop and enact appropriate incentives or adjustments.

Figure 3. Showcase State elements.

programs, and a brief description of their relevance to the Oregon Partners for Disaster Resistance program.

Interest in the partnership emerged from both the public and private sectors after the Rhode Island's Showcase State program was announced in late 1998. The Oregon Department of Geology and Mineral Industries and OEM continue to lead state agency interest from their missions in identifying hazards and reducing public safety risks. SAFECO Insurance Companies, and the Insurance Information Service of Oregon and Idaho lead private-sector interest in minimizing property damage and economic losses and expediting economic recovery after a disaster. The initiative is bolstered by the ongoing work of ONHW. The Partners for Disaster Resistance: Oregon Showcase State program provides a unique opportunity to demonstrate how state and local governments, the insurance industry, and academia can work together to promote awareness of natural hazard risks and associated risk reduction strategies.

A key objective of the partnership is to provide an integrated, cost-effective and systematic approach to prepare for and minimize natural disaster impacts. The state has, for many years, however, had another formal structure to maintain communications between agencies when disasters actually happen. The importance of the ability to maintain these communications became very apparent after the September 11th disaster.

Interagency Communication in the Post-disaster Environment

Interagency communication in Oregon during emergency response and post-disaster recovery is facilitated by the Oregon Emergency Response System (OERS), established by Governor Tom McCall of Oregon in 1972. Its mission is to improve communications and coordination between government agencies that respond to hazardous material incidents and to coordinate and manage state resources in response to natural and technological emergencies and civil unrest involving multi-jurisdictional cooperation between all levels of government and the private sector. OERS is the primary point of contact by which any public agency provides the state notification of an emergency or disaster, or requests access to state or federal resources (State of Oregon, 2002). The OERS is coordinated through the OERS Council, which is composed of staff from 21 state agencies and meets quarterly to discuss communications issues and conduct exercises.

Agencies participating on the OERS Council represent human resources, transportation, corrections, environment, building codes, utilities, and agriculture, among other areas. We distributed a survey designed to evaluate interagency communication and coordination in the post-disaster environment to the directors of the 22 member agencies of the OERS Council in October 2001, and received 20 responses for an 87% response rate.[4] The survey results and findings are organized in three core areas:

- Agency roles and responsibilities and agency understanding of OEM's role in coordinating emergency response and recovery;
- Emergency Coordination Center and OERS coordination roles in emergency response and recovery; and
- Communication and informational needs.

Agency Roles and Responsibilities

OERS is the primary point of contact for state notification of an emergency or disaster. Moreover, OERS coordinates 24-hour access to, and use of, personnel and equipment for all state agencies necessary to assess, alleviate, respond to, mitigate, or recover from conditions caused by an emergency or disaster. OERS provides service through OEM. Given the key role of OEM in directing and coordinating emergency activity, we were especially interested in the extent to which agency representatives understood and accepted this role.

The survey results suggest OEM is relatively effective in communicating its role in coordinating communication and assistance in emergency response and post-disaster recovery to other agencies. Consistent with its statutory mission, OEM is providing information and services to other state and local agencies during disasters and agencies are consulting OEM in post-disaster situations.

The survey findings, however, also suggest that not all agencies are clear on OEM's role in the OERS Council. For example, only 25% of respondents indicated that OEM is responsible for coordinating with FEMA during post-disaster recovery. Coordination with FEMA is one of the stated roles of OEM in disaster recovery. Survey results indicate that agency directors and their designees want to be kept apprised of the state's Emergency Communication Center's coordination and all agency-related emergency response plans and operations. Over 80% of respondents agreed that the nature of the emergency, potentially impacted areas of the state, activities being carried out by government officials to respond to the emergency or mitigate its effects, and actions the public should take for their protection should be included in information released from OEM during Emergency Communication Center activation.

Coordinating Roles in Emergency Response and Recovery

We were also interested in the extent to which agencies perceived that OEM provided effective help with needed services and communication, especially in the coordination of roles during periods of emergency response and recovery. Survey results suggest that up-to-date and accurate information on disaster incidents, and specific needs for response and recovery coordinated and disseminated through OEM and the Emergency Communication Center help agencies provide appropriate services and communicate an accurate message to the public, media, and other agencies. Over 50% of respondents agreed that OERS is very effective and an important means of communication. Additionally, respondents said that the OERS Council helps facilitate communication between state agencies by collecting and disseminating information, providing direct lines of communication, and increasing agency awareness.

Only 30% of the agencies surveyed have an agency or division emergency operations center. Of those that do have a center, 67% use the telephone and e-mail/internet to communicate with the state Emergency Communication Center. About 50% of respondents use the Oregon Emergency Response System. Notably, some agencies or divisions that do not have an emergency operations center do not have a plan for communicating and coordinating with OERS and the Emergency Communication Center. This suggests that communication issues may emerge in post-disaster situations.

Communication and Information Needs

Finally, we were interested in learning the extent to which agencies perceived that information had been effectively disseminated pre- and post-disaster. The distribution of timely and accurate information is a necessary function in emergency response and post-disaster recovery. OERS member agencies that understand OERS and Emergency Communication Center procedures can better facilitate the exchange of information and implementation of services during and after a disaster event. Responses concerning communication about pre-disaster planning, emergency response, and recovery information also suggest that improvements could be made. While 35% rated communication during the pre-disaster environment (in this instance, before September 11th) as somewhat or very effective, 40% of respondents rated such communication as average and 10% rated it as somewhat ineffective or not effective at all. When asked how effectively information has been disseminated among agencies during the post-disaster environment, however, 55% of respondents felt it had been somewhat or very effective in the past, 20% rated it as average, and 15% rated it as somewhat ineffective or not effective at all. While understanding of agency roles and responsibilities during post-disaster response seems relatively high, there is some difference in perceptions among the agencies we surveyed about the effectiveness of communication and information dissemination during pre-disaster and post-disaster situations.

The survey found that many agencies prefer to use communication methods that are dependant on external communication systems (e.g., telephone, e-mail and internet, and cellular phone). Alternative methods of accessing information before, during, and after disaster situations will help communication processes. Respondents suggested the internet, satellite telephones, radio, and other technology. This could present a problem in the event that these external systems are overwhelmed or fail during emergency response and post-disaster recovery activities.

Written comments suggest that information delivered to agencies in a concise and readily available manner would improve effectiveness. Moreover, about 70% of respondents said that communication between and among agencies could be improved by having increased education on current policies and operations. Over 50% felt that frequent exercises and increased training opportunities would further improve communication.

The OERS is a state-level mechanism for communication and coordination in disaster situations. The statutory requirement for this system provides assurance of delegated action in the post-disaster environment, and can serve as a model for communication and coordination. Furthermore, in partnership with state agency activity, and specifically, with the Governor's Interagency Hazard Mitigation Team, OERS provides an opportunity for an integrated and coordinated approach for emergency management. In summary, the OERS Council and operations have proven to be effective communication resources, although clarifying the purpose and use of the system and sending clear and more concise information to OERS Council members would lead to better communication.

Working Towards a Sustainable Collaborative Mitigation Model

The evidence of Oregon's efforts to minimize the impacts and potential loss from disasters is illustrated through a number of coordinated efforts among the public and private sector, policy and legislation, as well as the resources made available to communities statewide that help them to plan for disaster events. In the wake of September 11th, communities throughout the nation are struggling to build and maintain levels of preparedness for potential attacks against homeland security. Oregon is confronting this challenge by working to identify the gaps that exist between disaster response and mitigation, and focusing on long-term planning efforts that will build the capacity of community leaders to develop and implement programs that will reduce community risk from disasters.

Through its public policy framework, Oregon has made progress in natural hazard loss reduction. The state's land use planning laws, building code requirements, emergency management, hazard assessments, and other policies and programs provide a sound foundation on which to build. Engaging citizens and business owners in managing risk can be difficult, as it is challenging to change community and individual behavior. These efforts are not well coordinated or funded, and reduce the effectiveness of disaster safety messages. A recent statewide survey on household hazard preparedness indicated low levels of concern regarding natural hazards (Oregon Natural Hazards Workgroup, 2002). Even state agencies show a low level of preparedness activities, illustrated by the fact that only 30% of the agencies surveyed in the 2002 OERS Council survey indicated that they have emergency operations centers.

Building upon this understanding of the current level of preparedness among state agencies and Oregon citizens, we (ONHW) have continued to work with OEM to develop other resources and community mitigation plans and to coordinate the Partners for Disaster Resistance: Oregon Showcase State program to strengthen the state's risk and loss reduction efforts. OEM and ONHW are proposing to assist communities throughout Oregon to develop mitigation plans that will help them prepare for and reduce risk from natural hazards. The Partners for Disaster Resistance Program is a promising "next step" to bolster limited public resources and create partnerships that will generate activity that could not be accomplished by a single entity working independently.

Oregon's approach to natural hazard mitigation planning aims to build local capacity in developing and implementing risk reduction activities through technical assistance and training, partnership development, and resource sharing. To achieve this objective, the approach fosters partnerships among agencies, communities, and organizations to determine needs, identify issues and resources, and develop strategies for risk reduction. Below, we describe how our approach incorporates activities at state, regional, and local levels; how it builds on strategic partnerships between programs and, in summary, the opportunities and challenges that we see for the future.

Oregon's Tiered Mitigation Approach

Natural hazard mitigation activities in Oregon are organized in three inter-related levels: statewide activities, regional activities, and local activities (Figure 4). Each level of activity makes use of resources provided at the other levels and leads to more coordinated mitigation strategies and plans.

Figure 4: The Oregon mitigation approach.

Statewide Activities: Planning Tools, Resources, and Training

At the broadest level, Partners for Disaster Resistance provides a comprehensive framework for government and the private sector to work in coordination to reduce risk and prevent loss from natural hazards throughout the state. In 2002, Partners for Disaster Resistance developed a five-year strategic plan that provides a baseline of information on level of preparedness and types of mitigation activities among Oregon communities, organizations, and citizens, as well as recommended action items and strategies for implementation. Partners for Disaster Resistance is one example of how state agencies and organizations can come together to develop initiatives, and coordinate resources in a way that increases awareness and activities statewide.

Regional Activities: Planning, Partnerships, and Resources Sharing

Regional activities offer a narrower geographic scope and focus on planning and mitigation activities at the regional level. These planning activities can set baseline vulnerability data and regional mitigation goals and objectives for multiple jurisdictions. This can be accomplished through a comprehensive regional planning process that fosters partnership development, cooperation, and resource sharing among federal, state, and local governments and community and regional organizations. ONHW has found in prior projects that using a collaborative approach to mitigation planning promotes inter-governmental coordination, fosters public/private partnerships, and builds local capacity to develop risk reduction strategies. Since 2000, ONHW has been engaged in fostering regional activities through the development of several county natural hazard mitigation plans. These county plans included stakeholders from a broad range of sectors, and lay the foundation for jurisdictions in those areas to integrate local activities with those existing at the regional level.

One of the most recent examples of regional planning and partnerships is the Clackamas County Natural Hazards Mitigation Plan. Utilizing the collaborative planning process and framework developed by ONHW. Clackamas County, with assistance of ONHW and Showcase State partners, engaged in a year-long process to develop its natural hazard mitigation plan, which recently became the first plan in the nation to meet FEMA's new requirements for natural hazard mitigation plans. OEM also selected Clackamas County to be one of the Pre-Disaster Mitigation Communities for 2002–2003. In the next year, the county will assist its 14 cities in developing jurisdictional natural hazard mitigation strategies.

Local Activities: Plan Development and Implementation

Local activities focus on community-level activities and planning. Public participation processes are an important aspect of all planning and mitigation activities, as this participation feeds directly into action items and implementation strategies for the plans. Mitigation action and implementation strategies are the basis for local mitigation plan goals and objectives. At the local level, everyone involved in natural hazard planning and mitigation can draw on the resources and information at the regional and state levels. This cooperative aspect strengthens plans and leads to more disaster-resistant communities through an understanding of potential risk and methods for addressing the impacts. These plans are specific to a geographic area, yet they draw from statewide knowledge and strategies, and address vulnerability data and regional mitigation goals and objectives from regional plans where applicable.

This kind of coordination and resource sharing is illustrated by vulnerability assessments and other baseline information collected at a regional level that are then made available to local governments and organizations planning for natural hazards. Mitigation actions and implementation measures are most effective at the community level as local considerations drive the planning process. The tiered approach is best illustrated by the use of state resources to develop plans and activities at regional and local levels. In 2001, Washington County partnered with ONHW to develop a natural hazards mitigation plan. The *Planning for Natural Hazards: Oregon Technical Resource Guide* served as a primary basis for technical information for this plan, making use of an important statewide resource. Representatives from state and regional government, the private sector, and community organizations participated in the plan's development. In 2002, the cities of Beaverton and Hillsboro, two of Washington County's largest jurisdictions, are embarking on development of their natural hazard mitigation plans and will make use of both the county plan and state resources.

Addressing Disaster Mitigation Act of 2000 through Strategic Program Partnerships

In February 2002, FEMA published Interim Final Rule 44 *CFR* Part 201, part of the Disaster Mitigation Act of 2000 amendment to the Robert T. Stafford Disaster Assistance and Emergency Relief Act, which requires all states and communities to develop natural hazard mitigation plans by November 2004. These planning and mitigation requirements for states and communities will be accomplished through the Pre-Disaster Mitigation Program. We (ONHW) are working with Partners for Disaster Resistance partners, OEM, FEMA, and local governments statewide to coordinate Partners for Disaster Resistance activities in a manner consistent with the Pre-Disaster Mitigation Program and that will assist communities and the state in meeting the new requirements (Figure 5).

The program partnership between Partners for Disaster Resistance and the Pre-Disaster Mitigation Program is a step toward sustaining and institutionalizing Oregon's goals to reduce risk to natural disasters. Partnering the two programs will assist in achieving the broad goals of both programs, while assisting communities address the requirements of the new federal rule.

To accomplish these goals ONHW and Partners for Disaster Resistance are promoting a collaborative partnership approach to mitigation planning and activities that focuses on inter-governmental coordination, fosters public/private partnerships, and builds local capacity to develop risk reduction strategies and activities. The partnering of the Oregon Pre-Disaster Mitigation Program and Partners for Disaster Resistance is intended to result in an integrated, cost-effective, and systematic approach to prepare for and minimize natural disaster impacts.

The activities of both programs promise to provide measurable outcomes to institutionalize disaster protection into long-range policies, procedures, programs, designs, and plans and to take immediate action in reducing costs associated with disasters. Additionally, this planning process aims to

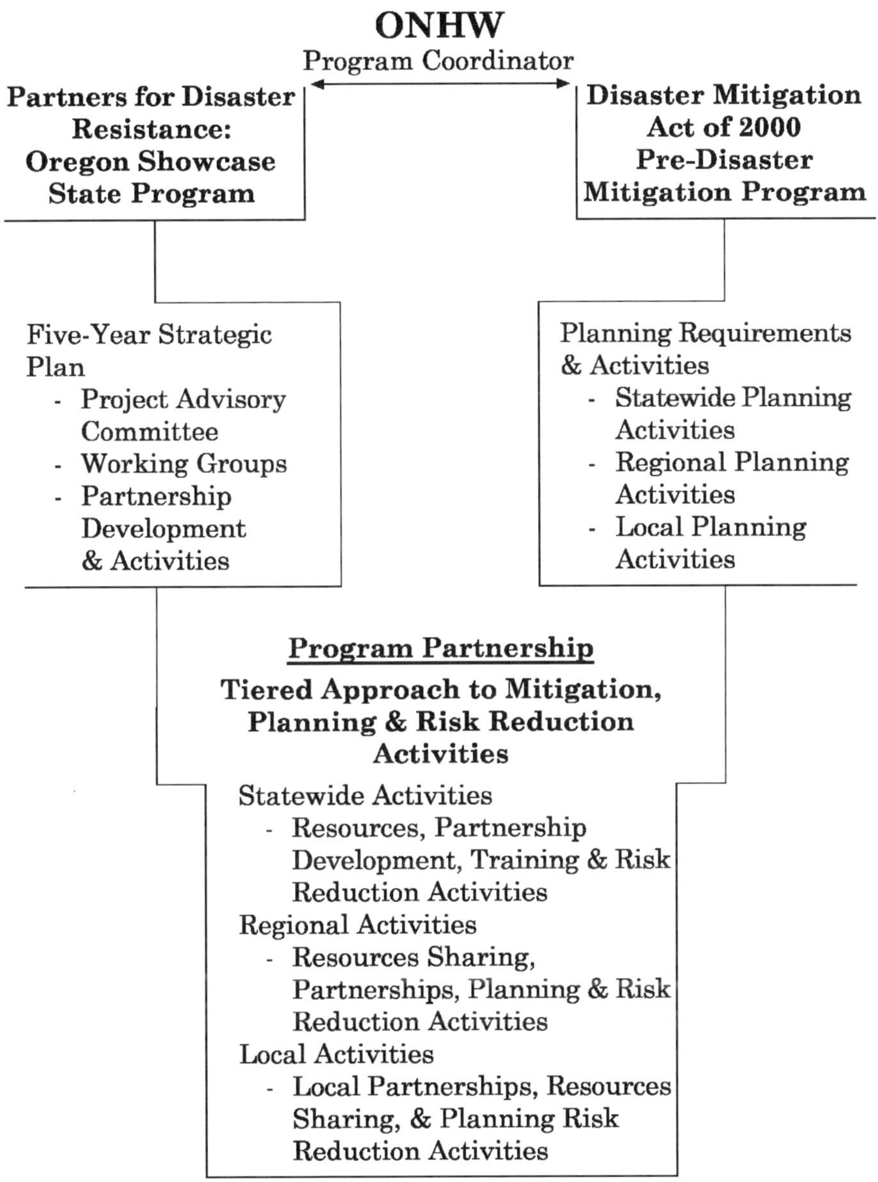

Figure 5: Pre-Disaster Mitigation Program partnerships.

incorporate economic, environmental, cultural, and historical considerations into natural hazard mitigation planning while adhering to state and federal requirements for community mitigation planning. These requirements include the Disaster Mitigation Act 2000, Oregon Statewide Land Use Planning Goal 7, and Senate Bill 360, among other federal and state requirements for mitigation planning.

Opportunities and Challenges

Findings from the agency survey illustrate the complex nature of interagency communication and coordination during disasters. Moreover, current world events demonstrate the importance of pre-disaster planning, and the crucial connection between preparing for, responding to, and recovering from disasters. Interagency communication is essential to all phases of disaster: response, recovery, mitigation, and preparedness, along with state government continuity. Developing disaster management strategies between agencies will lead to improved communication and coordination during disaster events.

Oregon addresses hazards through the Oregon Emergency Response System and the state Natural Hazard Mitigation Plan (Oregon Emergency Management, 2000), as well as through public policy and Statewide Land Use Planning Goals. Through its public policy framework, Oregon has made progress in hazard preparedness and loss reduction. The state's land use planning laws, building code requirements, emergency preparedness planning, hazards assessment, and other policies and programs that establish the basis for loss reduction provide direction for reducing risk and responding to natural hazard events. Moreover, in Oregon, two state organizations play central roles in communication and coordination for disaster management: the Oregon Emergency Response System Council and the Governor's Interagency Hazard Mitigation Team.

As previously stated, the objective of OERS is to provide and implement a plan for coordinated state agency action in cases involving natural or technological hazards or civil disorder that threaten the citizens or resources of Oregon. The Hazard Mitigation Team's broad focus is to understand losses arising from natural and technological hazards, and recommend strategies to mitigate loss of life, property, and natural resources by developing for promulgation a State Hazard Mitigation Plan. Furthermore, there are many other agencies, organizations, and programs throughout the state that are engaging in disaster response, recovery, mitigation, and preparedness. The Hazard Mitigation Team is a strong first step in coordinating mitigation activities across state agencies. Yet, as an initiative of Governor Kitzhaber, the Hazard Mitigation Team has no assurance of permanency during the transition to the next state administration. Thus, it needs to be institutionalized if Oregon is to continue its successes around natural hazard mitigation.

The events of September 11th aggravated an already weak U.S. economy. By December 2001, Oregon had the highest unemployment rate in the nation and faced a $1 billion state budget shortfall, which affected all state agencies, including OEM. All state agencies were asked to submit revised budgets as the Oregon legislature entered into a special session to address the problem. OEM, like many other state agencies, faces staff reductions as part of the budget balancing process. In short, OEM will need to find ways to meet the new demands of the Disaster Mitigation Act of 2000 with fewer state resources. Partnership and collaboration are logical ways to leverage limited resources.

A key challenge facing Oregon is how to integrate response and recovery with mitigation efforts. Our survey of state agencies that participate on the OERS Council suggests that Oregon is doing reasonably well in coordinating agency communications during response and recovery. This is due, in part, to a statutory mandate that agencies participate in the OERS. Mitigation has fared relatively well since the flood disasters of 1996. Discussions with representatives from the OERS and the Hazard Mitigation Team indicate that there is no communication between the two bodies. This disconnect between response and recovery and mitigation is an obvious area where partnerships could be developed that would result in long-term benefits.

Limited financial and human resources pose a challenge to the effective coordination in emergency situations during both the post-disaster recovery period and the pre-disaster mitigation environment. Agencies with direct missions to participate in hazard mitigation activities may be unable to do so, given budget constraints as well as a lack of accountability. One solution to dealing with limited authority would be the creation of a neutral statewide hazard mitigation coordinator who would work with both public and private interests and report to a governor-appointed body like Oregon's Hazard Mitigation Team (a role currently filled by the ONHW). This position would bring important recognition to the current threat of chronic and catastrophic natural disasters and the potential for technological or terrorist events, and bring about a level of coordination and activity among and between state agencies that does not currently exist. Based on the Oregon experience and the unique Showcase State partnership between the state agencies, the private sector, higher education, and communities, ONHW has proven that the coordination of limited resources can generate activity that could not be accomplished by any one group or organization working alone.

Collaboration among all levels of government and the private sector in natural hazards mitigation can result in better issue and resource identification, stronger political will, and more feasible strategies for implementing projects. Furthermore, the successes that ONHW has had as a neutral facilitator of hazard mitigation projects in Oregon has brought representatives from the business community, state agencies, and citizen groups to the same table. ONHW's coordination role illustrates the importance of a having a dedicated and impartial facilitator. ONHW has succeeded in this role by building local capacity, encouraging process and action, and fostering long-term, non-regulatory solutions to the challenges facing Oregon communities.

The success of Partners for Disaster Resistance and ONHW is transferable to other states. The natural divisions between different levels of government as well as businesses and communities makes the coordination role essential. Because these divisions exist in every state, the Partners for Disaster Resistance model can easily be implemented elsewhere. Success of the model, however, is not assured. It requires political and financial support. The political support in Oregon came from Governor John Kitzhaber's Executive Order declaring Oregon a Showcase State for Disaster Resistance and Resilience. Financial support has been more challenging. The coordination role provided by the ONHW has been funded through a grant from the Public Entity Risk Institute as well as state and local governments. Long-term, stable funding remains a barrier to the Partners for Disaster Resistance.

Our initial research focused on evaluating interagency communication in the post-disaster environment. Oregon's history of exposure to natural hazards, as well as the events of September 11th illustrate the essential role of public/private partnerships and demonstrate the successes of these partnerships. We should not lose sight, however, of the importance of planning for and reducing risks to natural hazards.

Further examination of communication and coordination in all phases of the disaster cycle, and among the various groups that engage in disaster management, can potentially lead to improved coordination and implementation of disaster management strategies for both natural and technological hazards. Activities to further understand and strengthen disaster-related interagency communication could include

- Examining the current state infrastructure for disaster management and how interagency communication is involved in coordinating disaster response, recovery, preparedness, and/or mitigation.

- Exploring the relationship between OERS and the Governor's Interagency Hazard Mitigation Team, and whether OERS's official state role makes it more efficient than the Hazard Mitigation Team, which is not a formal organization.

- Examine how the missions and composition of individual agencies affect their ability to communicate, coordinate, and respond during pre- and post-disaster periods.

- Examine the role that higher education programs such as the ONHW can play as a coordinating mechanism between federal, state, and local governments, business, and communities.

The public/private partnerships being established in Oregon are no better illustrated than by the coordination of the federal Pre-Disaster Mitigation Program and the Partners for Disaster Resistance: Oregon Showcase State program. This partnership will enable agencies and organizations to leverage resources for long-term planning and mitigation for natural hazards. Historically, there has been a focus on emergency response and recovery. Oregon's focus on mitigation is part of a paradigm shift highlighting risk reduction, thereby providing a cost-effective approach to reducing disaster loss. The coordination between the Pre-Disaster Mitigation Program and the Partners for Disaster Resistance Program, along with the proposed change in OEM's state agency status and the multitude of community mitigation activities statewide, has created a foundation for project implementation and future mitigation successes.

Notes

1. The Oregon Natural Hazards Workgroup is a program within the Community Service Center (CSC) at the University of Oregon. The CSC is a consortium of programs assisting Oregon communities and providing service-learning opportunities to University of Oregon students. Additional programs within the CSC include the Community Planning Workshop and Resource Assistance for Rural Environments. The authors of this paper are all on the staff of the Community Service Center.

2. Oregon Emergency Management, Department of Land Conservation and Development, Forestry, and Geologic and Mineral Industries were the key state agencies involved in the review.

3. For the purpose of this paper, we use the term "intergovernmental" to mean state agencies.

4. Of the 22 surveys distributed, there were 20 responses and three non-responses. One agency submitted responses from two different divisions.

References

Community Planning Workshop. 1998. "Implementation of Goal 7: An Evaluation and Discussion of Hazard Planning in Oregon." Eugene, OR: University of Oregon.

Federal Emergency Management Agency. 2002. Presidential Disaster Declarations. http://www.fema.gov/library/drcys.shtm.

Interagency Hazard Mitigation Team. 1996. *February 1996 Flooding, Landslide and Stream Erosion in the State of Oregon.*. FEMA-DR-1099-OR. Washington, D.C.: Federal Emergency Management Agency.

Oregon Cascades West, Community and Economic Development. 1996. *The Cascades West Region of Oregon and the February Flood of 1996: A Regional Flood Recovery Plan for Benton, Lane, Lincoln and Linn Counties.* Salem, OR: Oregon Economic Development Administration.

Oregon Emergency Management. 2000. *State of Oregon Emergency Management Plan: Natural Hazards Mitigation Plan.* Salem, OR: State of Oregon.

Oregon Natural Hazards Workgroup. 2002. *Household Disaster Preparedness in Oregon: Results of a Statewide Survey.* Eugene, OR: University of Oregon.

Oregon Natural Hazards Workgroup. 2001. *Planning for Natural Hazards: Oregon Technical Resource Guide.* Salem, OR: Oregon Department of Land Conservation and Development.

State of Oregon. n.d. Oregon Administrative Rule 660-015-0000(7), Planning Goal 7: Areas Subject to Natural Disasters and Hazards. Salem, OR: State of Oregon.

State of Oregon. 2002. Oregon Emergency Management website. http://www.osp.state.or.us/oem/.

State of Oregon. 2000. Executive Order No. 00-31, Oregon Showcase State Partnership for Natural Disaster Resistance and Resilience. Salem, OR: State of Oregon. http://www.governor.state.or.us/governor/legal/execords/eo00-31.pdf.

Steinberg, Michele and Raymond J. Burby. 2002. "Growing Safe: Is Your Comprehensive Plan Hazardous to Your Community's Health?" *Planning Magazine.* American Planning Association (April): 22–23.

Acknowledgments

We would like to acknowledge our funding sources for the Partners for Disaster Resistance and Resilience and research including the Public Entity Risk Institute, Oregon Emergency Management, and the Natural Hazards

Research and Application Center at the University of Colorado–Boulder for providing a quick response grant for the survey research.

We would also like to thank our state agency representatives for all of their assistance on this project. The authors would like to give a special thanks to the Governors Interagency Hazard Mitigation Team, Dennis Sigrist, Abby Kershaw and Dave Cassel, Oregon Emergency Management. An additional thanks to Skip McFarlane of the Community Service Center for graphic design work and Jean Stockard, Ph.D., Planning, Public Policy and Management, University of Oregon for editorial comments and support.

Correspondence should be directed to Robert Parker, Community Planning Workshop, University of Oregon, 1209 University of Oregon, Eugene, OR 97403-1209; (541) 346-3801; e-mail: rgp@darkwing.uoregon.edu or to Andre LeDuc, Community Service Center, Oregon Natural Hazards Workgroup, 1209 University of Oregon, Eugene, OR 97403-1209; (541) 346-5833; e-mail: onhw@darkwing.uoregon.edu.

Effects of the Terrorist Attacks of September 11, 2001, on Federal Emergency Management in the United States

Claire B. Rubin
Institute for Crisis, Disaster and Risk Management
The George Washington University

Irmak Renda-Tanali
Institute for Crisis, Disaster and Risk Management
The George Washington University

This report discusses the defining characteristics of the terrorist attacks on the United States on September 11, 2001, the role of the media, and the initial role and functions of two responding agencies—the U.S. Environmental Protection Agency (EPA) and the U.S. Coast Guard (USCG) in New York City. It also briefly describes the various impacts: economic and financial, damage to infrastructure, equipment losses, business interruption, human productivity, airline losses, insurance payouts, decreases in tourism, revenue losses, impacts on the stock exchanges, and donations and charities. The authors also evaluate the effects on public attitudes toward government, the new national public awareness of terrorism, public awareness of emergency management, and changes in public sector focus and workload. The authors describe anticipated changes in federal policy to better deal with such events in the future.

In the course of working on this report, the authors were stimulated to develop a related product: the *Terrorism Time Line: Major Milestone Events and Their U.S. Outcomes (1988-2001),* which was first published in March 2002 (Rubin and Renda-Tanali, 2002).

Introduction

The terrorist attacks on the World Trade Center and the Pentagon were horrific events, of a scale and type never before seen in the United States or the world. To our knowledge, no past terrorist disaster in the United States has resulted in both recovery and military actions to seek redress for the incident.

Given the timing, nature, and magnitude of the attacks, plus the immediate extensive media coverage, the topics of terrorism and emergency management received an unprecedented amount of attention not only in the United States but worldwide. Aspects of terrorism usually reserved to a small group of behind-the-scenes operational personnel suddenly became of interest and concern to citizens throughout the nation.

In researching and documenting the outcomes of the events in New York City and the Pentagon, the authors chose to focus primarily on emergency management at the federal level. Even with this limited focus, the effects of September 11th on the federal government involve a vast array of impacts and outcomes. This report briefly describes the events and their effects, primarily the early impacts and ramifications. It does not cover the many problems and issues connected with the public management of health and environment that began to emerge about four weeks after the attacks took place. Finally, our research relied mainly on secondary sources, because it was not possible to gain access to key actors for personal interviews in the first few weeks after the massive events.

The Unprecedented Role of the Public Sector

As noted by Waugh, "Emergency management is the quintessential governmental role. It is the role for which communities were formed and governments were constituted in the first place—to provide support and assistance when the resources of individuals and families are overwhelmed" (Waugh, 2000). For the emergency management community these vastly destructive terrorist attacks have a large number of aspects, impacts, and implications that are unprecedented. Clearly, the September 11th events will go down in history as a major milestone in emergency management and probably will result in major changes in the emergency management systems at each level of government in the future.

Given the vast scope of impacts and ramifications for government actions and policies, at every level of government, this paper can only outline or briefly discuss some of the impacts and outcomes of the September 11th event. This report should be viewed as an early step in what is likely to be a long-term sequence of analyses and reports about a milestone set of disaster events.

Approach

Our approach was to use the conceptual framework of the *Disaster Time Line: Selected Milestone Events and Outcomes (1965-2001)* as a starting point (Rubin and Renda-Tanali, 2001). The authors set out to research and document some of the political and policy impacts of the September 11th attacks and their ramifications at the federal level. While working on the *Disaster Time Line*, the authors discerned a predictable sequence of actions and outcomes from major defining disaster events since 1965. The key categories are major after-action reports and documents; legislation, regulations, and executive orders; response plans; and organizational changes.

Two issues arose while trying to apply this approach. Initially, it appeared that September 11th did not fit into the sequence observed previously. The authors later decided that although these events have some aberrations, they did fit into the basic pattern. The details of this finding will be discussed later. Second, the authors prepared a new graphic in order to focus on the details and underpinnings of the federal involvement in counter-terrorism, from 1988 through 2001. In the *Terrorism Time Line*, the authors documented the expected sequence of actions and determined, to the extent possible, the causal relationships between the events and major outcomes.

Events of September 11th

Many researchers and journalists have produced detailed descriptions of the events and the response efforts. Highlights of some of the most pertinent facts and some observations about their implications follow.

At 8:45 a.m. EDT on Tuesday, September 11, 2001, an American Airlines aircraft was hijacked by a group of terrorists after taking off from Boston and crashed into the north tower of the World Trade Center Complex in New York City. At that time, the severity of the incident, the numbers of people involved, and the reason for the crash were all unknown. At 9:03 a.m. a second plane, this time a United Airlines plane, hit the south tower of the World Trade Center.

During the period between the first and second crashes in New York City, the Washington Area Airport Authority had begun evacuating Reagan, BWI, and Dulles airports as a precaution. Immediately after the second crash, the Federal Aviation Administration issued a national "ground stop," which prevented all civil flights from taking off, thereby acknowledging that these actions were deliberate and that more attacks might be underway.

The roads were being closed in Washington, D.C., and the mayor had just given the order to evacuate the city, when another American Airlines plane hit the Pentagon office building in Arlington, Virginia, at approximately 9:40 a.m.. The FAA issued an immediate order to ground all planes flying in U.S. airspace. The news spread quickly through blanket media coverage that a fourth plane was heading towards Washington, D.C., with the expectation that it was aiming for the Congress or quite possibly the White House. Around 9:45 a.m. the decision was made to evacuate the White House.

At about 10:00 a.m., a fourth commercial plane went down in Somerset County, Pennsylvania, about 80 miles southeast of Pittsburgh. About the same time a partial collapse occurred at the Pentagon building in the area of impact. Shortly after 10:00 a.m. the south tower of the World Center collapsed. Within the next half-hour, the northern tower of the World Trade Center collapsed. At approximately 5:30 p.m. a third tower in the World Trade Center complex (Building 7) also collapsed.

Defining Characteristics

These attacks obviously were extraordinarily well planned and coordinated. They clearly had the goal of damaging the symbols of power in the United States, causing as many casualties as possible and spreading fear. Also, by hitting at the World Trade Center Complex in New York City, which is the heart of the international financial community, there is no doubt that the terrorists hoped for long-term negative economic consequences.

Not just the people living in New York City or in Washington, D.C., but also many millions of people all across the country felt they were potential targets, especially those living in other large cities. The local, state, and federal responses were immediate and massive amounts of resources were deployed to the attack sites. Initially, it was estimated that the casualties in the World Trade Center could be around 10,000 and 800 people were estimated to be dead in the Pentagon incident. Sadly, the initial fire fighting teams, including the New York City Fire Chief, deployed to the scene were among the dead and missing. The loss of about 300 skilled fire fighters and their chief was a major blow to the response force.

In addition to responding to the known disasters, prevention of further damage was a major concern. As these catastrophic series of events occurred, it was not—and it still is not—clear whether there were other attack plans and when the threat of further attacks would end. Both elected and appointed officials had to take immediate actions and make the kind of decisions that they had never made before to fulfill their duties to the citizens. No doubt the terrorists intended to shake the public trust towards the government. One immediate worry was how four commercial jetliners could have been successfully hijacked from different airports and their whereabouts while in the air remain unknown.

President Bush's mission changed profoundly in a matter of hours. He was forced to assume a defensive role for both himself and the country. And when the source of the attacks was determined, he mounted a war offensive against the perpetrators and other allied terrorists located in many countries. Within hours, measures were taken to ensure the continuity of the government, to avoid mass panic, and to protect the nation and its citizens from further attacks.

As thousands of members of urban search and rescue, emergency medical, emergency response teams, and tons of equipment were deployed, it became obvious that the debris removal would take months, if not years, and hopes of finding any survivors quickly faded. The Federal Bureau of Investigation (FBI), other federal teams, and the New York Police Department (NYPD) began the enormous task of sorting and sifting through debris for bodies and evidence, a task that also could take several months.

The Role of CNN and other Media

Given the time of day, and the fact that many governmental and financial workers have access to internet and television news, word and pictures of the events spread fast. Thanks to CNN and other media, many public officials could see the actual scenes of the events in New York City and at the Pentagon within minutes of their occurrence and were able to take action, such as opening emergency operations centers (EOCs) before being requested to do so officially.

Federal Response Actions

What follows are two brief examples of initial response actions on the part of federal and military organizations, U.S. EPA and the U.S. Coast Guard.

U.S. Environmental Protection Agency Headquarters

On Tuesday morning, September 11th, at the time of the first attack on the World Trade Center, at EPA headquarters in Washington, D.C., the Emergency Coordinator for the Agency, Jim Makris, and his deputy were engaged in a previously scheduled briefing for the EPA Administrator about EPA's emergency management system and capabilities. They received a call and were told to turn on the television to see the attack details. The officials then ended their meeting and opened the EOC immediately thereafter to begin disaster operations, according to Ed Terry, the manager of EPA's EOC. Shortly thereafter, EPA headquarters established links with all of its East Coast regional offices to begin coordination and support of the New York City response efforts.

EPA has the authorities and responsibilities needed to perform emergency response functions under the National Contingency Plan. In addition, when the Federal Response Plan is activated, EPA has the lead responsibility for Emergency Support Function #10: Hazardous Materials. In this case, no one waited for formal initiation of any of the emergency response plans, but went right to work with their existing authorities (Terry, 2001).

U.S. Coast Guard—Initial Response in New York City

Captain Dennis M. Egan, U.S. Coast Guard (USCG), who is the Director of the National Response Team (NRT), first learned about the New York City disaster on CNN television. He immediately ordered that the alarm to the FBI's Weapons of Mass Destruction WMD hotline be activated. Rescue helicopters were sent to New York City from USCG bases in Atlantic City and Cape Cod. When helicopters arrived one hour later, NYPD helicopters already were on scene. The USCG's Long Island helicopter facility was stocked for support of the NYPD for several days, but not used in the search and rescue. The New York City government immediately moved its resources from Staten Island to Manhattan.

Various ferry ships, under USCG direction, were used to evacuate civilians out of Manhattan. The ships involved were the Staten Island Ferry and three other ferries; there were no major USCG ships in the area. Captain Egan commented on the fact that the USCG ships were heading in, while the Navy ships were heading out of harbor. Many people fleeing the fires and destruction from the World Trade Center area ran toward the water, at the foot of Manhattan. The local police and USCG officials on board the ferries were armed and available for assistance. Egan commented that USCG was a counter-terrorism "node" in these actions. The USCG went quickly from the response to security phase when it began screening passenger vessels and putting armed guards on cargo ships.

When the second plane hit the World Trade Center, USCG area commanders were contacted. The Boston USCG Admiral invoked "'regional incident command," and was established as the senior USCG official in New York City. He was instructed not to be in charge of the entire incident. He joined in the governor's and the mayor's response activities, but returned shortly thereafter to his post in Boston.

The USCG Strike Teams set up in New York City to get the stock exchanges open again. They also did air sampling in the area. The USCG used the "Vessel Traffic System" for navigation around the city. Because the antenna on the World Trade Center tower was a major part of the system, range was reduced significantly. A new antenna was rigged on Staten Island as a backup.

The Coast Guard observed that the Federal Emergency Management Agency (FEMA) set up its Regional Operations Centers after a five- or six-day delay, due to the communications failures at the Federal Center in New York City, discussed elsewhere in this report. Egan commented that there were no major initial turf wars to report. The mayor was "significantly in charge."

Communication was perhaps the greatest problem. All cell phone lines were dead. Only two major phone trunk lines into New York City remained, and both were completely saturated (this problem persisted for several days). The National Response Center sent three portable communication units by van to New York City the night of September 11th. Those units were established at Battery Park, Staten Island, and on the USS *Comfort*. Nevertheless, the USCG had trouble setting up communications with those in charge in New York City.

Battery Park was taken over by the City of New York and by the FBI as a command center. The FBI Atlantic Strike Team had some initial trouble getting communications set up because their system depended on access by a self-contained van unit, which could not navigate the rubble-covered streets.

Within two hours of the start of the attacks, there was a National Response Team conference call. At about 1:00 p.m., there was another. The USCG established a liaison at the FBI Strategic Intelligence Operation Center; this post was filled for two weeks.

Captain Egan noted that the most valuable preparations for the actual response of the USCG on September 11th was due to TOPOFF, which was a major federal disaster exercise, mandated by Congress, held in 2000. This exercise apparently created many contacts that were vital in the September 11th response (Egan, 2001).

Emergency Management Considerations

In New York City, initial efforts on the part of local federal regional offices to deal with emergency response were hampered by damage to the city's EOCs. New York City had recently completed a multi-million dollar state-of-the-art EOC; but it was housed in one of the World Trade Center buildings that was totally destroyed. The State of New York seemed to fare better. The Federal Center in New York City was not physically damaged, but telecommunications were knocked out, which meant that FEMA Region II, EPA II and other federal agencies had to find other operational locations (Federal Emergency Management Agency, 2001).

In Arlington, Virginia, the response relationships appeared to be efficient and effective, since the Arlington County Fire Department and Pentagon officials had worked with each other and conducted response exercises before September 11th.

At the national level, things moved very quickly with Presidential declarations of emergency for the Pentagon and disaster for New York City. The conventional procedures for obtaining a Presidential declaration were not necessary: "self-initiating" requests, as allowed by the Stafford Act, occurred and the federal government as well as the military services began their response actions very rapidly.

Among the response actions that were highly unusual or unique to September 11th were the following:

- **Emergency and disaster declarations**: "self-initiating" declarations; use of an emergency and later a disaster declaration at the Pentagon.
- **Problems with EOCs** at the local and federal levels due to destruction and incapacitation, respectively.
- **White House involvement:** rapid creation and selection of a director for the Homeland Security Office. While fully operational, the White House and some federal agencies were making, or planning, major changes in processes, procedures, funding, and organizational arrangements for emergency management.

The Impacts

As of March 17, 2002, the latest information from federal and local officials, as reported in the *New York Times* (2002), gave the following totals for the number of people dead or missing from the September 11 attacks:

- In New York City, approximately 2,830 deaths had been confirmed. That number includes the 157 people on the two hijacked planes at the World Trade Center. Only 773 of the 2,830 people who died had been recovered and identified, although the remains of many are still being analyzed. Additional remains were recovered almost daily for more than nine months.
- At the Pentagon site, a total estimate of 189 persons died; 64 people, including the crew, died on board the hijacked plane; another 125 were dead or missing in the Pentagon building.
- At the Pennsylvania plane crash, 44 were confirmed dead on the hijacked plane initially.
- The number of injuries was relatively small, because all of the above events were so devastatingly deadly.

The Economic and Financial Impacts

It is a challenging task to calculate the overall costs of September 11th attacks. The destruction of the World Trade Center obliterated about 12 million square feet of Class A office space, which is the equivalent of all office space in Atlanta or Miami (Tully, 2001, p. 94). An additional 18 million square feet of office space in downtown Manhattan was damaged.

Infrastructure

In New York City, a significant amount of infrastructure was ruined in the neighborhood of the World Trade Center complex, including a crushed subway station, plus the loss of five phone-switching stations, two electrical substations, 300,000 telephone lines, and 33 miles of cable. It has been estimated that replacing the destroyed subway lines will cost around $3 billion and that utility repairs, including 300,000 telephone lines, one phone switching station, and six miles of electrical cable will cost $2 billion. Additionally, rebuilding the PATH NY/NJ station below World Trade Center would be about $2.4 billion. The estimated total cost for replacing the basic infrastructure is $7.4 billion (Coy et al., 2001, p. 114).

The Pentagon office building, which is owned by the Department of Defense, is estimated to have sustained $1 billion in damage. It was fortunate that the hijacked plane hit the Pentagon in the newly remodeled section, since relatively few people were in the not-yet-completed offices and the structure, windows, and other construction details were more attack-resistant than the rest of the building.

Equipment Losses

Going beyond the infrastructure costs in New York City, there were equipment and related losses—such as fire trucks, thousands of computers, furniture, and other equipment—that disappeared with the towers. Early estimates suggested that anywhere between $2 to $5 billion worth of telecom and computer equipment was destroyed. The total property loss was estimated at $34 million, according to the New York City Comptroller Alan Hevesi. That is nearly twice the $16.8 billion record set by 1992's Hurricane Andrew (Tully, 2001, p. 100). Similarly, but on a smaller scale, at the Pentagon computers, office equipment, and other unknown equipment and supplies were consumed in the fire after the plane hit.

Another unusually large cost in New York City was related to dealing with the immense amount of debris over the multi-acre area disaster site. The debris had to be sorted first for human remains and evidence, and later deposited in a landfill. The New York City Controller predicted that it would cost $14 billion just to clean up and police the site.

Business Interruption

The New York City site probably set an all-time record for business interruption costs, which were initially estimated at $21 billion; the most serious losses occurred in the downtown neighborhoods that were inaccessible for weeks after the attacks (*The Washington Post*, 2001a). Six months later, an official from the City of New York's Office of Emergency Management gave an estimate of $83 billion for the overall economic impact on the city from the attacks, based on her discussion with the business community (Moroccolo, 2002).

Built in 1970, the World Trade Center housed more than 430 companies from 28 countries. They were engaged in a wide variety of commercial activities, including banking and finance, insurance, transportation, import and export firms, customs brokerage, trade associations, and representatives of foreign governments. An estimated 50,000 people worked in the World Trade Center, and another 140,000 visited the complex daily. Estimates of how many people were in the World Trade Center when the attacks began vary from 15,000 to 40,000, according to an article in *The Washington Post* (2002). Thus, the ratio of people who safely got out of the many impacted buildings was many times higher than the number who died there on September 11th.

Companies like Morgan Stanley, which by far was the World Trade Center's largest tenant—with 3,700 employees (all but 15 unaccounted for)—was fully operational less than 48 hours after the tragedy. Remarkably, Cantor Fitzgerald lost 680 of its 1,000 employees but was operational for bond trading two days after the attacks.

Many Wall Street firms would have been inoperative for many more weeks after the attacks had it not been for the careful contingency planning they began after the 1987 stock market crash and accelerated after the 1993 World Trade Center bombing. These financial firms rely on two critical services to guarantee a quick rebound from natural and human-made disasters: (1) information backup services that collect computer tapes and store them in highly secure suburban facilities, and (2) alternative facilities that are fully equipped with mainframes and computer servers that replace lost computing power. For a subscription fee, plus a disaster assessment that may run into the millions of dollars, stricken firms were able to move their personnel to such a service provider's centers for up to six weeks (Tully, 2001). (After that the companies had to find their own space). Many companies have decided that it is prudent to spread operations over multiple locations on different electrical grids and telephone networks (Coy et al., 2001).

Human Productivity

Another sad but important indicator of loss is the loss of human lives and their future productivity as indicated in purely financial terms. Given the average age of the workers who lost their lives (40), the New York City Comptroller estimated the "lost human productive value" to be about $11 billion. Measured by payroll, New York City, with less than 3% of the country's workforce, accounts for 37% of the U.S. securities industry, 20% of advertising, and 18% of book publishing. The best and brightest from around the world are drawn to New York because it is where they can do their finest work and reap the highest rewards. In the short run, the September 11th attack would add a $500 billion blow to a city economy already stumbling from the bear market on Wall Street and the nationwide slump. More than 100,000 New Yorkers thus would eventually be thrown out of work by the attack, according to New York State Labor Department estimates (Coy et al., 2001, p. 104).

Airline Losses

The airline industry received a major blow due to the temporary shutdown of the air travel system and later widespread fear of flying by potential customers. Airlines and airfreight were down for weeks. People who chose to fly faced long lines due to increased security measures. Anything suspicious became a reason to ground planes. After the attacks, the airlines received a $15 billion government bailout, announced 100,000 layoffs, and slashed 20% of their flights (*The Washington Post*, 2001a).

In the Washington, D.C., area, Reagan National Airport and its businesses were the hardest hit in this ordeal. The airport was ordered to shut down immediately after the attacks and was not allowed to open until 23 days later due to its proximity to so many potential targets. The cost for closing was $330 million per day to the airport and northern Virginia businesses and $27 million to state and local tax revenue (Coy et al., 2001, p. 112).

Insurance Payout

The $126 billion commercial insurance industry is facing a $30 billion payout. This industry will never quite be the same, since insurers and reinsurers had never considered terrorism when pricing their premiums. The uncertainty about how to predict future attacks is a huge challenge for the insurance industry.

Tourism Income Losses

The tourism industry has been hit hardest in the Washington, D.C., area and New York, but with secondary and tertiary effects in Boston, Los Angeles,

Las Vegas and other major tourist destinations. About one-third of the nation's 265,000 unionized hotel and restaurant workers have been laid off. Hotel expansion plans have been on hold almost everywhere (Coy et al., 2001).

Revenue Losses

The U.S. economy, threatened by recession before September 11th, suffered a number of blows in the weeks since. The leading economic indicators dropped in September, yet the nation's financial markets have thus far weathered the uncertainty, making up the losses experienced in the days after reopening.

Former Mayor Rudolph Giuliani estimated the city would lose $1 billion in revenues this fiscal year—including a 20% decline in personal income taxes and more than 30% declines in hotel and real estate transfer taxes. Additional costs for additional police overtime, downtown cleanup, and other services would soar into the billions of dollars. Even with the help from Washington, New York was expecting a budget deficit of $4 billion in the next fiscal year. The city agencies would have to cut $1 billion from their spending plans. The federal government would reimburse the city for $11.4 billion in expenditures directly related to the attack, such as $5 billion for emergency construction at the World Trade Center site, and $3.8 billion for police, fire, and health services. Congress approved $20 billion in aid for New York, Virginia, and Pennsylvania (Coy et al., 2001).

Stock Exchanges

The New York and American Stock Exchanges were closed for a week until September 18. The stock market declined by double-digit percentages immediately after the terrorist attacks. The New York Stock Exchange dropped 1,369 points, the biggest point loss and the fifth worst week ever for the Dow Jones industrial average.

Charities and Donations

As a result of September 11th and all of the media attention given to the disaster, an unusually large number of charities formed, in addition to those already in existence—such as the Red Cross and the Salvation Army—and an unprecedented amount of donations were received. The resulting problems ultimately had to be straightened out by the Attorney General of New York City. As a sidebar to this topic, the current President of the American National Red Cross lost her job as a result of some disputes with the Board of Directors of that organization. It should be noted that donations related to the Pentagon disaster do not appear to have the same complications.

Health and Human Services Operations

According to a news release from the U.S. Department of Health and Human Services (HHS), the September 11th response in New York City constituted the largest National Disaster Medical System (NDMS) response ever. Of the more than 9,500 rescue workers, 1,364 were volunteer health and mortuary professionals who provided their services as part of the national NDMS, and more than 600 others were health professionals from HHS Commission Corps Readiness Forces and the Centers for Disease Control and Prevention. Disaster Mortuary Operations Response Teams supported the New York City Medical Examiner's Office, processing 15,528 human specimens and 270 bodies, and identifying 750 victims. On September 11, 2001, HHS declared a national health emergency; the Office of Emergency Preparedness immediately deployed NDMS and Commissioned Corps teams to the disaster site. The HHS funding totaled $301 million for response and recovery activities resulting from the September 11th attacks (Department of Health and Human Services, 2001).

Outcomes

It is not possible to overstate the dramatic changes in political culture, attitudes, and philosophy of the federal government regarding emergency management and counter-terrorism that have resulted from September 11th. Plus, many of these changes were immediate. Some elements of the emergency response went extremely well, such as the personal leadership of Mayor Giuliani, Governor Pataki, and the high level of competence of the Arlington County, Virginia, Police and Fire Services. But many concerns about weaknesses in the nation's ability to deal with a major terrorism event quickly surfaced, such as the need for better detection and warning systems for a terrorist attack, the need for improved central coordination at the federal level, weaknesses in the public health and disaster medical systems, and questions about the core capabilities of some states and localities to manage a massive disaster.

Other related systems were severely criticized for failures or weaknesses, such as the intelligence gathering and analysis capabilities of the international and domestic federal agencies. Lack of coordination among various federal agencies with information about suspected terrorists and problems in tracking foreign visitors and supposed students were also noted. The ramifications and implications are so substantial that it will take years of research and documentation to capture them.

A Major Sea Change

Within days after September 11th, the Bush Administration and the Congress rapidly made a major philosophical shift in their attitudes and willingness to combat terrorism, including major changes in national priorities, budget, and spending plans—all in a matter of a few weeks after the events.

Public Attitudes Toward Government

On September 30, a *New York Times* article titled, "Now Government is the Solution, Not the Problem," stated:

> After 20 years of exulting in the power of the private sector, in deregulation, tax cuts and reining in the Washington bureaucrats, Republicans and Democrats alike are talking about a muscular role for the government in the aftermath of the September 11th terrorist attacks. They are bailing out the airlines, establishing a new Office of Homeland Security, passing a big new aid package to rebuild the areas devastated by the attacks and pondering an even bigger effort to stimulate an ailing economy. When the chips are down, where do we turn? . . . To the government's firefighters, police officers, rescue teams. To the nonprofit sector's blood banks and shelters. And to big government's Army, Navy and Air Force.
>
> (*New York Times*, 2001a, p. 14)

Another perspective is that of the professional public administration community, which noted that the aftermath of September 11th provided a unique glimpse of public employees at work. In the newsletter of the American Society for Public Administration, it was noted:

> In a way unmatched in history, Americans had a chance to watch public administrators at work and, sometimes, under attack. They saw countless cases of unmatched bravery. The broadcast heroism, in fact, only hinted at the ways that government works rose to the challenges of their jobs.
>
> (*P.A. Times*, 2001)

The American Society for Public Administration further noted, "The real work—how to refashion the field to master the enormous new challenges facing it—begins now. Public administration will not only become more important, but its job has been dramatically transformed" (*P.A. Times*, 2001).

National Public Awareness of Terrorism

Given the timing, nature, and magnitude of the attacks, plus the immediate extensive media coverage, the topics of terrorism and emergency management received an unprecedented amount of attention not only in the United States but worldwide. Topics usually reserved to a small cadre of behind-the-scenes operational personnel suddenly were of interest and concern to citizens throughout the nation. This was captured in a *Washington Post* article entitled, "Think-Tank Presses are Suddenly Best-Selling Publishers." The article noted, "Across Washington, think tanks are finding their once obscure books, studies, and policy reports are hot with the general public" (*The Washington Post*, 2001b). Discussions of terrorism, bio-terrorism, and weapons of mass destruction are now commonplace among the general citizenry in the United States. The September 11th events provided a crash course on the topics. What was a somewhat esoteric technical area of interest, pursued by a relatively small group of responsible persons, is now discussed everywhere.

Public Awareness of Emergency Management

Citizens have become more aware of their public officials and how they conduct emergency management at each level of government. In New York City, Former Mayor Giuliani and Governor Pataki were directly involved in the response efforts and were highly visible doing their jobs on a daily basis. It should have been clear to most citizens that their local and state government officials were working ardently and effectively to help them.

One interesting indicator of the level of commitment and depth of the local emergency management effort is that at the third and final location of the city's EOC ultimately contained 350 workstations, according to newspaper accounts. That huge number is a crude indicator of the amount of coordination involved in the response and early recovery activities.

Similarly, public awareness of the key roles and functions of local public officials in Arlington, Virginia, was heightened by the attack on the Pentagon. Before the disaster, Pentagon staff had worked closely with the Arlington County Fire Department to prepare for a major fire in that building. The County Fire and Police Departments also were highly effective and committed to their jobs, according to two reports in *The Washington Post*. They too received great support and encouragement from the local citizens.

Changes in the Public Sector Focus and Workload

As noted above, the role of public practitioners in emergency management has changed and probably will continue to change as the United States goes into the recovery period. A related outcome is the effect on public officials, both elected and appointed, and the long-term burden on their workloads. For example, Senator Hillary Clinton (D– N.Y.) described the economic damage as "incalculable" and said ". . . [She has] been consumed with the details of organizing federal assistance for the city and expects that responding to the emergency on both the national and local levels will dominate her Senate career for the foreseeable future" (*New York Times*, 2001b, p. A7).

Major Policy and Program Outcomes

The five specific categories of observed outcomes of major disaster events that the authors developed and used in the *Disaster Time Line: Selected Major Milestone Events and Their U.S. Outcomes (1965-2001)* were applied to the September 11th events in order to capture some of the most frequently observed aspects of outcomes from a political and policy perspective.

(1) Major Reports and Documents

After examining dozens of major disaster events during the years 1965–2001, the authors noted that immediately after a major event, either the Congress or the White House initiated hearings, after-action reports, and/or studies to determine what the problems and deficiencies were in responding adequately to the disaster. This step occurred without exception in the 36 years examined (Rubin, 2000). Yet, in less than a week after the September 11th events, major national legislation was enacted and organizational changes occurred. There were two highly unusual aspects in the immediate aftermath of the terrorist attacks. First, no hearings or studies were ordered to determine what went wrong and what remedies were needed. Second, the speed and bipartisan nature of the legislative process were unprecedented.

The authors noted the sequence with great interest because it was an aberration from the pattern observed since 1965. After making a rough time line chart of the sequence, the authors surmised that because several major reports about terrorism had already been completed before September 11th, they were used rather than ordering new studies and reports. Some relevant ones that were quickly updated and issued are several GAO reports on counter-terrorism (U.S. General Accounting Office, 2001a, 2001b, 2001c) and on protecting critical infrastructure—Hart/Rudman Reports I & II, Gilmore Reports I & II, and the National Commission of Terrorism (Bremen Commission) Report.

It would appear that the information and knowledge about what to do already existed before September 11th. What was lacking was the political backing for change and the political will to act. A rapid sequence of actions regarding improved emergency management and protection of critical infrastructure then followed.

(2) Legislation

In a matter of about 16 weeks after the terrorist events, the degree of national attention and commitment to dealing with the outcome of the incidents led to the rapid enactment of four major pieces of legislation: the Supplemental Act for Response and Recovery; the U.S.A. Patriot Act of 2001; the Defense Authorization Act; and the Aviation and Transportation Security Act.

Other unusual characteristics of the aftermath of this disaster are (1) the speed with which the federal government and the New York state delegation met and agreed to create and pass congressional legislation and appropriation of $40 billion to finance the costs of response and recovery efforts, and (2) that major federal organizational and coordination changes occurred relatively rapidly, even before Congressional hearings were held or special task forces were formed.

Since September 11th, many new bills relating to terrorism are pending before Congress. The list of pending legislation is sizeable, and has been changing at a rapid rate.

(3) Executive Orders

Again, within about 16 weeks, three Executive Orders and two Homeland Security Presidential Directives (HSPD) were issued, including E.O. 13228, Homeland Security; E.O. 13231, Critical Infrastructure Protection; and E.O. 13234, Citizen Preparedness. HSPD1 deals with the Homeland Security Council and HSPD2 covers Immigration Policies.

(4) Key Federal Response Plans

It is expected that both the Federal Response Plan and the National Contingency Plans will be reviewed and revised, based on the September 11th attacks. It is too early to know the nature of these changes. The structural and organizational issues as well as the basic authorities for the Homeland Security Office probably will have to be clarified before the implementing mechanisms and response plans are changed.

(5) Major Organizational Changes

There were at least three new federal offices created, the Homeland Security Office and the Homeland Security Council in the Executive Office of the President, and the Transportation Security Administration in the Department of Transportation. Paramount among the changes here is the rapid creation of the Homeland Security Office. Other major changes pending include a wide array of security concerns, such as changes in airport and airline safety responsibilities, regulations, procedures; changes in immigration and naturalization laws and regulations; and changes in the transportation systems in the country.

It is too early to know just what the Homeland Security Office will do with regard to contributing to changes in response plans, systems, and even recovery. Given the breadth of the Executive Order mandating the formation of that office, it would be likely that major changes are in the offing. Some of the other changes that are likely to occur in the coming months: improved warning and alert systems, improved detection and treatment for chemical and biological agents, improved intelligence gathering and analysis from both domestic and international sources, changes in emergency management systems and personnel training, changes in FEMA's National Preparedness Office, changes in the Federal Response Plan and the National Contingency Plan, and more national counter-terrorism exercises.

Given the vast complexity of the attacks and their aftermath, the authors created the *Terrorism Time Line: Major Milestone Events and their U.S. Outcomes (1988-2001)*. Also under development are a narrative explanation of the chronology and a policy analysis of the major events and their outcomes.

In closing, in an article entitled, "Suddenly, Americans Trust Uncle Sam," noted author Francis Fukuyama is quoted as saying, "Trauma and war bring out communal solidarity and remind people of why we have government." Regarding the creation of trust in government, he said, ". . . a national crisis alone does not create trust in government. It's a combination of external threats and government effectiveness" (*New York Times*, 2001c).

References

Coy, P., D. Henry, N. Byrnes, R. McNatt, G. Khermouch, E. Thornton, and J. Rossant. 2001. "The Future of New York." *Business Week* (October 22): 96–126.

Egan, Dennis M., Captain, Director of the National Response Team, United States Coast Guard Headquarters, Washington D.C. 2001. Interview by John Harrald, Irmak Renda-Tanali, and Damon Coppola on October 4.

Federal Emergency Management Agency. http://www.fema.gov and http://www.fema.gov/priv/pao/clips.htm.

Moroccolo, Mary Ann, New York City Office of Emergency Services, 2002. Personal communication with Claire B. Rubin on March 19.

New York Times. 2001a. "Now Government is the Solution, Not the Problem." (September 30): 14.

New York Times. 2001b. (October 11): A7.

New York Times. 2001c. "Suddenly, Americans Trust Uncle Sam." (November 3).

New York Times. 2002. (March 17).

P.A. Times. 2001. "The Next Public Administration Revolution." *P.A. Times* 24 (October).

Rubin, Claire B. 2000. *Emergency Management in the 21st Century*. Working Paper #104. Boulder, CO; Natural Hazards Research and Applications Information Center, University of Colorado. http://www.colorado.edu/hazards/wp/wp104/wp104.html.

Rubin, Claire B. and Irmak Renda-Tanali. 2001. *Disaster Time Line: Selected Milestone Events (1965-2001)*. http://www.disaster-timeline.com.

Rubin, Claire B. and Irmak Renda-Tanali. 2002. *Terrorism Time Line: Major Milestone Events and Their U.S. Outcomes (1988-2001)*. http://www.disaster-timeline.com.

Terry, Ed, Manager of the Emergency Operations Center, U.S. Environmental Protection Agency Headquarters, Washington, D.C. 2001. Interview by Claire B. Rubin in November.

Tully, S. 2001. "Rebuilding Wall Street." *Fortune* (October 1 Special Report): 96–102.

U.S. Department of Health and Human Services. 2002. "Secretary Thompson Thanks 9/11 Rescue Workers." Press release issued March 11, 2002.

U.S. General Accounting Office. 2001b. *Homeland Security: A Framework for Addressing the Nation's Efforts*. GAO-01-1158T. Testimony presented on September 21. http://www.gao.gov.

U.S. General Accounting Office. 2001c. Combating Terrorism: Selected Challenges and Related Recommendations. GAO-01-822. 210pp. September 21. http://www.gao.gov/new.items/d01822.pdf.

U.S. General Accounting Office. 2001a. *Homeland Security: Key Elements of a Risk Management Approach*. GAO-02-150T. Testimony presented on October 12.

The Washington Post. 2001a. (October 28).

The Washington Post. 2001b. "Think-Tank Presses are Suddenly Best-Selling Publishers." (October 26).

The Washington Post. 2002. (March 10).

Waugh, William L. 2000. *Living with Hazards, Dealing with Disasters*. New York: M.E. Sharpe.

Acknowledgments

The authors acknowledge Quick Response grant funding support form the Natural Hazards Research and Applications Information Center, University of Colorado–Boulder. The authors appreciate the assistance of Damon Coppola, who shared his interview notes for another George Washington University research project.

This article was based on the authors' Quick Response Report #140, entitled *The Terrorist Attacks on September 11, 2001: Immediate Impacts and Their Ramifications for Federal Emergency Management*. 2002. Natural Hazards Research and Applications Information Center, University of Colorado, and on a subsequent article, "Federal Emergency Management in the U.S.: Implications of the Terrorist Attacks of September 11, 2001," which appeared in *The International Emergency Management Society, 9th Annual Conference Proceedings*. 2002. University of Waterloo.

Correspondence should be directed to Claire B. Rubin, President, Claire B. Rubin & Associates; e-mail: cbrubin@gwu.edu or to Irmak Renda-Tanali, Visiting Professor, The George Washington University, Institute for Crisis, Disaster and Risk Management, 1776 G St., N.W., Suite 150, Washington, D.C. 20052; e-mail: rendatan@seas.gwu.edu.

Appendix

Appendix

National Science Foundation Awards in Fiscal Year 2002 related to Homeland Security

Principal Investigator	Award	Institution	Title
AWARDS FROM THE DIRECTORATE FOR ENGINEERING			
Blatchley	BES 0210350	Purdue University	SGER: Inactivation of Biological Warfare Agents by Physical Disinfectants
Gurol	BES 0206861	San Diego State University	SGER: Application of Gaseous Ozone for Inactivation of Bacterial Spores
Mainelis	BES 0212007	Rutgers University	SGER: Inactivation of Airborne Microorganisms Using Electrical Fields: a Feasibility Study
Parlange	BES 0211625	John Hopkins University	SGER: Measurements of Particle Emissions from World Trade Center Site in New York City
Pereira da Cunha	BES 0233463	University of Maine	SGER: Detection of Bioterrorism-linked Microbial Pathogens Using Surface Acoustic Wave Liquid Sensors
Tepper	BES 0229026	Virginia Commonwealth University	SGER: Biosensing in the Gas Phase: A New Approach Based on Imprnted Nanoparticles of a Linear Polymer
Wei	BES 0228143	Purdue University	SGER: Antibody-cunjugated Nanoparticle Films as Spectroscopic Sensors of Chemical Agents
Astaneh	CMS 0139542	University of California Berkeley	SGER: World Trade Center Post-Disaster Reconnaissance and Perishable Structural Engineering Data Collection
Bloomquist	CMS 0139382	University of Florida	SGER: Infrastructural Damage Assessment Using Land-Based Laser Swath Mapping Technology
Frost	CMS 0139258	Georgia Institute of Technology	SGER: Digital Data Collection for Damage Assessment at World Trade Center
Garg	CMS 0228784	Duke University	SGER: A Multi-sensor-fused Intelligent Biomimetic Microrobot
Grigg	CMS 0139878	Colorado State University	Condition and Security Indicators for Interdependent Infrastructure Systems
Harrald	CMS 0139309	George Washington U	SGER: Observing and Documenting the Inter-Organizational Response to the 9/11 Terror Attacks

Harrald	CMS 0219953	George Washington U	Terrorism and Corporate Crisis Management: The Strategic Effect of the September 11 Attacks
Hendrickson	CMS 0223255	Carnegie Mellon University	Analyzing Critical Infrastructure Dependencies: Security & Survivability Effects in the Service Sector
Holguin-Veras	CMS 0205188	CUNY City College	SGER: Impacts of Extreme Events on Passenger Travel Behavior
Lee	CMS 0240691	SUNY Buffalo	US-PRC Workshop on Protection of Urban Infrastructure and Public Facilities Against Natural and Manmade Disasters
Lindell	CMS 0219155	Texas A&M University	Develop an Evacuation Management Decision Support System
Mileti	CMS 0139024	University of Colorado Boulder	Natural Hazards Research Application and Information Center (supplement)
Miller-Hooks	CMS 0218621	Pennsylvania State University	Emergency Preparedness Planning and On-Line Evacuation of Large Buildings
Mowrer	CMS 0139568	University of Maryland	SGER: World Trade Center Post-Disaster Fire Reconnaissance and Perishable Data Collection
Murphy	CMS 0229809	University of South Florida	SGER: Adaptive Shoring for Robot-assisted Search and Rescue
O'Rourke	CMS 0207266	Cornell University	SGER: Improved Security And Management Of Underground Infrastructure Systems: Lessons Learned From September 11, 2001
Phillips	CMS 0229389	Jacksonville State University	Organization-based Post Disaster Sheltering and Housing of Low Income and Minority Groups
Reid	CMS 0238991	National Academy of Engineering	Management of Precursors (w/DMII and SBE)
Rosenbloom	CMS 0215588	U of Southern California	Digital Government: Responding to the Unexpected (workshop through CISE)
Rubin	CMS 0231624	Claire Rubin & Assoc.	SGER: Analysis of Major Terrorist Disasters Affecting the U.S. (1988-2001)
Sozen	CMS 0204537	Purdue University	SGER: Gathering and Archiving Data on the Response of a Reinforced Concrete Building Subjected to Impact and Fire Loads
Stehr	CMS 0234100	Washington State University	The Implementation of Disaster Assistance in New York in the Aftermath of September 11: The Dynamics of Multiorganizational Response
Wallace	CMS 0139306	Rensselaer Polytech Institute	SGER: Impact of World Trade Center Disaster on Critical Infrastructure Interdependence
Yeh	CMS 0213686	University of Washington	International Travel Grant to Support U.S. Involvement in HAZARDS-2002
Zimmerman	CMS 0203371	New York University	Workshop on Research Related to the World Trade Center Disaster

Zimmerman	CMS 0204660	New York University	SGER: Urban Infrastructure Services in a Time of Crisis: Lessons from September 11th
Zimmie	CMS 0226864	Rensselaer Polytech Institute	SGER: Surface Blast Effects on Embankments and Dikes
Barat	CTS 0233582	New Jersey Institute of Technology	Terahertz Imaging System for Sensing of Chemical and Biological Agents
Janata	CTS 0207142	Georgia Institute of Technology	The New Challenges of Chemical and Biological Sensing (Workshop: January 9-10, 2002; funded through CHE in MPS)
Auerbach	DMI 0215158	EMTEL	SBIR Phase I: Remote Radio Frequency Measurements for Pipeline Monitoring - FloWatch911
Berg	DMII 0226831	Rensselaer Polytechnic Institute	SGER: Exploratory Research to Define Streamlining Opportunities for Construction Permitting in Disaster Recovery
Bier	DMII 0228204	U of Wisconsin Madison	Optimal and Near-Optimal Resource Allocation for Information Security and Critical Infrastructure Protection
Dong	DMI 0212035	NanoTek, Inc.	SBIR Phase I: Nanofabricated Gas Chromatography Column-Oven System
Drukier	DMII 0214853	BioTraces Inc.	SBIR Phase I: New Methods for Detecting Bioterrorism
Farquharson	DMII 0214280	RTA	SBIR Phase I: Anthrax Detector for Mail Sorting Systems
Fernandez	DMII 0215192	Ciencia Inc.	SBIR Phase I: Biosensor for Label-free, Real-time Monitoring of Environmental Pathogens
Forbes	DMII 0215159	VTK	SBIR Phase I: Flexible Displays for Radio Frequency Identification (RFID) Applications
Hunter	DMII 0214847	Physical Sciences Inc.	SBIR Phase I: Fast Response Sensor for Airborne Biological Particles
Jacobson	DMII 0114499	U of Illinois Urbana-Champaign	Collaborative Research: Aviation Access Control Security Systems
Kittrell	DMI 0215008	KSE Inc.	SBIR Phase I: Homeland Security: Photocatalytic Destruction of Air-Borne Bacteria
Kobza	DMII 0114046	Texas Tech University	Collaborative Research: Aviation Access Control Security Systems
Kosal	DMII 0215088	ChemSensing	SBIR Phase I: Colorimetric Sensor for Real-Time Detection of Nitroaromatic Explosives
Li	DMII 0213614	Genex Technologies, Inc.	SBIR Phase I: A Novel Large Depth of Field and High Resolution Imaging Scanner

Meisner	DMI 0213982	Titan	SBIR Phase I: Ultra-sensitive Charge-coupled Device Technology: A Photon Counting Camera
Mlcak	DMII 0215119	Boston Micro Systems, Inc.	SBIR Phase I: Rapid Reagent-less Multi-channel Biological Agent Detector
Morton	DMII 0228419	U of Texas Austin	Stochastic Network Interdiction Models for Homeland Security
Nelson	DMII 0215147	Aerodyne Research	SBIR Phase I: Development of a High Precision, Autonomous Quantum Cascade Laser-Based Detector for Methane and Nitrous Oxide
Rindova	DMII 0218029	U of Maryland, College Park	SGER: Evolving Dynamic Capabilities in Times of Dramatic Environmental Change
Schafer	DMII 0228246	Johns Hopkins University	Optimal Structural System Design for Catastrophic Unforeseen Events
Schlueter	DMII 0215299	Intellicon	SBIR Phase I: Preventive, Corrective, and Emergency Control for Equipment Outages Producing Voltage Collapse and Blackout
Thompson	DMI 0216574	RealTronics	SBIR Phase II: A High Frequency Beam Steered Electromagnetic Impulse Radar to Locate Human Targets Through Opaque Media
Vispute	DMI 0215175	Blue Wave	SBIR Phase I: Low Cost Visible Blind Ultra Violet Photodetectors on Glass and Polyimide
Wallace	DMII 0228402	Rensselaer Polytech Institute	Disruptions in Independent Infrastructures
Worsnop	DMII 0215061	Aerodyne Research	SBIR Phase I: Development of a Time of Flight Aerosol Mass Spectrometer for Atmospheric Aerosol Analysis
Zhang	DMII 0131228	JunTech, Inc.	SBIR Phase II: A New Digital Video Surveillance System
Lee	EEC 0139621	MCEER at SUNY Buffalo	Multidisciplinary Center for Earthquake Engineering Research (supplement)
Grimes	ECS 0225346	Pennsylvania State University	Integrated Sensing: Integrated Smart-sensor Networks for Monitoring Aqueous Environments

AWARDS FROM THE DIRECTORATE FOR SOCIAL, BEHAVIORAL AND ECONOMIC SCIENCES

Barrett	SBE 0204431	Boston College	SGER: Mechanisms of Resilience in the Face of On-Going Threat
Bearman	SBE 0140024	Columbia University	SGER: Narrative Networks: The World Trade Center Tragedy

Birnbaum	SBE 0202448	California State University Fullerton	SGER: Effects of Terrorism on Judgments and Decisions Concerning Civil Liberties
Bonanno	SBE 0202772	Columbia University	SGER: The Costs and Benefits of Self-Enhancement: Coping with the Terrorist Attack on the World Trade Center
Bozorgmehr	SBE 0140271	CUNY Graduate School	SGER: How Support Organizations Respond to Crises: Middle Eastern and South Asian American Organizations in the Aftermath of September 11
Brewer	SBE 0201511	George Washington U	SGER: Crisis and Confidence
Burkhardt	SES 0218007	U of Illinois at Urbana-Champaign	Dissertation Research: Reimagining Anthrax: Constructing the Threat and Science of Biological Warfare at Fort Dietrick, 1943-1969
Conway	SBE 0139292	University of Illinois Chicago	A Test of Flashbulb Memory: Tuesday, September 11, 2001
Goodwin	SBE 0203970	Purdue University	SGER: American Identity Under Siege: Re-categorization at Ground Zero
Huddy	SBE 0201650	SUNY Stonybrook	SGER: The Dynamic, Multi-Faceted Effects of Threat on U.S. Domestic and Foreign Policy Attitudes
Inglehart	SBE 0140566	University of Michigan	SGER: Islam and the West: Clash of Civilizations or Traumas of Modernization?
Jenkins-Smith	SBE 0234119	Texas A&M University	SGER: Public Responses to Terrorism
Krantz	SES 0136872	Columbia University	Understanding and Improving Protective Decision Making
Lecci	SBE 0204846	U of North Carolina Wilmington	SGER: A Social Cognitive Model for Processing Health Risk Information About Anthrax Fears
Lerner	SBE 0201525	Carnegie Mellon University	SGER: Emotional and Cognitive Carry Over from the September 11 Attacks
Lustick	SES 0218397	University of Pennsylvania	ITR: Development and Applications of the PS-I Computational Modeling Platform for Problems of Ethnic Conflict, Globalization, State Stability and Terrorism
Moaddel	SBE 0139908	Eastern Michigan University	SGER: A Post-Crisis Analysis of the Attitudes and Value Orientations of the Islamic Publics in Egypt, Iran and Morocco
Moore	SBE 0201149	University of Connecticut	SGER: Terror, Immigration, and Civil Liberties in the United States
Morrow-Jones	SBE 0202231	Ohio State University	SGER: The Impact of the September 11, 2001, Terrorist Attacks on American Metropolitan Household Decisions to Stay in the Same Residence, to Remodel, or to Move

Moskowitz	SBE 0213693	Lehigh University	Control over Stereotype Activation by Preconscious and Temporary Goals
O'Loughlin	SBE 0203087	U of Colorado Boulder	SGER: The Impacts of the September 11, 2001, Attacks and Response on Russian Geopolitical Culture
Rasinski	SBE 0139964	National Opinion Research Center	SGER: Public Response to a National Tragedy
Richardson	SBE 0200619	Association of American Geographers	SGER: The Geographical Dimensions of Terrorism: A Research Agenda for the Discipline
Rosenbloom	SBE 0215588	U of Southern California	Digital Government: Responding to the Unexpected (workshop through CISE)
Rubinstein	SBE 0129762	Syracuse University	Dissertation Research: Defending the Homeland: The Cultural Construction of U.S. Defense in Everyday Practice
Rubinstein	BCS 0129762	Syracuse University	Dissertation Research: Defending the Homeland: The Cultural Construction of U.S. Defense in Everyday Practice
Schuman	SBE 0206472	University of Michigan	SGER: Collective Memories Over The Life Course
Silver	SBE 0214143	Michigan State University	Attack on American Civil Liberties Trade-offs, and Ethnic Tolerance: Panel Survey
Silver	SBE 0211039	University of California Irvine	SGER: Coping with Community-based Traumatic Events: National Response to September 11, 2001
Silver	SBE 0140541	Michigan State University	SGER: The Attack on America, Civil Liberties Trade-offs and Ethnic Tolerance
Silver	SBE 9910223	University of California Irvine	SGER: Coping with Community-based Traumatic Events: The Columbine High School Shootings and the 9/11 Terrorist Attacks (supplement)
Skitka	SBE 0210053	University of Illinois Chicago	SGER: Understanding Political Intolerance: A Test of Deonance Theory
Thompson	SBE 0233978	Pomona College	SGER: Restoring Assumptions of Safety and Control Following the 2001 Terrorist Attacks
Wayment	SBE 0202652	Northern Arizona University	SGER: Predicting Affective Reactions to Collective Loss
Weisband	SBE 0218304	University of Arizona	The Role of Technology on Work Practices in Police Departments
Williams	SBE 0220979	Massachusetts Institute Technology	SGER: Workshop: Rethinking Technology After September 11 (March 2002)

AWARDS FROM OTHER DIRECTORATES

Cantley	BIO 0209454	Beth Israel, Harvard	SGER: Discovery of Substrates and Inhibitors of the Anthrax Lethal Factor using Peptide Libraries
Fraser	BIO 0202304	The Institute for Genomic Research	Genomic Diversity of Bacillus Anthracis Strains
Ammar	EIA 0131079	West Virginia University	Digital Government: Automated Dental Identification
Bhargava	IIS 0209059	Purdue University	Formalizing Evidence and Trust for User Authorization
Blumenthal	ANI 0238609	NAS	Telecommunications Research and Development
Dawes	CISE 0221927	SUNY Albany	SGER: Turning to Digital Government in a Crisis: Coordinating Government Business and Non-Profit Services in Response to the World Trade Center Attacks of September 11, 2001
DeFanti	EIA 0224306	University of Illinois Chicago	CISE Research Resources: Matching Advanced Visualization and Intelligent Data Mining to High-performance Experimental Networks
Fan	IIS 0208539	U of North Carolina Charlotte	A Novel Approach to Video Database Indexing via Semantic Classification
Goldberg	CISE 0204758	Rensselaer Polytechnic Institute	SGER: The Impact of Breakdowns of Physical Infrastructure on Social Networks
Hou	ANI 0221357	U of Illinois at Urbana-Champaign	Data Centric Sensor Networks
Markowsky	ANI 0210619	University of Maine	ANYWHERE, ANYTIME, ANYSIZE, ANY SIGNAL: Scalable Remote Information Sensing and Communications Systems
Murphy	EIA 0224401	University of South Florida	CISE Research Resources: R4: Rescue Robots for Research and Response
Rosenbloom	CISE 0215588	University of Southern California	Digital Government: Responding to the Unexpected (workshop with CMS and SBE)
Sastry	CISE 0239511	University of California Berkeley	Workshop on Critical Infrastructure Protection: September 23-24, 2002; Leesburg, VA
Sekar	CCR 0208877	SUNY Stony Brook	A New Approach for Securing Systems Using Automated Adaptive Intrusion Response
Silverstein	CCR 0208830	University of Virginia	Adaptive Digital Signal Processing for Spatial and Temporal Sampled Coherent Imaging Systems

Singh	ANI 0125728	Portland State University	Using Smart Antennas for Ad Hoc Networking
Sirer	EIA 0203449	Cornell University	CRCD: The Ad Hoc Classroom: Integrating Emerging Wireless Communications and Networking Technologies into Mainstream Computer Science and Electrical Engineering Curricula
Freiman	GEO 0229030	United Engineering Foundation	Conference: Risk-based Decision Making in Water Resources X
Olsen	GEO 0210001	University of Massachusetts Boston	SGER: Exploratory Research on the Environmental Impact of the World Trade Center: Attack on Sediment Quality and Dynamics in New York Harbor
Clarke	MPS 0139876	Purdue University	Kinetics and Mechanisms of Rapid Reactions in Solution
Collins	MPS 0211065	Carnegie Mellon University	SGER: Activated Hydrogen Peroxide Detoxification of Chemical-Biological Warfare Agents
Fenselau	MPS 0223645	University of Maryland College Park	Analytical Chemistry for Homeland Defense and National Security, ACS Meeting, Boston, MA, August 18-22, 2002
Janata	MPS 0207142	Georgia Institute of Technology	The New Challenges of Chemical and Biological Sensing (Workshop: January 9-10, 2002; funded with CTS in ENG)
Lehmann	MPS 0228797	Princeton	SGER: An Optical Fiber Resonator for Cavity Ring-down Spectroscopic Detection and Measurement of Trace Species
Becchetti	PHY 0214631	University of Michigan	2002 Symposium on Radiation Measurements and Applications, Ann Arbor, May 21-23, 2002
Friedman	DMS 0204029	Stanford University	Topics in Predictive and Descriptive Data Mining
Wang	CHE 0209707	New Mexico State University	Characterization of DNA-linked Nanoparticle Networks for Advanced Genetic Testing
Zoski	CHE 0210315	Georgia State University	GOALI: Addressable Multielectrode Arrays Based on Membrane Templates: Fabrication, Characterization and Instrumentation
Lewis	MPS 0228141	Caltech	Supplement: Achieving Molecular Level Control over the Chemical, Electrochemical, and Electrical Properties of Crystalline Si Surfaces
Lopez	MPS 0230818	U. New Mexico	SGER: Fluorescence Lifetime-Based Measurements of Biosensor Arrays Using Closed Loop Auto-Oscillating Systems
Majda	MPS 0228639	U. California-Berkeley	Supplement: Dynamics of Lateral Processes at the Air/Water Interface

Penner	MPS 0229434	U. California-Irvine	Supplement: Chemical Sensors Based on Electrodeposited Metal Nanowires
Rabolt/Chase	MPS 0228839	U. Delaware/Dupont	SGER: Ultra-Fast Infrared Spectroscopy Using a Focal Plane Array for the Real Time Detection of Chemical and Biological Agents
Sadik	MPS 0210968	SUNY Binghampton	SGER: Molecular Design of Intelligent Sensors for Selected Chemical Warfare Agents Using Support Vector machines
Solouki	MPS 0228971	U. Maine	SGER: Improving Performance Characteristics of Hyphenated Preconcentrator GC Fourier Transform Ion Cyclotron Resonance Mass Spectrometry: A "Gold Standard" for Sensor Development
Weidman	MPS 0215714	National Academy of Sciences	The Role of Mathematical Sciences in Homeland Defense
Kennedy	EHR 0236753	Morgan State University	Infrastructure Building to Develop a Center for Biological and Chemical Sensors Research
LaBar	EHR 0206350	Liberty Science Center	Skyscraper: Achievement and Impact - A Permanent Exhibition
Richardson	OIA 0236908	Potomac Institute for Policy Studies	Shaping Science and Technology to Serve National Security
Rippen	OIA 0234636	Rand Corporation	Data, Information and Recommendations Regarding Emergency Worker Safety and Health During Disaster Responses - CDC-NIOSH

AWARDS MADE THROUGH THE NSF/CMS-FUNDED NATURAL HAZARDS RESEARCH AND APPLICATIONS INFORMATION CENTER (NHRAIC), UNIVERSITY OF COLORADO–BOULDER

Cutter	University of South Carolina	Assessing the Use of Geographical Information Technology in Responding to the September 11 Terrorist Attack in New York City
Eads	University of Colorado Boulder	Marginalized Groups in Times of Crisis: Identity, Needs and Response
Edwards	East Carolina University	Social Capital and Advocacy Organization Response to a Suddenly Imposed National Disaster
Gilham	University of Colorado Boulder	Non-Profit Sector Responses to a Terrorist Attach While Preparing for Mass Demonstrations in Washington DC, September 2001
Grant	University of Akron	Downing of the UA Jet in PA: "Terrorism in Shanksville: A Study in Preparedness and Response
Hays	University of Colorado Boulder	The Role of Police Organizations During a Domestic Terrorist Attack

LeDuc	University of Oregon	Evaluation of Intergovernmental Communication in a Post Disaster Environment
Lowe	University of Colorado Boulder	Community Response in Terrorist Disasters
Michaels	University of Colorado Boulder	How Communications Technology Firms Responded in the Immediate Aftermath of the September 11, 2001 Explosions
O'Brien	California State University at Stanislaus	Institutional Warning Response Following the World Trade Center Attack
Peek-Gottschlich	University of Colorado Boulder	Ethnic Issues on University Campuses Following An Act of Terrorism: Middle Eastern Student Preparation and Response
Rodrigue	California State University Long Beach	Patterns of Media Coverage of The Terrorist Attacks on The United States In September of 2001
Rubin	George Washington University	The Events of Sept. 11, 2001: Immediate Ramifications for Federal Policy ($500 travel supplement)
Sattler	Western Washington University	Psychological Reactions following the Terrorist Attacks in New York City and Washington, D.C.
Simpson	University of Louisville	An Examination of the Process and Nature of Victim Identification Immediately Following the World Trade Center Collapse
Sutton	University of Colorado Boulder	Role of Faith Communities' Responses to Terrorist Attacks in New York City
Weber	University of North Texas, Denton	Public/Private Communication and Cooperation after a Recent Disaster Event: An Assessment of Coordination

Data provided by the National Science Foundation, June 2003